站在巨人的肩上

Standing on the Shoulders of Giants

图灵教育

站在巨人的肩上
Standing on the Shoulders of Giants

图灵程序设计丛书

ON JAVA 中文版 进阶卷

BY BRUCE ECKEL

[美] 布鲁斯·埃克尔(Bruce Eckel) —— 著　孙卓 陈德伟 臧秀涛 —— 译

人民邮电出版社

北京

图书在版编目(CIP)数据

On Java：中文版. 进阶卷 /（美）布鲁斯·埃克尔
(Bruce Eckel) 著；孙卓，陈德伟，臧秀涛译. -- 北京：
人民邮电出版社，2022.3（2022.9重印）
（图灵程序设计丛书）
ISBN 978-7-115-58502-8

Ⅰ. ①O… Ⅱ. ①布… ②孙… ③陈… ④臧… Ⅲ. ①
JAVA语言-程序设计 Ⅳ. ①TP312.8

中国版本图书馆CIP数据核字(2022)第005205号

Authorized translation from the English language edition, titled *On Java 8* by Bruce Eckel. Copyright © 2017, © 2021 by Bruce Eckel, President, MindView LLC.

Simplified Chinese-language edition published by Posts & Telecom Press Co., Ltd. Copyright © 2022. All rights reserved.

本书中文简体字版由Bruce Eckel授权人民邮电出版社有限公司独家出版。未经出版者事先书面许可，不得以任何方式或途径复制或传播本书内容。版权所有，侵权必究。

内 容 提 要

本书内容主要是对《On Java 中文版 基础卷》的拓展延伸，重点讲解Java的高级特性、并发、设计模式等相关进阶知识，对一些和开发密切相关的底层操作（如I/O系统、底层并发、数据压缩等）进行深入探讨，同时针对基础卷的重点章节进行了补充说明（如第3章增补了一些关于集合的高级特性）。在附录中，作者给出了67条关于低级程序设计和编写代码的建议，并分享了自己成为程序员的一些经验之谈。

本书适合有一定项目开发经验的Java程序员阅读。

◆ 著　　　[美]布鲁斯·埃克尔（Bruce Eckel）
　　译　　　孙　卓　陈德伟　臧秀涛
　　责任编辑　乐　馨
　　责任印制　周昇亮

◆ 人民邮电出版社出版发行　北京市丰台区成寿寺路11号
邮编　100164　　电子邮件　315@ptpress.com.cn
网址　https://www.ptpress.com.cn
三河市中晟雅豪印务有限公司印刷

◆ 开本：720×960　1/16
印张：35
字数：792千字　　　　　　2022年3月第1版
　　　　　　　　　　　　　2022年9月河北第5次印刷
著作权合同登记号　图字：01-2020-7186号

定价：129.80元
读者服务热线：(010)84084456-6009　印装质量热线：(010)81055316
反盗版热线：(010)81055315
广告经营许可证：京东市监广登字20170147号

致谢

1月16日是中文译稿完整交付的日子，距离本书翻译项目启动，时间恰好过去一年。感谢孙卓、陈德伟、臧秀涛、秦彬四位译者的辛勤付出。尤其感谢他们在翻译时秉承专业严谨的工作态度，校正了原稿中存在的诸多错漏。

特别感谢 DDD 专家张逸、服务端专家梁桂钊、软件系统架构专家王前明、译者陈德伟为本书重点章节录制了精讲视频，有效降低了本书的阅读门槛。

同时，非常感谢对本书译文给出诸多宝贵建议的审读专家们，他们包括：梁树桦、王明发、叶沄龙、杨光宇、施仪画、周剑飞、张伟、王前明、杨显波、史磊、张争龙、薛伟冬、张逸、陈鹜、小判、程锦、凌江涛、杜乐、龚泽龙。

还要感谢在本书"开放出版"阶段给出反馈建议的所有读者朋友们：陈明树、陈渝、丁春盈、冯治潍、葛瑞士、郭雅鑫、何金钢、刘一帆、刘梓桐、刘宗威、马康、冉浩、徐浩清、薛书猛、翟特特、赵希奥、郑炫坤、翟国平、吉宇晟、潘陈皓、苏超。

本书能够顺利出版，离不开以上所有人的帮助和贡献。

最后，感谢为 Java 中文社区做出卓越贡献的专家学者们，感谢 JCP 执行委员会委员周经森（Kingsum Chow）、阿里云程序语言与编译器技术总监李三红、腾讯 JDK 负责人杨晓峰、永源中间件总经理张建锋、《Java 并发编程的艺术》一书作者方腾飞为本书撰写推荐语。

中文版序

我听说这本书之所以要翻译为中文版，是因为读者的要求，我对此感到非常荣幸。谢谢你们——这让我知道，我在做正确的事情。也非常感谢人民邮电出版社图灵公司听到了读者的呼声，并投入了大量资源来促成本书中文版的翻译。

Thinking in Java, 4th Edition 涵盖了 Java 5 和部分 Java 6 的内容。自那以后，尽管还有其他的显著改进，但目前 Java 8 仍然是 Java 变化最大的一个版本。对许多程序员和公司来说，改用 Java 8 仍然是一个具有挑战性的选择，希望本书能对其有所帮助。

应图灵公司的要求，我在各个章节中添加了一些内容，用来讲解从 Java 9 到 Java 17 的新语言特性。

前言

本书基于 Java 8 的特性进行该语言的编程教学，同时根据 Java 11、17 等版本的新特性做了关键更新。

我的上一本 Java 书——*Thinking in Java, 4th Edition*，对于用 Java 5 编程仍然很有用，Android 编程用的就是这个语言版本。然而随着 Java 8 的到来，这门语言发生了许多显著的变化，编写和阅读新版本 Java 代码的感受都与以往有了明显的不同。于是，花费两年时间编写一本新书也就在情理之中了。

Java 8 最大的改进是引入了函数式编程的一些长处，简单来说包括 lambda 表达式、流（stream），以及"函数式基本类型"（functional primitive）等。即便如此，Java 依然是一门受 Smalltalk 启发而设计的面向对象编程语言。由于受制于向后兼容性，Java 无法彻底翻新为一门函数式编程语言。但我还是要为 Brian Goetz 和他的团队在重重限制之下所做出的贡献喝彩。毋庸置疑，Java 8 让这门语言获得了升华，也有助于你学习 Java 语言。同时，我希望这本书能够让你的 Java 学习之旅变得轻松和愉悦。

关于 Java 8 后续版本的新特性

就在本书的编写过程中，Java 17 发布了。本书的内容原本是基于 Java 8 的，但是应人民邮电出版社图灵公司（本书中文版的出版商）的要求，我也会在本书中向读者介绍 Java 9 到 Java 17 的新特性，同时会在对应的章节标题中用"新特性："后加该特性的说明来予以标识。此外，通过目录也可以很容易地找到这些新特性。如果你只能使用 Java 8，你大可跳过这些章节，这样做并不会影响你阅读本书的其他内容。也就是说，新特性只会在对应的章节中使用，而不会出现在本书其他关于 Java 8 的主要内容中。

本书所使用的位于 GitHub 代码仓库的示例都包含一个基于 Java 8 环境的构建文件（使用 Gradle 构建工具生成）。只要你安装的是 Java 8（或者是 Java 8 后续的新版本），这些构建文件就可以正常运行。此外，所有用于演示 Java 8 后续版本新特性的代码示例，其顶部的注释都会包含一个特殊标签"{NewFeature}"以及实现了该特性的 JDK 版本号（极少数情况下，你会见到一些未完成的特性）。同时，Gradle 构建时会自动排除带有"{NewFeature}"标签的示例。如果想要测试这些代码示例，你需要先安装对应的 JDK 版本，然后就可以通过命令行来编译对应的代码示例了。

出版说明

本书使用了自动化的构建过程，同样的自动化过程还有解压、编译以及测试所有示例代码。我使用 Python 3 编写了大量的应用程序来处理所有的自动化过程。

封面设计

本书的封面插图来自美国公共事业振兴署（Works Progress Administration，简称 WPA，是 1935—1943 年美国大萧条时期所创建的一个大型公共事业项目，其目标是援助失业人口重新返回工作岗位）。此外，它也让我想起了《绿野仙踪》系列丛书的插图。我的设计师朋友 Daniel Will-Harris 和我都十分喜爱这张图片。

致谢

Thinking in Java 一书面世至今，我很感谢它带给我的诸多益处，尤其是让我有机会在世界各地进行演讲。借此机会，我才得以与更多的人和公司建立联系，这是无价的。

感谢 Eric Evans（《领域驱动设计》一书的作者）针对本书书名提供了宝贵意见，也感谢所有在讨论组里帮助我确定书名的人们。

感谢 James Ward，他使我得以为本书使用 Gradle 构建工具，感谢他一直以来提供的帮助以及跟我的友谊。感谢 Ben Muschko 对构建文件所做的优化，同时也要感谢 Hans Dockter 给予 Ben 时间来做这件事。

感谢 Jeremy Cerise 与 Bill Frasure 参与本书的开发者活动，并提供了有价值的帮助。

感谢所有抽出宝贵时间莅临科罗拉多州克雷斯特德比特市，参加我所组织的会议、研讨活动、开发者活动以及其他活动的嘉宾们。即便你们做出的贡献不易让人察觉，但我想说的是，这些贡献依然是至关重要的。

献词

谨献给我敬爱的父亲 E. Wayne Eckel，他生于 1924 年 4 月 1 日，卒于 2016 年 11 月 23 日。

本书导读

"我的语言之局限，即我的世界之局限。"
——Ludwig Wittgenstein（1889—1951）

这句话不仅适用于我们日常读写的语言，也适用于编程语言。很微妙的一件事是，一门语言会悄然无息地引导你进入某种思维模式，同时远离其他思维模式。Java 尤其如此。

Java 是一门派生语言。当时的情况是，早期的语言设计师不想用 C++ 来开发项目，于是创建了一门和 C++ 极为相似的新编程语言，不过也做出了一些改进。这种新编程语言最主要的改动是加入了**虚拟机**和**垃圾收集**机制，本书后续章节会对这两点进行详细介绍。此外，Java 还在其他方面推动着行业的持续发展。

Java 最主要的概念之一来自 SmallTalk，这门语言强调"对象"（详见《On Java 中文版 基础卷》第 1 章）是编程的基本单位，所以任何东西都必须是对象。经历过长时间的洗礼之后，这个概念被证明是有些激进的，有些人甚至断定对象的概念是彻头彻尾的失败，应该果断丢弃。我个人认为，把所有内容都封装为对象不仅是一种负担，而且还会将许多程序设计推向错误的方向。然而不可否认的是，在一些情况下对象依然十分有用。所以，将一切都封装为对象（尤其是深入到最底层的时候）是一种设计失误，但完全抛弃对象同样太过极端。

Java 还有一些设计决策也没有达成预期目标。关于这一点，本书中会陆续加以说明，以便你能够理解这些语言特性如鲠在喉的原因何在。但是，我并不是要将 Java 盖棺定论为一门优秀或拙劣的语言。我想表达的是，如果你了解了一门语言的不足之处和局限性，当你遇到某个语言特性不可用时，就不会被卡住，以致无法继续。同时，因为你已经知晓其局限性，所以就可以更好地进行程序设计。

编程是一门管理复杂性的艺术，而问题的复杂程度取决于机器的复杂程度。由于这种复杂性的存在，导致了大多数编程项目的失败。

许多编程语言在设计时充分考虑了复杂性的问题，然而有时候，其他问题才是更为本质的问题。几乎不可避免的是，那些"其他问题"才是让使用该语言的程序员最终碰壁的原因。例如，C++ 语言不得不向后兼容 C 语言（这是为了让 C 语言程序员更容易上手），同时还要保证运行效率。不可否认的是，这两者都是非常实用的设计目标，并且成了 C++ 语言获得成功的功臣，但是随之也带来了大量额外的复杂性。

Visual Basic（VB）语言依赖于 BASIC 语言，而 BASIC 语言本身并不是一种扩展性良好的语言。这导致 VB 在扩展时经常出现各种非常难以维护的语法。Perl 语言能够向后兼容 awk、sed、grep 以及其他 UNIX 工具，然而这些旧时代的工具本身就是需要被替换和更新的。结果就是，Perl 程序里面充斥着大量的"只写代码"（意思是你自己都读不懂自己写的代码是什么意思）。不过话说回来，C++、VB、Perl 以及其他一些语言（比如 SmallTalk）都提供了一些能够处理复杂性的设计方案，并且从解决特定问题的角度来看，它们做得还相当不错。

信息革命让我们所有人可以更为便捷地交流，不管是一对一、在群组之内还是在全球范围内。我听说下一次革命将促生一个由足够多的人和连接组合而成的全球化的大脑。Java 会不会成为这种革命所需的工具之一呢？一切皆有可能。

本书目标

本书每一章都会介绍一个或者一组互相关联的概念，同时这些概念不依赖于当前章节没有介绍的特性。因此，你可以结合当前获取的知识来充分理解上下文，然后再阅读下一章。

我个人为本书设定的目标如下。

1. 循序渐进地呈现相关知识点，以便你充分理解每一个理念，之后再继续前行。同时，精心编排语言特性的介绍顺序，以便你在看到某个特性的运用之前，先对该特性的概念有所了解。然而我并不能保证百分之百可以做到这一点,当出现意外情况时,我也会提供一些简要的相关说明。

2. 所使用的示例尽可能地浅显易懂。有时候我会因为这一条原则而放弃引入所谓"现实世界"的问题,然而我发现对于初学者而言,相比于因为示例解决了一个范围很大的问题而感到惊讶,当他们理解示例中所有细节的时候会觉得更有收获。对于这一点,也许有人会批评我只热衷于"简单示例",但是为了产生更为明显的教育成效,我依然乐于接受目前的做法。

3. 我相信有些细节对于 95% 的程序员而言是无关紧要的。这些细节只会让人们感到困惑,并且增加他们对于语言复杂度的认知。

4. 为你打下坚实的编程语言基础,以便你之后学习难度更大的课程和图书时,可以充分理解自己所遇到的问题。

语言设计缺陷

每一种语言都存在设计缺陷。屡屡让新手程序员感到不安和挫败的是,他们必须"周旋"于各种语言特性之中,不断猜测应该用什么、不应该用什么。承认错误总是让人感到不快,但是相比承认错误所带来的不适感,这种糟糕的新手体验要严重得多。令人尴尬的是,所有失败的语言/库设计一直存在于 Java 的发布版本里。

诺贝尔经济学奖得主 Joseph Stiglitz 有一句生活哲言十分应景,也叫作"承诺升级理论"(The Theory of Escalating Commitment):

持续犯错的代价由别人承担,而承认错误的代价由你自己来承担。

当我发现编程语言的设计缺陷时,我倾向于指出这些问题。Java 发展到今天,已拥有了许多热心的拥护者,其中有些人甚至将 Java 视为自家"孩子",而非一种语言工具。因为我编写了一些关于 Java 的著作,所以他们以为我也会像他们一样袒护 Java。于是,当我发现了某个语言缺陷并进行批判时,经常会出现以下两种情况:

1. 起初会引起一阵类似于"我的孩子无论对错"的愤怒,到了最后(也许会经过许多年),该缺陷逐渐被大家广泛承认,从此被视为 Java 的历史遗留问题。

2. 更为关键的是,新手程序员并没有经历过"想不通为什么会这样"的痛苦挣扎,尤其是发现了某个看起来不对劲儿的地方之后所产生的自我怀疑,在这种情况下人们会很自然地认为**要么是自己做错了,要么就是自己还没有搞明白**。更糟糕的是,有些教授该语言的人会直接引用一些错误的概念,而不是对问题进行更加深入的研究和分析。而如果能够理解语言的设计缺陷，即使是新手程序员也能够理解不对劲儿的地方是一个错误，从而绕过它继续前行。

我认为，理解语言和库的设计缺陷是必要的，因为它们会影响程序员的生产力。有些语言特性非常具有吸引力，但可能会在你毫无准备之时突然卡住你的工作进程。此外，设计缺陷也会影响新语言的采用。探索一门语言能做什么的过程十分有趣，然而设计缺陷能够告诉你该语言**不能**做什么。

多年以来，我真切地感受到 Java 语言的设计者不够关心用户。有些语言缺陷可谓太过明显，根本没有经过深思熟虑，看起来像是设计者的思绪早已飞到了九霄云外，对自己的用户不管不顾。而这种对程序员看似不尊重的态度，也是我当初放弃 Java 选择其他语言，并且在相当长的一段时间内都不想回头的主要原因。

而当我重新回过头来审视 Java 的时候，Java 8 给我的感觉焕然一新，就好像是该语言的设计者对于语言和用户的态度发生了 180 度大转弯。比如，许多被用户诟病已久，甚至被视为语言毒瘤的特性和库都得到了修正。新引入的特性也让人耳目一新，就好像是设计团队中新加入了几位极其关注程序员使用体验的设计者。这些设计者终于行动了起来并致力于让 Java 语言变得更为出众，这明显好过在没有深入探究一个理念的本质时就急不可待地把它添加进来。此外，部分新特性十分优雅（至少可以说在考虑到 Java 局限性的情况下，已经尽可能地优雅了）。

得益于语言设计者的良苦用心（其实我并没有料想到这一点），编写本书的过程相比以往要顺利得多。Java 8 包含了许多基础和重要的改进，而由于 Java 一直严格遵守自己的向后兼容性承诺，做出这些改进无疑需要花费相当多的精力。因此可以预料的是，将来也很难再见到如此重大的改进了（关于这一点，希望我是错的）。话虽如此，我依然要为那些把 Java 重新带入正确航道的人献上掌声。当终于能够用 Java 8 编写出某段代码时，我第一次下意识地喊出："我爱死这个了！"

普及程度

Java 的普及具有重要意义。我的意思是，如果你学会了 Java，也许找工作会容易一些，而且市面上有大量的 Java 培训材料、课程以及其他学习资源等。另外，如果你开一家公司并且选择 Java 作为工作语言，招募 Java 程序员时也会容易一些。Java 的这一点优势确实无可争辩。

话虽如此，目光短浅终归不是好事。如果你并不是真心喜爱 Java，建议你还是远离它为好。我的意思是，如果学习 Java 只是为了找工作，无异于选择了一种不幸福的人生。而对于公司来说，如果你选择 Java 只是为了降低招聘难度，请务必三思而后行。根据你的实际需求，也许采用其他语言的话，你可以雇用更少的员工，但能达到更高的生产力（比如通过我的另一本书 *Atomic Kotlin* 学习 Kotlin 语言）。此外，使用一种更新也更激动人心的编程语言也许更容易吸引有志之士的加盟。

不过，如果你真的喜爱 Java 这门语言，那么欢迎你加入。同时，我希望本书可以进一步丰富你的编程经验。

Java 新的"发布节奏"

Java 的版本号总是显得十分怪异。比如 Java 早期的 1.1~1.4 版本使用带小数点的数字代表主版本号，到了 Java 5 则变成使用整数代表主版本号。

现在 Java 拥有了一套新的版本号规则，也可以称之为"发布节奏"，内容如下。

1. 每隔 6 个月发布一个新版本，使用整数作为版本号。

2. 发布的版本会包含一些试用功能，让用户可以体验和指出问题。而这种 6 个月的版本节奏，其主要目的可能就在于让用户尽早发现功能试用的相关问题。不过，由于无法保证这些功能之后能够长期存在，一旦这些功能出于某些原因没有达成预期的效果，它们就会被取缔。所以，你不应该依赖这些试用性质的功能。

3. 区分清楚短期支持（Short-Term-Support, STS）版本和长期支持（Long-Term-Support, LTS）版本。Java 8、11、17 都是 LTS 版本，其他版本则是支持周期只有 6 个月的 STS 版本。具体而言，只要有新版本问世，对 STS 版本的支持即宣告终止。类似地，一旦有新的 LTS 版本问世，（通常在一年以内）很快也会停止对原 LTS 版本的支持（这里指的是 Oracle 所提供的免费支持，也就是说，OpenJDK 可能会支持更长时间）。

值得一提的是，STS 版本和 LTS 版本都可能包含一些试用性质的功能。

此外，每一个 Java 版本都会包含不同类型的功能试用，举例如下。

- 实验（Experimental）：代表该功能仍处于早期阶段，可以认为完成度只有 25% 左右。
- 预览（Preview）：该功能已经完全实现，但是在最终确定之前仍然可能会有所调整。可以认为这些功能达到了 beta 版本，甚至是候选发布（Release Candidate,

RC）版本的标准。有时候会看到，某些功能带有标注"预览 2"（Preview 2），这大概表示此功能已经做出了一些修改，同时希望之后可以获得一些相关的反馈。

- 孵化中（Incubating）：代表一个 API 或工具（相对于语言的核心功能而言）还不是 Java 发布内容的一部分。因为 Java 的标准下载包并不会包含这些内容，所以必须主动获取这些 API 或工具才能使用它们。比如 jshell [1]，在 Java 8 里依然是一个孵化中的功能，然而从 Java 9 开始，它就成了正式发布版本的一部分。

实验和孵化中的功能被统称为"非正式"功能。非正式功能默认不会被启用，需要通过命令行或者 IDE 的设置菜单手动启用它们。本书之后的内容也会介绍一些 Java 17 所包含的非正式功能，并且相关示例也会指引如何通过命令行编译这些功能。

对于大多数公司和程序员来说，关注 STS 版本不仅可能需要付出额外的精力，而且使用这种生命周期较短的版本究竟有多少回报也让人存疑，所以我只推荐使用 LTS 版本。如果你只更新 LTS 版本，那就没什么问题了，而且无须担心 STS 版本的快速更新所带来的影响。

图形用户界面

对于 Java 而言，图形用户界面（GUI）和桌面编程代表着一段动荡甚至有些悲惨的历史。

在 Java 1.0 时代，GUI 库最初的设计目的是让程序员可以创建一种在所有平台上看起来都光鲜亮丽的 GUI。遗憾的是，这个目标并没有达成。取而代之的是，Java 1.0 通过**抽象窗口工具集**（Abstract Windowing Toolkit, AWT）创建了一种在所有平台上都表现平平的 GUI。不仅如此，这套 GUI 还有一些局限性。比如，你最多只能使用 4 种字体，而且你不能调用操作系统中任何成熟的 GUI 组件。此外，Java 1.0 AWT 的编程模式最令人尴尬的是，它甚至不支持面向对象编程。我的研讨班中的一名学生（他曾经在 Sun 公司经历过最初创造 Java 语言的那段时光）曾经解释过这一情况：最初的 AWT 是在一个月之内构想、设计和实现出来的。这样的产能效率纵然让人称奇，却也是体现框架设计重要性的一份反面教材。

随后发展到 Java 1.1 AWT 事件模型的时期，情况终于有所改善。这次的 AWT 使用一种更为清晰且面向对象的编程方式，同时添加了一种名为 JavaBeans 的组件编程模式（现已不复存在），其目的是可以轻松创建可视化的编程环境。到了 Java 2（也叫 Java 1.2）

---

[1] 一个交互式的编程工具（Read-Evaluate-Print-Loop, REPL），通过命令行可以直接输入 Java 语句并查看其输出结果。

时期，Java 不再继续改进 Java 1.0 AWT，而是用 Java 基础类（Java Foundation Classes, JFC）重写了一切，其中 GUI 部分称为"Swing"。通过 JavaBeans 及其丰富的代码库，用户可以创建出效果不错的 GUI。

然而，Swing 也不是 Java 语言 GUI 库的最终解决方案，随后 Sun 公司又做出了最后一次努力，推出了 JavaFX。当 Oracle 公司收购 Sun 公司后，他们将这个曾经野心勃勃的项目（其中甚至还包含了一种脚本语言）调整为 Java 的一个库，现在它似乎是唯一一个得以继续开发的 UI 工具包（详情请参考维基百科关于 JavaFX 的文章）。然而即便是这种程度的开发力度也难以为继，于是 JavaFX 和它的几个前辈一样，最终也难逃覆灭的命运。

现如今，Swing 依然是 Java 的一部分（不过只是维护，没有再开发新内容）。因为 Java 现在已经是开源项目，所以也可以轻松获取 Swing。此外，Swing 和 JavaFX 之间存在一些有限的交互，因其原本的目的是将 Swing 的功能移植到 JavaFX 中。

归根结底，Java 在桌面领域从未真正强大过，甚至从未触及设计师的雄心壮志。至于其他，比如 JavaBeans，也总是雷声大、雨点小（不幸的是，有不少人花费了大量心血来编写关于 Swing 的书，甚至是仅仅关于 JavaBeans 的书），始终没有获得大众的青睐。结果就是，Java 在桌面领域的大多数应用场景是 IDE 以及一些企业内部的应用程序。虽然人们确实也会用 Java 开发用户界面，但要清楚地意识到，这只是 Java 语言的一个小众需求。

JDK HTML 文档

Oracle 公司为 Java 开发工具集（Java Development Kit, JDK）提供了电子文档，用 Web 浏览器即可查看。除非必要，本书不会重复文档的内容，因为你用浏览器查看一个类的详细说明要比在本书中查找快得多（此外，在线文档的内容还是即时更新的）。所以在本书中，通常我只会提及某处需要参考"JDK 文档"。如果 JDK 文档的内容不足以让你理解某个特定的示例，我也会提供额外的说明。

经过测试的示例

本书提供的示例使用的是 Java 8 环境和 Gradle 编译工具。虽然我也使用新版本的 Java 测试过这些示例，但我依然推荐使用该语言的 LTS 版本：在我写这本书时，对应的是 Java 11 或 Java 17。此外，本书所有示例都可以从 GitHub 仓库免费获取。

每当构建一个应用程序时，如果没有一套内置测试流程来测试你的代码，就无法判断该代码是否坚实可信。因此，我为本书创建了一套测试系统，用于展示和验证大多数示例的输出结果。具体而言，运行示例代码后的输出结果会包含一段注释，附加在代码的末尾处。有时候注释并不显示全部内容，而是只显示开头的几行，或者开头和末尾的几行。这种嵌入式的输出方式提升了代码可读性，降低了学习门槛，同时也提供了一种验证代码正确性的方式。

代码规范

在本书中，各种标识符（关键字、方法名、变量名、类名等）会以等宽字体显示。而例如"类"（class）等频繁出现的关键字，如果使用特殊字体的话反而可能会让人感到不适。因此，具备足够辨识度的词语将采用常规字体显示。

本书示例会采用一种特定的编程风格。在尽可能满足本书格式要求的前提下，这种编程风格和 Oracle 网站上提供的编程风格几乎完全一致，同时能够兼容大多数 Java 开发环境。鉴于编程风格这个话题足以引发长达数小时的激烈争论，我需要在此澄清的是，我并没有试图通过我的代码示例来表明何为正确的编程风格，我使用的编程风格完全只是根据自己的意愿而为之。由于 Java 是一种形态自由的编程语言，所以你可以按照自己的喜好选择编程风格。此外，在使用诸如 IntelliJ IDEA 或者 Visual Studio Code（VSCode）等 IDE（Integrated Development Environment，集成开发环境）时，你可以设置自己熟悉的编程风格，以此解决编程风格不一致的问题。

本书的源代码都通过了自动化测试，最新版本的 Java 应该可以正常运行这些源代码（除了被特别标识的内容）。

bug 反馈

即使作者本人用尽各种办法来检测编程错误，依然可能会有漏网之鱼，通常新的读者可能会有所发现。在阅读本书的过程中，只要你确信自己发现了某处错误，不管是文字还是代码示例问题，请第一时间将该错误以及你修正后的内容提交到：https://github.com/BruceEckel/Onjava8-examples/issues。[①] 感谢你的帮助！

[①] 本书中文版勘误请提交到 ituring.cn/book/2935。——编者注

源代码

本书所有源代码都可以在 GitHub 网站上获取：https://github.com/BruceEckel/Onjava8-examples。这些源代码可以用于在校学习或者其他教育类场景。

源代码的版权保护主要是为了确保这些源代码可以被正确地引用，以及防止在未经授权的情况下被随意发布。（只要是本书中引用了版权信息的源代码，在大多数情况下，使用是没有问题的。）

在所有源代码文件里，你都会发现类似以下的版权信息说明：

```
// Copyright.txt
This computer source code is Copyright ©2021 MindView LLC.
All Rights Reserved.

Permission to use, copy, modify, and distribute this
computer source code (Source Code) and its documentation
without fee and without a written agreement for the
purposes set forth below is hereby granted, provided that
the above copyright notice, this paragraph and the
following five numbered paragraphs appear in all copies.

1. Permission is granted to compile the Source Code and to
include the compiled code, in executable format only, in
personal and commercial software programs.

2. Permission is granted to use the Source Code without
modification in classroom situations, including in
presentation materials, provided that the book "On
Java 8" is cited as the origin.

3. Permission to incorporate the Source Code into printed
media may be obtained by contacting:

MindView LLC, PO Box 969, Crested Butte, CO 81224
MindViewInc@gmail.com

4. The Source Code and documentation are copyrighted by
MindView LLC. The Source code is provided without express
or implied warranty of any kind, including any implied
warranty of merchantability, fitness for a particular
purpose or non-infringement. MindView LLC does not
warrant that the operation of any program that includes the
Source Code will be uninterrupted or error-free. MindView
LLC makes no representation about the suitability of the
Source Code or of any software that includes the Source
Code for any purpose. The entire risk as to the quality
and performance of any program that includes the Source
Code is with the user of the Source Code. The user
```

```
understands that the Source Code was developed for research
and instructional purposes and is advised not to rely
exclusively for any reason on the Source Code or any
program that includes the Source Code. Should the Source
Code or any resulting software prove defective, the user
assumes the cost of all necessary servicing, repair, or
correction.

5. IN NO EVENT SHALL MINDVIEW LLC, OR ITS PUBLISHER BE
LIABLE TO ANY PARTY UNDER ANY LEGAL THEORY FOR DIRECT,
INDIRECT, SPECIAL, INCIDENTAL, OR CONSEQUENTIAL DAMAGES,
INCLUDING LOST PROFITS, BUSINESS INTERRUPTION, LOSS OF
BUSINESS INFORMATION, OR ANY OTHER PECUNIARY LOSS, OR FOR
PERSONAL INJURIES, ARISING OUT OF THE USE OF THIS SOURCE
CODE AND ITS DOCUMENTATION, OR ARISING OUT OF THE INABILITY
TO USE ANY RESULTING PROGRAM, EVEN IF MINDVIEW LLC, OR
ITS PUBLISHER HAS BEEN ADVISED OF THE POSSIBILITY OF SUCH
DAMAGE. MINDVIEW LLC SPECIFICALLY DISCLAIMS ANY
WARRANTIES, INCLUDING, BUT NOT LIMITED TO, THE IMPLIED
WARRANTIES OF MERCHANTABILITY AND FITNESS FOR A PARTICULAR
PURPOSE. THE SOURCE CODE AND DOCUMENTATION PROVIDED
HEREUNDER IS ON AN "AS IS" BASIS, WITHOUT ANY ACCOMPANYING
SERVICES FROM MINDVIEW LLC, AND MINDVIEW LLC HAS NO
OBLIGATIONS TO PROVIDE MAINTENANCE, SUPPORT, UPDATES,
ENHANCEMENTS, OR MODIFICATIONS.

Please note that MindView LLC maintains a Web site which
is the sole distribution point for electronic copies of the
Source Code, https://github.com/BruceEckel/OnJava8-examples,
where it is freely available under the terms stated above.

If you think you've found an error in the Source Code,
please submit a correction at:
https://github.com/BruceEckel/OnJava8-examples/issues
```

在编程过程中，只要你在每一个源代码文件里都保留了上面提及的版权信息，这些源代码就可以用于你的项目以及在校学习等教育用途（包括幻灯片演示等文件）。

获取随书资源

扫描下方二维码，获取"随书源码"和"导读指南"。

目录

01 枚举类型 —— 001

1.1 枚举类型的基本特性 / 001
    静态导入枚举类型 / 002
1.2 在枚举类型中增加自定义方法 / 003
    重写枚举类型中的方法 / 004
1.3 在 switch 语句中使用枚举 / 005
1.4 values() 方法的神秘之处 / 006
1.5 实现，而不是继承 / 009
1.6 随机选择 / 010
1.7 使用接口来组织枚举 / 011
1.8 用 EnumSet 来代替标识 / 015
1.9 使用 EnumMap / 017
1.10 常量特定方法 / 018
    1.10.1 用枚举实现职责链模式 / 023
    1.10.2 用枚举实现状态机 / 027
1.11 多路分发 / 032
    1.11.1 使用枚举类型分发 / 034
    1.11.2 使用常量特定方法 / 036
    1.11.3 使用 EnumMap 分发 / 038
    1.11.4 使用二维数组 / 039
1.12 支持模式匹配的新特性 / 041
1.13 新特性：switch 中的箭头语法 / 041
1.14 新特性：switch 中的 case null / 042
1.15 新特性：将 switch 作为表达式 / 044
1.16 新特性：智能转型 / 046
1.17 新特性：模式匹配 / 048
    1.17.1 违反里氏替换原则 / 049
    1.17.2 守卫 / 053
    1.17.3 支配性 / 055
    1.17.4 覆盖范围 / 057
1.18 总结 / 058

02 对象传递和返回 —— 060

2.1 传递引用 / 061
    引用别名 / 061
2.2 创建本地副本 / 063
    2.2.1 值传递 / 063
    2.2.2 克隆对象 / 064
    2.2.3 为类增加可克隆能力 / 065
    2.2.4 成功的克隆 / 067
    2.2.5 Object.clone() 的效果 / 068
    2.2.6 克隆组合对象 / 070
    2.2.7 深拷贝 ArrayList / 073

2.2.8 通过序列化进行深拷贝 / 074

2.2.9 在继承层次结构中增加可克隆性并向下覆盖 / 076

2.2.10 为什么用这种奇怪的设计 / 077

2.3 控制可克隆性 / 078

复制构造器 / 082

2.4 不可变类 / 086

2.4.1 创建不可变类 / 088

2.4.2 不可变性的缺点 / 089

2.4.3 String 很特殊 / 091

2.5 总结 / 091

03 集合主题 — 093

3.1 样例数据 / 093

3.2 List 的行为 / 099

3.3 Set 的行为 / 102

3.4 在 Map 上使用函数式操作 / 104

3.5 选择 Map 的部分元素 / 105

3.6 填充集合 / 107

3.6.1 使用 Suppliers 来填充 Collection / 108

3.6.2 使用 Suppliers 来填充 Map / 109

3.7 使用享元自定义 Collection 和 Map / 112

3.8 Collection 的功能 / 123

3.9 可选的操作 / 125

不支持的操作 / 127

3.10 Set 与存储顺序 / 129

SortedSet / 132

3.11 Queue / 134

3.11.1 优先级队列 / 135

3.11.2 Deque / 136

3.12 理解 Map / 137

3.12.1 性能 / 139

3.12.2 SortedMap / 141

3.12.3 LinkedHashMap / 143

3.13 工具函数 / 144

3.13.1 List 上的排序和查找 / 147

3.13.2 创建不可修改的 Collection 或 Map / 148

3.13.3 同步 Collection 或 Map / 150

3.14 持有引用 / 151

WeakHashMap / 154

3.15 Java 1.0/1.1 的集合类 / 155

3.15.1 Vector 和 Enumeration / 155

3.15.2 Hashtable / 156

3.15.3 Stack / 157

3.15.4 BitSet / 158

3.16 总结 / 160

04 注解 — 162

4.1 基本语法 / 163
    4.1.1 定义注解 / 164
    4.1.2 元注解 / 165

4.2 编写注解处理器 / 166
    4.2.1 注解元素 / 167
    4.2.2 默认值的限制 / 168
    4.2.3 生成外部文件 / 168
    4.2.4 注解不支持继承 / 172
    4.2.5 实现处理器 / 172

4.3 用 javac 处理注解 / 174
    4.3.1 最简单的处理器 / 175
    4.3.2 更复杂的处理器 / 178

4.4 基于注解的单元测试 / 182
    4.4.1 在 @Unit 中使用泛型 / 190
    4.4.2 实现 @Unit / 192

4.5 总结 / 202

05 并发编程 — 203

5.1 令人迷惑的术语 / 204
    并发的新定义 / 206

5.2 并发的超能力 / 207

5.3 并发为速度而生 / 209

5.4 Java 并发四定律 / 211
    5.4.1 不要使用并发 / 211
    5.4.2 一切都不可信，一切很重要 / 212
    5.4.3 能运行并不代表没有问题 / 212
    5.4.4 你终究要理解并发 / 213

5.5 残酷的事实 / 214

5.6 本章剩余部分 / 216

5.7 并行流 / 218
    5.7.1 parallel() 并非灵丹妙药 / 219
    5.7.2 parallel() 和 limit() 的作用 / 224
    5.7.3 并行流只是看起来很简单 / 228

5.8 创建和运行任务 / 228
    5.8.1 Task 和 Executor / 228
    5.8.2 使用更多的线程 / 232
    5.8.3 生成结果 / 234
    5.8.4 作为任务的 lambda 与方法引用 / 236

5.9 终止长时间运行的任务 / 237

5.10 CompletableFuture / 240
    5.10.1 基本用法 / 241
    5.10.2 其他操作 / 244
    5.10.3 合并多个 CompletableFuture / 247
    5.10.4 模拟场景应用 / 251
    5.10.5 异常 / 253

5.11 死锁 / 259

5.12 构造器并不是线程安全的 / 264

5.13 工作量、复杂性、成本 / 268

5.14 总结 / 273

    5.14.1 缺点 / 274

    5.14.2 Java 核心设计的失败之处 / 275

    5.14.3 其他的库 / 275

    5.14.4 设想一种为并发而设计的语言 / 276

    5.14.5 延伸阅读 / 276

06 底层并发 —— 277

6.1 什么是线程？/ 277

    6.1.1 最佳线程数 / 279

    6.1.2 我可以创建多少线程 / 280

6.2 捕获异常 / 282

6.3 共享资源 / 286

    6.3.1 资源竞争 / 286

    6.3.2 解决资源竞争 / 290

    6.3.3 将 EvenProducer 同步化 / 292

6.4 volatile 关键字 / 293

    6.4.1 字分裂 / 293

    6.4.2 可见性 / 294

    6.4.3（指令）重排序和先行发生 / 295

    6.4.4 何时使用 volatile / 296

6.5 原子性 / 296

    6.5.1 Josh 的序列号 / 300

    6.5.2 原子类 / 303

6.6 临界区 / 304

    6.6.1 在其他对象上进行同步 / 307

    6.6.2 使用显式 Lock 对象 / 308

6.7 库组件 / 310

    6.7.1 延迟队列 DelayQueue / 311

    6.7.2 优先级阻塞队列 PriorityBlockingQueue / 313

    6.7.3 无锁集合 / 316

6.8 总结 / 317

07 Java I/O 系统 —— 319

7.1 I/O 流 / 319

    7.1.1 各种 InputStream 类型 / 321

    7.1.2 各种 OutputStream 类型 / 321

    7.1.3 添加属性和有用的接口 / 322

    7.1.4 各种 Reader 和 Writer / 324

    7.1.5 自成一家的 RandomAccessFile / 327

    7.1.6 I/O 流的典型用法 / 328

    7.1.7 小结 / 335

7.2 标准 I/O / 335

    7.2.1 从标准输入中读取 / 336

    7.2.2 将 System.out 转换为 PrintWriter / 336

    7.2.3 标准 I/O 重定向 / 337

    7.2.4 进程控制 / 338

7.3 新 I/O 系统 / 340

    7.3.1 字节缓冲区 ByteBuffer / 340

    7.3.2 转换数据 / 344

    7.3.3 获取基本类型 / 347

    7.3.4 视图缓冲区 / 348

7.3.5 用缓冲区操纵数据 / 352

7.3.6 内存映射文件 / 357

7.3.7 文件加锁 / 361

08 设计模式

364

8.1 设计模式的概念 / 364

8.2 单例模式 / 366

8.3 设计模式的分类 / 370

8.4 模板方法 / 370

8.5 封装实现 / 371

    8.5.1 代理模式 / 372

    8.5.2 状态模式 / 373

    8.5.3 状态机模式 / 376

8.6 工厂模式：封装对象的创建 / 377

    8.6.1 动态工厂模式 / 380

    8.6.2 多态工厂模式 / 381

    8.6.3 抽象工厂模式 / 382

8.7 函数对象模式 / 384

    8.7.1 命令模式 / 385

    8.7.2 策略模式 / 386

    8.7.3 职责链模式 / 388

8.8 改变接口 / 391

    8.8.1 适配器模式 / 391

    8.8.2 外观模式 / 392

8.9 解释器模式：运行时的灵活性 / 394

8.10 回调 / 394

    8.10.1 观察者模式 / 394

    8.10.2 示例：观察花朵 / 395

    8.10.3 一个可视化的观察者示例 / 398

8.11 多路分发 / 401

8.12 模式重构 / 405

    8.12.1 Trash 和它的子类 / 405

    8.12.2 信使对象 / 410

    8.12.3 使工厂通用化 / 411

    8.12.4 从文件解析 Trash / 412

    8.12.5 用 DynaFactory 实现回收 / 415

    8.12.6 将用法抽象化 / 416

    8.12.7 用多路分发重新设计 / 420

    8.12.8 访问者模式 / 427

    8.12.9 反射是有害的？ / 431

8.13 总结 / 435

A 编程指南 / 436

B Javadoc / 446

C 理解 equals() 和 hashCode() / 451

D 数据压缩 / 475

E 对象序列化 / 481

F 静态类型检查的利与弊 / 505

枚举类型

> enum 关键字用于创建一个新类型，其中包含一组数量有限的命名变量，并视这些变量为常规程序组件。实践表明这是一种非常有用的类型。[1]

《On Java 中文版 基础卷》[2]第 6 章 6.9 节曾对枚举类型做过简短的介绍。不过，鉴于你现在已经了解了 Java 的一些进阶知识，我们可以对 Java 枚举类型做一些更深入的讲解。你会看到枚举类型非常有用，而且还会领略更多的语言特性，比如泛型和反射等。同时你还会了解更多设计模式方面的知识。

1.1 枚举类型的基本特性

正如在基础卷第 6 章中所看到的，你可以调用枚举类型中的 `values()` 方法来遍历枚举常量列表。`values()` 方法生成一个由枚举常量组成的数组，其中常量的顺序和常量声明的顺序保持一致，这样你就可以方便地（比如通过 for-in 循环）使用结果数组了。

---

[1] Joshua Bloch 为本章的写作提供了极大的帮助。
[2] 后简称"基础卷"。——编者注

当创建枚举类型时，编译器会为你生成一个辅助类，这个类自动继承自 java.lang.Enum。java.lang.Enum 提供了下例所示的一些功能：

```java
// enums/EnumClass.java
// Enum 类的能力

enum Shrubbery { GROUND, CRAWLING, HANGING }

public class EnumClass {
  public static void main(String[] args) {
    for(Shrubbery s : Shrubbery.values()) {
      System.out.println(
        s + " ordinal: " + s.ordinal());
      System.out.print(
        s.compareTo(Shrubbery.CRAWLING) + " ");
      System.out.print(
        s.equals(Shrubbery.CRAWLING) + " ");
      System.out.println(s == Shrubbery.CRAWLING);
      System.out.println(s.getDeclaringClass());
      System.out.println(s.name());
      System.out.println("********************");
    }
    // 根据字符串名生成一个枚举值：
    for(String s :
        "HANGING CRAWLING GROUND".split(" ")) {
      Shrubbery shrub =
        Enum.valueOf(Shrubbery.class, s);
      System.out.println(shrub);
    }
  }
}
/* 输出：
GROUND ordinal: 0
-1 false false
class Shrubbery
GROUND
********************
CRAWLING ordinal: 1
0 true true
class Shrubbery
CRAWLING
********************
HANGING ordinal: 2
1 false false
class Shrubbery
HANGING
********************
HANGING
CRAWLING
GROUND
*/
```

ordinal() 方法返回一个从 0 开始的 int 值，代表每个枚举实例的声明顺序。你可以放心地使用 == 来比较枚举实例（equals() 和 hashCode() 方法会由编译器自动为你生成）。Enum 类实现了 Comparable 接口（因此可比较），所以自动包含了 compareTo() 方法，另外它还实现了 Serializable 接口（因此可序列化）。

如果调用枚举实例的 getDeclaringClass() 方法，则会得到该枚举实例所属的外部包装类。

name() 方法返回枚举实例被声明的名称，使用 toString() 同样也可以返回该名称。valueOf() 方法是 Enum 类中的静态方法，它根据传入的 String，返回名称与该 String 匹配的枚举实例。如果匹配的实例不存在，则抛出异常。

静态导入枚举类型

下面是基础卷第 6 章中 Burrito.java 类的一个变体：

```
// enums/SpicinessEnum.java
package enums;

public enum SpicinessEnum {
  NOT, MILD, MEDIUM, HOT, FLAMING
}

// enums/Burrito2.java
// {java enums.Burrito2}
package enums;
import static enums.SpicinessEnum.*;

public class Burrito2 {
  SpicinessEnum degree;
  public Burrito2(SpicinessEnum degree) {
    this.degree = degree;
  }
  @Override public String toString() {
    return "Burrito is "+ degree;
  }
  public static void main(String[] args) {
    System.out.println(new Burrito2(NOT));
    System.out.println(new Burrito2(MEDIUM));
    System.out.println(new Burrito2(HOT));
  }
}
/* 输出：
Burrito is NOT
Burrito is MEDIUM
Burrito is HOT
*/
```

　　`static import` 将所有的枚举实例标识符都引入了本地命名空间，因此它们不需要显式地使用枚举类型来限定。相较于显式地用枚举类型来限定枚举实例，哪种方式更好呢？这很大程度上要视代码的复杂程度而定。编译器肯定会保障类型的正确性，所以你唯一要关心的就是代码的可读性如何。一般来说不会有大问题，但还是要根据具体情况评估。

　　注意，如果枚举定义在同一个文件中，或者定义在默认包中，则无法使用该方式（显然在 Sun 公司的内部，对于是否允许这种情况有过一些争论）。

1.2　在枚举类型中增加自定义方法

　　对于枚举类型来说，除了无法继承它以外，基本可以将它看作一个普通的类。这意味着你可以在里面增加自定义的方法，甚至可以增加一个 main() 方法。

　　正如你所见，默认的 `toString()` 方法只会返回枚举实例的名称，而你很可能想为枚举实例生成不同于该默认方式的描述。为此，你可以实现一个构造方法，以获取额外的信息，然后再用额外的方法来提供扩展描述，如下例所示：

```java
// enums/OzWitch.java
// 《绿野仙踪》中的女巫

public enum OzWitch {
  // 实例必须在方法之前定义：
  WEST("Miss Gulch, aka the Wicked Witch of the West"),
  NORTH("Glinda, the Good Witch of the North"),
  EAST("Wicked Witch of the East, wearer of the Ruby " +
    "Slippers, crushed by Dorothy's house"),
  SOUTH("Good by inference, but missing");
  private String description;
  // 构造器的访问权限必须是包级或 private：
  private OzWitch(String description) {
    this.description = description;
  }
  public String getDescription() { return description; }
  public static void main(String[] args) {
    for(OzWitch witch : OzWitch.values())
      System.out.println(
        witch + ": " + witch.getDescription());
  }
}
/* 输出：
WEST: Miss Gulch, aka the Wicked Witch of the West
NORTH: Glinda, the Good Witch of the North
EAST: Wicked Witch of the East, wearer of the Ruby
Slippers, crushed by Dorothy's house
SOUTH: Good by inference, but missing
*/
```

如果你想增加自定义方法，则必须先用分号结束枚举实例的序列。同时，Java 会强制你在枚举中先定义实例。如果在定义实例之前定义了任何方法或字段，则会抛出编译时错误。

枚举类型的构造器和方法的写法与普通类一样，因为除了少量特殊限制外，它就是一个普通的类。你几乎可以对它做任何你想做的事（虽然你通常只会使用最简单的枚举类型）。

虽然本例中的构造器是私有的，但使用哪种访问权限实际上区别并不大：构造器只能用来创建你在枚举定义中声明的枚举实例；在枚举定义完成后，编译器不会允许你用它来创建任何新的类型。

重写枚举类型中的方法

还有另一种为枚举生成不同的 String 值的方式：重写 toString() 方法。在下面的示例中，实例名没什么问题，但我们希望换一种格式来显示。重写 enum 的 toString() 方法和重写任何普通类的方法相同：

```java
// enums/SpaceShip.java
import java.util.stream.*;

public enum SpaceShip {
  SCOUT, CARGO, TRANSPORT,
  CRUISER, BATTLESHIP, MOTHERSHIP;
  @Override public String toString() {
    String id = name();
    String lower = id.substring(1).toLowerCase();
    return id.charAt(0) + lower;
  }
  public static void main(String[] args) {
    Stream.of(values)
      .forEach(System.out::println);
  }
}
/* 输出：
Scout
Cargo
Transport
Cruiser
Battleship
Mothership
*/
```

toString() 方法通过调用 name() 方法获取 SpaceShip 的名称，并且修改了结果，使得结果中的英文单词仅首字母为大写。

1.3 在 switch 语句中使用枚举

枚举类型的一个非常方便之处就是可以用在 switch 语句中。通常，switch 语句只能使用整型或字符串类型的值，但是由于 enum 内部已经构建了一个整型序列，并且可以通过 ordinal() 方法来得到枚举实例的顺序（显然编译器做了相应的工作），所以枚举类型可以用在 switch 语句中。

虽然通常要使用枚举实例，就必须用枚举的类型名来限定它，但在 case 语句中你无须这么做。以下示例中使用了枚举来创建一个简单的状态机：

```java
// enums/TrafficLight.java
// 在 switch 语句中使用枚举

// 定义一个枚举类型：
enum Signal { GREEN, YELLOW, RED, }

public class TrafficLight {
  Signal color = Signal.RED;
  public void change() {
    switch(color) {
      // 注意在 case 语句中，无须使用 Signal.RED
      case RED:    color = Signal.GREEN;
                   break;
      case GREEN:  color = Signal.YELLOW;
/* 输出：
The traffic light is RED
The traffic light is GREEN
The traffic light is YELLOW
The traffic light is RED
The traffic light is GREEN
The traffic light is YELLOW
The traffic light is RED
*/
```

```
                        break;
      case YELLOW: color = Signal.RED;
                        break;
    }
  }
  @Override public String toString() {
    return "The traffic light is " + color;
  }
  public static void main(String[] args) {
    TrafficLight t = new TrafficLight();
    for(int i = 0; i < 7; i++) {
      System.out.println(t);
      t.change();
    }
  }
}
```

编译器没有因为该 switch 中没有 default 语句而报错，但这并不是因为它注意到了你为每个 Signal 实例都编写了 case 分支。即使你注释掉某个 case 分支，也仍然不会报错。这意味着你必须要小心，确保你手动覆盖了所有的分支。但是，如果此时你在 case 语句中调用了 return（而且没有编写 dafault），则编译器**会**报错，即使你已经覆盖到了枚举中的所有值。

1.4　values() 方法的神秘之处

本书之前提到过，所有的枚举类型都是由编译器通过继承 Enum 类来创建的。然而，如果仔细查看 Enum 类的代码，你会发现里面并没有 values() 方法，我们却已经能直接使用它了。是哪里有其他"隐藏"的方法吗？我们可以编写一个小的反射程序来一探究竟：

```
// enums/Reflection.java
// 使用反射分析枚举类
import java.lang.reflect.*;
import java.util.*;
import onjava.*;

enum Explore { HERE, THERE }

public class Reflection {
  public static
  Set<String> analyze(Class<?> enumClass) {
    System.out.println(
      "_____ Analyzing " + enumClass + " _____");
    System.out.println("Interfaces:");
    for(Type t : enumClass.getGenericInterfaces())
      System.out.println(t);
    System.out.println(
      "Base: " + enumClass.getSuperclass());
    System.out.println("Methods: ");
```

```
    Set<String> methods = new TreeSet<>();
    for(Method m : enumClass.getMethods())
      methods.add(m.getName());
    System.out.println(methods);
    return methods;
  }
  public static void main(String[] args) {
    Set<String> exploreMethods =
      analyze(Explore.class);
    Set<String> enumMethods = analyze(Enum.class);
    System.out.println(
      "Explore.containsAll(Enum)? " +
      exploreMethods.containsAll(enumMethods));
    System.out.print("Explore.removeAll(Enum): ");
    exploreMethods.removeAll(enumMethods);
    System.out.println(exploreMethods);
    // 反编译 enum：
    OSExecute.command(
      "javap -cp build/classes/java/main Explore");
  }
}
```

```
/* 输出：
_____ Analyzing class Explore _____
Interfaces:
Base: class java.lang.Enum
Methods:
[compareTo, equals, getClass, getDeclaringClass,
hashCode, name, notify, notifyAll, ordinal, toString,
valueOf, values, wait]
_____ Analyzing class java.lang.Enum _____
Interfaces:
java.lang.Comparable<E>
interface java.io.Serializable
Base: class java.lang.Object
Methods:
[compareTo, equals, getClass, getDeclaringClass,
hashCode, name, notify, notifyAll, ordinal, toString,
valueOf, wait]
Explore.containsAll(Enum)? true
Explore.removeAll(Enum): [values]
Compiled from "Reflection.java"
final class Explore extends java.lang.Enum<Explore> {
  public static final Explore HERE;
  public static final Explore THERE;
  public static Explore[] values();
  public static Explore valueOf(java.lang.String);
  static {};
}
*/
```

答案揭晓，values() 方法是由编译器添加的一个静态方法。注意在创建枚举的过程中，valueOf() 方法同样也被添加到了 Explore 枚举中。这有点让人糊涂，Enum 类中同样也有一个 valueOf() 方法，但是该方法有 2 个参数，而新加入的方法则只有 1 个。然而，这里的 Set 方法只关心方法名，并不关心方法签名，所以在调用 Explore.removeAll(Enum) 后，只剩下了 [values]。

打印结果显示 Explore 枚举被编译器限定为 final 类，所以你无法继承一个枚举类。此外还有一个 static 的初始化子句，你稍后会看到它可以被重定义。

由于类型擦除（相关介绍参见基础卷第 20 章）的缘故，反编译器得不到 Enum 类的完整信息，因此只能将 Explore 类的基类作为一个原始的 Enum 类来显示，而不是实际上的 Enum<Explore>。

由于 values() 方法是由编译器在枚举类的定义中插入的一个静态方法，因此如果你将枚举类型向上转型为 Enum，则 values() 方法将不可用。然而要注意的是，Class 中有个 getEnumConstants() 方法，所以即使 Enum 的接口中没有 values() 方法，仍然可以通过 Class 对象来得到 enum 的实例：

```java
// enums/UpcastEnum.java
// 如果向上转型枚举，便会丢失 values() 方法

enum Search { HITHER, YON }

public class UpcastEnum {
  public static void main(String[] args) {
    Search[] vals = Search.values();
    Enum e = Search.HITHER; // 向上转型
    // e.values(); // Enum 中没有 values() 方法
    for(Enum en : e.getClass().getEnumConstants())
      System.out.println(en);
  }
}
/* 输出：
HITHER
YON
*/
```

由于 getEnumConstants() 是 Class 类中的一个方法，因此对一个没有枚举的类也可以调用该方法。

```java
// enums/NonEnum.java

public class NonEnum {
  public static void main(String[] args) {
    Class<Integer> intClass = Integer.class;
    try {
      for(Object en : intClass.getEnumConstants())
        System.out.println(en);
    } catch(Exception e) {
```

```
      System.out.println("Expected: " + e);
    }
  }
}
```

```
/* 输出:
Expected: java.lang.NullPointerException
*/
```

该方法会返回 null，因此如果你尝试引用该结果，就会抛出异常。

1.5 实现，而不是继承

我们已经确认所有的 enum 对象都继承自 java.lang.Enum。Java 不支持多重继承，这意味着你无法通过以下这样的继承方式创建一个枚举对象：

```
enum NotPossible extends Pet { ... // 无法执行
```

不过，可以创建实现了一个或多个接口的枚举类型：

```
// enums/cartoons/EnumImplementation.java
// 枚举类型可以实现接口
// {java enums.cartoons.EnumImplementation}
package enums.cartoons;
import java.util.*;
import java.util.function.*;

enum CartoonCharacter
implements Supplier<CartoonCharacter> {
  SLAPPY, SPANKY, PUNCHY,
  SILLY, BOUNCY, NUTTY, BOB;
  private Random rand =
    new Random(47);
  @Override public CartoonCharacter get() {
    return values()[rand.nextInt(values().length)];
  }
}

public class EnumImplementation {
  public static <T> void printNext(Supplier<T> rg) {
    System.out.print(rg.get() + ", ");
  }
  public static void main(String[] args) {
    // 选择任一实例:
    CartoonCharacter cc = CartoonCharacter.BOB;
    for(int i = 0; i < 10; i++)
      printNext(cc);
  }
}
/* 输出:
BOB, PUNCHY, BOB, SPANKY, NUTTY, PUNCHY, SLAPPY, NUTTY,
NUTTY, SLAPPY,
*/
```

结果看起来有点奇怪,因为你必须先得有一个枚举实例,才能在其上调用某个方法。但是此处的 CartoonCharacter 对象可以传入任何将 Supplier 对象作为参数的方法,比如 printNext()。

1.6 随机选择

本章中的许多示例需要从 enum 中随机选择实例,正如你在 CartoonCharacter.get() 中看到的一样。可以使用泛型将这项任务的实现抽象成公共能力,并放到公共库中:

```
// onjava/Enums.java
package onjava;
import java.util.*;

public class Enums {
  private static Random rand = new Random(47);
  public static
  <T extends Enum<T>> T random(Class<T> ec) {
    return random(ec.getEnumConstants());
  }
  public static <T> T random(T[] values) {
    return values[rand.nextInt(values.length)];
  }
}
```

这个看起来相当奇怪的语法 <T extends Enum<T>>,声明了 T 是一个枚举的实例。通过传入 Class<T>,使得这个 class 对象变得可用,从而可以生成枚举实例的数组。重载的 random() 方法只需要知道传给自己的参数是一个 T[],因为它并不执行具体的对 Enum 的操作,只是随机地选择一个数组元素即可。而返回的对象类型则是确切的 enum 类型。

下面对 random() 方法做个简单的测试:

```
// enums/RandomTest.java
import onjava.*;

enum Activity { SITTING, LYING, STANDING, HOPPING,
  RUNNING, DODGING, JUMPING, FALLING, FLYING }

public class RandomTest {
  public static void main(String[] args) {
    for(int i = 0; i < 20; i++)
      System.out.print(
        Enums.random(Activity.class) + " ");
  }
}
```

```
/* 输出:
STANDING FLYING RUNNING STANDING RUNNING STANDING LYING
DODGING SITTING RUNNING HOPPING HOPPING HOPPING RUNNING
STANDING LYING FALLING RUNNING FLYING LYING
*/
```

虽然 Enum 是个很简单的类，但在本章中你可以看到它避免了相当多的重复操作。重复操作有带来错误的风险，因此需要尽量避免。

1.7 使用接口来组织枚举

枚举类型无法被继承，这一点可能有时会让人沮丧。想要继承枚举的动机，一部分源自希望扩充原始枚举中的元素，另一部分源自想要使用子类型来创建不同的子分组。

你可以在一个接口内对元素进行分组，然后基于这个接口生成一个枚举，通过这样的方式来实现元素的分类。举个例子，假如你有一些类型互不相同的 food（食物），想创建若干 enum 来组织它们，但又希望它们仍然是 Food 类型。你可以这么做：

```java
// enums/menu/Food.java
// 在接口内对枚举进行子归类
package enums.menu;

public interface Food {
  enum Appetizer implements Food {
    SALAD, SOUP, SPRING_ROLLS;
  }
  enum MainCourse implements Food {
    LASAGNE, BURRITO, PAD_THAI,
    LENTILS, HUMMUS, VINDALOO;
  }
  enum Dessert implements Food {
    TIRAMISU, GELATO, BLACK_FOREST_CAKE,
    FRUIT, CREME_CARAMEL;
  }
  enum Coffee implements Food {
    BLACK_COFFEE, DECAF_COFFEE, ESPRESSO,
    LATTE, CAPPUCCINO, TEA, HERB_TEA;
  }
}
```

实现接口是唯一可子类化枚举的方式，因此所有嵌套在 Food 中的枚举类型都实现了 Food 接口。现在我们基本可以说"一切都是某种类型的 Food"，如下例所示：

```java
// enums/menu/TypeOfFood.java
// {java enums.menu.TypeOfFood}
package enums.menu;
import static enums.menu.Food.*;

public class TypeOfFood {
  public static void main(String[] args) {
    Food food = Appetizer.SALAD;
    food = MainCourse.LASAGNE;
    food = Dessert.GELATO;
    food = Coffee.CAPPUCCINO;
  }
}
```

对于每个实现了 Food 接口的枚举类型，都可以向上转型为 Food，因此它们全都是 Food 类型。

但是，当你要处理一组类型时，接口往往就不如枚举有用。如果要创建"由枚举组成的枚举"，你可以为 Food 中的每个枚举类型都创建一个外部枚举类型：

```java
// enums/menu/Course.java
package enums.menu;
import onjava.*;

public enum Course {
  APPETIZER(Food.Appetizer.class),
  MAINCOURSE(Food.MainCourse.class),
  DESSERT(Food.Dessert.class),
  COFFEE(Food.Coffee.class);
  private Food[] values;
  private Course(Class<? extends Food> kind) {
    values = kind.getEnumConstants();
  }
  public Food randomSelection() {
    return Enums.random(values);
  }
}
```

上面的代码中，每个枚举类型都接收相应的 Class 对象以作为构造参数，从而可以使用 getEnumConstants() 来提取出所有的枚举实例并存储起来，这些实例稍后会在 randomSelection() 中被用到。好了，现在我们可以从每个 Course（菜项，如 "前菜 / 主菜 / 甜点"）中选择一种 Food（食物），生成一份随机配好的午餐了。

```java
// enums/menu/Meal.java
// {java enums.menu.Meal}
package enums.menu;

public class Meal {
```

```java
  public static void main(String[] args) {
    for(int i = 0; i < 5; i++) {
      for(Course course : Course.values()) {
        Food food = course.randomSelection();
        System.out.println(food);
      }
      System.out.println("***");
    }
  }
}
```

```
/* 输出:
SPRING_ROLLS
VINDALOO
FRUIT
DECAF_COFFEE
***
SOUP
VINDALOO
FRUIT
TEA
***
SALAD
BURRITO
FRUIT
```
（转右栏）

```
TEA
***
SALAD
BURRITO
CREME_CARAMEL
LATTE
***
SOUP
BURRITO
TIRAMISU
ESPRESSO
***
*/
```

此处，创建一个由枚举类型组成的枚举类型的意义在于，可以方便地遍历每个 Course。在稍后的例子 VendingMachine.java 中，你会看到另一种由不同约束条件指定的分类方法。

另一种更简洁的分类方法是在枚举内嵌套枚举，如下例所示：

```java
// enums/SecurityCategory.java
// 更简洁的枚举子归类
import onjava.*;

enum SecurityCategory {
  STOCK(Security.Stock.class),
  BOND(Security.Bond.class);
  Security[] values;
  SecurityCategory(Class<? extends Security> kind) {
    values = kind.getEnumConstants();
  }
  interface Security {
    enum Stock implements Security {
      SHORT, LONG, MARGIN
    }
    enum Bond implements Security {
```

```
/* 输出:
BOND: MUNICIPAL
BOND: MUNICIPAL
STOCK: MARGIN
STOCK: MARGIN
BOND: JUNK
STOCK: SHORT
STOCK: LONG
STOCK: LONG
BOND: MUNICIPAL
BOND: JUNK
*/
```

```java
      MUNICIPAL, JUNK
  }
}
public Security randomSelection() {
  return Enums.random(values);
}
public static void main(String[] args) {
  for(int i = 0; i < 10; i++) {
    SecurityCategory category =
      Enums.random(SecurityCategory.class);
    System.out.println(category + ": " +
      category.randomSelection());
  }
}
}
```

Security 接口用于将内部的枚举类型作为公共类型聚合到一起，然后再将它们归类到 SecurityCategory 中的枚举中。

如果将此方法应用到前面的 Food 示例，那么结果如下：

```java
// enums/menu/Meal2.java
// {java enums.menu.Meal2}
package enums.menu;
import onjava.*;

public enum Meal2 {
  APPETIZER(Food.Appetizer.class),
  MAINCOURSE(Food.MainCourse.class),
  DESSERT(Food.Dessert.class),
  COFFEE(Food.Coffee.class);
  private Food[] values;
  private Meal2(Class<? extends Food> kind) {
    values = kind.getEnumConstants();
  }
  public interface Food {
    enum Appetizer implements Food {
      SALAD, SOUP, SPRING_ROLLS;
    }
    enum MainCourse implements Food {
      LASAGNE, BURRITO, PAD_THAI,
      LENTILS, HUMMUS, VINDALOO;
    }
    enum Dessert implements Food {
      TIRAMISU, GELATO, BLACK_FOREST_CAKE,
      FRUIT, CREME_CARAMEL;
    }
    enum Coffee implements Food {
      BLACK_COFFEE, DECAF_COFFEE, ESPRESSO,
      LATTE, CAPPUCCINO, TEA, HERB_TEA;
    }
```

```
/* 输出：
SPRING_ROLLS
VINDALOO
FRUIT
DECAF_COFFEE
***
SOUP
VINDALOO
FRUIT
TEA
***
SALAD
BURRITO
FRUIT
TEA
***
SALAD
BURRITO
CREME_CARAMEL
LATTE
***
SOUP
BURRITO
TIRAMISU
ESPRESSO
***
*/
```

```
    }
    public Food randomSelection() {
      return Enums.random(values);
    }
    public static void main(String[] args) {
      for(int i = 0; i < 5; i++) {
        for(Meal2 meal : Meal2.values()) {
          Food food = meal.randomSelection();
          System.out.println(food);
        }
        System.out.println("***");
      }
    }
  }
```

最后，虽然这只是对代码重新组织了一下，但在某些情况下，这样可以使结构更加清晰。

1.8 用 EnumSet 来代替标识

Set 是一种不允许有重复元素存在的集合。enum 要求每个内部成员都是唯一的，因此看起来很像 Set，但是由于无法添加或移除元素，它并不如 Set 那么好用。于是 EnumSet 被引入，用来配合 enum 的使用，以替代传统的基于 int 的"位标识"用法。这种标识通常被用来表明某些开 - 关状态信息，但最终你实际上是在操作各种位状态，而不是业务逻辑，所以非常容易写出难以理解的代码。

速度是 EnumSet 的设计目标之一，因为它需要和位标识竞争（位操作的性能通常远高于 HashSet）。其内部实现其实是一个被用作位数组的 long 型变量，所以它非常高效。这样做的好处是，你现在拥有了一种更具表现力的方式来表达二进制特征的存在与否，而且无须担心速度。

EnumSet 中的元素必须来自某个枚举类型。下面是一个假想的例子，使用枚举来表示一栋大楼中警报感应器的安装位置：

```
// enums/AlarmPoints.java
package enums;
public enum AlarmPoints {
  STAIR1, STAIR2, LOBBY, OFFICE1, OFFICE2, OFFICE3,
  OFFICE4, BATHROOM, UTILITY, KITCHEN
}
```

EnumSet 持续跟踪警报的状态：

```
// enums/EnumSets.java
// EnumSets 的操作
```

```java
// {java enums.EnumSets}
package enums;
import java.util.*;
import static enums.AlarmPoints.*;

public class EnumSets {
  public static void main(String[] args) {
    EnumSet<AlarmPoints> points =
      EnumSet.noneOf(AlarmPoints.class); // 为空
    points.add(BATHROOM);
    System.out.println(points);
    points.addAll(
      EnumSet.of(STAIR1, STAIR2, KITCHEN));
    System.out.println(points);
    points = EnumSet.allOf(AlarmPoints.class);
    points.removeAll(
      EnumSet.of(STAIR1, STAIR2, KITCHEN));
    System.out.println(points);
    points.removeAll(
      EnumSet.range(OFFICE1, OFFICE4));
    System.out.println(points);
    points = EnumSet.complementOf(points);
    System.out.println(points);
  }
}
/* 输出:
[BATHROOM]
[STAIR1, STAIR2, BATHROOM, KITCHEN]
[LOBBY, OFFICE1, OFFICE2, OFFICE3, OFFICE4, BATHROOM, UTILITY]
[LOBBY, BATHROOM, UTILITY]
[STAIR1, STAIR2, OFFICE1, OFFICE2, OFFICE3, OFFICE4, KITCHEN]
*/
```

静态导入（static import）的作用是简化枚举常量的使用。方法名的自解释性相当好，你还可以在 JDK 的文档里找到完整的细节。如果你仔细看文档，会发现 of() 方法分别以可变参数和接收 3~5 个显式参数的方式进行了重载。这体现了 EnumSet 对性能的关注，因为本来只需要一个接收可变参数的 of() 方法就能解决问题，但是这样会比通过显式参数的方式略为低效。因此，如果用 2~5 个参数来调用 of() 方法，实际起作用的是显式调用（速度稍快）；如果用 1 个或 5 个以上的参数来调用，则会是可变参数的版本。注意，如果用 1 个参数来调用，编译器并不会构建可变参数数组，因此这种情况下不会产生额外的开销。

EnumSet 是基于 64 位的 long 构建的，每个枚举实例需要占用 1 位来表达是否存在的状态，这意味着在单个 long 的支撑范围内，1 个 EnumSet 最多可支持包含 64 个元素的枚举类型。如果枚举类型中的元素超过了 64 个，会发生什么呢？

```java
// enums/BigEnumSet.java
import java.util.*;

public class BigEnumSet {
  enum Big { A0, A1, A2, A3, A4, A5, A6, A7, A8, A9,
    A10, A11, A12, A13, A14, A15, A16, A17, A18, A19,
    A20, A21, A22, A23, A24, A25, A26, A27, A28, A29,
    A30, A31, A32, A33, A34, A35, A36, A37, A38, A39,
    A40, A41, A42, A43, A44, A45, A46, A47, A48, A49,
    A50, A51, A52, A53, A54, A55, A56, A57, A58, A59,
    A60, A61, A62, A63, A64, A65, A66, A67, A68, A69,
    A70, A71, A72, A73, A74, A75 }
  public static void main(String[] args) {
    EnumSet<Big> bigEnumSet = EnumSet.allOf(Big.class);
    System.out.println(bigEnumSet);
  }
}
/* 输出：
[A0, A1, A2, A3, A4, A5, A6, A7, A8, A9, A10, A11, A12,
A13, A14, A15, A16, A17, A18, A19, A20, A21, A22, A23,
A24, A25, A26, A27, A28, A29, A30, A31, A32, A33, A34,
A35, A36, A37, A38, A39, A40, A41, A42, A43, A44, A45,
A46, A47, A48, A49, A50, A51, A52, A53, A54, A55, A56,
A57, A58, A59, A60, A61, A62, A63, A64, A65, A66, A67,
A68, A69, A70, A71, A72, A73, A74, A75]
*/
```

显然 EnumSet 可以支持包含超过 64 个元素的枚举类型，所以我们可以推测，它在必要的时候会引入新的 long 型变量。

1.9 使用 EnumMap

EnumMap 是一种特殊的 Map，它要求自身所有的键来自某个枚举类型。由于枚举的约束（元素和索引的映射关系与数组相似），EnumMap 的内部可以作为一个数组来实现，因此它们的性能非常好，你可以放心地用 EnumMap 来实现基于枚举的查询。

你只能用枚举中的元素作为键来调用 put() 方法，除此之外，就和调用一个普通的 Map 没什么区别了。

下面的例子演示了设计模式中的**命令**模式。这种模式由一个（通常）只包含一个方法的接口开始，然后为该方法创建多个具有不同行为的实现。只需要配置好这些命令对象，程序就会根据需要来调用它们。

```java
// enums/EnumMaps.java
// EnumMaps 基础
// {java enums.EnumMaps}
```

```
package enums;
import java.util.*;
import static enums.AlarmPoints.*;

interface Command { void action(); }

public class EnumMaps {
  public static void main(String[] args) {
    EnumMap<AlarmPoints,Command> em =
      new EnumMap<>(AlarmPoints.class);
    em.put(KITCHEN,
      () -> System.out.println("Kitchen fire!"));
    em.put(BATHROOM,
      () -> System.out.println("Bathroom alert!"));
    for(Map.Entry<AlarmPoints,Command> e:
         em.entrySet()) {
      System.out.print(e.getKey() + ": ");
      e.getValue().action();
    }
    try { // 如果指定的 key 没有对应值：
      em.get(UTILITY).action();
    } catch(Exception e) {
      System.out.println("Expected: " + e);
    }
  }
}
/* 输出：
BATHROOM: Bathroom alert!
KITCHEN: Kitchen fire!
Expected: java.lang.NullPointerException
*/
```

和 EnumSet 一样，EnumMap 中的元素顺序由它们在枚举中定义的顺序决定。

通过 main() 函数中的末尾部分，可以看到每个枚举都有个键，但是值都是 null，除非在该键上调用过 put() 方法。

相较于**常量特定方法**（constant-specific method，参见本书第 4 章），EnumMap 的优势在于：利用 EnumMap，你可以改变值对象。而你可以看到常量特定方法在编译时是不可变的。

正如你在本章稍后会看到的，EnumMap 支持**多路分发**（multiple dispatching），以应对多个类型的枚举共存且相互影响的各种场景。

1.10 常量特定方法

Java 的枚举机制可以通过为每个枚举实例编写不同的方法，来赋予它们不同的行为。要实现这一点，你可以在枚举类型中定义一个或多个抽象方法，然后为每个枚举实例编写不同的实现。例如：

```java
// enums/ConstantSpecificMethod.java
import java.util.*;
import java.text.*;

public enum ConstantSpecificMethod {
  DATE_TIME {
    @Override String getInfo() {
      return
        DateFormat.getDateInstance()
          .format(new Date());
    }
  },
  CLASSPATH {
    @Override String getInfo() {
      return System.getenv("CLASSPATH");
    }
  },
  VERSION {
    @Override String getInfo() {
      return System.getProperty("java.version");
    }
  };
  abstract String getInfo();
  public static void main(String[] args) {
    for(ConstantSpecificMethod csm : values())
      System.out.println(csm.getInfo());
  }
}
/* 输出:
Jan 24, 2021
C:\Git\OnJava8\ExtractedExamples\\gradle\wrapper\gradle
-wrapper.jar
1.8.0_41
*/
```

你可以通过关联的枚举实例来查找和调用方法。这通常叫作**表驱动模式**（注意，和前面的命令模式很相似）。

在面向对象编程中，不同的行为和不同的类相关联。通过常量特定方法，枚举类型的各种实例可以拥有各自的行为，这表明每个实例都是不同的类型。在上面的例子中，每个枚举实例都被视同于 "基类" ConstantSpecificMethod 的实例，但调用 getInfo() 方法时的行为是多态的。

然而，这两者的相似性也只能到此为止，你无法将 enum 实例等同于类类型：

```java
// enums/NotClasses.java
// {ExcludeFromGradle}
// javap -c LikeClasses
```

```java
enum LikeClasses {
  WINKEN {
    @Override void behavior() {
      System.out.println("Behavior1");
    }
  },
  BLINKEN {
    @Override void behavior() {
      System.out.println("Behavior2");
    }
  },
  NOD {
    @Override void behavior() {
      System.out.println("Behavior3");
    }
  };
  abstract void behavior();
}

public class NotClasses {
  // void f1(LikeClasses.WINKEN instance) {} // 不行
}
/* 输出（前12行）:
Compiled from "NotClasses.java"
abstract class LikeClasses extends
java.lang.Enum<LikeClasses> {
  public static final LikeClasses WINKEN;

  public static final LikeClasses BLINKEN;

  public static final LikeClasses NOD;

  public static LikeClasses[] values();
    Code:
       0: getstatic       #2                  // 字段
$VALUES:[LLikeClasses;
       3: invokevirtual #3                    // 方法
"[LLikeClasses;".clone:()Ljava/lang/Object;
                   ...
*/
```

在 f1() 方法中，编译器不允许将枚举实例作为类类型来使用。只要设想一下编译器是如何生成代码的，这一点就说得通了——每个枚举元素都是 LikeClasses 的一个 static final 的实例。

同样，由于它们是静态的，内部枚举中的枚举实例表现得并不像普通的内部类，你无法从外部类访问非静态域或方法。

再来看一个洗车的例子。每个用户都拿到了一个洗车选项的菜单，每个选项都代表不同的操作。每个选项都可以分配一个常量特定方法,然后用一个 EnumSet 来持有用户的选择。

```java
// enums/CarWash.java
import java.util.*;

public class CarWash {
  public enum Cycle {
    UNDERBODY {
      @Override void action() {
        System.out.println("Spraying the underbody");
      }
    },
    WHEELWASH {
      @Override void action() {
        System.out.println("Washing the wheels");
      }
    },
    PREWASH {
      @Override void action() {
        System.out.println("Loosening the dirt");
      }
    },
    BASIC {
      @Override void action() {
        System.out.println("The basic wash");
      }
    },
    HOTWAX {
      @Override void action() {
        System.out.println("Applying hot wax");
      }
    },
    RINSE {
      @Override void action() {
        System.out.println("Rinsing");
      }
    },
    BLOWDRY {
      @Override void action() {
        System.out.println("Blowing dry");
      }
    };
    abstract void action();
  }
  EnumSet<Cycle> cycles =
    EnumSet.of(Cycle.BASIC, Cycle.RINSE);
  public void add(Cycle cycle) {
    cycles.add(cycle);
  }
  public void washCar() {
    for(Cycle c : cycles)
      c.action();
```

```
/* 输出:
[BASIC, RINSE]
The basic wash
Rinsing
[BASIC, HOTWAX,
RINSE, BLOWDRY]
The basic wash
Applying hot wax
Rinsing
Blowing dry
*/
```

```java
  }
  @Override public String toString() {
    return cycles.toString();
  }
  public static void main(String[] args) {
    CarWash wash = new CarWash();
    System.out.println(wash);
    wash.washCar();
    // 添加的顺序并不重要:
    wash.add(Cycle.BLOWDRY);
    wash.add(Cycle.BLOWDRY); // 重复添加会被忽略
    wash.add(Cycle.RINSE);
    wash.add(Cycle.HOTWAX);
    System.out.println(wash);
    wash.washCar();
  }
}
```

定义一个常量特定方法的语法和定义匿名内部类差不多，但是要更简洁一些。

这个例子还展示了 EnumSet 的更多特点。它是一种 Set，所以每个可选洗车项目都只能持有一个选项，用相同的参数对 add() 方法的重复调用会被忽略（这很好理解，因为打开开关一次和多次是一样的）。另外，添加枚举实例的顺序并不重要——输出顺序由枚举声明的顺序决定。

重写常量特定方法，而不是实现一个抽象方法，这是可能的吗？是的，如以下示例所示。

```java
// enums/OverrideConstantSpecific.java

public enum OverrideConstantSpecific {
  NUT, BOLT,
  WASHER {
    @Override void f() {
      System.out.println("Overridden method");
    }
  };
  void f() {
    System.out.println("default behavior");
  }
  public static void main(String[] args) {
    for(OverrideConstantSpecific ocs : values()) {
      System.out.print(ocs + ": ");
      ocs.f();
    }
  }
}
/* 输出:
NUT: default behavior
BOLT: default behavior
WASHER: Overridden method
*/
```

虽然枚举确实阻止了某些类型的代码，但是通常来说，可以把它们当作类来实验。

1.10.1 用枚举实现职责链模式

职责链（Chain of Responsibility）设计模式先创建了一批用于解决目标问题的不同方法，然后将它们连成一条"链"。当一个请求到达时，会顺着这条链传递下去，直到遇到链上某个可以处理该请求的方法。

可以很容易地用常量特定方法实现一条简单的**职责链**。考虑一个邮局模型，它对每一封邮件都会尝试用最常见的方式来处理，（如果行不通）并不断尝试别的方式，直到该邮件最终被视为"死信"（无法投递）。每种尝试都可以看作一个策略（另一种设计模式），而整个策略列表放在一起就是一条职责链。

我们从一封邮件开始说起。它所有的重要特征都可以用枚举来表达。由于 Mail 对象是随机创建的，想要减小一封邮件的 GeneralDelivery 被赋予 YES 的可能性，最简单的方法是创建更多的非 YES 的实例，因此枚举的定义一开始可能看起来有点好笑。

在 Mail 中，你会看到 randomMail() 方法，用来随机创建测试邮件。generator() 方法生成了一个 Iterable 对象，它使用 randomMail() 方法来生成一定数量的 Mail 对象，每通过迭代器调用一次 next() 就会生成一个。这种结构允许通过调用 Mail.generator() 方法实现 for-in 循环的简单创建能力。

```java
// enums/PostOffice.java
// 邮局建模
import java.util.*;
import onjava.*;

class Mail {
  // NO 减少了随机选择的可能性：
  enum GeneralDelivery {YES,NO1,NO2,NO3,NO4,NO5}
  enum Scannability {UNSCANNABLE,YES1,YES2,YES3,YES4}
  enum Readability {ILLEGIBLE,YES1,YES2,YES3,YES4}
  enum Address {INCORRECT,OK1,OK2,OK3,OK4,OK5,OK6}
  enum ReturnAddress {MISSING,OK1,OK2,OK3,OK4,OK5}
  GeneralDelivery generalDelivery;
  Scannability scannability;
  Readability readability;
  Address address;
  ReturnAddress returnAddress;
  static long counter = 0;
  long id = counter++;
  @Override public String toString() {
    return "Mail " + id;
  }
  public String details() {
    return toString() +
      ", General Delivery: " + generalDelivery +
```

```java
        ", Address Scannability: " + scannability +
        ", Address Readability: " + readability +
        ", Address Address: " + address +
        ", Return address: " + returnAddress;
  }
  // 生成测试邮件:
  public static Mail randomMail() {
    Mail m = new Mail();
    m.generalDelivery =
      Enums.random(GeneralDelivery.class);
    m.scannability =
      Enums.random(Scannability.class);
    m.readability =
      Enums.random(Readability.class);
    m.address = Enums.random(Address.class);
    m.returnAddress =
      Enums.random(ReturnAddress.class);
    return m;
  }
  public static
  Iterable<Mail> generator(final int count) {
    return new Iterable<Mail>() {
      int n = count;
      @Override public Iterator<Mail> iterator() {
        return new Iterator<Mail>() {
          @Override public boolean hasNext() {
            return n-- > 0;
          }
          @Override public Mail next() {
            return randomMail();
          }
          @Override
          public void remove() { // 未实现
            throw new UnsupportedOperationException();
          }
        };
      }
    };
  }
}

public class PostOffice {
  enum MailHandler {
    GENERAL_DELIVERY {
      @Override boolean handle(Mail m) {
        switch(m.generalDelivery) {
          case YES:
            System.out.println(
              "Using general delivery for " + m);
            return true;
          default: return false;
        }
      }
```

```java
    },
    MACHINE_SCAN {
      @Override boolean handle(Mail m) {
        switch(m.scannability) {
          case UNSCANNABLE: return false;
          default:
            switch(m.address) {
              case INCORRECT: return false;
              default:
                System.out.println(
                  "Delivering "+ m + " automatically");
                return true;
            }
        }
      }
    },
    VISUAL_INSPECTION {
      @Override boolean handle(Mail m) {
        switch(m.readability) {
          case ILLEGIBLE: return false;
          default:
            switch(m.address) {
              case INCORRECT: return false;
              default:
                System.out.println(
                  "Delivering " + m + " normally");
                return true;
            }
        }
      }
    },
    RETURN_TO_SENDER {
      @Override boolean handle(Mail m) {
        switch(m.returnAddress) {
          case MISSING: return false;
          default:
            System.out.println(
              "Returning " + m + " to sender");
            return true;
        }
      }
    };
    abstract boolean handle(Mail m);
  }
  static void handle(Mail m) {
    for(MailHandler handler : MailHandler.values())
      if(handler.handle(m))
        return;
    System.out.println(m + " is a dead letter");
  }
  public static void main(String[] args) {
    for(Mail mail : Mail.generator(10)) {
      System.out.println(mail.details());
```

```
        handle(mail);
        System.out.println("*****");
    }
  }
}
/* 输出:
Mail 0, General Delivery: NO2, Address Scannability:
UNSCANNABLE, Address Readability: YES3, Address
Address: OK1, Return address: OK1
Delivering Mail 0 normally
*****
Mail 1, General Delivery: NO5, Address Scannability:
YES3, Address Readability: ILLEGIBLE, Address Address:
OK5, Return address: OK1
Delivering Mail 1 automatically
*****
Mail 2, General Delivery: YES, Address Scannability:
YES3, Address Readability: YES1, Address Address: OK1,
Return address: OK5
Using general delivery for Mail 2
*****
Mail 3, General Delivery: NO4, Address Scannability:
YES3, Address Readability: YES1, Address Address:
INCORRECT, Return address: OK4
Returning Mail 3 to sender
*****
Mail 4, General Delivery: NO4, Address Scannability:
UNSCANNABLE, Address Readability: YES1, Address
Address: INCORRECT, Return address: OK2
Returning Mail 4 to sender
*****
Mail 5, General Delivery: NO3, Address Scannability:
YES1, Address Readability: ILLEGIBLE, Address Address:
OK4, Return address: OK2
Delivering Mail 5 automatically
*****
Mail 6, General Delivery: YES, Address Scannability:
YES4, Address Readability: ILLEGIBLE, Address Address:
OK4, Return address: OK4
Using general delivery for Mail 6
*****
Mail 7, General Delivery: YES, Address Scannability:
YES3, Address Readability: YES4, Address Address: OK2,
Return address: MISSING
Using general delivery for Mail 7
*****
Mail 8, General Delivery: NO3, Address Scannability:
YES1, Address Readability: YES3, Address Address:
INCORRECT, Return address: MISSING
Mail 8 is a dead letter
*****
Mail 9, General Delivery: NO1, Address Scannability:
UNSCANNABLE, Address Readability: YES2, Address
Address: OK1, Return address: OK4
Delivering Mail 9 normally
*****
*/
```

职责链模式的作用体现在了 `MailHandler` 枚举中，枚举的定义顺序则决定了各个策略在每封邮件上被应用的顺序。该模式会按顺序尝试应用每个策略，直到某个策略执行成功，或者全部策略都执行失败（即邮件无法投递）。

1.10.2　用枚举实现状态机

枚举类型很适合用来实现**状态机**。状态机可以处于有限数量的特定状态。它们通常根据输入，从一个状态移动到下一个状态，但同时也会存在**瞬态**。当任务执行完毕后，状态机会立即跳出所有状态。

每个状态都有某些可接受的输入，不同的输入会使状态机从当前状态切换到新的状态。由于枚举限制了可能出现的状态集大小（即状态数量），因此很适合表达（枚举）不同的状态和输入。

每种状态一般也会有某种对应的输出。

自动售货机是个很好的状态机应用的例子。首先，在一个枚举中定义一系列输入：

```java
// enums/Input.java
import java.util.*;

public enum Input {
  NICKEL(5), DIME(10), QUARTER(25), DOLLAR(100),
  TOOTHPASTE(200), CHIPS(75), SODA(100), SOAP(50),
  ABORT_TRANSACTION {
    @Override public int amount() { // 不允许
      throw new RuntimeException("ABORT.amount()");
    }
  },
  STOP { // 这必须是最后一个实例
    @Override public int amount() { // 不允许
      throw new
        RuntimeException("SHUT_DOWN.amount()");
    }
  };
  int value; // 单位为美分（cent）
  Input(int value) { this.value = value; }
  Input() {}
  int amount() { return value; }; // 单位为美分（cent）
  static Random rand = new Random(47);
  public static Input randomSelection() {
    // 不包括 STOP：
    return
      values()[rand.nextInt(values().length - 1)];
  }
}
```

注意其中两个 Input 有着对应的金额，所以在接口中定义了 amount() 方法。然而，对另外两个 Input 调用 amount() 是不合适的，如果调用就会抛出异常。尽管这是个有点奇怪的机制（在接口中定义一个方法，然后如果在某些具体实现中调用它的话就会抛出异常），但这是枚举的限制所导致的。

VendingMachine（自动售货机）接收到输入后，首先通过 Category（类别）枚举来对这些输入进行分类，这样就可以在各个类别间切换了。下例演示了枚举是如何使代码变得更清晰、更易于管理的。

```java
// enums/VendingMachine.java
// {java VendingMachine VendingMachineInput.txt}
import java.util.*;
import java.io.IOException;
import java.util.function.*;
import java.nio.file.*;
import java.util.stream.*;

enum Category {
  MONEY(Input.NICKEL, Input.DIME,
        Input.QUARTER, Input.DOLLAR),
  ITEM_SELECTION(Input.TOOTHPASTE, Input.CHIPS,
                 Input.SODA, Input.SOAP),
  QUIT_TRANSACTION(Input.ABORT_TRANSACTION),
  SHUT_DOWN(Input.STOP);
  private Input[] values;
  Category(Input... types) { values = types; }
  private static EnumMap<Input,Category> categories =
    new EnumMap<>(Input.class);
  static {
    for(Category c : Category.class.getEnumConstants())
      for(Input type : c.values)
        categories.put(type, c);
  }
  public static Category categorize(Input input) {
    return categories.get(input);
  }
}

public class VendingMachine {
  private static State state = State.RESTING;
  private static int amount = 0;
  private static Input selection = null;
  enum StateDuration { TRANSIENT } // 标识 enum
  enum State {
    RESTING {
      @Override void next(Input input) {
        switch(Category.categorize(input)) {
          case MONEY:
            amount += input.amount();
```

```
        state = ADDING_MONEY;
        break;
      case SHUT_DOWN:
        state = TERMINAL;
      default:
    }
  }
},
ADDING_MONEY {
  @Override void next(Input input) {
    switch(Category.categorize(input)) {
      case MONEY:
        amount += input.amount();
        break;
      case ITEM_SELECTION:
        selection = input;
        if(amount < selection.amount())
          System.out.println(
            "Insufficient money for " + selection);
        else state = DISPENSING;
        break;
      case QUIT_TRANSACTION:
        state = GIVING_CHANGE;
        break;
      case SHUT_DOWN:
        state = TERMINAL;
      default:
    }
  }
},
DISPENSING(StateDuration.TRANSIENT) {
  @Override void next() {
    System.out.println("here is your " + selection);
    amount -= selection.amount();
    state = GIVING_CHANGE;
  }
},
GIVING_CHANGE(StateDuration.TRANSIENT) {
  @Override void next() {
    if(amount > 0) {
      System.out.println("Your change: " + amount);
      amount = 0;
    }
    state = RESTING;
  }
},
TERMINAL {@Override
void output() { System.out.println("Halted"); } };
private boolean isTransient = false;
State() {}
State(StateDuration trans) { isTransient = true; }
void next(Input input) {
  throw new RuntimeException("Only call " +
```

```java
        "next(Input input) for non-transient states");
    }
    void next() {
      throw new RuntimeException(
        "Only call next() for " +
        "StateDuration.TRANSIENT states");
    }
    void output() { System.out.println(amount); }
  }
  static void run(Supplier<Input> gen) {
    while(state != State.TERMINAL) {
      state.next(gen.get());
      while(state.isTransient)
        state.next();
      state.output();
    }
  }
  public static void main(String[] args) {
    Supplier<Input> gen = new RandomInputSupplier();
    if(args.length == 1)
      gen = new FileInputSupplier(args[0]);
    run(gen);
  }
}

// 基本的稳健性检查:
class RandomInputSupplier implements Supplier<Input> {
  @Override public Input get() {
    return Input.randomSelection();
  }
}

// 从以;分隔的字符串的文件创建输入
class FileInputSupplier implements Supplier<Input> {
  private Iterator<String> input;
  FileInputSupplier(String fileName) {
    try {
      input = Files.lines(Paths.get(fileName))
        .skip(1) // 跳过注释行
        .flatMap(s -> Arrays.stream(s.split(";")))
        .map(String::trim)
        .collect(Collectors.toList())
        .iterator();
    } catch(IOException e) {
      throw new RuntimeException(e);
    }
  }
  @Override public Input get() {
    if(!input.hasNext())
      return null;
    return Enum.valueOf(
      Input.class, input.next().trim());
  }
}
```

```
/* 输出:
25
50
75
here is your CHIPS
0
100
200
here is your TOOTHPASTE
0
25
35
Your change: 35
0
25
35
Insufficient money for SODA
35
60
70
75
Insufficient money for SODA
75
Your change: 75
0
Halted
*/
```

因为通过 switch 语句在枚举实例中进行选择操作是最常见的方式（注意，为了使 switch 便于操作枚举，语言层面需要付出额外的代价），所以在组织多个枚举类型时，最常问的问题之一就是"我需要什么东西之上（即以什么粒度）进行 switch"。这里最简单的办法是，回头梳理一遍 VendingMachine，就会发现在每种 State 下，你需要针对输入操作的基本类别进行 switch 操作：投入钱币、选择商品、退出交易、关闭机器。并且在这些类别内，你还可以投入不同类别的货币，选择不同类别的商品。Category 枚举会对不同的 Input 类型进行分类，因此 categorize() 方法可以在 switch 中生成恰当的 Category。这种方法用一个 EnumMap 实现了高效且安全的查询。

如果你研究一下 VendingMachine 类，便会发现每个状态的区别，以及对输入的响应区别。同时还要注意那两个瞬态：在 run() 方法中，售货机等待一个 Input，并且会一直在状态间移动，直到它不再处于某个瞬态中。

VendingMachine 可以通过两种不同的 Supplier 对象，以两种方法来测试。RandomInputSupplier 只需要持续生成除 SHUT_DOWN 以外的任何输入。通过一段较长时间的运行后，就相当于做了一次健康检查，以确定售货机不会偏离到某些无效状态。FileInputSupplier 接收文本形式的输入描述文件，并将它们转换为 enum 实例，然后创建 Input 对象。下面是用于生成以上输出的文本文件：

```
// enums/VendingMachineInput.txt
QUARTER; QUARTER; QUARTER; CHIPS;
DOLLAR; DOLLAR; TOOTHPASTE;
QUARTER; DIME; ABORT_TRANSACTION;
QUARTER; DIME; SODA;
QUARTER; DIME; NICKEL; SODA;
ABORT_TRANSACTION;
STOP;
```

FileInputSupplier 的构造器将这个文件转换为行级的 Stream 流，并忽略注释行。然后它通过 String.split() 方法将每一行都根据分号拆开。这样就能生成一个字符串数组，可以通过先将该数组转化为 Steam，然后执行 flatMap()，来将其注入（前面 FileInputSupplier 中生成的）Stream 中。结果将删除所有的空格，并转换为 List<String>，并从中得到 Iterator<String>。

上述设计有个限制：VendingMachine 中会被 State 枚举实例访问到的字段都**必须**是静态的，这意味着只能存在一个 VendingMachine 实例。这可能不会是个大问题——你可以想想一个实际的（嵌入式 Java）实现，每台机器可能就只有一个应用程序。

1.11 多路分发

当你处理多个交互类型时,程序可能会变得相当混乱。举例来说,考虑一个解析并执行数学表达式的系统。里面可能包括 Number.plus(Number)、Number.multiply(Number) 等,此处的 Number 是数值对象家族的基类。但是当你要执行 a.plus(b),并且不知道 a 或 b 的具体类型时,如何保证它们间的相互作用是正确的?

答案一开始可能让你意外:Java 只能进行**单路分发**。也就是说,如果你想对多个类型未知的对象进行操作,Java 只会对其中一个类型调用动态绑定机制。这并不能解决当前的问题,所以你最终只能手动检测类型,然后再实现你自己的动态绑定行为。

这个问题的解决方法就是**多路分发**(此处称为**双路分发**,因为只有两路)。多态只能在方法调用时发生,所以如果你要使用双路分发,就必须执行两次方法调用:第一次用来确定第一个未知类型,第二次用来确定第二个未知类型。要使用多路分发,就必须对每个类型都进行虚拟调用——如果是在操作两个不同的交互类型层次结构,则需要在每个层次结构都执行虚拟调用。这通常会启用一个配置,使得一次方法调用可以产生多次虚拟调用,从而服务于过程中的多个类型。要达到这个目的,你需要不止一个方法互相配合:每个分发都需要一个方法调用。以下示例(实现了一个"猜拳"游戏)中对应的方法是 compete() 和 eval(),它们都是相同类型的成员,会生成三个可能结果中的一个:

```
// enums/Outcome.java
package enums;
public enum Outcome { WIN, LOSE, DRAW }
```

Item 是多路分发类型的接口:

```
// enums/Item.java
package enums;

public interface Item {
  Outcome compete(Item it);
  Outcome eval(Paper p);
  Outcome eval(Scissors s);
  Outcome eval(Rock r);
}
```

每个特定的 Item 都提供了这些方法的对应实现:

```
// enums/Paper.java
package enums;
import static enums.Outcome.*;
```

```java
public class Paper implements Item {
  @Override public Outcome compete(Item it) {
    return it.eval(this);
  }
  @Override
  public Outcome eval(Paper p) { return DRAW; }
  @Override
  public Outcome eval(Scissors s) { return WIN; }
  @Override
  public Outcome eval(Rock r) { return LOSE; }
  @Override public String toString() {
    return "Paper";
  }
}
// enums/Scissors.java
package enums;
import static enums.Outcome.*;

public class Scissors implements Item {
  @Override public Outcome compete(Item it) {
    return it.eval(this);
  }
  @Override
  public Outcome eval(Paper p) { return LOSE; }
  @Override
  public Outcome eval(Scissors s) { return DRAW; }
  @Override
  public Outcome eval(Rock r) { return WIN; }
  @Override public String toString() {
    return "Scissors";
  }
}
// enums/Rock.java
package enums;
import static enums.Outcome.*;

public class Rock implements Item {
  @Override public Outcome compete(Item it) {
    return it.eval(this);
  }
  @Override
  public Outcome eval(Paper p) { return WIN; }
  @Override
  public Outcome eval(Scissors s) { return LOSE; }
  @Override
  public Outcome eval(Rock r) { return DRAW; }
  @Override public String toString() {
    return "Rock";
  }
}
```

RoShamBo1.match() 接收两个 Item 对象作为参数，然后通过调用 Item.compete() 开始

执行双路分发：

```java
// enums/RoShamBo1.java
// 多路分发示例
// {java enums.RoShamBo1}
package enums;
import java.util.*;

public class RoShamBo1 {
  static final int SIZE = 20;
  private static Random rand = new Random(47);
  public static Item newItem() {
    switch(rand.nextInt(3)) {
      default:
      case 0: return new Scissors();
      case 1: return new Paper();
      case 2: return new Rock();
    }
  }
  public static void match(Item a, Item b) {
    System.out.println(
      a + " vs. " + b + ": " +  a.compete(b));
  }
  public static void main(String[] args) {
    for(int i = 0; i < SIZE; i++)
      match(newItem(), newItem());
  }
}
/* 输出：
Rock vs. Rock: DRAW
Paper vs. Rock: WIN
Paper vs. Rock: WIN
Paper vs. Rock: WIN
Scissors vs. Paper: WIN
Scissors vs. Scissors: DRAW
Scissors vs. Paper: WIN
Rock vs. Paper: LOSE
Paper vs. Paper: DRAW
Rock vs. Paper: LOSE
Paper vs. Scissors: LOSE
Paper vs. Scissors: LOSE
Rock vs. Scissors: WIN
Rock vs. Paper: LOSE
Paper vs. Rock: WIN
Scissors vs. Paper: WIN
Paper vs. Scissors: LOSE
Paper vs. Scissors: LOSE
Paper vs. Scissors: LOSE
Paper vs. Scissors: LOSE
*/
```

虚拟机制决定了 a 的类型，因此它在 a 的实际类型的 compete() 函数中唤醒。compete() 函数通过对剩余类型调用 eval() 方法，调用执行了第二次分发。将自身（this）作为参数传入 eval()，会产生一个对重载 eval() 的调用，由此保留了第一次分发的类型信息。当第二次分发结束后，你就知道了两个 Item 对象的准确类型。

构建多路分发需要很多道工序，但是记住它所带来的好处：在调用中保持了语法的优雅，而不是在调用中编写丑陋的代码来确定一个又一个的对象类型。你只需要说："你们两个！我不关心你们是什么类型，你们自己处理好彼此的交互！"当然，在决定使用多路分发前，要确定这种优雅的语法确实对你很重要。

1.11.1　使用枚举类型分发

如果将 RoShamBo1.java 直接转换为基于枚举的实现版本，则会出现问题。因为枚举实例并不是类型，所以无法重载 eval() 方法……你无法将枚举实例作为参数类型。不过，还有别的方法可以利用枚举来实现多路分发。

一种方法是通过构造方法初始化每个枚举实例,并以一组结果作为参数,最终组成类似查询表的结构:

```java
// enums/RoShamBo2.java
// 将一个枚举切换到另一个枚举
// {java enums.RoShamBo2}
package enums;
import static enums.Outcome.*;

public enum RoShamBo2 implements Competitor<RoShamBo2> {
  PAPER(DRAW, LOSE, WIN),
  SCISSORS(WIN, DRAW, LOSE),
  ROCK(LOSE, WIN, DRAW);
  private Outcome vPAPER, vSCISSORS, vROCK;
  RoShamBo2(Outcome paper,
    Outcome scissors, Outcome rock) {
    this.vPAPER = paper;
    this.vSCISSORS = scissors;
    this.vROCK = rock;
  }
  @Override public Outcome compete(RoShamBo2 it) {
    switch(it) {
      default:
      case PAPER: return vPAPER;
      case SCISSORS: return vSCISSORS;
      case ROCK: return vROCK;
    }
  }
  public static void main(String[] args) {
    RoShamBo.play(RoShamBo2.class, 20);
  }
}
/* 输出:
ROCK vs. ROCK: DRAW
SCISSORS vs. ROCK: LOSE
SCISSORS vs. ROCK: LOSE
SCISSORS vs. ROCK: LOSE
PAPER vs. SCISSORS: LOSE
PAPER vs. PAPER: DRAW
PAPER vs. SCISSORS: LOSE
ROCK vs. SCISSORS: WIN
SCISSORS vs. SCISSORS: DRAW
ROCK vs. SCISSORS: WIN
SCISSORS vs. PAPER: WIN
SCISSORS vs. PAPER: WIN
ROCK vs. PAPER: LOSE
ROCK vs. SCISSORS: WIN
SCISSORS vs. ROCK: LOSE
PAPER vs. SCISSORS: LOSE
SCISSORS vs. PAPER: WIN
SCISSORS vs. PAPER: WIN
SCISSORS vs. PAPER: WIN
SCISSORS vs. PAPER: WIN
*/
```

在 compete() 方法中,一旦两个类型都被确定了,那么唯一的动作就是返回得到的 Outcome。然而,你也可以调用其他方法,甚至是(比如)构造方法中分配的**命令**对象中的方法。

RoShamBo2.java 比最初的版本更为简单明了,因此也更容易理解。注意,你仍然在使用两路分发来确定这两个对象的类型。在 RoShamBo1.java 中,两路的分发都是通过虚拟方法调用的。在这里,只有第一次分发使用了虚拟方法调用,第二次分发用的是 switch。不过这样做是安全的,因为枚举限制了 switch 中的状态可选范围。

用来操作枚举的代码是独立的,因此也可以放到其他例子中使用。首先,Competitor(竞争者)接口定义了一个和另一个 Competitor 竞争的类型:

```
// enums/Competitor.java
// 将一个枚举切换到另一个枚举
package enums;

public interface Competitor<T extends Competitor<T>> {
  Outcome compete(T competitor);
}
```

然后我们定义了两个静态方法（使用静态是为了避免显式地指定参数类型）。首先，match() 方法中调用了 compete() 方法，让两个 Competitor 竞争，可以看到此处类型参数只需要是 Competitor<T>。但是在 play() 方法中，类型参数必须同时是 Enum<T>（因为要放在 Enums.random() 中使用）和 Competitor<T>（因为要传给 match()）：

```
// enums/RoShamBo.java
// RoShamBo 示例集的通用工具
package enums;
import onjava.*;

public class RoShamBo {
  public static <T extends Competitor<T>>
  void match(T a, T b) {
    System.out.println(
      a + " vs. " + b + ": " +  a.compete(b));
  }
  public static <T extends Enum<T> & Competitor<T>>
  void play(Class<T> rsbClass, int size) {
    for(int i = 0; i < size; i++)
      match(
        Enums.random(rsbClass),Enums.random(rsbClass));
  }
}
```

play() 方法没有将类型参数 T 作为返回值，因此，似乎你应该在 Class<T> 中用通配符来取代主要的参数声明。然而通配符无法继承多个基类，因此我们必须使用上面的表达式。

1.11.2　使用常量特定方法

由于常量特定方法允许为不同的枚举实例提供不同的方法实现，所以看起来似乎是实现多路分发的完美解决方案。但是即使可以通过这种方式赋予枚举实例不同的行为，枚举实例也仍然不是类型，因此你无法将它们作为方法签名中的参数类型。对于这个例子，最好的办法是构建一个 switch 语句。

```
// enums/RoShamBo3.java
// 使用常量特定方法
// {java enums.RoShamBo3}
package enums;
import static enums.Outcome.*;
```

```java
public enum RoShamBo3 implements Competitor<RoShamBo3> {
  PAPER {
    @Override public Outcome compete(RoShamBo3 it) {
      switch(it) {
        default: // 为了安抚编译器
        case PAPER: return DRAW;
        case SCISSORS: return LOSE;
        case ROCK: return WIN;
      }
    }
  },
  SCISSORS {
    @Override public Outcome compete(RoShamBo3 it) {
      switch(it) {
        default:
        case PAPER: return WIN;
        case SCISSORS: return DRAW;
        case ROCK: return LOSE;
      }
    }
  },
  ROCK {
    @Override public Outcome compete(RoShamBo3 it) {
      switch(it) {
        default:
        case PAPER: return LOSE;
        case SCISSORS: return WIN;
        case ROCK: return DRAW;
      }
    }
  };
  @Override
  public abstract Outcome compete(RoShamBo3 it);
  public static void main(String[] args) {
    RoShamBo.play(RoShamBo3.class, 20);
  }
}
/* 输出：
ROCK vs. ROCK: DRAW
SCISSORS vs. ROCK: LOSE
SCISSORS vs. ROCK: LOSE
SCISSORS vs. ROCK: LOSE
PAPER vs. SCISSORS: LOSE
PAPER vs. PAPER: DRAW
PAPER vs. SCISSORS: LOSE
ROCK vs. SCISSORS: WIN
SCISSORS vs. SCISSORS: DRAW
ROCK vs. SCISSORS: WIN
SCISSORS vs. PAPER: WIN
SCISSORS vs. PAPER: WIN
ROCK vs. PAPER: LOSE
ROCK vs. SCISSORS: WIN
SCISSORS vs. ROCK: LOSE
PAPER vs. SCISSORS: LOSE
SCISSORS vs. PAPER: WIN
SCISSORS vs. PAPER: WIN
SCISSORS vs. PAPER: WIN
SCISSORS vs. PAPER: WIN
*/
```

虽然以上代码可以运行，看起来也没什么问题，但相比之下，RoShamBo2.java 的方案似乎在新增类型时需要的代码更少，因此看起来也更简洁易懂。

不过，RoShamBo3.java 也可以简化为以下版本：

```java
// enums/RoShamBo4.java
// {java enums.RoShamBo4}
package enums;

public enum RoShamBo4 implements Competitor<RoShamBo4> {
  ROCK {
    @Override
```

```java
    public Outcome compete(RoShamBo4 opponent) {
      return compete(SCISSORS, opponent);
    }
  },
  SCISSORS {
    @Override
    public Outcome compete(RoShamBo4 opponent) {
      return compete(PAPER, opponent);
    }
  },
  PAPER {
    @Override
    public Outcome compete(RoShamBo4 opponent) {
      return compete(ROCK, opponent);
    }
  };
  Outcome compete(RoShamBo4 loser, RoShamBo4 opponent) {
    return ((opponent == this) ? Outcome.DRAW
        : ((opponent == loser) ? Outcome.WIN
                               : Outcome.LOSE));
  }
  public static void main(String[] args) {
    RoShamBo.play(RoShamBo4.class, 20);
  }
}
/* 输出:
PAPER vs. PAPER: DRAW
SCISSORS vs. PAPER: WIN
SCISSORS vs. PAPER: WIN
SCISSORS vs. PAPER: WIN
ROCK vs. SCISSORS: WIN
ROCK vs. ROCK: DRAW
ROCK vs. SCISSORS: WIN
PAPER vs. SCISSORS: LOSE
SCISSORS vs. SCISSORS: DRAW
PAPER vs. SCISSORS: LOSE
SCISSORS vs. ROCK: LOSE
SCISSORS vs. ROCK: LOSE
PAPER vs. ROCK: WIN
PAPER vs. SCISSORS: LOSE
SCISSORS vs. PAPER: WIN
ROCK vs. SCISSORS: WIN
SCISSORS vs. ROCK: LOSE
SCISSORS vs. ROCK: LOSE
SCISSORS vs. ROCK: LOSE
SCISSORS vs. ROCK: LOSE
*/
```

此处的第二个分发是由两个参数版本的 compete() 执行的，它执行了一系列比较操作，因此看起来和 switch 的行为很相似。它更简短，却更难理解。如果将其用于大型系统的开发，这一混淆可能会削弱系统的可维护性。

1.11.3 使用 EnumMap 分发

使用 EnumMap 可以实现"真正的"双路分发，它是专门为 enum 设计的高效 Map。我们的目标是在两个未知类型中切换，因此由 EnumMap 组成的 EnumMap（嵌套 EnumMap）可以实现双路分发。

```java
// enums/RoShamBo5.java
// 使用由 EnumMap 组成的 EnumMap 进行多路分发
// {java enums.RoShamBo5}
package enums;
import java.util.*;
import static enums.Outcome.*;

enum RoShamBo5 implements Competitor<RoShamBo5> {
  PAPER, SCISSORS, ROCK;
  static EnumMap<RoShamBo5,EnumMap<RoShamBo5,Outcome>>
    table = new EnumMap<>(RoShamBo5.class);
  static {
    for(RoShamBo5 it : RoShamBo5.values())
      table.put(it, new EnumMap<>(RoShamBo5.class));
    initRow(PAPER, DRAW, LOSE, WIN);
    initRow(SCISSORS, WIN, DRAW, LOSE);
    initRow(ROCK, LOSE, WIN, DRAW);
  }
  static void initRow(RoShamBo5 it,
    Outcome vPAPER, Outcome vSCISSORS, Outcome vROCK) {
    EnumMap<RoShamBo5,Outcome> row =
      RoShamBo5.table.get(it);
    row.put(RoShamBo5.PAPER, vPAPER);
    row.put(RoShamBo5.SCISSORS, vSCISSORS);
    row.put(RoShamBo5.ROCK, vROCK);
  }
  @Override public Outcome compete(RoShamBo5 it) {
    return table.get(this).get(it);
  }
  public static void main(String[] args) {
    RoShamBo.play(RoShamBo5.class, 20);
  }
}
/* 输出:
ROCK vs. ROCK: DRAW
SCISSORS vs. ROCK: LOSE
SCISSORS vs. ROCK: LOSE
SCISSORS vs. ROCK: LOSE
PAPER vs. SCISSORS: LOSE
PAPER vs. PAPER: DRAW
PAPER vs. SCISSORS: LOSE
ROCK vs. SCISSORS: WIN
SCISSORS vs. SCISSORS: DRAW
ROCK vs. SCISSORS: WIN
SCISSORS vs. PAPER: WIN
SCISSORS vs. PAPER: WIN
ROCK vs. PAPER: LOSE
ROCK vs. SCISSORS: WIN
SCISSORS vs. ROCK: LOSE
PAPER vs. SCISSORS: LOSE
SCISSORS vs. PAPER: WIN
SCISSORS vs. PAPER: WIN
SCISSORS vs. PAPER: WIN
SCISSORS vs. PAPER: WIN
*/
```

EnumMap 是用一个 static{} 子句进行初始化的，可以看到对 initRow() 的调用是类似表格的结构。注意 compete() 方法，其在内部的一条语句内发生了两次分发。

1.11.4 使用二维数组

我们发现每个枚举实例都持有一个固定的值（基于它的声明顺序），该值由 ordinal()

方法生成，因此可以进一步简化该方案。使用一个二维数组将竞争者映射到结果，便可以实现最简单易懂的解决方案（而且有可能是最快的，尽管 EnumMap 用了内部数组）：

```java
// enums/RoShamBo6.java
// 基于 "tables"（表）而非多路分发的 Enum
// {java enums.RoShamBo6}
package enums;
import static enums.Outcome.*;

enum RoShamBo6 implements Competitor<RoShamBo6> {
  PAPER, SCISSORS, ROCK;
  private static Outcome[][] table = {
    { DRAW, LOSE, WIN }, // 布
    { WIN, DRAW, LOSE }, // 剪刀
    { LOSE, WIN, DRAW }, // 石头
  };
  @Override public Outcome compete(RoShamBo6 other) {
    return table[this.ordinal()][other.ordinal()];
  }
  public static void main(String[] args) {
    RoShamBo.play(RoShamBo6.class, 20);
  }
}
/* 输出：
ROCK vs. ROCK: DRAW
SCISSORS vs. ROCK: LOSE
SCISSORS vs. ROCK: LOSE
SCISSORS vs. ROCK: LOSE
PAPER vs. SCISSORS: LOSE
PAPER vs. PAPER: DRAW
PAPER vs. SCISSORS: LOSE
ROCK vs. SCISSORS: WIN
SCISSORS vs. SCISSORS: DRAW
ROCK vs. SCISSORS: WIN
SCISSORS vs. PAPER: WIN
SCISSORS vs. PAPER: WIN
ROCK vs. PAPER: LOSE
ROCK vs. SCISSORS: WIN
SCISSORS vs. ROCK: LOSE
PAPER vs. SCISSORS: LOSE
SCISSORS vs. PAPER: WIN
SCISSORS vs. PAPER: WIN
SCISSORS vs. PAPER: WIN
SCISSORS vs. PAPER: WIN
*/
```

table（表）的顺序和前面的例子中调用 initRow() 的顺序一致。

和之前的示例相比，本示例代码的简短让其看起来非常有吸引力，部分原因是它看起来更易于理解和修改，也可能只是因为它看起来更直接。然而，它并没有之前的示例那么"安全"，因为它使用了数组。如果处理一个更大的数组，则可能会得到错误的数组大小，并且如果测试未覆盖到所有的可能情况，则也许会出现一些不易发现的问题。

所有这些解决方案其实都是各种不同类型的表，但仍然值得去探索各种表的表现方式，从而找到最合适的那一种。注意，虽然上述解决方案是最紧凑的，但它也相当死板，因为在给定若干常量输入的情况下，它只能生成一个常量输出。不过，也没有什么可以阻

止你通过 table 来生成函数对象（function object）。对于某些特定类型的问题，"表驱动模式"（table-driven code）的概念是非常强大的。

1.12 支持模式匹配的新特性

你可以认为**模式匹配**（pattern matching）是在 switch 关键字上进行了显著的功能扩充。它是分成了多个模块、历经了 Java 的多个版本持续实现的。这保证了每个模块在其他模块加入前都可以安全地运行。最后，所有的模块集中到一起，就产生了这个新特性（这种形式在本书完成时可用，更多子特性会在未来加入到 Java 中）。

这种拆分实现的方式使得人们在面对跨多个版本增加的多个"子特性"时会有点困惑。试图通过逐个分析各个单独模块来理解模式匹配并不是明智的选择。在接下来的章节中，我将所有支持模式匹配的模块都汇总到了一起——包括最新的 Java 17 中增加的部分——希望能够帮你减少理解上的困惑。

虽然模式匹配和枚举类型并不是绑定的关系，但是它解决了一些曾经由枚举类型来解决的问题。本节内容也能让你有足够的知识来理解模式匹配所解决的所有问题。

1.13 新特性：switch 中的箭头语法

JDK 14 增加了在 switch 中使用一种不同语法的 case 的能力。在下面的示例中，colons() 用的是旧方式，arrows() 用的是新方式：

```
// enumerations/ArrowInSwitch.java
// {NewFeature} Since JDK 14
import static java.util.stream.IntStream.range;

public class ArrowInSwitch {
  static void colons(int i) {
    switch(i) {
      case 1: System.out.println("one");
              break;
      case 2: System.out.println("two");
              break;
      case 3: System.out.println("three");
              break;
      default: System.out.println("default");
    }
  }
  static void arrows(int i) {
    switch(i) {
      case 1 -> System.out.println("one");
```

```
/* 输出：
default
one
two
three
default
one
two
three
*/
```

```
      case 2 -> System.out.println("two");
      case 3 -> System.out.println("three");
      default -> System.out.println("default");
    }
  }
  public static void main(String[] args) {
    range(0, 4).forEach(i -> colons(i));
    range(0, 4).forEach(i -> arrows(i));
  }
}
```

在 `colons()` 中,你可以看到为了防止继续向下执行,每个 `case` 语句(除了最后一个 `default` 以外)后面都有必要加上 `break`。而在我们将 `arrows()` 中的冒号替换为箭头后,便不再需要 `break` 语句了。然而这只是箭头的一部分功能而已。

你不能在同一个 `switch` 中同时使用冒号和箭头。

1.14 新特性:`switch` 中的 `case null`

JDK 17 新增了(预览)功能,可以在 `switch` 中引入原本非法的 `case null`。以前只能在 `switch` 的外部检查是否为 `null`,如 `old()` 中所示:

```
// enumerations/CaseNull.java
// {NewFeature} Preview in JDK 17
// 带上此标志用 javac 编译:
//    --enable-preview --source 17
// 带上此标志运行: --enable-preview
import java.util.*;
import java.util.function.*;

public class CaseNull {
  static void old(String s) {
    if(s == null) {
      System.out.println("null");
      return;
    }
    switch(s) {
      case "XX" -> System.out.println("XX");
      default   -> System.out.println("default");
    }
  }
  static void checkNull(String s) {
    switch(s) {
      case "XX" -> System.out.println("XX");
      case null -> System.out.println("null");
      default   -> System.out.println("default");
    }
  }
  // 也可以用于冒号语法:
```

```java
    switch(s) {
      case "XX": System.out.println("XX");
                 break;
      case null: System.out.println("null");
                 break;
      default  : System.out.println("default");
    }
  }
  static void defaultOnly(String s) {
    switch(s) {
      case "XX" -> System.out.println("XX");
      default   -> System.out.println("default");
    }
  }
  static void combineNullAndCase(String s) {
    switch(s) {
      case "XX", null -> System.out.println("XX|null");
      default -> System.out.println("default");
    }
  }
  static void combineNullAndDefault(String s) {
    switch(s) {
      case "XX" -> System.out.println("XX");
      case null, default -> System.out.println("both");
    }
  }
  static void test(Consumer<String> cs) {
    cs.accept("XX");
    cs.accept("YY");
    try {
      cs.accept(null);
    } catch(NullPointerException e) {
      System.out.println(e.getMessage());
    }
  }
  public static void main(String[] args) {
    test(CaseNull::old);
    test(CaseNull::checkNull);
    test(CaseNull::defaultOnly);
    test(CaseNull::combineNullAndCase);
    test(CaseNull::combineNullAndDefault);
  }
}
/* 输出：
XX
default
null
XX
XX
default
default
```

```
null
null
XX
default
Cannot invoke "String.hashCode()" because "<local1>" is null
XX|null
default
XX|null
XX
both
both
*/
```

你可以在 checkNull() 中看到，null 现在在 switch 中可以作为合法的 case 了，而且同时适用于箭头和冒号这两种语法。

你可能会好奇 default 是否会覆盖值为 null 的情况。从 defaultOnly() 可以看出 default 并不会捕获 null，而如果你又没有使用 case null，那么就会抛出 NullPointerException 异常。Java 声称 switch 会覆盖所有可能的值，即使它实际并未覆盖 null。这是向后兼容性导致的问题——如果 Java 突然开始强制进行 null 检查，那么大部分已有代码会无法通过编译。

总体来说，多种模式可以用逗号合并在一个 case 中。也可以将 null 和其他模式合并，如 combineNullAndCase() 中所示。为方便起见，还可以像 combineNullAndDefault() 中那样将 case null 和 default 合并。

1.15　新特性：将 switch 作为表达式

switch 一直以来都只是一个**语句**，并不会生成结果。JDK 14 使得 switch 还可以作为一个**表达式**来使用，因此它可以得到一个值：

```java
// enumerations/SwitchExpression.java
// {NewFeature} Since JDK 14
import java.util.*;

public class SwitchExpression {
  static int colon(String s) {
    var result = switch(s) {
      case "i": yield 1;
      case "j": yield 2;
      case "k": yield 3;
      default:  yield 0;
```

```
/* 输出：
i 1 1
j 2 2
k 3 3
z 0 0
*/
```

```
    };
    return result;
  }
  static int arrow(String s) {
    var result = switch(s) {
      case "i" -> 1;
      case "j" -> 2;
      case "k" -> 3;
      default  -> 0;
    };
    return result;
  }
  public static void main(String[] args) {
    for(var s: new String[]{"i", "j", "k", "z"})
      System.out.format(
        "%s %d %d%n", s, colon(s), arrow(s));
  }
}
```

如 colon() 中所示，在使用旧的冒号语法的同时，可以使用新的 yield 关键字从 switch 中返回结果。注意在使用 yield 的时候，并不需要用到 break——实际上如果加上了 break，反而会在编译时产生错误消息"试图跳出 switch 表达式"（attempt to break out of a switch expression）。

如果试图在 switch 语句中使用 yield，编译器会产生错误消息"在 switch 表达式外部 yield"（yield outside of switch expression）。

arrow() 会产生和 colon() 相同的效果，但是要注意这样的语法会更清晰、更紧凑，也更易读。可以看看这种方式为 TrafficLight.java 带来的改进（相较于在本章前面实现的版本）：

```
// enumerations/EnumSwitch.java
// {NewFeature} Since JDK 14

public class EnumSwitch {
  enum Signal { GREEN, YELLOW, RED, }
  Signal color = Signal.RED;
  public void change() {
    color = switch(color) {
      case RED -> Signal.GREEN;
      case GREEN -> Signal.YELLOW;
      case YELLOW -> Signal.RED;
    };
  }
}
```

同样，如果在不加上 case BLUE 的情况下将 BLUE 加入到 enum Signal 中，Java 的反应则是"switch 表达式并未覆盖所有可能的输入值"（the switch expression does not cover all possible input values）。编译器会确保在修改代码的时候不会缺少 case。

如果一个 case 需要多个语句或表达式，那就将它们放在一对花括号中。

```java
// enumerations/Planets.java
// {NewFeature} Since JDK 14

enum CelestialBody {
  MERCURY, VENUS, EARTH, MARS, JUPITER,
  SATURN, URANUS, NEPTUNE, PLUTO
}

public class Planets {
  public static String classify(CelestialBody b) {
    var result = switch(b) {
      case MERCURY, VENUS, EARTH,
           MARS, JUPITER,
           SATURN, URANUS, NEPTUNE -> {
             System.out.print("A planet: ");
             yield b.toString();
           }
      case PLUTO -> {
             System.out.print("Not a planet: ");
             yield b.toString();
           }
    };
    return result;
  }
  public static void main(String[] args) {
    System.out.println(classify(CelestialBody.MARS));
    System.out.println(classify(CelestialBody.PLUTO));
  }
}
/* 输出：
A planet: MARS
Not a planet: PLUTO
*/
```

注意，在从多行代码组成的 case 表达式中生成结果时，即便使用了箭头语法，也要使用 yield。

1.16　新特性：智能转型

JDK 16 最终完成了 JEP（Java 增强建议）394 的定稿，但它有个糟糕的名字，即"针对 instanceof 的模式匹配"（Pattern Matching for instanceof）。如果你看到了完整的模式匹配，便会明白这个名字不是很合理。称其为"模式匹配支持"（support for pattern matching）可能会更恰当。在 Kotlin 语言中，该特性被简单地称为"智能转型"（smart casting），因为一旦确定类型后，就永远不需要对其转型了。

下面的示例中可以看到如何分别用旧方法和新方法实现同样的目的：

```java
// enumerations/SmartCasting.java
// {NewFeature} Since JDK 16

public class SmartCasting {
  static void dumb(Object x) {
    if(x instanceof String) {
      String s = (String)x;
      if(s.length() > 0) {
        System.out.format(
          "%d %s%n", s.length(), s.toUpperCase());
      }
    }
  }
  static void smart(Object x) {
    if(x instanceof String s && s.length() > 0) {
      System.out.format(
        "%d %s%n", s.length(), s.toUpperCase());
    }
  }
  static void wrong(Object x) {
    // "||"永远不会生效：
    // if(x instanceof String s || s.length() > 0) {}
    // error: cannot find symbol       ^
  }
  public static void main(String[] args) {
    dumb("dumb");
    smart("smart");
  }
}
/* 输出：
4 DUMB
5 SMART
*/
```

在 dumb() 中，一旦 instanceof 确定了 x 是 String 后，就必须显式地将它转型为 String s。否则就得在该函数中其余的地方到处插入转型操作。但是在 smart() 中，注意 x instanceof String s 自动用 String 类型创建了一个新的变量 s。s 在整个作用域中都可用，即使是在剩余的 if 条件中，如 && s.length() > 0 中所示。这样可以得到更紧凑的代码。

从 wrong() 可以看出，在 if 智能转型表达式中只能使用 &&。使用 || 则意味着可能 x 是个 instanceof String，也可能 s.length() > 0。但这也就意味着 x 也可能不是 String，在这种情况下，Java 就不会将 x 智能转型以生成 s，因此 s 在 || 右侧是不可用的。

JEP 394 将 s 称为**模式变量**（pattern variable）。

虽然这个特性并未完全消除某些混乱的 if 语句，但这并不是加入该特性的目的。该特性是作为模式匹配的构建块而引入的，你很快就会看到。

该特性可以产生某些奇怪的作用域行为：

```java
// enumerations/OddScoping.java
// {NewFeature} Since JDK 16

public class OddScoping {
  static void f(Object o) {
    if(!(o instanceof String s)) {
      System.out.println("Not a String");
      throw new RuntimeException();
    }
    // 此处 s 仍在作用域中！
    System.out.println(s.toUpperCase());  // [1]
  }
  public static void main(String[] args) {
    f("Curiouser and Curiouser");
    f(null);
  }
}
/* 输出：
CURIOUSER AND CURIOUSER
Not a String
Exception in thread "main" java.lang.RuntimeException
        at OddScoping.f(OddScoping.java:8)
        at OddScoping.main(OddScoping.java:15)
*/
```

[1] 只有在不引入会引发异常的语句时，s 才在作用域中。如果注释掉 throw new RuntimeException()，编译器便会告诉你它无法在行 [1] 中找到 s——这是你通常所预期的行为。

乍一看这像个 bug，但它就是这么设计的——该行为在 JEP 394 中有着明确的描述。尽管这可以说是一个极端情况，但你能想象追踪由这种行为引起的 bug 会有多困难。

1.17 新特性：模式匹配

现在所有的基础模块都已准备好，我们可以以更全局的视角来看看模式匹配了。不过要记住，Java 的模式匹配仍在开发中，和成熟的模式匹配系统（如 Kotlin、Scala，以及 Python 中的那些）相比，功能还相当有限。

在编写本书时，模式匹配还是 JDK 17 中的预览特性——虽然不太可能再有变化了，但也仍未定稿。当你读到此处时，这些特性可能已经不再是预览状态了，因此可能不再需要用到示例注释中给出的编译器和运行时标志了。

Java 团队还打算加入更多的特性。可能要到好几年后，才会看到完整形式的 Java 模式匹配。

1.17.1 违反里氏替换原则

基于继承的多态性实现了基于类型的行为，但是要求这些类型都在同一个继承层次结构中。而模式匹配也实现了基于类型的行为，却并不要求类型全都具有相同的接口，或都处于相同的继承层次结构中。

这是一种不同的反射使用方式。你仍然需要在运行时确定类型，但该方式比反射更加正式、更加结构化。

如果感兴趣的类型全都具有共同的基类，并且只会使用公共基类中定义的方法，那么就遵循**里氏替换原则**（Liskov Substitution Principle, LSP）。在这种情况下，模式匹配就不是必需的了——只需使用普通的继承多态性即可。

```java
// enumerations/NormalLiskov.java
import java.util.stream.*;

interface LifeForm {
  String move();
  String react();
}

class Worm implements LifeForm {
  @Override public String move() {
    return "Worm::move()";
  }
  @Override public String react() {
    return "Worm::react()";
  }
}

class Giraffe implements LifeForm {
  @Override public String move() {
    return "Giraffe::move()";
  }
  @Override public String react() {
    return "Giraffe::react()";
  }
}

public class NormalLiskov {
  public static void main(String[] args) {
    Stream.of(new Worm(), new Giraffe())
      .forEach(lf -> System.out.println(
        lf.move() + " " + lf.react()));
  }
}
/* 输出：
Worm::move()
Worm::react()
Giraffe::move()
Giraffe::react()
*/
```

所有的方法都完全定义在 LifeForm 接口中，实现类中也并没有增加任何新的方法。

但如果需要增加难以放在基类中的方法呢？比如某些蠕虫身体被切断后可以再生，而长颈鹿肯定做不到。长颈鹿可以踢人，但是很难想象怎样可以在基类中表现出该行为，同时又不会使 Worm（蠕虫）错误地实现（该行为）。Java 集合库遇到了这种问题，并试图以一种笨拙的方式解决：在基类中增加"可选"方法，某些子类可以实现该方法，而另一些子类则不实现。这种方法遵循了里氏替换原则，但造成了混乱的设计。

Java 的基本灵感来自一种叫作 SmallTalk 的动态语言，它会利用已有的类并增加方法，以此实现代码复用。以 Smalltalk 的设计方式来实现一套"pet"（宠物）继承层次结构，可能最终看起来是这样的：

```java
// enumerations/Pet.java

public class Pet {
  void feed() {}
}

class Dog extends Pet {
  void walk() {}
}

class Fish extends Pet {
  void changeWater() {}
}
```

我们利用了基本 Pet 的功能性，并通过增加我们需要的方法扩展了该类。这不同于本书中一贯的主张（正如 NormalLiskov.java 所示）：应该仔细地设计基类，以包含所有继承层次结构中可能需要的方法，因此遵循了里氏替换原则。虽然这是个很有抱负的目标，但也有些不切实际。试图将基于动态类型的 SmallTalk 模型强制引入基于静态类型的 Java 系统，这必然会造成妥协。在某些情况下，这些妥协可能是行不通的。模式匹配允许使用 SmallTalk 的方式向子类添加新方法，同时仍然保持里氏替换原则的大部分形式。基本上，模式匹配允许违反里氏替换原则，而不产生不可控的代码。

有了模式匹配，就可以通过为每种可能的类型进行检查并编写不同的代码，来处理 Pet 继承层次结构的非里氏替换原则的性质：

```java
// enumerations/PetPatternMatch.java
// {NewFeature} Preview in JDK 17
// 用以下标志，使用 javac 编译：
//     --enable-preview --source 17
import java.util.*;
```

```
public class PetPatternMatch {
  static void careFor(Pet p) {
    switch(p) {
      case Dog d -> d.walk();
      case Fish f -> f.changeWater();
      case Pet sp -> sp.feed();
    };
  }
  static void petCare() {
    List.of(new Dog(), new Fish())
      .forEach(p -> careFor(p));
  }
}
```

switch(p) 中的 p 称为**选择器表达式**（selector expression）。在模式匹配诞生之前，选择器表达式只能是完整的基本类型（char、byte、short 或 int），对应的包装类形式（Character、Byte、Short 或 Integer）、String 或 enum 类型。有了模式匹配，选择器表达式扩展到可以支持任何引用类型。此处的选择器表达式可以是 Dog、Fish 或 Pet。

注意，这和继承层次结构中的动态绑定类似，但是并没有将不同类型的代码放在重写的方法中，而是将其放在了不同的 case 表达式中。

编译器强制增加了 case Pet，因为该类可以在不是 Dog 或 Fish 的情况下仍然合法地存在。如果没有增加 case Pet，switch 就无法覆盖所有可能的输入值。为基类使用接口可以消除该约束，但会增加另一个约束。下面这个示例被放置在自己的 package 中，以防止命名冲突：

```
// enumerations/PetPatternMatch2.java
// {NewFeature} Preview in JDK 17
// 用以下标志，使用 javac 编译：
//    --enable-preview --source 17
package sealedpet;
import java.util.*;

sealed interface Pet {
  void feed();
}

final class Dog implements Pet {
  @Override public void feed() {}
  void walk() {}
}

final class Fish implements Pet {
  @Override public void feed() {}
  void changeWater() {}
}
```

```java
public class PetPatternMatch2 {
  static void careFor(Pet p) {
    switch(p) {
      case Dog d -> d.walk();
      case Fish f -> f.changeWater();
    };
  }
  static void petCare() {
    List.of(new Dog(), new Fish())
      .forEach(p -> careFor(p));
  }
}
```

如果 Pet 没有用 sealed（密封，Java 15 引入的关键字，用于限制类或方法的随意继承扩充）修饰，编译器会再次发出警告"switch 语句没有覆盖所有可能的输入值"。本例中的具体原因则是 interface Pet 可以被其他任何文件中的任何类实现，所以破坏了 switch 语句覆盖的完整性。通过将 Pet 修饰为 sealed 的，编译器就能确保 switch 覆盖到了所有可能的 Pet 类型。

模式匹配不会像继承多态性那样将你约束在单一继承层次结构中——你可以匹配任意类型。想要这样做，就需要将 Object 传入 switch：

```java
// enumerations/ObjectMatch.java
// {NewFeature} Preview in JDK 17
// 用以下标志，使用 javac 编译：
//    --enable-preview --source 17
// 用以下 java 标志运行：--enable-preview
import java.util.*;

record XX() {}

public class ObjectMatch {
  static String match(Object o) {
    return switch(o) {
      case Dog d -> "Walk the dog";
      case Fish f -> "Change the fish water";
      case Pet sp -> "Not dog or fish";
      case String s -> "String " + s;
      case Integer i -> "Integer " + i;
      case String[] sa -> String.join(", ", sa);
      case null, XX xx -> "null or XX: " + xx;
      default -> "Something else";
    };
  }
  public static void main(String[] args) {
    List.of(new Dog(), new Fish(), new Pet(),
      "Oscar", Integer.valueOf(12),
      Double.valueOf("47.74"),
      new String[]{ "to", "the", "point" },
```

```
/* 输出：
Walk the dog
Change the fish water
Not dog or fish
String Oscar
Integer 12
Something else
to, the, point
null or Object: XX[]
*/
```

```
      new XX()
    ).forEach(
      p -> System.out.println(match(p))
    );
  }
}
```

将 Object 参数传入 switch 的时候，编译器会要求有 default 存在，也是为了覆盖所有可能的输入值（不过 null 除外，编译器并不要求存在为 null 的 case，尽管这种情况也可能发生）。

可以将为 null 的 case 和另一种模式合并到一起，如 case null, XX xx。这样是可以的，因为对象引用也可以是 null。

1.17.2 守卫

守卫（guard）使你可以进一步细化匹配条件，而不只是简单地匹配类型。它是出现在类型判断和 && 后的一项测试。守卫可以是任何布尔表达式。如果选择器表达式和 case 的类型相同，并且守卫判断为 true，那么模式就匹配上了：

```
// enumerations/Shapes.java
// {NewFeature} Preview in JDK 17
// 用以下标志，使用 javac 编译：
//   --enable-preview --source 17
// 用以下 java 标志运行：--enable-preview
import java.util.*;

sealed interface Shape {
  double area();
}

record Circle(double radius) implements Shape {
  @Override public double area() {
    return Math.PI * radius * radius;
  }
}

record Rectangle(double side1, double side2)
  implements Shape {
  @Override public double area() {
    return side1 * side2;
  }
}

public class Shapes {
  static void classify(Shape s) {
    System.out.println(switch(s) {
      case Circle c && c.area() < 100.0
```

```
/* 输出：
Small Circle: Circle[radius=5.0]
Large Circle: Circle[radius=25.0]
Square: Rectangle[side1=12.0, side2=12.0]
Rectangle: Rectangle[side1=12.0, side2=15.0]
*/
```

```
      -> "Small Circle: " + c;
    case Circle c -> "Large Circle: " + c;
    case Rectangle r && r.side1() == r.side2()
      -> "Square: " + r;
    case Rectangle r -> "Rectangle: " + r;
  });
}
public static void main(String[] args) {
  List.of(
    new Circle(5.0),
    new Circle(25.0),
    new Rectangle(12.0, 12.0),
    new Rectangle(12.0, 15.0)
  ).forEach(t -> classify(t));
}
}
```

第一个针对 Circle 的守卫用于确定该 Circle 是否是"小"的。第二个针对 Rectangle 的守卫用于确定 Rectangle 是否是正方形。

下面是个更复杂的守卫应用示例。Tank（罐子）可以持有不同类型的液体，而液体的 Level（浓度）必须在 0%~100% 范围内。

```
// enumerations/Tanks.java
// {NewFeature} Preview in JDK 17
// 用以下标志，使用javac编译:
//   --enable-preview --source 17
// 用以下java标志运行: --enable-preview
import java.util.*;

enum Type { TOXIC, FLAMMABLE, NEUTRAL }

record Level(int percent) {
  Level {
    if(percent < 0 || percent > 100)
      throw new IndexOutOfBoundsException(
        percent + " percent");
  }
}

record Tank(Type type, Level level) {}

public class Tanks {
  static String check(Tank tank) {
    return switch(tank) {
      case Tank t && t.type() == Type.TOXIC
        -> "Toxic: " + t;
      case Tank t && (                    // [1]
        t.type() == Type.TOXIC &&
        t.level().percent() < 50
      ) -> "Toxic, low: " + t;
```

```
      case Tank t && t.type() == Type.FLAMMABLE
        -> "Flammable: " + t;
      // 相当于 default：
      case Tank t -> "Other Tank: " + t;
    };
  }
  public static void main(String[] args) {
    List.of(
      new Tank(Type.TOXIC, new Level(49)),
      new Tank(Type.FLAMMABLE, new Level(52)),
      new Tank(Type.NEUTRAL, new Level(75))
    ).forEach(
      t -> System.out.println(check(t))
    );
  }
}
```

[1] 如果守卫包含多个表达式，简单地将其放在圆括号内即可。

由于我们是在对 Tank（而不是 Object）做 switch 操作，所以最后 case Tank 的作用就相当于 default，因为它会捕捉到所有没有匹配到任何其他模式的 Tank。

1.17.3 支配性

switch 中 case 语句的顺序很重要。如果基类先出现，就会支配任何出现在后面的 case[1]：

```
// enumerations/Dominance.java
// {NewFeature} Preview in JDK 17
// 用以下标志，使用 javac 编译
//    --enable-preview --source 17
import java.util.*;

sealed interface Base {}
record Derived() implements Base {}

public class Dominance {
  static String test(Base base) {
    return switch(base) {
      case Derived d -> "Derived";
      case Base b -> "B";           // [1]
    };
  }
}
```

基类 Base 处于最后一个位置，即行 [1]。但是如果将该行上移，便会出现在 case

---

[1] 支配即 dominate，如果前面 case 的判断条件所覆盖的范围包含后面的 case，即可认为前者支配了后者，也就是说后者实际上不会得到执行的机会了。——译者注

Derived 的前面，这就意味着 switch 将永远没有机会测试到 Derived 了，因为任何子类都将被 case Base 捕获。如果尝试这么做，编译器就会报错，"该 case 标签被前面的 case 标签支配了"（this case label is dominated by a preceding case label）。

在使用守卫时，顺序的敏感性常常会体现出来。在 Tanks.java 中，将 switch 里处于末尾的 case 移动到更高的位置会导致"支配性"的错误消息。如果在同一个模式上有多个守卫，更具体的模式必须出现在更泛化的模式之前，否则更泛化的模式会在更具体的模式之前进行匹配，后者就永远没有机会被检查了。所幸编译器会报告支配性的问题。

编译器只有在一个模式中的**类型**支配了另一个模式中的类型时，才能检测出支配性问题。它无法知道守卫中的逻辑是否会导致问题：

```java
// enumerations/People.java
// {NewFeature} Preview in JDK 17
// 用以下标志，使用 javac 编译：
//   --enable-preview --source 17
// 用以下 java 标志运行：--enable-preview
import java.util.*;

record Person(String name, int age) {}

public class People {
  static String categorize(Person person) {
    return switch(person) {
      case Person p && p.age() > 40           // [1]
        -> p + " is middle aged";
      case Person p &&
        (p.name().contains("D") || p.age() == 14)
        -> p + " D or 14";
      case Person p && !(p.age() >= 100)      // [2]
        -> p + " is not a centenarian";
      case Person p -> p + " Everyone else";
    };
  }
  public static void main(String[] args) {
    List.of(
      new Person("Dorothy", 15),
      new Person("John Bigboote", 42),
      new Person("Morty", 14),
      new Person("Morty Jr.", 1),
      new Person("Jose", 39),
      new Person("Kane", 118)
    ).forEach(
      p -> System.out.println(categorize(p))
    );
  }
}
```

```
/* 输出：
Person[name=Dorothy, age=15] D or 14
Person[name=John Bigboote, age=42] is middle aged
Person[name=Morty, age=14] D or 14
Person[name=Morty Jr., age=1] is not a centenarian
Person[name=Jose, age=39] is not a centenarian
Person[name=Kane, age=118] is middle aged
*/
```

模式 [2] 中的守卫似乎可以匹配到 118 岁的 Kane，但是 Kane 被模式 [1] 匹配到了。不能依赖编译器来帮你处理守卫表达式的逻辑。

如果没有最后的 case Person p，编译器会认为 "switch 表达式没有覆盖所有可能的输入"。而有了 case Person p，也仍然不需要 default。因此最泛化的 case 便成了 default。由于 switch 的参数是 Person，因此所有的 case 都被覆盖了（null 仍然除外）。

1.17.4 覆盖范围

模式匹配会引导你逐渐使用 sealed 关键字，这有助于确保你已覆盖了所有可能传入选择器表达式的类型。不过接下来再看一个示例：

```java
// enumerations/SealedPatternMatch.java
// {NewFeature} Preview in JDK 17
// 用以下标志，使用 javac 编译：
//    --enable-preview --source 17
// 用以下 java 标志运行：--enable-preview
import java.util.*;

sealed interface Transport {};
record Bicycle(String id) implements Transport {};
record Glider(int size) implements Transport {};
record Surfboard(double weight) implements Transport {};

// 如果取消下面这行的注释：
// record Skis(int length) implements Transport {};
// 便会得到错误消息 "switch 表达式未覆盖所有可能的输入值"

public class SealedPatternMatch {
  static String exhaustive(Transport t) {
    return switch(t) {
      case Bicycle b -> "Bicycle " + b.id();
      case Glider g -> "Glider " + g.size();
      case Surfboard s -> "Surfboard " + s.weight();
    };
  }
  public static void main(String[] args) {
    List.of(
```

```
      new Bicycle("Bob"),
      new Glider(65),
      new Surfboard(6.4)
    ).forEach(
      t -> System.out.println(exhaustive(t))
    );
    try {
      exhaustive(null); // 永远有可能发生！  // [1]
    } catch(NullPointerException e) {
      System.out.println("Not exhaustive: " + e);
    }
  }
}
/* 输出：
Bicycle Bob
Glider 65
Surfboard 6.4
Not exhaustive: java.lang.NullPointerException
*/
```

  `sealed interface Transport` 是通过 `record` 实现的，`record` 自动就是 `final` 的。`switch` 覆盖了所有可能的 `Transport` 类型，而且如果增加一个新的类型，编译器就会检测出来，并告知你没有完全覆盖所有可能的模式。但是从 [1] 可以看出，仍然有一个 `case`，编译器并未坚持让你覆盖：`null`。如果你没忘记加上 `case null`，便可以防止出现异常。但是编译器在此处帮不了你，显然是因为这会影响到太多已有的 `switch` 代码。

  虽然模式匹配对于 Java 来说肯定是一项改进，但它也遭遇着和其他 Java 新增特性相同的问题：由于向后兼容性的缘故，它只能是相对更强大的 Scala 及 Kotlin 版本的某种不完全的实现。另外，由于匹配模式进入 Java 那缓慢而谨慎的步伐（这一点可以理解），可能需要过好几年才能看到完整版本的 Java 模式匹配。

  理想情况下，Java 的模式匹配足以解决问题了。不过，如果你发现自己对模式匹配的依赖非常强烈，那么你可能需要一个比 Java 所提供的功能的更完整的功能集。Kotlin 语言有着更全面的模式匹配能力，而且能生成可以无缝放入现有 Java 程序的类文件。

1.18 总结

  虽然枚举类型本身并不是特别复杂，但由于 `enum` 可以与多态、泛型和反射等特性结合使用，所以本章被安排在了本书比较靠后的位置。

  虽然 Java 的枚举明显比 C 或 C++ 中的 `enum` 更复杂、更先进，但它仍然是一个"小"特性（功能），Java 在没有它的情况下也"生存"了很多年（虽然有点尴尬）。然而本章告诉你，一个"小"特性也可以有相当大的价值——有时它恰恰给了你最合适的杠杆来优雅而清晰地解决问题。正如本书中常常提到的，优雅非常重要，而清晰则可以决定一个解决方案是成功的还是失败的（失败是因为人们无法看懂）。

对于清晰性这一点，一个困惑之源是 Java 1.0 的失败选择：使用术语"enumeration"，而不是常见且已被广泛接受的术语"iterator"，来表示一个从序列中依次选取元素的对象（如基础卷第 12 章所述）。有些语言甚至将枚举的数据类型称为"enumerator"！这个错误后来被 Java 修正了，但是 Enumeration 接口显然无法简单地直接移除，因此它仍然会存在于在一些老旧代码（甚至是新代码）、库以及文档中。

02

对象传递和返回

> 现在，你应该已经对对象传递的理念习以为常了：当你在"传递"一个对象的时候，你实际上是在传递它的引用。

在许多编程语言中，你可以使用该语言的"常规"方式来传递对象，而且大部分时候一切都很正常。但总有些时候，你需要用到一些反常规的方式，而这时事情就会突然变得有点复杂。Java 也不例外，理解在传递和操纵对象时究竟具体发生了些什么，这一点非常重要。本章会为你揭示这背后的真相。

还有另外一种理解本章核心问题的角度。如果你之前使用的是带有指针的语言，那你就可以这样问："Java 有指针吗？"Java 中的所有对象标识（除了基本类型）都是指针，但是它们的使用都受到了严格的限制和保护，这些限制和保护不仅来自编译器，而且还来自运行时系统。换句话说，Java 有指针，但是没有指针运算。Java 的指针即我一直所说的"引用"，你可以认为它是"安全的指针"，这和小学生使用的安全剪刀没什么不同——它并不锋利，因此你不费吹灰之力就可以保证自己的安全，但有时用起来会显得烦琐。

2.1 传递引用

当你将一个引用传给方法后，该引用指向的仍然是原来的对象。可以通过下面这个简单的实验看出来这一点：

```java
// references/PassReferences.java

public class PassReferences {
  public static void f(PassReferences h) {
    System.out.println("h inside f(): " + h);
  }
  public static void main(String[] args) {
    PassReferences p = new PassReferences();
    System.out.println("p inside main(): " + p);
    f(p);
  }
}
/* 输出：
p inside main(): PassReferences@19e0bfd
h inside f(): PassReferences@19e0bfd
*/
```

print 打印语句会自动调用 toString() 方法，同时 PassReferences 直接从 Object 继承了 toString() 方法，并未对该方法进行重新定义。因此，这里使用了 Object 自身的 toString() 方法，打印出了对象的类名，并在后面附上对象在内存中的地址（并非引用，而是实际的对象所保存的地址）。

从输出可以看出，p 和 h 引用都指向了同一个对象。这比只是为了能够向方法传递参数，而专门复制一个新的 PassReferences 对象要高效得多。但是这会带来一个重要的问题。

引用别名

引用别名指的是不止一个引用被绑定到了同一个对象上的情况，如前面的示例所示。别名导致的问题主要发生在有人对对象进行**写**操作时。如果该对象的其他引用的持有者并不希望对象发生变更，那么结果将使其大感意外。示例如下：

```java
// references/Alias1.java
// 一个对象有两个引用别名

public class Alias1 {
  private int i;
  public Alias1(int ii) { i = ii; }
  public static void main(String[] args) {
    Alias1 x = new Alias1(7);
    Alias1 y = x; // 分配引用    // [1]
    System.out.println("x: " + x.i);
    System.out.println("y: " + y.i);
/* 输出：
x: 7
y: 7
Incrementing x
x: 8
y: 8
*/
```

```
    System.out.println("Incrementing x");
    x.i++;                                       // [2]
    System.out.println("x: " + x.i);
    System.out.println("y: " + y.i);
  }
}
```

[1] 此处创建了一个新的 Alias1 引用，但并未将其分配给某个用 new 创建的新对象，而是分配给了一个已有对象的引用。所以引用 x 的内容，也就是对象 x 所指向的地址，被分配给了 y，因此 x 和 y 都指向了同一个对象。

[2] 当 x 的 i 自增后，y 的 i 也被影响，如输出中所示。

最好的解决方案其实很简单，就是不要这么做。不要在同一个作用域内，有意识地对一个对象使用一个以上的引用别名。这样你的代码就会很大程度上简化理解和调试的难度。不过，在你将引用作为参数传入的时候——这正是 Java 的工作方式——你实际上已经自动命名引用别名了，因为本地创建的引用可以修改"外部的对象"（在方法作用域外创建的对象）：

```
// references/Alias2.java
// 方法调用隐式地对其参数命名了引用别名

public class Alias2 {
  private int i;
  public Alias2(int i) { this.i = i; }
  public static void f(Alias2 reference) {
    reference.i++;
  }
  public static void main(String[] args) {
    Alias2 x = new Alias2(7);
    System.out.println("x: " + x.i);
    System.out.println("Calling f(x)");
    f(x);
    System.out.println("x: " + x.i);
  }
}
/* 输出：
x: 7
Calling f(x)
x: 8
*/
```

上面的示例中，方法修改了其参数，即外部的对象。当这种情况出现时，你必须判断这是否合理，是否符合用户的预期，以及是否会导致问题。这三个问题的答案通常是不合理、不符合预期、会导致问题，这也是纯粹的函数式语言并不允许这种行为的原因。

通常来说，你调用方法是为了得到返回值，以及 / 或者改变**调用方法**的对象的状态（内部属性）。很少有为了操作方法的参数而调用方法的情况，这是"调用方法只是为了利用其**副作用**"的一种形式。因此，如果你要创建会修改参数的方法，就必须清楚地让使用者知道该方法的用法并警惕其潜在风险。因为这种方式会带来不确定性及隐患，所以最好避免修改参数。

如果你在方法调用的过程中必须修改参数，同时又不想改变该参数在方法外部的状态，那么就要通过在方法内部复制出一个参数的副本的方式对其进行保护。这是本章大部分内容的主题。

2.2 创建本地副本

Java 中所有的参数传递都是通过传递引用实现的。也就是说，当你传递"一个对象"时，你实际上传递的是存活于方法外部的指向这个对象的一个引用。因此，如果你通过该引用执行了任何修改，你也将修改外部的对象。此外：

- 引用别名会在传递参数时自动发生；
- 并没有本地对象，只有本地引用；
- 引用是有作用域的，对象则没有；
- Java 中的对象生命周期从来就不是个问题。

语言层面上，Java 并未提供防止对象被修改和消除别名负面影响的相关支持（例如 "const"）。你不能简单地在参数列表中使用 final 关键字，这只能阻止你将该引用重新绑定到另一个不同的对象上。

如果你严格地只读取对象的相关信息，并不修改它，那么传递对象引用就是最高效的传递参数的方式。这样很不错，默认的方式就是最高效的方式。然而，有时需要将对象当作"本地"的，这样你所做的修改就只会影响本地的副本，而不会修改外部的对象。许多编程语言支持在方法内自动创建外部对象的本地副本的能力。[1]Java 并不支持这样做，但是它可以让你实现相同的效果。

2.2.1 值传递

这就引出了术语方面的问题，这对参数来说似乎总是好的。这个术语就是"值传递"（pass by value），其含义取决于你如何理解程序的操作。这个概念是指不论你传递的是什么，你都会得到一个本地的副本。而真正的问题是，你如何看待你所传递的东西。说到"值传递"的含义，有如下两个截然不同的阵营。

---

[1] 在 C 中，通常只处理少量的数据，默认是按值传递。C++ 必须遵循这个形式，但是对于对象，按值传递通常不是最高效的方法。另外，在 C++ 中通过手写代码的方式支持按值传递是件非常令人头疼的事。

1. Java 传递任何事物时，都是在传递该事物的值。当你向方法传入基本类型时，得到的是该基本类型的一份单独的副本。而当你向方法传入一个引用时，得到的则是该引用的副本。因此，一切都是通过值在传递。该观点所基于的假设是：你总是认为并且在意传递的就是引用（和对象不同），但 Java 的设计似乎已经在很大程度上让你（在大部分时候）忽略你是在使用引用。也就是说，Java 似乎允许你将引用视为"对象"，因为每当你调用方法时，它都会隐式地反向解析引用（为对象）。

2. Java 在传递基本类型时，传递的是值，但在传递对象时，传递的则是引用。在通行的观点中，引用是对象的一个别名，因此你**不会**认为是在传递引用，而会认为"传递的是对象"。在将对象传入方法时，你不会得到该对象的本地副本，因此很明显，传递的并不是对象的值。在 Sun 公司的内部似乎对该观点有一定的支持，因为曾经有一个"保留但未实现"的关键字 byvalue（它永远不会被实现）。

以上详细介绍了两个不同阵营的观点，我也表达了自己的观点："这取决于你如何看待引用"，我感觉这个问题可以告一段落了。总之，这并没有**那么**重要——重要的是，你要理解传递引用会使得调用者的对象被意外地修改。

2.2.2 克隆对象

在深入研究克隆之前，请查看本章 2.4 节末尾的备选方案。

创建对象的本地副本，最可能的原因是你要修改该对象，但并不想修改调用者的原有对象。一种方法是使用 clone() 方法。clone() 是在基类 Object 中定义的，并且访问权限为 protected。你需要在任何你想要克隆的子类中将 clone() 方法重写为 public 的方法。例如，标准库中的 ArrayList 类重写了 clone() 方法，这样就可以对 ArrayList 调用 clone() 方法了：

```java
// references/CloneArrayList.java
// clone() 操作仅可用于标准 Java 库中的少数项
import java.util.*;
import java.util.stream.*;

class Int {
  private int i;
  Int(int ii) { i = ii; }
  public void increment() { i++; }
  @Override public String toString() {
    return Integer.toString(i);
  }
}

public class CloneArrayList {
```

```
/* 输出：
v: [0, 1, 2, 3, 4, 5, 6, 7, 8, 9]
v: [1, 2, 3, 4, 5, 6, 7, 8, 9, 10]
*/
```

```
    public static void main(String[] args) {
      ArrayList<Int> v = IntStream.range(0, 10)
        .mapToObj(Int::new)
        .collect(Collectors
          .toCollection(ArrayList::new));
      System.out.println("v: " + v);
      @SuppressWarnings("unchecked")
      ArrayList<Int> v2 = (ArrayList<Int>)v.clone();
      // 对 v2 中的所有元素进行自增：
      v2.forEach(Int::increment);
      // 看看是否修改了 v 中的元素：
      System.out.println("v: " + v);
    }
  }
```

clone() 方法生成了一个 Object，然后该 Object 必须被转换为合适的类型。本例演示了 ArrayList 的 clone() 方法如何**不去**自动尝试克隆 ArrayList 中的每个对象——原有的 ArrayList 和克隆的 ArrayList 都是同一个对象的不同引用别名。这是一种**浅拷贝**（shallow copy），因为只复制了对象的"表层"部分。实际对象的组成部分包括该"表层"（引用）、该引用指向的所有对象，以及所有**这些**对象所指向的所有对象，以此类推。这通常称为"对象网络"。创建所有这些内容的完整副本，称为**深拷贝**（deep copy）。

你可以在输出中看到浅拷贝的效果，对 v2 执行的操作影响了 v。不对 ArrayList 中的所有对象进行 clone() 可能是个合理的假设，因为无法保证这些对象都是可克隆（cloneable[①]）的。

2.2.3 为类增加可克隆能力

尽管 clone() 是在所有类的基类 Object 中定义的，但并不是所有类就因此自动具备克隆能力了。这看起来似乎有点反常，理论上基类中的方法应该在子类中也总是可用的。Java 中的克隆确实违反了这个概念，如果你希望某个类具备克隆能力，就必须专门添加代码来使其生效。

1. 利用 protected 的技巧

为了避免你创建的所有类都具有默认的可克隆能力，基类 Object 中的 clone() 方法是 protected 的。这意味着对于那些只是简单使用某个类（并不继承出子类）的调用方程序员来说，可克隆能力并不是默认可用的。这同时也意味着，你无法通过基类的引用来调用 clone()。这实际上是在编译期给你的一个信号，让你知道你的对象是不可克隆的——并且

[①] 这并不是该单词在字典中的拼写方式，但它在 Java 库中就是这么用的，因此我在这里也这样使用了，希望能避免混淆。

奇怪的是，Java 标准库中的大部分类是不可克隆的。因此，如果你声明：

```
Integer x = 1;
x = x.clone();
```

你就会在编译期得到一条错误消息，指出没有调用 clone() 方法的权限（Integer 并未重写该方法，该方法仍然是默认的 protected 版本）。

然而，如果是在一个 Object 的子类（就像所有的类一样）中，那么你就有调用 Object.clone() 的权限，因为该方法是 protected 的，而你是继承者。基类中的 clone() 的功能很有用，它会执行实际**子类对象**的按位复制，因此充当了公用的克隆操作。不过，为了使你的克隆操作可被访问，你必须将它设为 public 的。因此，在你克隆时，有两个关键步骤：

- 调用 super.clone()
- 将你的克隆操作设为 public 的

你很可能会在后续的子类中重写 clone() 方法，否则执行的将是你现有的 clone()（现在是 public 的了），这或许并非你想要的操作（虽然由于 Object.clone() 会创建实际对象的副本，它也有可能就是你想要的操作）。protected 技巧只有一次使用时机，即你首次继承一个不具备可克隆能力的类，且希望继承出的类具备可克隆性时。对于从该继承类再度继承出的任何类，clone() 方法都是可用的了，因为在 Java 中是不可能在派生过程中缩小方法访问权限的。也就是说，一旦一个类成为可克隆的，一切从其派生的类就都是可克隆的，除非你使用了语言提供的某种机制（稍后会描述）来"关闭"克隆能力。

2. 实现 Cloneable 接口

要让对象实现完整的可克隆能力，还需要进一步操作：实现 Cloneable 接口。这是一个空（标记）接口。

Cloneable 接口的存在有两个原因。第一个原因是，你可能有个对某父类型的向上转型的引用，但并不知道是否可以克隆该对象。这时可以使用 instanceof 关键字（在基础卷第 19 章中描述过）来找出该引用是否和某个可克隆的对象有关联：

```
if(myReference instanceof Cloneable) // ...
```

第二个原因是，可克隆能力的设计背后有一种考虑，即可能你并不希望所有类型的对象都是可克隆的。因此 Object.clone() 可以验证一个类是否实现了 Cloneable 接口，如果没有，便会抛出 CloneNotSupportedException 异常。所以一般来说，你是被迫加上

implement Cloneable 的，以作为支持克隆的必要操作。

2.2.4 成功的克隆

一旦你理解了实现 clone() 方法的细节，就能够创建出可轻松复制的类，以提供生成本地副本的能力了：

```java
// references/LocalCopy.java
// 用 clone() 创建本地副本

class Duplo implements Cloneable {
  private int n;
  Duplo(int n) { this.n = n; }
  @Override public Duplo clone() {            // [1]
    try {
      return (Duplo)super.clone();
    } catch(CloneNotSupportedException e) {
      throw new RuntimeException(e);
    }
  }
  public int getValue() { return n; }
  public void setValue(int n) { this.n = n; }
  public void increment() { n++; }
  @Override public String toString() {
    return Integer.toString(n);
  }
}

public class LocalCopy {
  public static Duplo g(Duplo v) {
    // 传递引用，修改了外部的对象:
    v.increment();
    return v;
  }
  public static Duplo f(Duplo v) {
    v = v.clone(); // 本地副本              // [2]
    v.increment();
    return v;
  }
  public static void main(String[] args) {
    Duplo a = new Duplo(11);
    Duplo b = g(a);
    // 引用相等，并不是对象相等：
    System.out.println("a == b: " + (a == b) +
      "\na = " + a + "\nb = " + b);
    Duplo c = new Duplo(47);
    Duplo d = f(c);
    System.out.println("c == d: " + (c == d) +
      "\nc = " + c + "\nd = " + d);
  }
}
/* 输出:
a == b: true
a = 12
b = 12
c == d: false
c = 47
d = 48
*/
```

首先，为了使 clone() 能被访问，必须将它设为 public 的。其次，在你的 clone() 操作的开始部分，调用基类版本的 clone()。这里调用的 clone() 即在 Object 中预先定义好的那个 clone()，你可以调用该方法，是因为它是 protected 的，可以在子类中被访问。

Object.clone() 会检测出该对象有多大，并为新对象创建足够的内存空间，然后将旧对象所有的二进制位都复制到新对象中。这称为**按位复制**，通常也是你希望 clone() 方法做的事。但是 Object.clone() 在执行操作之前，会先检查（在继承层次结构中）是否有类是 Cloneable 的，也就是它是否实现了 Cloneable 接口。如果没有，Object.clone() 会抛出 CloneNotSupportedException 异常，表明你无法对它进行克隆。因此，你必须用 try 块，将对 super.clone() 的调用包裹起来，以捕获那个永远不会发生的异常（因为你已经实现了 Cloneable 接口）。

[1] 注意此处用到的协变返回类型。基类 Object 的 clone() 方法只能返回 Object，但是子类的 clone() 可以返回更具体的类型。在协变返回类型出现之前，你必须将返回类型向下转型为合适的类型，但现在它可以在编译期进行验证了。

[2] 此处，你可以看到对 clone() 的调用返回了 Duplo，并不需要转型。

在 LocalCopy 中，方法 g() 和 f() 演示了参数传递的两种方式。方法 g() 演示了引用传递，它改变了外部对象，并返回了一个该外部对象的引用。f() 则克隆了参数，从而实现了解耦，不再影响原始的外部对象，然后它就可以继续做任何想要的操作了，甚至是返回该新对象的引用，而不会对原有对象产生任何副作用。

在 main() 中，针对参数传递的两种方法的效果进行了测试。要注意的是，Java 中的相等判定并不会深入对象的内部去看两者的值是否相等。== 和 != 操作只是简单地比较引用。如果引用内部的地址相同，那么指向的就是同一个对象，因此就是"相等"的。所以操作符真正验证的是两个引用是否是同一个对象的两个别名！

2.2.5 Object.clone() 的效果

调用 Object.clone() 时究竟发生了些什么，以至于在重写 clone() 时调用 super.clone() 的这一步如此重要呢？根类中的 clone() 方法负责创建正确大小的存储空间，并执行了从原始对象中的所有二进制位到新对象存储中的按位复制。也就是说，该方法并不只是创建存储空间和复制一个 Object，它实际上会判断要复制的实际对象（不只是基类对象，还包括派生对象）的大小，然后再进行复制。所有这些都依靠在根类中定义的 clone() 方法的代码来实现，而根类并不知道具体哪个对象会被继承，因此你可以猜到，这个过程中用到了反射来确定要克隆的实际对象。这样，clone() 方法就可以创建大小合适的存储空间，并正确地对该类型执行按位复制。

克隆过程通常应该从调用 super.clone() 开始，这样便通过创建精确的副本，为克隆操作奠定了基础。然后就可以执行所需的其他操作来完成克隆了。

要清楚地知道其他操作是什么，就需要理解 Object.clone() 到底给你带来了什么好处，特别是该方法是否会自动克隆所有引用指向的目标。我们可以通过测试来了解：

```java
// references/Snake.java
// 测试克隆，看看是否引用的所有目标都被克隆了
public class Snake implements Cloneable {
  private Snake next;
  private char c;
  // i 的值 == 蛇身的段数
  public Snake(int i, char x) {
    c = x;
    if(--i > 0)
      next = new Snake(i, (char)(x + 1));
  }
  public void increment() {
    c++;
    if(next != null)
      next.increment();
  }
  @Override public String toString() {
    String s = ":" + c;
    if(next != null)
      s += next.toString();
    return s;
  }
  @Override public Snake clone() {
    try {
      return (Snake)super.clone();
    } catch(CloneNotSupportedException e) {
      throw new RuntimeException(e);
    }
  }
  public static void main(String[] args) {
    Snake s = new Snake(5, 'a');
    System.out.println("s = " + s);
    Snake s2 = s.clone();
    System.out.println("s2 = " + s2);
    s.increment();
    System.out.println(
      "after s.increment, s2 = " + s2);
  }
}
/* 输出：
s = :a:b:c:d:e
s2 = :a:b:c:d:e
after s.increment,
s2 = :a:c:d:e:f
*/
```

一条 Snake（蛇）由多个段组成，每一段都是 Snake 类型。因此这是一个单向链表。

这些段通过递归的方式创建，每一段都会对第一个构造参数做递减操作，直到值为 0。为了给每一段都定义一个唯一的标识，第二个 char 类型的参数会在每次递归调用构造方法时递增。

increment() 方法递归地对标识自增，以体现出变化，而 toString() 则递归地打印出每一个标识。从结果可以看出，只有第一段被 Object.clone() 复制了，因此它执行的是浅拷贝。如果想要复制整条蛇，也就是深拷贝，你需要在重写的 clone() 方法中执行额外的操作。

在任何派生自可克隆类的类中，你通常需要调用 super.clone()，以确保所有基类的操作（包括 Object.clone()）都执行了。然后对对象中的所有引用都显式地执行 clone()，否则这些引用都将只是原始对象中对应引用的别名。这和调用构造器的方式相似：最开始调用基类的构造器，然后是下一层派生的构造器，以此类推，直到最后一层派生的构造器。区别在于，clone() 并不是构造器，因此无法自动实现这种调用机制。你必须确保自己实现这种机制。

2.2.6 克隆组合对象

在你试图深拷贝组合对象时，会遇到一个问题。你必须假设所有成员对象中的 clone() 方法都会按顺序对各自的引用执行深拷贝，并照此进行下去。这一点是必须确保的。它实际上意味着为了正确执行深拷贝，你要么需要控制所有类的所有代码，要么至少对深拷贝涉及的所有类都足够了解，以确定它们都能正确地执行各自的深拷贝。

下面便是你在处理组合对象的深拷贝时必须做的事情：

```
// references/DepthReading.java
// 克隆组合对象
package references;

public class DepthReading implements Cloneable {
  private double depth;
  public DepthReading(double depth) {
    this.depth = depth;
  }
  @Override public DepthReading clone() {
    try {
      return (DepthReading)super.clone();
    } catch(CloneNotSupportedException e) {
      throw new RuntimeException(e);
    }
  }
  public double getDepth() { return depth; }
  public void setDepth(double depth) {
    this.depth = depth;
```

```java
  }
  @Override public String toString() {
    return String.valueOf(depth);
  }
}
// references/TemperatureReading.java
// 克隆组合对象
package references;

public class TemperatureReading implements Cloneable {
  private long time;
  private double temperature;
  public TemperatureReading(double temperature) {
    time = System.currentTimeMillis();
    this.temperature = temperature;
  }
  @Override public TemperatureReading clone() {
    try {
      return (TemperatureReading)super.clone();
    } catch(CloneNotSupportedException e) {
      throw new RuntimeException(e);
    }
  }
  public double getTemperature() {
    return temperature;
  }
  public void setTemperature(double temp) {
    this.temperature = temp;
  }
  @Override public String toString() {
    return String.valueOf(temperature);
  }
}
// references/OceanReading.java
// 克隆组合对象
package references;

public class OceanReading implements Cloneable {
  private DepthReading depth;
  private TemperatureReading temperature;
  public
  OceanReading(double tdata, double ddata) {
    temperature = new TemperatureReading(tdata);
    depth = new DepthReading(ddata);
  }
  @Override public OceanReading clone() {
    OceanReading or = null;
    try {
      or = (OceanReading)super.clone();
    } catch(CloneNotSupportedException e) {
      throw new RuntimeException(e);
    }
    // 必须克隆引用:
```

```java
    or.depth = (DepthReading)or.depth.clone();
    or.temperature =
      (TemperatureReading)or.temperature.clone();
    return or;
  }
  public TemperatureReading getTemperatureReading() {
    return temperature;
  }
  public void
  setTemperatureReading(TemperatureReading tr) {
    temperature = tr;
  }
  public DepthReading getDepthReading() {
    return depth;
  }
  public void setDepthReading(DepthReading dr) {
    this.depth = dr;
  }
  @Override public String toString() {
    return "temperature: " + temperature +
      ", depth: " + depth;
  }
}
```

现在我们可以用 JUnit 来测试了：

```java
// references/tests/DeepCopyTest.java
package references;
import org.junit.jupiter.api.*;
import static org.junit.jupiter.api.Assertions.*;

public class DeepCopyTest {
  @Test
  public void testClone() {
    OceanReading reading =
      new OceanReading(33.9, 100.5);
    // 进行克隆：
    OceanReading clone = reading.clone();
    TemperatureReading tr =
      clone.getTemperatureReading();
    tr.setTemperature(tr.getTemperature() + 1);
    clone.setTemperatureReading(tr);
    DepthReading dr = clone.getDepthReading();
    dr.setDepth(dr.getDepth() + 1);
    clone.setDepthReading(dr);
    assertEquals(reading.toString(),
      "temperature: 33.9, depth: 100.5");
    assertEquals(clone.toString(),
      "temperature: 34.9, depth: 101.5");
  }
}
```

DepthReading 和 TemperatureReading 很相似，它们都只包含基本类型。因此，clone()
方法可以很简单：调用 super.clone()，然后返回结果。注意，这两者的 clone() 代码是完
全相同的。

OceanReading 是由 DepthReading 和 TemperatureReading 对象组合而成的，因此，要实现
深拷贝，它的 clone() 就必须克隆 OceanReading 内部的所有引用。要完成这项任务，super.
clone() 的结果必须转型为 OceanReading 对象（这样才可以访问 depth 和 temperature 的
引用）。

2.2.7 深拷贝 ArrayList

再回头看看本章早些时候出现的 CloneArrayList.java。这一次 Int2 类是可克隆的了，
因此 ArrayList 可以进行深拷贝：

```java
// references/AddingClone.java
// 要将克隆能力添加到自己的类中，你需要绕点弯路
import java.util.*;
import java.util.stream.*;

class Int2 implements Cloneable {
  private int i;
  Int2(int ii) { i = ii; }
  public void increment() { i++; }
  @Override public String toString() {
    return Integer.toString(i);
  }
  @Override public Int2 clone() {
    try {
      return (Int2)super.clone();
    } catch(CloneNotSupportedException e) {
      throw new RuntimeException(e);
    }
  }
}

// 继承不会移除可克隆性:
class Int3 extends Int2 {
  private int j; // 自动创建了副本
  Int3(int i) { super(i); }
}

public class AddingClone {
  @SuppressWarnings("unchecked")
  public static void main(String[] args) {
    Int2 x = new Int2(10);
    Int2 x2 = x.clone();
    x2.increment();
```

```
/* 输出:
x = 10, x2 = 11
v: [0, 1, 2, 3, 4, 5, 6, 7, 8, 9]
v2: [1, 2, 3, 4, 5, 6, 7, 8, 9, 10]
v: [0, 1, 2, 3, 4, 5, 6, 7, 8, 9]
*/
```

```
    System.out.println(
      "x = " + x + ", x2 = " + x2);
    // 继承出的任何事物同样也是可克隆的：
    Int3 x3 = new Int3(7);
    x3 = (Int3)x3.clone();
    ArrayList<Int2> v = IntStream.range(0, 10)
      .mapToObj(Int2::new)
      .collect(Collectors
        .toCollection(ArrayList::new));
    System.out.println("v: " + v);
    ArrayList<Int2> v2 =
      (ArrayList<Int2>)v.clone();
    // 现在克隆每个元素：
    IntStream.range(0, v.size())
      .forEach(i -> v2.set(i, v.get(i).clone()));
    // 对 v2 的所有元素进行累加：
    v2.forEach(Int2::increment);
    System.out.println("v2: " + v2);
    // 看看是否改变了 v 中的所有元素：
    System.out.println("v: " + v);
  }
}
```

Int3 继承自 Int2，并增加了一个新的基本类型成员 int j。你可能会认为需要再次重写 clone() 以确保 j 被复制了，但并不是这样。在调用 Int3 的 clone() 时，其内部所调用的 Int2 的 clone()，实际上调用的是 Object.clone()，它检测到此处起作用的是 Int3，并对 Int3 执行了按位复制。只要你不增加需要克隆的引用，对 Object.clone() 的一次调用便可以执行所有必要的复制，不论 clone() 定义在继承层次结构中多么深的位置。

要深拷贝 ArrayList，需要这么做：在克隆 ArrayList 后，还需要进一步克隆 ArrayList 所指向的每一个对象。如果要深拷贝如 HashMap 这样的内容，你也需要执行类似的操作。

本示例余下的部分证明克隆确实发生了：一旦克隆了一个对象，你就可以对副本进行修改了，而原对象不会受到影响。

2.2.8 通过序列化进行深拷贝

如果仔细想想 Java 的对象序列化（见附录 E），你可能会发现，如果对一个对象先进行序列化，再将其反序列化，那么它实际上就是被克隆了。

那么为什么不用序列化来实现深拷贝呢？下面这个示例对这两种方式分别计时，以比较两者有什么不同：

```java
// references/Compete.java
import java.io.*;
import onjava.Timer;

class Thing1 implements Serializable {}
class Thing2 implements Serializable {
  Thing1 t1 = new Thing1();
}

class Thing3 implements Cloneable {
  @Override public Thing3 clone() {
    try {
      return (Thing3)super.clone();
    } catch(CloneNotSupportedException e) {
      throw new RuntimeException(e);
    }
  }
}

class Thing4 implements Cloneable {
  private Thing3 t3 = new Thing3();
  @Override public Thing4 clone() {
    Thing4 t4 = null;
    try {
      t4 = (Thing4)super.clone();
    } catch(CloneNotSupportedException e) {
      throw new RuntimeException(e);
    }
    // 对字段也进行克隆:
    t4.t3 = t3.clone();
    return t4;
  }
}

public class Compete {
  public static final int SIZE = 100000;
  public static void
  main(String[] args) throws Exception {
    Thing2[] a = new Thing2[SIZE];
    for(int i = 0; i < SIZE; i++)
      a[i] = new Thing2();
    Thing4[] b = new Thing4[SIZE];
    for(int i = 0; i < SIZE; i++)
      b[i] = new Thing4();
    Timer timer = new Timer();
    try(
      ByteArrayOutputStream buf =
        new ByteArrayOutputStream();
      ObjectOutputStream oos =
        new ObjectOutputStream(buf)
    ) {
      for(Thing2 a1 : a) {
```

/* 输出:
Duplication via serialization: 385 Milliseconds
Duplication via cloning: 38 Milliseconds
*/

```
      oos.writeObject(a1);
    }
    // 现在获取副本:
    try(
      ObjectInputStream in =
        new ObjectInputStream(
          new ByteArrayInputStream(
            buf.toByteArray()))
    ) {
      Thing2[] c = new Thing2[SIZE];
      for(int i = 0; i < SIZE; i++)
        c[i] = (Thing2)in.readObject();
    }
  }
  System.out.println(
    "Duplication via serialization: " +
    timer.duration() + " Milliseconds");

  // 现在试试克隆:
  timer = new Timer();
  Thing4[] d = new Thing4[SIZE];
  for(int i = 0; i < SIZE; i++)
    d[i] = b[i].clone();
  System.out.println(
    "Duplication via cloning: " +
    timer.duration() + " Milliseconds");
  }
}
```

Thing2 和 Thing4 包含了成员对象，所以需要进行一些深拷贝操作。Serializable 类很容易构建，但是需要大量的额外操作来复制它们。另外，在克隆所需要的操作中，类的构建工作更多，但是实际的对象复制操作相对简单。

注意，序列化至少要比克隆慢一个数量级。

2.2.9 在继承层次结构中增加可克隆性并向下覆盖

如果你创建了一个新的类，则其基类默认是 Object，因此默认是不具备可克隆性的。只要你不显式地增加可克隆性，它就不会具备该能力。但你可以在继承层次结构中的任意一层增加该能力，而且从该层开始向下的所有层次都会具备该能力，就像这样：

```
// references/HorrorFlick.java
// 在继承层次结构中的任意一层插入可克隆性

class Person {}

class Hero extends Person {}

class Scientist extends Person implements Cloneable {
```

```java
  @Override public Scientist clone() {
    try {
      return (Scientist)super.clone();
    } catch(CloneNotSupportedException e) {
      throw new RuntimeException(e);
    }
  }
}

class MadScientist extends Scientist {}

public class HorrorFlick {
  public static void main(String[] args) {
    Person p = new Person();
    Hero h = new Hero();
    Scientist s = new Scientist();
    MadScientist m = new MadScientist();
    //- p = (Person)p.clone(); // 编译错误
    //- h = (Hero)h.clone(); // 编译错误
    s = s.clone();
    m = (MadScientist)m.clone();
  }
}
```

在向继承层次结构中增加可克隆性之前，编译器会阻止你进行克隆。在对 Scientist 加入可克隆性后，Scientist 及它的所有后代就都是可克隆的了。注意 Scientist 的 clone() 返回的是 Scientist，而在克隆 MadScientist 时，它并未创建自己的克隆方法，而是使用了继承自 Scientist 的 clone()，因此需要转型。

2.2.10　为什么用这种奇怪的设计

如果这种设计整体看起来很奇怪，那是因为它确实很奇怪。你可能会好奇，为什么会这样设计，这种设计背后的意义又是什么呢？

最初，Java 是为了控制硬件设备而设计的一种语言，而且当时肯定没想到将来会用在互联网中。对于这样的通用型语言，让程序员能够克隆任何对象是很合理的。因此 clone() 被放到了根类 Object 中，**但是**它之前是 public 方法，因此你永远可以克隆任何对象。这看起来是最灵活的方式，但为什么终究会带来坏处呢？

当 Java 被视为互联网编程语言后，事情就变得不同了。突然，安全问题开始出现，并且这些问题是用对象处理的，而你并不一定想让任何人都可以复制你的安全相关的对象。因此你可以看到原本简单易懂的设计上增加了许多补丁：Object 中的 clone() 现在是 protected 的了，你必须重写它，并且实现 Cloneable 接口，还要处理异常。

值得注意的是，你只有在需要调用 Object 的 clone() 方法时才必须实现 Cloneable 接口,因为该方法会在运行时进行检查,确保你的类实现了 Cloneable 接口。

2.3 控制可克隆性

你可能会建议通过简单地将 clone() 方法权限改为 private 来移除可克隆性,但这样做行不通,因为你无法降低一个取自于基类的方法在子类中的可访问性。然而,控制一个对象的克隆性是有必要的。这里有一些可供你在编写类时采取的方案。

1. 不关心(可克隆性)。你不做任何和克隆相关的处理,这意味着你的类无法被克隆,但是它的继承类可以在需要的时候增加克隆能力。这只有在默认的 Object.clone() 可以对类中的所有字段执行合理操作的时候才有用。

2. 支持 clone()。实现 Cloneable 接口并重写 clone() 方法。在重写的 clone() 方法中,要调用 super.clone() 并捕获所有的异常(这样你重写的 clone() 才不会抛出任何异常)。

3. 视情况支持克隆。如果你的类持有指向其他对象的引用,该对象可能是可克隆的,也可能不是(例如容器类),那么你的 clone() 方法可以尽量克隆所有持有引用的对象,并且在它们抛出异常时将那些异常传给程序员。例如,考虑一种会试图克隆自身所持有的所有对象的特殊 ArrayList。在编写这样的 ArrayList 时,你并不知道调用方程序员会向其放入何种对象,因此你并不知道这些对象是否可以克隆。

4. 不实现 Cloneable 接口,但是将 clone() 重写为 protected 的,并实现适用于所有字段的正确复制行为。这样,从该类继承的任何类都可以重写 clone() 并调用 super.clone() 来实现正确的复制行为。注意你的实现可以且应该调用 super.clone(),尽管该方法期待的是 Cloneable 的对象(否则就会抛出异常),因为没有人会直接在你当前编写类型的对象上调用该方法,而只会通过某个子类来调用,并且如果要正常工作,子类就需要实现 Cloneable 接口。

5. 通过不实现 Cloneable 接口并重写 clone(),使其抛出异常,来尽量避免克隆。这只有在该类的子类在自身重定义的 clone() 中调用 super.clone() 时才会成功。否则,程序员可能会设法绕过限制。

6. 将类定义为 final 的,以此阻止克隆。如果 clone() 没有被任何父类重写,那么它就不能被重写。如果被重写过,那么就再次重写,并抛出 CloneNotSupportException 异常。将类定义为 final 是唯一保证可以阻止克隆的方法。此外,在处理安全对象时,或其他你需要控制创建对象数量的场景下,将所有的构造器都设为 private 的,并提供一个或多个专门的方法来创建对象。这样,这些方法就可以限制创建对象的数量和创建对象的条件。

下面这个示例演示了实现克隆,然后在继承层次结构中将其"关闭"的多种方法:

```java
// references/CheckCloneable.java
// 检查一个引用是否可以克隆

// 无法克隆——未重写clone()：
class Ordinary {}

// 重写clone()，未实现Cloneable接口：
class WrongClone extends Ordinary {
  @Override public Object clone()
  throws CloneNotSupportedException {
    return super.clone(); // 抛出异常
  }
}

// 完美克隆：
class IsCloneable extends Ordinary
implements Cloneable {
  @Override public Object clone()
  throws CloneNotSupportedException {
    return super.clone();
  }
}

// 通过抛出异常来关闭克隆：
class NoMore extends IsCloneable {
  @Override public Object clone()
  throws CloneNotSupportedException {
    throw new CloneNotSupportedException();
  }
}

class TryMore extends NoMore {
  @Override public Object clone()
  throws CloneNotSupportedException {
    // 调用NoMore.clone()，抛出异常：
    return super.clone();
  }
}

class BackOn extends NoMore {
  private BackOn duplicate(BackOn b) {
    // 以某种方式生成b的副本，并返回该副本。这个副本毫无作用，只是为了举例：
    return new BackOn();
  }
  @Override public Object clone() {
    // 未调用NoMore.clone()：
    return duplicate(this);
  }
}

// 你无法继承该类，因此无法像在BackOn中一样重写clone()：
final class ReallyNoMore extends NoMore {}
```

```
/* 输出：
Attempting IsCloneable
Cloned IsCloneable
Attempting WrongClone
Doesn't implement Cloneable
Attempting NoMore
Could not clone NoMore
Attempting TryMore
Could not clone TryMore
Attempting BackOn
Cloned BackOn
Attempting ReallyNoMore
Could not clone ReallyNoMore
*/
```

```java
public class CheckCloneable {
  public static
  Ordinary tryToClone(Ordinary ord) {
    String id = ord.getClass().getName();
    System.out.println("Attempting " + id);
    Ordinary x = null;
    if(ord instanceof Cloneable) {
      try {
        x = (Ordinary)((IsCloneable)ord).clone();
        System.out.println("Cloned " + id);
      } catch(CloneNotSupportedException e) {
        System.out.println(
          "Could not clone " + id);
      }
    } else {
      System.out.println("Doesn't implement Cloneable");
    }
    return x;
  }
  public static void main(String[] args) {
    // 向上转型：
    Ordinary[] ord = {
      new IsCloneable(),
      new WrongClone(),
      new NoMore(),
      new TryMore(),
      new BackOn(),
      new ReallyNoMore(),
    };
    Ordinary x = new Ordinary();
    // 这样无法编译，因为 Object 中的 clone() 是 protected 的：
    //- x = (Ordinary)x.clone();
    // 先检查该类是否实现了 Cloneable 接口：
    for(Ordinary ord1 : ord) {
      tryToClone(ord1);
    }
  }
}
```

第一个类 Ordinary 代表了我们在本书各处可见的那种类：未支持克隆，但事实证明，也没有阻止克隆。但是如果你持有一个可能在后续派生的类中被向上转型的 Ordinary 引用，则你并不知道它能否被克隆。

WrongClone 类演示了一种不正确的实现克隆的方式。它确实重写了 Object.clone() 并将其改为 public 的，但是并未实现 Cloneable 接口，因此在调用 super.clone() 时（会导致调用 Object.clone()），会抛出 CloneNotSupportedException 异常，所以这样克隆是不行的。

IsCloneable 实现了克隆所需的所有正确操作，重写了 clone() 并实现了 Cloneable 接口。不过，该 clone() 方法以及本例后面的若干其他 clone() 方法并未捕获 CloneNotSupportedException

异常，而是将异常传递给了调用者，调用者必须用 try-catch 块将方法包裹起来。在你自己实现的 clone() 中，一般需要在该方法内部捕获 CloneNotSupportedException 异常，而不是将它传递下去。你稍后会看到，在本例中将异常传递下去是为了得到更多有用的信息。

NoMore 试图关闭克隆能力，它使用的是 Java 设计者们预期的方法：在子类的 clone() 方法中抛出 CloneNotSupportedException 异常。TryMore 类中的 clone() 方法正确地调用了 super.clone()，这指向了 NoMore.clone()，它抛出了异常，阻止了克隆。

但是如果程序员并没有遵循在重写的 clone() 中调用 super.clone() 这个"正确"的途径，又会怎样呢？在 BackOn 中，可以看到这是如何发生的。该类通过一个单独的 duplicate() 方法创建了一个当前对象的副本，并在 clone() 中调用了这个方法，而不是 super.clone() 方法。异常永远不会被抛出，而新的类依旧是可克隆的。你无法依靠抛出异常来阻止生成可克隆的类。唯一万无一失的办法是像 ReallyNoMore 那样，用 final 来修饰类，使其无法被继承。这意味着如果 final 类中的 clone() 方法抛出了异常，该方法便无法通过继承来修改，确保阻止了克隆（在继承层次结构中，你无法在任何一层的类上显式地调用 Object.clone()。你被限制于仅能调用 super.clone()，该方法只能访问直接的上一层基类）。因此，如果你要生成任何涉及安全问题的类，就将这些类设为 final 的。

CheckCloneable 中的第一个方法是 tryToClone()，其接收任意 Ordinary 对象作为参数，并通过 instaceof 检查该对象是否为可克隆的。如果是，便会将该对象向上转型为 IsCloneable，调用 clone()，并将结果转型回 Ordinary 类型，而且会捕获任何异常。注意这里用到了反射（见基础卷第 19 章）来显示类名和具体过程。

在 main() 中的数组定义里，创建了不同类型的 Ordinary 对象并它们向上转型为 Ordinary。随后的两行代码创建了一个复杂的 Ordinary 对象并试图克隆它。不过，这段代码无法编译，因为在 Object 中，clone() 是个 protected 的方法。这段代码的剩余部分对数组进行了遍历，并尝试克隆每个对象，以及报告各自的结果是成功还是失败。

因此作为总结，如果你想要一个可克隆的类，则需要：

1. 实现 Cloneable 接口；
2. 重写 clone() 方法；
3. 在 clone() 方法中调用 super.clone() 方法；
4. 在 clone() 方法中捕获异常。

这样能够达到最方便的效果。

复制构造器

克隆操作的构建过程可能会十分复杂,是否还有其他解决方案呢?一种(很慢的)方案是使用序列化,如本章 2.2 节所示。另一种方案是实现一个专门用于创建对象副本的构造器。在 C++ 中,这叫作**复制构造器**(copy constructor,又称拷贝构造器,即会创建对象副本的构造器)。下面进行第一次尝试,看起来似乎应该奏效,但是没有成功:

```java
// references/CopyConstructor.java
// 尝试对一个对象创建相同类型的本地副本的构造器
import java.lang.reflect.*;

class FruitQualities {
  private int weight;
  private int color;
  private int firmness;
  private int ripeness;
  private int smell;
  // (省略代码)
  // 无参构造器:
  FruitQualities() {
    // 此处省略具体的处理逻辑
  }
  // 其他构造器:
  // ...
  // 复制构造器:
  FruitQualities(FruitQualities f) {
    weight = f.weight;
    color = f.color;
    firmness = f.firmness;
    ripeness = f.ripeness;
    smell = f.smell;
    // (省略代码)
  }
}

class Seed {
  // 成员列表(略)
  Seed() { /* 无参构造器 */ }
  Seed(Seed s) { /* 复制构造器 */ }
}

class Fruit {
  private FruitQualities fq;
  private int seeds;
  private Seed[] s;
  Fruit(FruitQualities q, int seedCount) {
    fq = q;
    seeds = seedCount;
    s = new Seed[seeds];
    for(int i = 0; i < seeds; i++)
      s[i] = new Seed();
```

```java
  }
  // 其他构造器：
  // ...
  // 复制构造器：
  Fruit(Fruit f) {
    fq = new FruitQualities(f.fq);
    seeds = f.seeds;
    s = new Seed[seeds];
    // 调用所有的 Seed 复制构造器：
    for(int i = 0; i < seeds; i++)
      s[i] = new Seed(f.s[i]);
    // 复制构造器的其他行为……
  }
  // 这样可以使派生的构造器（或其他方法）放入不同的 qualities（品质参数）：
  protected void addQualities(FruitQualities q) {
    fq = q;
  }
  protected FruitQualities getQualities() {
    return fq;
  }
}

class Tomato extends Fruit {
  Tomato() {
    super(new FruitQualities(), 100);
  }
  Tomato(Tomato t) { // 复制构造器
    super(t); // 向上转型为父复制构造器
    // 复制构造器的其他行为……
  }
}

class ZebraQualities extends FruitQualities {
  private int stripedness;
  // 无参构造器：
  ZebraQualities() {
    super();
    // 此处省略具体的处理逻辑……
  }
  ZebraQualities(ZebraQualities z) {
    super(z);
    stripedness = z.stripedness;
  }
}

class GreenZebra extends Tomato {
  GreenZebra() {
    addQualities(new ZebraQualities());
  }
  GreenZebra(GreenZebra g) {
    super(g); // 调用 Tomato(Tomato)
    // 恢复正确的 qualities
    addQualities(new ZebraQualities());
```

```
  }
  public void evaluate() {
    ZebraQualities zq =
      (ZebraQualities)getQualities();
    // 省略进一步处理 qualities 的逻辑
    // ...
  }
}

public class CopyConstructor {
  public static void ripen(Tomato t) {
    // 使用"复制构造器":
    t = new Tomato(t);                        // [1]
    System.out.println("In ripen, t is a " +
      t.getClass().getName());
  }
  public static void slice(Fruit f) {
    f = new Fruit(f); // 嗯......这样行得通吗?   // [2]
    System.out.println("In slice, f is a " +
      f.getClass().getName());
  }
  @SuppressWarnings("unchecked")
  public static void ripen2(Tomato t) {
    try {
      Class c = t.getClass();
      // 使用"复制构造器":
      Constructor ct =
        c.getConstructor(new Class[] { c });
      Object obj =
        ct.newInstance(new Object[] { t });
      System.out.println("In ripen2, t is a " +
        obj.getClass().getName());
    } catch(NoSuchMethodException |
            SecurityException |
            InstantiationException |
            IllegalAccessException |
            IllegalArgumentException |
            InvocationTargetException e) {
      System.out.println(e);
    }
  }
  @SuppressWarnings("unchecked")
  public static void slice2(Fruit f) {
    try {
      Class c = f.getClass();
      Constructor ct =
        c.getConstructor(new Class[] { c });
      Object obj =
        ct.newInstance(new Object[] { f });
      System.out.println("In slice2, f is a " +
        obj.getClass().getName());
    } catch(NoSuchMethodException |
            SecurityException |
```

```
                    InstantiationException |
                    IllegalAccessException |
                    IllegalArgumentException |
                    InvocationTargetException e) {
      System.out.println(e);
    }
  }
  public static void main(String[] args) {
    Tomato tomato = new Tomato();
    ripen(tomato); // OK
    slice(tomato); // OOPS!
    ripen2(tomato); // OK
    slice2(tomato); // OK
    GreenZebra g = new GreenZebra();
    ripen(g); // OOPS!
    slice(g); // OOPS!
    ripen2(g); // OK
    slice2(g); // OK
    g.evaluate();
  }
}
/* 输出：
In ripen, t is a Tomato
In slice, f is a Fruit
java.lang.NoSuchMethodException: Tomato.<init>(Tomato)
java.lang.NoSuchMethodException: Tomato.<init>(Tomato)
In ripen, t is a Tomato
In slice, f is a Fruit
java.lang.NoSuchMethodException:
GreenZebra.<init>(GreenZebra)
java.lang.NoSuchMethodException:
GreenZebra.<init>(GreenZebra)
*/
```

这个示例一开始看起来会有点奇怪。当然，水果（fruit）都有品质参数（qualities），但是为什么不直接将代表这些品质参数的字段直接放到 Fruit 类中呢？主要有如下两个潜在原因。

1. 为了易于插入或修改这些品质参数。注意 Fruit 中有个 protected 的方法 addQualities()，该方法可以让子类来完成这些工作。（你可能会觉得合理的做法应该是在 Fruit 中实现一个 protected 的构造器，并以 FruitQualities 作为参数，但是构造器无法继承，所以它无法在继承层次结构的第二层及更深的层次中使用）。通过将水果品质参数放在一个单独的类中，再使用组合的方式加入到代码中，可以提升代码的灵活性，包括在某个特定的 Fruit 对象生命周期中途改变品质参数的能力。

2. 将 FruitQualities 实现为一个单独的类，可以使你通过继承和多态增加新的 qualities 或改变其行为。注意，对于 GreenZebra（绿斑马，这真的是一个番茄品种）来说，构造器调用了 addQualities() 方法，并向其传入由 FruitQualities 派生而来的 ZebraQualities 对象，因此该对象可以被赋值到基类定义的 FruitQualities 引用上。GreenZebra 在使用这个 FruitQualities 的时候，必须将其向下转型为正确的类型（正如你在 evaluate() 中所见），但你总能知道实际的类型是 ZebraQualities。

你还会看到 Seed 类，而 Fruit（其定义中持有自己的 seeds）则持有一个 Seed 数组。

最后，注意每个类都有一个复制构造器，它们各自都需要调用基类及成员对象的复制构造器，以实现深拷贝。CopyConstructor 中则对复制构造器进行了测试。

[1] ripen() 以 Tomato 为参数，并为其实现了复制构造器以生成该对象的副本。

[2] slice() 以更泛化的 Fruit 对象为构造器，并同样生成了该对象的副本。

在 main() 中，我们使用不同类型的 Fruit 对代码进行了测试。从输出可以看出问题所在。slice() 中的复制构造器对 Tomato 发生作用后，输出的结果不再是 Tomato 对象，而只是一个 Fruit。它已经丢失了所有 tomato 的特征。并且，使用 GreenZebra 的时候，ripen() 和 slice() 分别将其转换成了 Tomato 和 Fruit。因此很遗憾，在 Java 中使用复制构造器的方案来实现本地对象副本的创建，对我们而言并不是一个好的选择。

为什么 C++ 可以，而 Java 就不行？

复制构造器是 C++ 的一个重要组成部分，因为它为对象自动创建了本地副本。然而之前的示例表明，这对 Java 来说行不通。这是为什么呢？在 Java 中，我们操纵的一切都是引用，但在 C++ 中，你可以操纵类似引用的实体，也可以直接传递对象。这就是 C++ 的复制构造器的目的：接收对象并将其按值传递，从而复制该对象。因此在 C++ 中它可以正常工作，但是要记住这种方法在 Java 中会失败，所以不要使用。

2.4 不可变类

虽然 clone() 生成的本地副本在合适的场景下可以给出想要的结果，但这也意味着程序员（该方法的作者）被迫对引用别名导致的副作用负责。如果你要创建一个非常通用而且应用广泛的库，以至于无法确保它永远都会在正确的地方被克隆，又会怎样呢？你更有可能面临的问题是，怎样保证可以只利用引用别名的高效性（即避免不必要的对象复制），同时避开其带来的副作用呢？

一种解决方案（已在纯函数式编程语言中使用）是创建属于只读类的**不可变对象**（immutable object）。你可以定义一个类，其中没有方法会去修改对象的内部状态。引用别名在这样的类中不会造成影响，因为你只能读取内部状态（而无法修改），所以如果各处代码对同一个对象都只会进行读取，那么就不会出现问题。

Java 标准库中对所有基本类型都提供了"包装"类，这可以作为不可变对象的简单示例。你可能已经发现了这一点，要在诸如 ArrayList（只能接收 Object 引用）这样的容器中存储 int，你需要将 int 包装在标准库中的 Integer 类中。此处的包装是通过自动装箱来实现的：

```java
// references/ImmutableInteger.java
// Integer 类不可被改变
import java.util.*;
import java.util.stream.*;

public class ImmutableInteger {
  public static void main(String[] args) {
    @SuppressWarnings("deprecation")
    List<Integer> v = IntStream.range(0, 10)
      .mapToObj(Integer::new)
      .collect(Collectors.toList());
    System.out.println(v);
    // 但是你怎样修改 Integer 内部的 int 呢？
  }
}
/* 输出：
[0, 1, 2, 3, 4, 5, 6, 7, 8, 9]
*/
```

Integer 类（其他所有基本类型的"包装"类也一样）以一种很简单的方式实现了不可变性：对象内部不提供允许你修改该对象的方法。

如果你确实需要一个类来持有一个可以被修改的基本类型，就必须自己创建。幸好这很简单：

```java
// references/MutableInteger.java
// 可变包装类
import java.util.*;
import java.util.stream.*;

class IntValue {
  private int n;
  IntValue(int x) { n = x; }
  public int getValue() { return n; }
  public void setValue(int n) { this.n = n; }
  public void increment() { n++; }
  @Override public String toString() {
    return Integer.toString(n);
  }
}

public class MutableInteger {
  public static void main(String[] args) {
    List<IntValue> v = IntStream.range(0, 10)
      .mapToObj(IntValue::new)
      .collect(Collectors.toList());
    System.out.println(v);
    v.forEach(IntValue::increment);
    System.out.println(v);
  }
}
/* 输出：
[0, 1, 2, 3, 4, 5, 6, 7, 8, 9]
[1, 2, 3, 4, 5, 6, 7, 8, 9, 10]
*/
```

如果不用考虑隐私问题，IntValue 甚至可以更简单：

```java
// references/SimplerMutableInteger.java
// 一个简单的包装类
import java.util.*;
import java.util.stream.*;

class IntValue2 {
  public int n;
  IntValue2(int n) { this.n = n; }
}

public class SimplerMutableInteger {
  public static void main(String[] args) {
    List<IntValue2> v = IntStream.range(0, 10)
      .mapToObj(IntValue2::new)
      .collect(Collectors.toList());
    v.forEach(iv2 ->
      System.out.print(iv2.n + " "));
    System.out.println();
    v.forEach(iv2 -> iv2.n += 1);
    v.forEach(iv2 ->
      System.out.print(iv2.n + " "));
  }
}
/* 输出：
0 1 2 3 4 5 6 7 8 9
1 2 3 4 5 6 7 8 9 10
*/
```

不过，直接调用 n 的值，这有点不太优雅。

2.4.1 创建不可变类

下面是一种创建自有的不可变类的方法：

```java
// references/Immutable1.java
// 不可变对象可以对引用别名的风险免疫

public class Immutable1 {
  private int data;
  public Immutable1(int initVal) {
    data = initVal;
  }
  public int read() { return data; }
  public boolean nonzero() { return data != 0; }
  public Immutable1 multiply(int multiplier) {
    return new Immutable1(data * multiplier);
  }
  public static void f(Immutable1 i1) {
    Immutable1 quad = i1.multiply(4);
    System.out.println("i1 = " + i1.read());
    System.out.println("quad = " + quad.read());
  }
  public static void main(String[] args) {
```

/* 输出：
x = 47
i1 = 47
quad = 188
x = 47
*/

```
    Immutable1 x = new Immutable1(47);
    System.out.println("x = " + x.read());
    f(x);
    System.out.println("x = " + x.read());
  }
}
```

各处的 data 都是 private 的，并且你可以发现没有任何 public 方法可以修改这些 data。确实，multiply() 看起来像是修改了对象，但实际上它只是创建了一个新的 Immutable1，并未影响原有对象。

f() 方法接收 Immutable1 对象为参数，并在其上实现了多种操作，从 main() 的输出可以看出 x 并没有任何变化。因此，x 的对象可以无风险地使用任意多个引用别名，因为 Immutable1 类从设计上就确保了对象不会被改变。

2.4.2 不可变性的缺点

创建不可变类，乍一看似乎为我们提供了一种优雅的解决方案。不过，每当你确实需要修改这种新类型的对象时，就必须忍受创建新对象的开销，并可能导致更频繁的垃圾收集。对有些类来说，这不是问题（而且函数式编程语言依赖这种设计）。但是对于另一些类（如 String 类）来说，这种开销是非常昂贵的——牢记格言，切勿过早优化。（注意，尽管存在明显的开销，但那些只提供了不变性、未经其他优化的语言已经做得相当不错了。）

解决办法是创建一个可以被修改的伴生类。然后，如果你要做很多的修改工作，就可以切换使用该可被修改的伴生类，并在完成后切换回不可变类。

我们可以修改 Immutable1.java，以证明这一点：

```
// references/Immutable2.java
// 用于修改不可变类的伴生类

class Mutable {
  private int data;
  Mutable(int initVal) {
    data = initVal;
  }
  public Mutable add(int x) {
    data += x;
    return this;
  }
  public Mutable multiply(int x) {
    data *= x;
    return this;
  }
  public Immutable2 makeImmutable2() {
```

```
/* 输出：
i2 = 47
r1 = 376
r2 = 376
*/
```

```java
    return new Immutable2(data);
  }
}

public class Immutable2 {
  private int data;
  public Immutable2(int initVal) {
    data = initVal;
  }
  public int read() { return data; }
  public boolean nonzero() {
    return data != 0;
  }
  public Immutable2 add(int x) {
    return new Immutable2(data + x);
  }
  public Immutable2 multiply(int x) {
    return new Immutable2(data * x);
  }
  public Mutable makeMutable() {
    return new Mutable(data);
  }
  public static
  Immutable2 modify1(Immutable2 y) {
    Immutable2 val = y.add(12);
    val = val.multiply(3);
    val = val.add(11);
    val = val.multiply(2);
    return val;
  }
  // 这样会得到相同的结果:
  public static
  Immutable2 modify2(Immutable2 y) {
    Mutable m = y.makeMutable();
    m.add(12).multiply(3).add(11).multiply(2);
    return m.makeImmutable2();
  }
  public static void main(String[] args) {
    Immutable2 i2 = new Immutable2(47);
    Immutable2 r1 = modify1(i2);
    Immutable2 r2 = modify2(i2);
    System.out.println("i2 = " + i2.read());
    System.out.println("r1 = " + r1.read());
    System.out.println("r2 = " + r2.read());
  }
}
```

和之前一样，Immutable2 内部的方法在需要修改对象的时候创建了一个新的对象，以保护对象的不可变性。此处指的便是 add() 和 multiply() 方法。Mutable 则是所实现的协同类，它同样有 add() 和 multiply() 方法，但是这两个方法会修改 Mutable 对象，而不是创建一个新对象。另外，Mutable 中还有一个使用类中的 data 生成 Immutable2 对象的方法，反之亦然。

`modify1()` 和 `modify2()` 这两个静态方法演示了生成相同结果的两种方法。在 `modify1()` 中，一切都是在 Immutable2 类中完成的，你可以看到此过程中创建了四个新的 Immutable2 对象（并且每当 val 被重新赋值的时候，之前的对象就变成了垃圾）。

`modify2()` 中的第一个行为是接收 Immutable2 y 参数，并用该参数生成了 Mutable 对象（这就和你之前看到的 `clone()` 调用一样，不过这次是创建了一个不同类型的对象）。然后便用 Mutable 对象执行了很多修改操作，并不需要创建那么多新对象。最后，它又变回了 Immutable2 对象。此处创建了两个新对象（Mutable 和结果对象 Immutable2），而不是四个。

因此这种方法在以下情况下是合理的选择：

1. 你需要不可变对象；
2. 并且你经常需要做很多修改操作；
3. 或者创建新的不可变对象的开销很大。

2.4.3　String 很特殊

String 本质上是不可变的——在基础卷第 18 章曾做过相关介绍。如果回顾这一章，你会看到当 String 的内容发生变化时，每个 String 的方法都会谨慎地返回一个新的 String 对象。如果内容不需要更改，则该方法只返回指向原始 String 的引用，这样便节省了存储空间和开销。

Java 中的 String 类有些与众不同。String 中有很多特殊情况，最重要的一点是，它是内建的类，也是 Java 的重要基础。因此，你会看到编译器将双引号包围的字符转换为 String，以及特殊的重载操作符 + 和 += 的使用。剩下的特例便是使用配套的 StringBuilder 和编译器中一些附加的神奇技术所精心构建的不变性（详见基础卷第 18 章）。

2.5　总结

由于 Java 中所有的对象标识符都是引用，同时由于所有的对象都是在堆上创建的，并且只有在不再使用的时候才会被垃圾收集，所以对象操作的风格发生了变化，特别是在传递和返回对象的时候。在 C 或 C++ 方法中初始化一段内存空间的时候，你可能会要求用户向方法中传递该段内存的地址。否则，你就必须确保存储的正确销毁。因此，这类方法的接口和所需的理解都会更加复杂。但在 Java 中，你不必担心这些职责，也不必担心对象在需要时是否仍然存在，因为这些总会自动为你处理好。你什么时候需要对象，就什么时候创建（而不用提前），并且永远不用担心该对象的职责传递机制，只需简单地传递引用即可。这种简化有时不会引起你的注意，但有时则会令你大吃一惊。

这些潜在的巧妙操作有如下两方面的缺点。

1. 额外的内存管理总是会导致运行效率降低，尽管这通常不是问题，此外，一段代码的执行时间还有一些不确定性，因为每当内存不足时，垃圾收集器就会运行。对于大多数应用程序来说，这些操作利大于弊，且 Java 的改进已经将性能提升到了不会造成任何影响的地步。

2. 引用别名：有时你最终会遇到两个指向同一个对象的引用，这只有在这两个引用都应该指向**不同**的对象时，才会成为问题。你需要在这里多加小心，必要时可以使用 clone() 或者其他方式复制对象，来防止其他引用被意外地修改。或者，你可以通过创建不可变的对象来支持引用别名，以提高效率，这些对象的操作可以返回相同类型或某些不同类型的新对象，但绝不会更改原始对象，因此任何指向该对象的引用别名都不会看到任何更改。

有些人会说 Java 中的克隆是一项拙劣的设计，不应该使用，因此他们选择实现自己版本的克隆。Doug Lea 帮助解决了此问题，他向我提出了这个建议，说他只是为每个类都创建了一个名为 duplicate() 的函数。这样，你就永远不用调用 Object.clone() 了，也就不再需要实现 Cloneable 并捕获 CloneNotSupportedException 异常了。这肯定是个合理的方法，并且由于 Java 标准库中对于 clone() 的支持是如此罕见，因此这显然也是个安全的方法。

相较于自行编写支持克隆的代码，你更应该考虑使用 Apache **通用序列化实用工具类**（Apache Commons Serialization Utility Classes），或深克隆库（deep cloning library），或者一些其他的克隆库。

03 集合主题

> 本章将介绍一些相较于基础卷第 12 章更为高级的集合主题。

3.1 样例数据

我们先来创建一些样例数据,以便与集合示例一起使用。下面的数据将颜色名与 HTML 中颜色对应的 RGB 值关联了起来。请注意每个键和值都是唯一的:

```java
// onjava/HTMLColors.java
// 用于集合示例的样例数据
package onjava;
import java.util.*;
import java.util.stream.*;
import java.util.concurrent.*;

public class HTMLColors {
  public static final Object[][] ARRAY = {
    { 0xF0F8FF, "AliceBlue" },
    { 0xFAEBD7, "AntiqueWhite" },
    { 0x7FFFD4, "Aquamarine" },
    { 0xF0FFFF, "Azure" },
    { 0xF5F5DC, "Beige" },
    { 0xFFE4C4, "Bisque" },
    { 0x000000, "Black" },
```

```
{ 0xFFEBCD, "BlanchedAlmond" },
{ 0x0000FF, "Blue" },
{ 0x8A2BE2, "BlueViolet" },
{ 0xA52A2A, "Brown" },
{ 0xDEB887, "BurlyWood" },
{ 0x5F9EA0, "CadetBlue" },
{ 0x7FFF00, "Chartreuse" },
{ 0xD2691E, "Chocolate" },
{ 0xFF7F50, "Coral" },
{ 0x6495ED, "CornflowerBlue" },
{ 0xFFF8DC, "Cornsilk" },
{ 0xDC143C, "Crimson" },
{ 0x00FFFF, "Cyan" },
{ 0x00008B, "DarkBlue" },
{ 0x008B8B, "DarkCyan" },
{ 0xB8860B, "DarkGoldenRod" },
{ 0xA9A9A9, "DarkGray" },
{ 0x006400, "DarkGreen" },
{ 0xBDB76B, "DarkKhaki" },
{ 0x8B008B, "DarkMagenta" },
{ 0x556B2F, "DarkOliveGreen" },
{ 0xFF8C00, "DarkOrange" },
{ 0x9932CC, "DarkOrchid" },
{ 0x8B0000, "DarkRed" },
{ 0xE9967A, "DarkSalmon" },
{ 0x8FBC8F, "DarkSeaGreen" },
{ 0x483D8B, "DarkSlateBlue" },
{ 0x2F4F4F, "DarkSlateGray" },
{ 0x00CED1, "DarkTurquoise" },
{ 0x9400D3, "DarkViolet" },
{ 0xFF1493, "DeepPink" },
{ 0x00BFFF, "DeepSkyBlue" },
{ 0x696969, "DimGray" },
{ 0x1E90FF, "DodgerBlue" },
{ 0xB22222, "FireBrick" },
{ 0xFFFAF0, "FloralWhite" },
{ 0x228B22, "ForestGreen" },
{ 0xDCDCDC, "Gainsboro" },
{ 0xF8F8FF, "GhostWhite" },
{ 0xFFD700, "Gold" },
{ 0xDAA520, "GoldenRod" },
{ 0x808080, "Gray" },
{ 0x008000, "Green" },
{ 0xADFF2F, "GreenYellow" },
{ 0xF0FFF0, "HoneyDew" },
{ 0xFF69B4, "HotPink" },
{ 0xCD5C5C, "IndianRed" },
{ 0x4B0082, "Indigo" },
{ 0xFFFFF0, "Ivory" },
{ 0xF0E68C, "Khaki" },
{ 0xE6E6FA, "Lavender" },
{ 0xFFF0F5, "LavenderBlush" },
{ 0x7CFC00, "LawnGreen" },
{ 0xFFFACD, "LemonChiffon" },
{ 0xADD8E6, "LightBlue" },
{ 0xF08080, "LightCoral" },
{ 0xE0FFFF, "LightCyan" },
{ 0xFAFAD2, "LightGoldenRodYellow" },
{ 0xD3D3D3, "LightGray" },
{ 0x90EE90, "LightGreen" },
{ 0xFFB6C1, "LightPink" },
{ 0xFFA07A, "LightSalmon" },
{ 0x20B2AA, "LightSeaGreen" },
{ 0x87CEFA, "LightSkyBlue" },
{ 0x778899, "LightSlateGray" },
{ 0xB0C4DE, "LightSteelBlue" },
{ 0xFFFFE0, "LightYellow" },
{ 0x00FF00, "Lime" },
{ 0x32CD32, "LimeGreen" },
{ 0xFAF0E6, "Linen" },
{ 0xFF00FF, "Magenta" },
{ 0x800000, "Maroon" },
{ 0x66CDAA, "MediumAquaMarine" },
{ 0x0000CD, "MediumBlue" },
{ 0xBA55D3, "MediumOrchid" },
{ 0x9370DB, "MediumPurple" },
{ 0x3CB371, "MediumSeaGreen" },
{ 0x7B68EE, "MediumSlateBlue" },
{ 0x00FA9A, "MediumSpringGreen" },
{ 0x48D1CC, "MediumTurquoise" },
{ 0xC71585, "MediumVioletRed" },
{ 0x191970, "MidnightBlue" },
{ 0xF5FFFA, "MintCream" },
{ 0xFFE4E1, "MistyRose" },
{ 0xFFE4B5, "Moccasin" },
{ 0xFFDEAD, "NavajoWhite" },
{ 0x000080, "Navy" },
{ 0xFDF5E6, "OldLace" },
{ 0x808000, "Olive" },
{ 0x6B8E23, "OliveDrab" },
{ 0xFFA500, "Orange" },
{ 0xFF4500, "OrangeRed" },
{ 0xDA70D6, "Orchid" },
{ 0xEEE8AA, "PaleGoldenRod" },
{ 0x98FB98, "PaleGreen" },
{ 0xAFEEEE, "PaleTurquoise" },
{ 0xDB7093, "PaleVioletRed" },
{ 0xFFEFD5, "PapayaWhip" },
{ 0xFFDAB9, "PeachPuff" },
{ 0xCD853F, "Peru" },
{ 0xFFC0CB, "Pink" },
{ 0xDDA0DD, "Plum" },
{ 0xB0E0E6, "PowderBlue" },
{ 0x800080, "Purple" },
```

（上接第94页）

```
    { 0xFF0000, "Red" },
    { 0xBC8F8F, "RosyBrown" },
    { 0x4169E1, "RoyalBlue" },
    { 0x8B4513, "SaddleBrown" },
    { 0xFA8072, "Salmon" },
    { 0xF4A460, "SandyBrown" },
    { 0x2E8B57, "SeaGreen" },
    { 0xFFF5EE, "SeaShell" },
    { 0xA0522D, "Sienna" },
    { 0xC0C0C0, "Silver" },
    { 0x87CEEB, "SkyBlue" },
    { 0x6A5ACD, "SlateBlue" },
    { 0x708090, "SlateGray" },
    { 0xFFFAFA, "Snow" },
    { 0x00FF7F, "SpringGreen" },
    { 0x4682B4, "SteelBlue" },
    { 0xD2B48C, "Tan" },
    { 0x008080, "Teal" },
    { 0xD8BFD8, "Thistle" },
    { 0xFF6347, "Tomato" },
    { 0x40E0D0, "Turquoise" },
    { 0xEE82EE, "Violet" },
    { 0xF5DEB3, "Wheat" },
    { 0xFFFFFF, "White" },
    { 0xF5F5F5, "WhiteSmoke" },
    { 0xFFFF00, "Yellow" },
    { 0x9ACD32, "YellowGreen" },
};
public static final Map<Integer,String> MAP =
    Arrays.stream(ARRAY)
        .collect(Collectors.toMap(
            element -> (Integer)element[0],
            element -> (String)element[1],
            (v1, v2) -> { // 合并函数
                throw new IllegalStateException();
            },
            LinkedHashMap::new
        ));
// 只有值唯一的情况下才能将键和值反过来：
public static <V, K> Map<V, K>
invert(Map<K, V> map) {
    return map.entrySet().stream()
        .collect(Collectors.toMap(
            Map.Entry::getValue,
            Map.Entry::getKey,
            (v1, v2) -> {
                throw new IllegalStateException();
            },
            LinkedHashMap::new
        ));
}
public static final Map<String,Integer>
```

```java
    INVMAP = invert(MAP);
  // 给定颜色名，查找对应的RGB值：
  public static Integer rgb(String colorName) {
    return INVMAP.get(colorName);
  }
  public static final List<String> LIST =
    Arrays.stream(ARRAY)
      .map(item -> (String)item[1])
      .collect(Collectors.toList());
  public static final List<Integer> RGBLIST =
    Arrays.stream(ARRAY)
      .map(item -> (Integer)item[0])
      .collect(Collectors.toList());
  public static
  void show(Map.Entry<Integer,String> e) {
    System.out.format(
      "0x%06X: %s%n", e.getKey(), e.getValue());
  }
  public static void
  show(Map<Integer,String> m, int count) {
    m.entrySet().stream()
      .limit(count)
      .forEach(e -> show(e));
  }
  public static void show(Map<Integer,String> m) {
    show(m, m.size());
  }
  public static
  void show(Collection<String> lst, int count) {
    lst.stream()
      .limit(count)
      .forEach(System.out::println);
  }
  public static void show(Collection<String> lst) {
    show(lst, lst.size());
  }
  public static
  void showrgb(Collection<Integer> lst, int count) {
    lst.stream()
      .limit(count)
      .forEach(n -> System.out.format("0x%06X%n", n));
  }
  public static void showrgb(Collection<Integer> lst) {
    showrgb(lst, lst.size());
  }
  public static
  void showInv(Map<String,Integer> m, int count) {
    m.entrySet().stream()
      .limit(count)
      .forEach(e ->
        System.out.format(
          "%-20s  0x%06X%n", e.getKey(), e.getValue()));
  }
```

```
    public static void showInv(Map<String,Integer> m) {
      showInv(m, m.size());
    }
    public static void border() {
      System.out.println(
        "*****************************");
    }
  }
```

  MAP 是使用基础卷第 14 章介绍的流创建的。我们通过流将二维的 ARRAY 转换到了一个 Map 中，但是请注意，我们并没有直接使用 Collectors.toMap() 的简单版本。这个版本会产生一个 HashMap，它所使用的哈希函数会打乱键的顺序。为了保留顺序，必须将键值对直接放入一个 LinkedHashMap 中，这意味着我们使用的是 Collectors.toMap() 的复杂版本。它接受的两个函数分别负责从每个被流化的元素中提取键和值，就像简单的 Collectors.toMap() 一样。它还需要一个**合并函数**（merge function），负责在有两个值关联到同一个键的情况下解决冲突。因为我们的数据经过预先检查，所以不会出现这样的情况；如果出现，我们就抛出异常。最后传入的函数负责生成一个指定类型的空映射，随后由流来填充。

  rgb() 方法是一个便捷函数，接受表示颜色名的字符串，返回对应的 RGB 值。为了实现该函数，我们需要 COLORS 的一个**逆向**版本，它接受一个 String 类型的键，查找 Integer 类型的 RGB 值。逆向是通过 invert() 方法实现的，如果 COLORS 中有任何值是不唯一的，它就会抛出异常。

  我们还创建了包含所有颜色名的 LIST，以及包含所有十六进制 RGB 值的 RGBLIST。

  第一个 show() 方法只接受一个 Map.Entry，并以十六进制显示键，这样我们很容易针对原始的 ARRAY 进行二次检查。每个以 show 开头的方法都重载一个接受 count 参数的版本，这个参数用来说明我们想显示多少个元素。第二个版本会显示序列中的所有元素。

  下面是一个基本的测试：

```
// collectiontopics/HTMLColorTest.java
import static onjava.HTMLColors.*;

public class HTMLColorTest {
  static final int DISPLAY_SIZE = 20;
  public static void main(String[] args) {
    show(MAP, DISPLAY_SIZE);
    border();
    showInv(INVMAP, DISPLAY_SIZE);
    border();
    show(LIST, DISPLAY_SIZE);
    border();
```

```
    showrgb(RGBLIST, DISPLAY_SIZE);
  }
}
/* 输出：
0xF0F8FF: AliceBlue
0xFAEBD7: AntiqueWhite
0x7FFFD4: Aquamarine
0xF0FFFF: Azure
0xF5F5DC: Beige
0xFFE4C4: Bisque
0x000000: Black
0xFFEBCD: BlanchedAlmond
0x0000FF: Blue
0x8A2BE2: BlueViolet
0xA52A2A: Brown
0xDEB887: BurlyWood
0x5F9EA0: CadetBlue
0x7FFF00: Chartreuse
0xD2691E: Chocolate
0xFF7F50: Coral
0x6495ED: CornflowerBlue
0xFFF8DC: Cornsilk
0xDC143C: Crimson
0x00FFFF: Cyan
****************************
AliceBlue              0xF0F8FF
AntiqueWhite           0xFAEBD7
Aquamarine             0x7FFFD4
Azure                  0xF0FFFF
Beige                  0xF5F5DC
Bisque                 0xFFE4C4
Black                  0x000000
BlanchedAlmond         0xFFEBCD
Blue                   0x0000FF
BlueViolet             0x8A2BE2
Brown                  0xA52A2A
BurlyWood              0xDEB887
CadetBlue              0x5F9EA0
Chartreuse             0x7FFF00
Chocolate              0xD2691E
Coral                  0xFF7F50
CornflowerBlue         0x6495ED
Cornsilk               0xFFF8DC
Crimson                0xDC143C
Cyan                   0x00FFFF
****************************
（转右栏）
```

```
AliceBlue
AntiqueWhite
Aquamarine
Azure
Beige
Bisque
Black
BlanchedAlmond
Blue
BlueViolet
Brown
BurlyWood
CadetBlue
Chartreuse
Chocolate
Coral
CornflowerBlue
Cornsilk
Crimson
Cyan
****************************
0xF0F8FF
0xFAEBD7
0x7FFFD4
0xF0FFFF
0xF5F5DC
0xFFE4C4
0x000000
0xFFEBCD
0x0000FF
0x8A2BE2
0xA52A2A
0xDEB887
0x5F9EA0
0x7FFF00
0xD2691E
0xFF7F50
0x6495ED
0xFFF8DC
0xDC143C
0x00FFFF
*/
```

使用 LinkedHashMap，我们确实能保留 HTMLColors.ARRAY 的顺序。

3.2 List 的行为

List 是除数组之外最基本的对象存储和检索方式。基本的列表操作包括：

- add() 用于插入元素
- get() 用于随机访问元素，注意这个操作在特定的 List 实现上成本不同
- iterator() 用于返回该序列上的 Iterator
- stream() 用于生成序列中元素的 Stream

List 的构造器总是保留元素插入的顺序。

下列示例中的方法分别涵盖了一组不同的活动：每个 List 都能做的事情（basicTest()），在 Iterator 上移动（iterMotion()）与用 Iterator 修改元素（iterManipulation()），查看 List 操作的效果（testVisual()），以及只能用于 LinkedList 的操作。

```java
// collectiontopics/ListOps.java
// 可以在 List 上做的事情
import java.util.*;
import onjava.HTMLColors;

public class ListOps {
  // 创建一个用于测试的短列表
  static final List<String> LIST =
    HTMLColors.LIST.subList(0, 10);
  private static boolean b;
  private static String s;
  private static int i;
  private static Iterator<String> it;
  private static ListIterator<String> lit;
  public static void basicTest(List<String> a) {
    a.add(1, "x"); // 在位置 1 处插入
    a.add("x"); // 在末尾插入
    // 插入一个集合：
    a.addAll(LIST);
    // 从位置 3 处开始，插入一个集合：
    a.addAll(3, LIST);
    b = a.contains("1"); // 列表中是否包含该元素？
    // 作为参数的这个集合是否包含在列表中？
    b = a.containsAll(LIST);
    // List 支持随机访问，该操作对于 ArrayList 而言
    // 成本很低，但是对于 LinkedList 而言成本很高
    s = a.get(1); // 在位置 1 处获取（有类型的）对象
    i = a.indexOf("1"); // 确定对象的索引
    b = a.isEmpty(); // 列表中是否存在元素？
    it = a.iterator(); // 普通的迭代器
    lit = a.listIterator(); // ListIterator
    lit = a.listIterator(3); // 从位置 3 处开始
    i = a.lastIndexOf("1"); // 最后一个与参数匹配的元素的索引
```

```java
    a.remove(1); // 移除位置1处的元素
    a.remove("3"); // 移除该对象
    a.set(1, "y"); // 将位置1处的元素设置为"y"
    // 保留所有与参数中元素匹配的元素
    // (对两个集合求交集)：
    a.retainAll(LIST);
    // 移除参数中的所有元素：
    a.removeAll(LIST);
    i = a.size(); // 这个集合有多大？
    a.clear(); // 移除所有元素
  }
  public static void iterMotion(List<String> a) {
    ListIterator<String> it = a.listIterator();
    b = it.hasNext();
    b = it.hasPrevious();
    s = it.next();
    i = it.nextIndex();
    s = it.previous();
    i = it.previousIndex();
  }
  public static void iterManipulation(List<String> a) {
    ListIterator<String> it = a.listIterator();
    it.add("47");
    // 在调用add()之后，必须将迭代器后移：
    it.next();
    // 移除新元素后面的元素：
    it.remove();
    // 在调用remove()之后，必须将迭代器后移：
    it.next();
    // 修改被移除元素后面的元素：
    it.set("47");
  }
  public static void testVisual(List<String> a) {
    System.out.println(a);
    List<String> b = LIST;
    System.out.println("b = " + b);
    a.addAll(b);
    a.addAll(b);
    System.out.println(a);
    // 使用 ListIterator
    // 插入、移除和替换元素：
    ListIterator<String> x =
      a.listIterator(a.size()/2);
    x.add("one");
    System.out.println(a);
    System.out.println(x.next());
    x.remove();
    System.out.println(x.next());
    x.set("47");
    System.out.println(a);
    // 反向遍历列表：
    x = a.listIterator(a.size());
    while(x.hasPrevious())
```

```java
      System.out.print(x.previous() + " ");
    System.out.println();
    System.out.println("testVisual finished");
  }
  // 有些只有 LinkedList 才有的操作:
  public static void testLinkedList() {
    LinkedList<String> ll = new LinkedList<>();
    ll.addAll(LIST);
    System.out.println(ll);
    // 将其看作一个栈, 压栈:
    ll.addFirst("one");
    ll.addFirst("two");
    System.out.println(ll);
    // 类似于在栈顶执行的 peek 操作:
    System.out.println(ll.getFirst());
    // 类似于从栈顶弹出:
    System.out.println(ll.removeFirst());
    System.out.println(ll.removeFirst());
    // 将其看作一个队列, 从队尾取出元素:
    System.out.println(ll.removeLast());
    System.out.println(ll);
  }
  public static void main(String[] args) {
    // 每次创建并填充一个新的列表:
    basicTest(new LinkedList<>(LIST));
    basicTest(new ArrayList<>(LIST));
    iterMotion(new LinkedList<>(LIST));
    iterMotion(new ArrayList<>(LIST));
    iterManipulation(new LinkedList<>(LIST));
    iterManipulation(new ArrayList<>(LIST));
    testVisual(new LinkedList<>(LIST));
    testLinkedList();
  }
}
/* 输出:
[AliceBlue, AntiqueWhite, Aquamarine, Azure, Beige,
Bisque, Black, BlanchedAlmond, Blue, BlueViolet]
b = [AliceBlue, AntiqueWhite, Aquamarine, Azure, Beige,
Bisque, Black, BlanchedAlmond, Blue, BlueViolet]
[AliceBlue, AntiqueWhite, Aquamarine, Azure, Beige,
Bisque, Black, BlanchedAlmond, Blue, BlueViolet,
AliceBlue, AntiqueWhite, Aquamarine, Azure, Beige,
Bisque, Black, BlanchedAlmond, Blue, BlueViolet,
AliceBlue, AntiqueWhite, Aquamarine, Azure, Beige,
Bisque, Black, BlanchedAlmond, Blue, BlueViolet]
[AliceBlue, AntiqueWhite, Aquamarine, Azure, Beige,
Bisque, Black, BlanchedAlmond, Blue, BlueViolet,
AliceBlue, AntiqueWhite, Aquamarine, Azure, Beige, one,
Bisque, Black, BlanchedAlmond, Blue, BlueViolet,
AliceBlue, AntiqueWhite, Aquamarine, Azure, Beige,
Bisque, Black, BlanchedAlmond, Blue, BlueViolet]
```

```
Bisque
Black
[AliceBlue, AntiqueWhite, Aquamarine, Azure, Beige,
Bisque, Black, BlanchedAlmond, Blue, BlueViolet,
AliceBlue, AntiqueWhite, Aquamarine, Azure, Beige, one,
47, BlanchedAlmond, Blue, BlueViolet, AliceBlue,
AntiqueWhite, Aquamarine, Azure, Beige, Bisque, Black,
BlanchedAlmond, Blue, BlueViolet]
BlueViolet Blue BlanchedAlmond Black Bisque Beige Azure
Aquamarine AntiqueWhite AliceBlue BlueViolet Blue
BlanchedAlmond 47 one Beige Azure Aquamarine
AntiqueWhite AliceBlue BlueViolet Blue BlanchedAlmond
Black Bisque Beige Azure Aquamarine AntiqueWhite
AliceBlue
testVisual finished
[AliceBlue, AntiqueWhite, Aquamarine, Azure, Beige,
Bisque, Black, BlanchedAlmond, Blue, BlueViolet]
[two, one, AliceBlue, AntiqueWhite, Aquamarine, Azure,
Beige, Bisque, Black, BlanchedAlmond, Blue, BlueViolet]
two
two
one
BlueViolet
[AliceBlue, AntiqueWhite, Aquamarine, Azure, Beige,
Bisque, Black, BlanchedAlmond, Blue]
*/
```

在 basicTest() 和 iterMotion() 中，方法调用都是为了演示正确的语法，尽管保存了返回值，但是并没有使用。在某些情况下，根本没有保存返回值。在使用每个方法之前，要查看 JDK 文档，充分了解其用法。

3.3 Set 的行为

Set 的意义在于测试成员身份，也可以用于删除重复元素。如果不关心元素的顺序或并发，HashSet 总是最好的选择，因为它就是为实现尽可能快的查找而设计的（使用了附录 C 中介绍的**哈希函数**）。

对于其他 Set 实现，元素顺序会不同：

```
// collectiontopics/SetOrder.java
import java.util.*;
import onjava.HTMLColors;

public class SetOrder {
  static String[] sets = {
    "java.util.HashSet",
    "java.util.TreeSet",
```

```
    "java.util.concurrent.ConcurrentSkipListSet",
    "java.util.LinkedHashSet",
    "java.util.concurrent.CopyOnWriteArraySet",
  };
  static final List<String> RLIST =
    new ArrayList<>(HTMLColors.LIST);
  static {
    Collections.reverse(RLIST);
  }
  public static void
  main(String[] args) throws Exception {
    for(String type: sets) {
      System.out.format("[-> %s <-]%n",
        type.substring(type.lastIndexOf('.') + 1));
      @SuppressWarnings("unchecked")
      Set<String> set = (Set<String>)
        Class.forName(type).getConstructor().newInstance();
      set.addAll(RLIST);
      set.stream()
        .limit(10)
        .forEach(System.out::println);
    }
  }
}
```

```
/* 输出:
[-> HashSet <-]
MediumOrchid
PaleGoldenRod
Sienna
LightSlateGray
DarkSeaGreen
Black
Gainsboro
Orange
LightCoral
DodgerBlue
[-> TreeSet <-]
AliceBlue
AntiqueWhite
Aquamarine
Azure
Beige
Bisque
Black
BlanchedAlmond
Blue
BlueViolet
[-> ConcurrentSkipListSet <-]
AliceBlue
AntiqueWhite
Aquamarine
Azure
Beige
```

```
Bisque
Black
BlanchedAlmond
Blue
BlueViolet
[-> LinkedHashSet <-]
YellowGreen
Yellow
WhiteSmoke
White
Wheat
Violet
Turquoise
Tomato
Thistle
Teal
[-> CopyOnWriteArraySet <-]
YellowGreen
Yellow
WhiteSmoke
White
Wheat
Violet
Turquoise
Tomato
Thistle
Teal
*/
```

（转右栏）

这里需要 @SuppressWarnings("unchecked")，因为不确定我们向 Class.forName(type).newInstance() 传递的 String 到底是什么。编译器不能确保这个操作一定成功。

RLIST 是 HTMLColors.LIST 的逆向版本。因为 Collections.reverse() 通过修改参数来实现逆向，而不是返回一个包含逆向排列元素的新 List，所以这个调用在 static 子句内执行。RLIST 可以防止我们误认为 Set 会对结果进行排序。

HashSet 的输出看上去没有明显的顺序（因为它是基于哈希函数的）。TreeSet 和 ConcurrentSkipListSet 都会对其元素进行排序，而且都实现了 SortedSet 接口来表明这一点。因为这样的 Set 是有序的，所以它们还提供了更多操作。LinkedHashSet 和 CopyOnWriteArraySet 会保留元素插入的顺序，尽管没有接口表明这一点。

ConcurrentSkipListSet 和 CopyOnWriteArraySet 是线程安全的。

在本章最后，我们会了解非 HashSet 实现执行排序所带来的性能开销，以及不同实现中其他功能的开销。

3.4　在 Map 上使用函数式操作

和 Collection 接口一样，Map 接口也内置了 forEach()。但是如果我们想执行像 map()、flatMap()、reduce() 或 filter() 等其他基本操作，又该怎么做呢？看一下 Map 接口，并没有与这些操作相关的线索。

我们通过 entrySet() 连接到这些方法，它会生成一个由 Map.Entry 对象组成的 Set。这个 Set 又包含了 stream() 和 parallelStream() 方法。只需要记住：我们在使用 Map.Entry 对象。

```java
// collectiontopics/FunctionalMap.java
// 在 Map 上执行函数式操作
import java.util.*;
import java.util.stream.*;
import java.util.concurrent.*;
import static onjava.HTMLColors.*;

public class FunctionalMap {
  public static void main(String[] args) {
    MAP.entrySet().stream()
      .map(Map.Entry::getValue)
      .filter(v -> v.startsWith("Dark"))
      .map(v -> v.replaceFirst("Dark", "Hot"))
      .forEach(System.out::println);
```

```
      }
    }
    /* 输出：              HotOrchid
    HotBlue               HotRed
    HotCyan               HotSalmon
    HotGoldenRod          HotSeaGreen
    HotGray               HotSlateBlue
    HotGreen              HotSlateGray
    HotKhaki              HotTurquoise
    HotMagenta            HotViolet
    HotOliveGreen         */
    HotOrange
（转右栏）
```

生成 Stream 之后，所有的基本函数式方法（当然还有更多方法）就都可以使用了。

3.5 选择 Map 的部分元素

TreeMap 和 ConcurrentSkipListMap 都实现了 NavigableMap 接口。这个接口的目的是解决需要选择某个 Map 中部分元素的问题。下面是利用 HTMLColors 实现的一个示例：

```java
// collectiontopics/NavMap.java
// NavigableMap 可以生成 Map 的片段
import java.util.*;
import java.util.concurrent.*;
import static onjava.HTMLColors.*;

public class NavMap {
  public static final
  NavigableMap<Integer,String> COLORS =
    new ConcurrentSkipListMap<>(MAP);
  public static void main(String[] args) {
    show(COLORS.firstEntry());
    border();
    show(COLORS.lastEntry());
    border();
    NavigableMap<Integer, String> toLime =
      COLORS.headMap(rgb("Lime"), true);
    show(toLime);
    border();
    show(COLORS.ceilingEntry(rgb("DeepSkyBlue") - 1));
    border();
    show(COLORS.floorEntry(rgb("DeepSkyBlue") - 1));
    border();
    show(toLime.descendingMap());
    border();
    show(COLORS.tailMap(rgb("MistyRose"), true));
    border();
```

```
    show(COLORS.subMap(
      rgb("Orchid"), true,
      rgb("DarkSalmon"), false));
  }
}
/* 输出:
0x000000: Black
***************************
0xFFFFFF: White
***************************
0x000000: Black
0x000080: Navy
0x00008B: DarkBlue
0x0000CD: MediumBlue
0x0000FF: Blue
0x006400: DarkGreen
0x008000: Green
0x008080: Teal
0x008B8B: DarkCyan
0x00BFFF: DeepSkyBlue
0x00CED1: DarkTurquoise
0x00FA9A: MediumSpringGreen
0x00FF00: Lime
***************************
0x00BFFF: DeepSkyBlue
***************************
0x008B8B: DarkCyan
***************************
0x00FF00: Lime
0x00FA9A: MediumSpringGreen
0x00CED1: DarkTurquoise
0x00BFFF: DeepSkyBlue
0x008B8B: DarkCyan
0x008080: Teal
0x008000: Green
0x006400: DarkGreen

0x0000FF: Blue
0x0000CD: MediumBlue
0x00008B: DarkBlue
0x000080: Navy
0x000000: Black
***************************
0xFFE4E1: MistyRose
0xFFEBCD: BlanchedAlmond
0xFFEFD5: PapayaWhip
0xFFF0F5: LavenderBlush
0xFFF5EE: SeaShell
0xFFF8DC: Cornsilk
0xFFFACD: LemonChiffon
0xFFFAF0: FloralWhite
0xFFFAFA: Snow
0xFFFF00: Yellow
0xFFFFE0: LightYellow
0xFFFFF0: Ivory
0xFFFFFF: White
***************************
0xDA70D6: Orchid
0xDAA520: GoldenRod
0xDB7093: PaleVioletRed
0xDC143C: Crimson
0xDCDCDC: Gainsboro
0xDDA0DD: Plum
0xDEB887: BurlyWood
0xE0FFFF: LightCyan
0xE6E6FA: Lavender
*/
```

（转右栏）

我们可以在 main() 中看到 NavigableMap 的各种功能。因为 NavigableMap 的键是有顺序的，所以它支持 firstEntry() 和 lastEntry() 的概念。在 COLORS 上调用 headMap()，生成的 NavigableMap 包含了从 COLORS 开头到第一个参数所指元素的所有元素，第二个参数（即 boolean 值）用于指示在结果中是否包含第一个参数所指的元素。在 COLORS 上调用 tailMap() 会执行类似的操作，但范围是从第一个参数所指的元素开始，直到 COLORS 的末尾。subMap() 支持生成该 Map 中间的某段。

ceilingEntry() 从指定的键值向后查找下一个键值对，而 floorEntry() 则按反方向查找。descendingMap() 会将 NavigableMap 的顺序反过来。

如果要解决的问题可以通过切分 Map 来简化,那么用 NavigableMap 就能解决。其他集合实现也有类似的功能,可以帮助我们解决问题。

3.6　填充集合

和 Arrays 一样,集合也有一个叫作 Collections 的伴生类,包含很多静态的工具方法,其中有一个叫作 fill()。fill() 会将集合中的所有元素都替换为同一个对象引用。虽然它只能用于 List 对象,但是它生成的列表可以传递给构造器或 addAll() 方法:

```java
// collectiontopics/FillingLists.java
// Collections.fill() 与 Collections.nCopies()
import java.util.*;

class StringAddress {
  private String s;
  StringAddress(String s) { this.s = s; }
  @Override public String toString() {
    return super.toString() + " " + s;
  }
}

public class FillingLists {
  public static void main(String[] args) {
    List<StringAddress> list = new ArrayList<>(
      Collections.nCopies(4,
        new StringAddress("Hello")));
    System.out.println(list);
    Collections.fill(list,
      new StringAddress("World!"));
    System.out.println(list);
  }
}
/* 输出:
[StringAddress@19e0bfd Hello, StringAddress@19e0bfd Hello, StringAddress@19e0bfd Hello, StringAddress@19e0bfd Hello]
[StringAddress@139a55 World!, StringAddress@139a55 World!, StringAddress@139a55 World!, StringAddress@139a55 World!]
*/
```

这个示例演示了两种使用指向某个对象的引用来填充 Collection 的方法。第一种是使用 Collections.nCopies()。它会创建一个 List,该 List 被传递给 ArrayList 的构造器,从而填充这个 ArrayList。

第二种是使用 Collections.fill()。StringAddress 中的 toString() 方法会调用 Object.toString()，产生的信息是类名加该对象的哈希码（由 hashCode() 方法生成）的无符号十六进制表示。输出表明，所有的引用都被设置为了同一个对象。在调用 Collections.fill() 之后，情况也是如此。因为 fill() 方法只能替换 List 中已有的元素，而不能添加新元素，所以这个方法不是很有用。

3.6.1 使用 Suppliers 来填充 Collection

基础卷第 20 章引入的 onjava.Suppliers 类提供了用于填充 Collection 的通用解决方案。下面的示例使用 Suppliers 初始化了几类不同的 Collection：

```java
// collectiontopics/SuppliersCollectionTest.java
import java.util.*;
import java.util.function.*;
import java.util.stream.*;
import onjava.*;

class Government implements Supplier<String> {
  static String[] foundation = (
    "strange women lying in ponds " +
    "distributing swords is no basis " +
    "for a system of government").split(" ");
  private int index;
  @Override public String get() {
    return foundation[index++];
  }
}

public class SuppliersCollectionTest {
  public static void main(String[] args) {
    // 基础卷第 20 章引入的 Suppliers 类：
    Set<String> set = Suppliers.create(
      LinkedHashSet::new, new Government(), 15);
    System.out.println(set);
    List<String> list = Suppliers.create(
      LinkedList::new, new Government(), 15);
    System.out.println(list);
    list = new ArrayList<>();
    Suppliers.fill(list, new Government(), 15);
    System.out.println(list);

    // 或者使用流：
    set = Arrays.stream(Government.foundation)
      .collect(Collectors.toSet());
    System.out.println(set);
    list = Arrays.stream(Government.foundation)
      .collect(Collectors.toList());
    System.out.println(list);
    list = Arrays.stream(Government.foundation)
```

```
      .collect(Collectors
        .toCollection(LinkedList::new));
    System.out.println(list);
    set = Arrays.stream(Government.foundation)
      .collect(Collectors
        .toCollection(LinkedHashSet::new));
    System.out.println(set);
  }
}
/* 输出：
[strange, women, lying, in, ponds, distributing,
swords, is, no, basis, for, a, system, of, government]
[strange, women, lying, in, ponds, distributing,
swords, is, no, basis, for, a, system, of, government]
[strange, women, lying, in, ponds, distributing,
swords, is, no, basis, for, a, system, of, government]
[ponds, no, a, in, swords, for, is, basis, strange,
system, government, distributing, of, women, lying]
[strange, women, lying, in, ponds, distributing,
swords, is, no, basis, for, a, system, of, government]
[strange, women, lying, in, ponds, distributing,
swords, is, no, basis, for, a, system, of, government]
[strange, women, lying, in, ponds, distributing,
swords, is, no, basis, for, a, system, of, government]
*/
```

LinkedHashSet 中的元素会按照插入顺序排列，因为它会维护一个链表来保存顺序信息。

不过请注意该示例的第二部分：大部分时间可以使用 Stream 来创建并填充 Collection。和 Suppliers 方案不同的是，Stream 方案没有要求我们注明想创建的元素个数，它会处理所有的 Stream 元素。

在可能的情况下，应该首选 Stream 方案。

3.6.2 使用 Suppliers 来填充 Map

要使用 Suppliers 向 Map 中填充数据，需要一个 Pair 类，因为每次调用 Supplier 的 get() 方法，必定会产生一对对象（一个键，一个值）。

```
// onjava/Pair.java
package onjava;

public class Pair<K, V> {
  public final K key;
  public final V value;
  public Pair(K k, V v) {
    key = k;
    value = v;
  }
}
```

```
  public K key() { return key; }
  public V value() { return value; }
  public static <K,V> Pair<K, V> make(K k, V v) {
    return new Pair<K,V>(k, v);
  }
}
```

Pair 是只读的**数据传输对象**（Data Transfer Object）或**信使**（Messenger）。它和基础卷第 20 章的 Tuple2 基本是一样的，不过名字更适合用于 Map 的初始化。为了简化 Pair 对象的创建，这里还添加了 static make() 方法。

利用 Java 8 的 Stream，可以很方便地生成填充好的 Map：

```
// collectiontopics/StreamFillMaps.java
import java.util.*;
import java.util.function.*;
import java.util.stream.*;
import onjava.*;

class Letters
implements Supplier<Pair<Integer,String>> {
  private int number = 1;
  private char letter = 'A';
  @Override public Pair<Integer,String> get() {
    return new Pair<>(number++, "" + letter++);
  }
}

public class StreamFillMaps {
  public static void main(String[] args) {
    Map<Integer,String> m =
      Stream.generate(new Letters())
        .limit(11)
        .collect(Collectors
          .toMap(Pair::key, Pair::value));
    System.out.println(m);

    // 分别提供键和值：
    Rand.String rs = new Rand.String(3);
    Count.Character cc = new Count.Character();
    Map<Character,String> mcs = Stream.generate(
      () -> Pair.make(cc.get(), rs.get()))
      .limit(8)
      .collect(Collectors
        .toMap(Pair::key, Pair::value));
    System.out.println(mcs);

    // 用一个类来生成键，它们使用同一个值：
    Map<Character,String> mcs2 = Stream.generate(
      () -> Pair.make(cc.get(), "Val"))
      .limit(8)
```

```
        .collect(Collectors
          .toMap(Pair::key, Pair::value));
    System.out.println(mcs2);
  }
}
      /* 输出:
      {1=A, 2=B, 3=C, 4=D, 5=E, 6=F, 7=G, 8=H, 9=I, 10=J,
      11=K}
      {b=btp, c=enp, d=ccu, e=xsz, f=gvg, g=mei, h=nne,
      i=elo}
      {p=Val, q=Val, j=Val, k=Val, l=Val, m=Val, n=Val,
      o=Val}
      */
```

以上示例体现了一个模式，我们可以用它来编写一个自动创建和填充 Map 的工具：

```
// onjava/FillMap.java
package onjava;
import java.util.*;
import java.util.function.*;
import java.util.stream.*;

public class FillMap {
  public static <K, V> Map<K,V>
  basic(Supplier<Pair<K,V>> pairGen, int size) {
    return Stream.generate(pairGen)
      .limit(size)
      .collect(Collectors
        .toMap(Pair::key, Pair::value));
  }
  public static <K, V> Map<K,V>
  basic(Supplier<K> keyGen,
        Supplier<V> valueGen, int size) {
    return Stream.generate(
      () -> Pair.make(keyGen.get(), valueGen.get()))
      .limit(size)
      .collect(Collectors
        .toMap(Pair::key, Pair::value));
  }
  public static <K, V, M extends Map<K,V>>
  M create(Supplier<K> keyGen,
           Supplier<V> valueGen,
           Supplier<M> mapSupplier, int size) {
    return Stream.generate( () ->
      Pair.make(keyGen.get(), valueGen.get()))
        .limit(size)
        .collect(Collectors
          .toMap(Pair::key, Pair::value,
            (k, v) -> k, mapSupplier));
  }
}
```

basic() 方法可以生成一个默认的 Map，而 create() 支持指定确切的映射类型，而且会返回这个确切的类型。

这是一个测试：

```java
// collectiontopics/FillMapTest.java
import java.util.*;
import java.util.function.*;
import java.util.stream.*;
import onjava.*;

public class FillMapTest {
  public static void main(String[] args) {
    Map<String,Integer> mcs = FillMap.basic(
      new Rand.String(4), new Count.Integer(), 7);
    System.out.println(mcs);
    HashMap<String,Integer> hashm =
      FillMap.create(new Rand.String(4),
        new Count.Integer(), HashMap::new, 7);
    System.out.println(hashm);
    LinkedHashMap<String,Integer> linkm =
      FillMap.create(new Rand.String(4),
        new Count.Integer(), LinkedHashMap::new, 7);
    System.out.println(linkm);
  }
}
/* 输出：
{npcc=1, ztdv=6, gvgm=3, btpe=0, einn=4, eelo=5, uxsz=2}
{npcc=1, ztdv=6, gvgm=3, btpe=0, einn=4, eelo=5, uxsz=2}
{btpe=0, npcc=1, uxsz=2, gvgm=3, einn=4, eelo=5, ztdv=6}
*/
```

3.7 使用享元自定义 Collection 和 Map

本节将演示如何创建自定义的 Collection 和 Map 实现。每个 java.util 集合都有自己的抽象类，提供了该集合的部分实现。因此，只需要实现必要的方法就能生成想要的集合。你会看到通过继承 java.util.Abstract 类来创建自定义的 Map 和 Collection 多么简单。比如，要创建一个只读的 Set，只需要继承 AbstractSet 并实现 iterator() 和 size() 方法。上面的示例就是生成测试数据的另一种方式。这样生成的集合通常是只读的，而且只需提供最少的方法。

这一解决方案也展示了**享元**（Flyweight）设计模式。当普通的解决方案需要太多对象时，或者当生成正常的对象会占用太多空间时，就可以使用享元。享元模式将对象的一部分外化了：我们不再将对象的所有内容置于该对象之内，而是通过一个更高效的外部表来查找部分或整个对象（或者通过其他一些可以节省空间的计算来生成）。

下面定义了一个可以为任意大小的 List，从效果上看，它相当于用 Integer 预先初始化了。要从 AbstractList 创建一个只读的 List，必须实现 get() 和 size() 两个方法：

```java
// onjava/CountingIntegerList.java
// 包含样例数据、不限长度的 List
// {java onjava.CountingIntegerList}
package onjava;
import java.util.*;

public class CountingIntegerList
extends AbstractList<Integer> {
  private int size;
  public CountingIntegerList() { size = 0; }
  public CountingIntegerList(int size) {
    this.size = size < 0 ? 0 : size;
  }
  @Override public Integer get(int index) {
    return index;
  }
  @Override public int size() { return size; }
  public static void main(String[] args) {
    List<Integer> cil =
      new CountingIntegerList(30);
    System.out.println(cil);
    System.out.println(cil.get(500));
  }
}
/* 输出：
[0, 1, 2, 3, 4, 5, 6, 7, 8, 9, 10, 11, 12, 13, 14, 15,
16, 17, 18, 19, 20, 21, 22, 23, 24, 25, 26, 27, 28, 29]
500
*/
```

只有当我们想限制这个 List 的长度时（正如我们在 main() 中所做的那样），size 的值才是重要的。即使在这种情况下，get() 也可能会产生任意的值。

这是享元模式的一个很好的示例。get() 会在我们请求时"计算"值，所以并没有需要存储和初始化的实际底层 List 结构。

在大多数程序中，这里节省的存储空间不会产生什么影响。然而，它允许我们使用一个非常大的 index 来调用 List.get()，不需要一个填充好所有这些值的 List。另外，我们可以在自己的程序中使用大量的 CountingIntegerList，而不用担心存储问题。确实，享元的一大好处就是让我们无须考虑资源问题就能使用更好的抽象。

我们可以使用享元来实现其他"已初始化"的、数据集大小不限的自定义集合。下面是一个 Map，它会为每个 Integer 键生成一个独一无二的值：

```java
// onjava/CountMap.java
// 包含样例数据、不限长度的 Map
// {java onjava.CountMap}
package onjava;
import java.util.*;
import java.util.stream.*;

public class CountMap
extends AbstractMap<Integer,String> {
  private int size;
  private static char[] chars =
    "ABCDEFGHIJKLMNOPQRSTUVWXYZ".toCharArray();
  private static String value(int key) {
    return
      chars[key % chars.length] +
      Integer.toString(key / chars.length);
  }
  public CountMap(int size) {
    this.size = size < 0 ? 0 : size;
  }
  @Override public String get(Object key) {
    return value((Integer)key);
  }
  private static class Entry
  implements Map.Entry<Integer,String> {
    int index;
    Entry(int index) { this.index = index; }
    @Override   public boolean equals(Object o) {
      return o instanceof Entry &&
        Objects.equals(index, ((Entry)o).index);
    }
    @Override public Integer getKey() { return index; }
    @Override public String getValue() {
      return value(index);
    }
    @Override public String setValue(String value) {
      throw new UnsupportedOperationException();
    }
    @Override public int hashCode() {
      return Objects.hashCode(index);
    }
  }
  @Override
  public Set<Map.Entry<Integer,String>> entrySet() {
    // LinkedHashSet 会记住初始化顺序:
    return IntStream.range(0, size)
      .mapToObj(Entry::new)
      .collect(Collectors
        .toCollection(LinkedHashSet::new));
  }
  public static void main(String[] args) {
```

```
        final int size = 6;
        CountMap cm = new CountMap(60);
        System.out.println(cm);
        System.out.println(cm.get(500));
        cm.values().stream()
          .limit(size)
          .forEach(System.out::println);
        System.out.println();
        new Random(47).ints(size, 0, 1000)
          .mapToObj(cm::get)
          .forEach(System.out::println);
    }
}
/* 输出:
{0=A0, 1=B0, 2=C0, 3=D0, 4=E0, 5=F0, 6=G0, 7=H0, 8=I0,
9=J0, 10=K0, 11=L0, 12=M0, 13=N0, 14=O0, 15=P0, 16=Q0,
17=R0, 18=S0, 19=T0, 20=U0, 21=V0, 22=W0, 23=X0, 24=Y0,
25=Z0, 26=A1, 27=B1, 28=C1, 29=D1, 30=E1, 31=F1, 32=G1,
33=H1, 34=I1, 35=J1, 36=K1, 37=L1, 38=M1, 39=N1, 40=O1,
41=P1, 42=Q1, 43=R1, 44=S1, 45=T1, 46=U1, 47=V1, 48=W1,
49=X1, 50=Y1, 51=Z1, 52=A2, 53=B2, 54=C2, 55=D2, 56=E2,
57=F2, 58=G2, 59=H2}
G19
A0
B0
C0
D0
E0
F0

Y9
J21
R26
D33
Z36
N16
*/
```

为了创建只读的 Map，我们要继承 AbstractMap 并实现 entrySet() 方法。private value() 方法负责计算任意键的值，并会在 get() 和 Entry.getValue() 方法内使用。CountMap 的大小可以忽略不计。

这里使用了 LinkedHashSet，并没有创建自定义的 Set 类，所以并没有完全实现享元。只有当我们调用 entrySet() 时，这个对象才会生成。

现在我们来创建一个更复杂的享元。在这个示例中，数据集是一个由国家及其首都组成的 Map。capitals() 方法可以生成这样的 Map。[1] names() 方法则会生成包含国家名的 List。

---

[1] 生成的数据和表示国家分类的代码注释可能有不准确之处。——编者注

这两个方法都可以通过一个指定了所需大小的 int 参数来获得由部分元素组成的列表：

```java
// onjava/Countries.java
// 用享元模式设计的 Map 和 List，包含样例数据
// {java onjava.Countries}
package onjava;
import java.util.*;

public class Countries {
  public static final String[][] DATA = {
    // 非洲
    {"ALGERIA","Algiers"},
    {"ANGOLA","Luanda"},
    {"BENIN","Porto-Novo"},
    {"BOTSWANA","Gaberone"},
    {"BURKINA FASO","Ouagadougou"},
    {"BURUNDI","Bujumbura"},
    {"CAMEROON","Yaounde"},
    {"CAPE VERDE","Praia"},
    {"CENTRAL AFRICAN REPUBLIC","Bangui"},
    {"CHAD","N'djamena"},
    {"COMOROS","Moroni"},
    {"CONGO","Brazzaville"},
    {"DJIBOUTI","Dijibouti"},
    {"EGYPT","Cairo"},
    {"EQUATORIAL GUINEA","Malabo"},
    {"ERITREA","Asmara"},
    {"ETHIOPIA","Addis Ababa"},
    {"GABON","Libreville"},
    {"THE GAMBIA","Banjul"},
    {"GHANA","Accra"},
    {"GUINEA","Conakry"},
    {"BISSAU","Bissau"},
    {"COTE D'IVOIR (IVORY COAST)","Yamoussoukro"},
    {"KENYA","Nairobi"},
    {"LESOTHO","Maseru"},
    {"LIBERIA","Monrovia"},
    {"LIBYA","Tripoli"},
    {"MADAGASCAR","Antananarivo"},
    {"MALAWI","Lilongwe"},
    {"MALI","Bamako"},
    {"MAURITANIA","Nouakchott"},
    {"MAURITIUS","Port Louis"},
    {"MOROCCO","Rabat"},
    {"MOZAMBIQUE","Maputo"},
    {"NAMIBIA","Windhoek"},
    {"NIGER","Niamey"},
    {"NIGERIA","Abuja"},
    {"RWANDA","Kigali"},
    {"SAO TOME E PRINCIPE","Sao Tome"},
    {"SENEGAL","Dakar"},
    {"SEYCHELLES","Victoria"},
```

```
{"SIERRA LEONE","Freetown"},
{"SOMALIA","Mogadishu"},
{"SOUTH AFRICA","Pretoria/Cape Town"},
{"SUDAN","Khartoum"},
{"SWAZILAND","Mbabane"},
{"TANZANIA","Dodoma"},
{"TOGO","Lome"},
{"TUNISIA","Tunis"},
{"UGANDA","Kampala"},
{"DEMOCRATIC REPUBLIC OF THE CONGO (ZAIRE)",
 "Kinshasa"},
{"ZAMBIA","Lusaka"},
{"ZIMBABWE","Harare"},
// 亚洲
{"AFGHANISTAN","Kabul"},
{"ARMENIA","Yerevan"},
{"AZERBAIJAN","Baku"},
{"BAHRAIN","Manama"},
{"BANGLADESH","Dhaka"},
{"BHUTAN","Thimphu"},
{"BRUNEI","Bandar Seri Begawan"},
{"CAMBODIA","Phnom Penh"},
{"CHINA","Beijing"},
{"CYPRUS","Nicosia"},
{"GEORGIA","Tbilisi"},
{"INDIA","New Delhi"},
{"INDONESIA","Jakarta"},
{"IRAN","Tehran"},
{"IRAQ","Baghdad"},
{"ISRAEL","Jerusalem"},
{"JAPAN","Tokyo"},
{"JORDAN","Amman"},
{"KAZAKSTAN","Almaty"},
{"KUWAIT","Kuwait City"},
{"KYRGYZSTAN","Alma-Ata"},
{"LAOS","Vientiane"},
{"LEBANON","Beirut"},
{"MALAYSIA","Kuala Lumpur"},
{"THE MALDIVES","Male"},
{"MONGOLIA","Ulan Bator"},
{"MYANMAR (BURMA)","Rangoon"},
{"NEPAL","Katmandu"},
{"NORTH KOREA","P'yongyang"},
{"OMAN","Muscat"},
{"PAKISTAN","Islamabad"},
{"PHILIPPINES","Manila"},
{"QATAR","Doha"},
{"SAUDI ARABIA","Riyadh"},
{"SINGAPORE","Singapore"},
{"SOUTH KOREA","Seoul"},
{"SRI LANKA","Colombo"},
{"SYRIA","Damascus"},
{"TAJIKISTAN","Dushanbe"},
```

```
{"THAILAND","Bangkok"},
{"TURKEY","Ankara"},
{"TURKMENISTAN","Ashkhabad"},
{"UNITED ARAB EMIRATES","Abu Dhabi"},
{"UZBEKISTAN","Tashkent"},
{"VIETNAM","Hanoi"},
{"YEMEN","Sana'a"},
// 大洋洲
{"AUSTRALIA","Canberra"},
{"FIJI","Suva"},
{"KIRIBATI","Bairiki"},
{"MARSHALL ISLANDS","Dalap-Uliga-Darrit"},
{"MICRONESIA","Palikir"},
{"NAURU","Yaren"},
{"NEW ZEALAND","Wellington"},
{"PALAU","Koror"},
{"PAPUA NEW GUINEA","Port Moresby"},
{"SOLOMON ISLANDS","Honaira"},
{"TONGA","Nuku'alofa"},
{"TUVALU","Fongafale"},
{"VANUATU","Port Vila"},
{"WESTERN SAMOA","Apia"},
// 欧洲
{"ALBANIA","Tirana"},
{"ANDORRA","Andorra la Vella"},
{"AUSTRIA","Vienna"},
{"BELARUS (BYELORUSSIA)","Minsk"},
{"BELGIUM","Brussels"},
{"BOSNIA-HERZEGOVINA","Sarajevo"},
{"BULGARIA","Sofia"},
{"CROATIA","Zagreb"},
{"CZECH REPUBLIC","Prague"},
{"DENMARK","Copenhagen"},
{"ESTONIA","Tallinn"},
{"FINLAND","Helsinki"},
{"FRANCE","Paris"},
{"GERMANY","Berlin"},
{"GREECE","Athens"},
{"HUNGARY","Budapest"},
{"ICELAND","Reykjavik"},
{"IRELAND","Dublin"},
{"ITALY","Rome"},
{"LATVIA","Riga"},
{"LIECHTENSTEIN","Vaduz"},
{"LITHUANIA","Vilnius"},
{"LUXEMBOURG","Luxembourg"},
{"MACEDONIA","Skopje"},
{"MALTA","Valletta"},
{"MOLDOVA","Chisinau"},
{"MONACO","Monaco"},
{"MONTENEGRO","Podgorica"},
{"THE NETHERLANDS","Amsterdam"},
{"NORWAY","Oslo"},
```

```
    {"POLAND","Warsaw"},
    {"PORTUGAL","Lisbon"},
    {"ROMANIA","Bucharest"},
    {"RUSSIA","Moscow"},
    {"SAN MARINO","San Marino"},
    {"SERBIA","Belgrade"},
    {"SLOVAKIA","Bratislava"},
    {"SLOVENIA","Ljuijana"},
    {"SPAIN","Madrid"},
    {"SWEDEN","Stockholm"},
    {"SWITZERLAND","Berne"},
    {"UKRAINE","Kyiv"},
    {"UNITED KINGDOM","London"},
    {"VATICAN CITY","Vatican City"},
    // 中北美洲
    {"ANTIGUA AND BARBUDA","Saint John's"},
    {"BAHAMAS","Nassau"},
    {"BARBADOS","Bridgetown"},
    {"BELIZE","Belmopan"},
    {"CANADA","Ottawa"},
    {"COSTA RICA","San Jose"},
    {"CUBA","Havana"},
    {"DOMINICA","Roseau"},
    {"DOMINICAN REPUBLIC","Santo Domingo"},
    {"EL SALVADOR","San Salvador"},
    {"GRENADA","Saint George's"},
    {"GUATEMALA","Guatemala City"},
    {"HAITI","Port-au-Prince"},
    {"HONDURAS","Tegucigalpa"},
    {"JAMAICA","Kingston"},
    {"MEXICO","Mexico City"},
    {"NICARAGUA","Managua"},
    {"PANAMA","Panama City"},
    {"ST. KITTS AND NEVIS","Basseterre"},
    {"ST. LUCIA","Castries"},
    {"ST. VINCENT AND THE GRENADINES","Kingstown"},
    {"UNITED STATES OF AMERICA","Washington, D.C."},
    // 南美洲
    {"ARGENTINA","Buenos Aires"},
    {"BOLIVIA","Sucre (legal)/La Paz(administrative)"},
    {"BRAZIL","Brasilia"},
    {"CHILE","Santiago"},
    {"COLOMBIA","Bogota"},
    {"ECUADOR","Quito"},
    {"GUYANA","Georgetown"},
    {"PARAGUAY","Asuncion"},
    {"PERU","Lima"},
    {"SURINAME","Paramaribo"},
    {"TRINIDAD AND TOBAGO","Port of Spain"},
    {"URUGUAY","Montevideo"},
    {"VENEZUELA","Caracas"},
};
// 通过实现 entrySet() 来使用 AbstractMap
```

```java
private static class FlyweightMap
extends AbstractMap<String,String> {
  private static class Entry
  implements Map.Entry<String,String> {
    int index;
    Entry(int index) { this.index = index; }
    @Override public boolean equals(Object o) {
      return o instanceof FlyweightMap &&
        Objects.equals(DATA[index][0], o);
    }
    @Override public int hashCode() {
      return Objects.hashCode(DATA[index][0]);
    }
    @Override
    public String getKey() { return DATA[index][0]; }
    @Override public String getValue() {
      return DATA[index][1];
    }
    @Override public String setValue(String value) {
      throw new UnsupportedOperationException();
    }
  }
  // 实现 AbstractSet 的 size() 和 iterator() 方法:
  static class EntrySet
  extends AbstractSet<Map.Entry<String,String>> {
    private int size;
    EntrySet(int size) {
      if(size < 0)
        this.size = 0;
      // 不能比 DATA 数组大:
      else if(size > DATA.length)
        this.size = DATA.length;
      else
        this.size = size;
    }
    @Override public int size() { return size; }
    private class Iter
    implements Iterator<Map.Entry<String,String>> {
      // 每个 Iterator 只有一个 Entry 对象:
      private Entry entry = new Entry(-1);
      @Override public boolean hasNext() {
        return entry.index < size - 1;
      }
      @Override
      public Map.Entry<String,String> next() {
        entry.index++;
        return entry;
      }
      @Override public void remove() {
        throw new UnsupportedOperationException();
      }
    }
    @Override public
```

```java
      Iterator<Map.Entry<String,String>> iterator() {
        return new Iter();
      }
    }
    private static
    Set<Map.Entry<String,String>> entries =
      new EntrySet(DATA.length);
    @Override
    public Set<Map.Entry<String,String>> entrySet() {
      return entries;
    }
  }
  // 创建一个由 'size' 个国家组成的部分映射:
  static Map<String,String> select(final int size) {
    return new FlyweightMap() {
      @Override
      public Set<Map.Entry<String,String>> entrySet() {
        return new EntrySet(size);
      }
    };
  }
  static Map<String,String> map = new FlyweightMap();
  public static Map<String,String> capitals() {
    return map; // 整个映射
  }
  public static Map<String,String> capitals(int size) {
    return select(size); // 部分映射
  }
  static List<String> names =
    new ArrayList<>(map.keySet());
  // 所有的国名:
  public static List<String> names() { return names; }
  // 部分列表:
  public static List<String> names(int size) {
    return new ArrayList<>(select(size).keySet());
  }
  public static void main(String[] args) {
    System.out.println(capitals(10));
    System.out.println(names(10));
    System.out.println(new HashMap<>(capitals(3)));
    System.out.println(
      new LinkedHashMap<>(capitals(3)));
    System.out.println(new TreeMap<>(capitals(3)));
    System.out.println(new Hashtable<>(capitals(3)));
    System.out.println(new HashSet<>(names(6)));
    System.out.println(new LinkedHashSet<>(names(6)));
    System.out.println(new TreeSet<>(names(6)));
    System.out.println(new ArrayList<>(names(6)));
    System.out.println(new LinkedList<>(names(6)));
    System.out.println(capitals().get("BRAZIL"));
  }
}
```

```
/* 输出：
{ALGERIA=Algiers, ANGOLA=Luanda, BENIN=Porto-Novo,
BOTSWANA=Gaberone, BURKINA FASO=Ouagadougou,
BURUNDI=Bujumbura, CAMEROON=Yaounde, CAPE VERDE=Praia,
CENTRAL AFRICAN REPUBLIC=Bangui, CHAD=N'djamena}
[ALGERIA, ANGOLA, BENIN, BOTSWANA, BURKINA FASO,
BURUNDI, CAMEROON, CAPE VERDE, CENTRAL AFRICAN
REPUBLIC, CHAD]
{BENIN=Porto-Novo, ANGOLA=Luanda, ALGERIA=Algiers}
{ALGERIA=Algiers, ANGOLA=Luanda, BENIN=Porto-Novo}
{ALGERIA=Algiers, ANGOLA=Luanda, BENIN=Porto-Novo}
{ALGERIA=Algiers, ANGOLA=Luanda, BENIN=Porto-Novo}
[BENIN, BOTSWANA, ANGOLA, BURKINA FASO, ALGERIA,
BURUNDI]
[ALGERIA, ANGOLA, BENIN, BOTSWANA, BURKINA FASO,
BURUNDI]
[ALGERIA, ANGOLA, BENIN, BOTSWANA, BURKINA FASO,
BURUNDI]
[ALGERIA, ANGOLA, BENIN, BOTSWANA, BURKINA FASO,
BURUNDI]
[ALGERIA, ANGOLA, BENIN, BOTSWANA, BURKINA FASO,
BURUNDI]
Brasilia
*/
```

二维数组 String DATA 是公共（public）的，在其他地方也可以用。FlyweightMap 必须实现 entrySet() 方法，而该方法需要一个自定义的 Set 实现和一个自定义的 Map.Entry 类。这里用了另一种方式来实现享元：每个 Map.Entry 对象保存的是其索引，而非实际的键和值。当我们调用 getKey() 或 getValue() 时，它使用这个索引来返回相应的 DATA 元素。EntrySet 确保它的 size 不大于 DATA。

享元的另一部分是在 EntrySet.Iterator 中实现的。这里没有为 DATA 中的每个数据对创建一个 Map.Entry，而是**每个迭代器**只有一个 Map.Entry。Entry 对象被用作进入 DATA 数据的一个窗口，它只包含一个用于静态 String 数组的 index。每当我们调用迭代器的 next() 方法时，Entry 中的 index 会自增，也就指向了下一个元素对，然后 next() 方法就会返回 Iterator 中的这一个 Entry 对象。[①]

select() 方法会生成一个 FlyweightMap，其中包含了一个指定大小的 EntrySet。该方法被用在了重载的 capitals() 和 names() 方法中，可以在 main() 中看到。

---

① java.util 中的 Map 会使用 getKey() 和 getValue() 方法来实现批量复制，所以这里的做法行得通。如果某个自定义的 Map 只是简单地复制整个 Map.Entry，这么做就会出问题。

3.8 Collection 的功能

表 3-1 列出了我们可以在 Collection 上做的所有事情（没有包括从 Object 类自动继承的方法），因此这些事情也都可以在 List、Set、Queue 或 Deque 上做（当然，这些接口可能还会提供额外的功能）。Map 并非继承自 Collection，所以需要单独处理。

表 3-1

操作	作用
boolean add(T)	确保将参数（其类型为泛型类型 T）指向的对象添加到集合中。如果没有添加成功，则返回 false（这是一个"可选"的方法，下一节会讲解）
boolean addAll(Collection<? extends T>)	将参数中的所有元素添加进集合。如果有任何元素被添加进来，则返回 true（"可选"）
void clear()	移除集合中的所有元素（"可选"）
boolean contains(Object o)	如果集合中包含参数 o 所引用的对象，则返回 true
boolean containsAll(Collection<?>)	如果集合中有参数中的所有元素，则返回 true
boolean isEmpty()	如果集合中没有元素，则返回 true
Iterator<T> iterator() Spliterator<T> spliterator()	返回一个可以在集合的元素之间移动的迭代器。Spliterator 更为复杂，用于并发场景
boolean remove(Object)	如果参数指向的对象在集合中，则移除该元素的一个实例。如果发生了移除，则返回 true（"可选"）
boolean removeAll(Collection<?>)	移除参数包含的所有元素。如果发生了任何移除，则返回 true（"可选"）
boolean retainAll(Collection<?>)	只保留参数包含的元素（从集合论的角度看，就是求"交集"）。如果发生了任何变化，则返回 true（"可选"）
boolean removeIf(Predicate<? super E>)	移除这个集合中满足给定谓词条件的每一个元素
Stream<E> stream() Stream<E> parallelStream()	返回一个由这个集合中的元素组成的流
int size()	返回集合中元素的数量
Object[] toArray()	返回一个包含集合中所有元素的数组
<T> T[] toArray(T[] a)	返回一个包含集合中所有元素的数组。结果的运行时类型是参数数组的类型，而不是简单的 Object

这里没有用于随机访问元素的 get() 方法，因为 Collection 也包含 Set，而 Set 维护着自己的内部排序方式（这就使随机访问查找没有了意义）。因此，要检查 Collection 中的元素，必须使用迭代器。

这里展示了所有的 Collection 方法。ArrayList 被用作其"最小公分母"的一个 Collection：

```
// collectiontopics/CollectionMethods.java
// 我们可以在 Collection 上做的所有操作
import java.util.*;
import static onjava.HTMLColors.*;

public class CollectionMethods {
```

```java
public static void main(String[] args) {
  Collection<String> c =
    new ArrayList<>(LIST.subList(0, 4));
  c.add("ten");
  c.add("eleven");
  show(c);
  border();
  // 从 List 创建一个数组:
  Object[] array = c.toArray();
  // 从 List 创建一个 String 数组:
  String[] str = c.toArray(new String[0]);
  // 寻找最大和最小元素, 这意味着
  // 不同之处取决于 Comparable 接口
  // 是如何实现的:
  System.out.println(
    "Collections.max(c) = " + Collections.max(c));
  System.out.println(
    "Collections.min(c) = " + Collections.min(c));
  border();
  // 将一个集合添加到另一个集合中
  Collection<String> c2 =
    new ArrayList<>(LIST.subList(10, 14));
  c.addAll(c2);
  show(c);
  border();
  c.remove(LIST.get(0));
  show(c);
  border();
  // 移除包含在参数所指向集合中的所有元素:
  c.removeAll(c2);
  show(c);
  border();
  c.addAll(c2);
  show(c);
  border();
  // 某个元素是不是在该集合中?
  String val = LIST.get(3);
  System.out.println(
    "c.contains(" + val + ") = " + c.contains(val));
  // 某个集合是不是在该集合中?
  System.out.println(
    "c.containsAll(c2) = " + c.containsAll(c2));
  Collection<String> c3 =
    ((List<String>)c).subList(3, 5);
  // 保留同时存在于 c2 和 c3 中的所有元素
  //（集合的交集）:
  c2.retainAll(c3);
  show(c2);
  // 移除 c2 中所有在 c3 中出现过的元素:
  c2.removeAll(c3);
  System.out.println(
    "c2.isEmpty() = " +  c2.isEmpty());
  border();
  // 函数式操作:
  c = new ArrayList<>(LIST);
```

```
        c.removeIf(s -> !s.startsWith("P"));
        c.removeIf(s -> s.startsWith("Pale"));
        // 流操作:
        c.stream().forEach(System.out::println);
        c.clear(); // 移除所有元素
        System.out.println("after c.clear():" + c);
    }
}
/* 输出:
AliceBlue
AntiqueWhite
Aquamarine
Azure
ten
eleven
****************************
Collections.max(c) = ten
Collections.min(c) = AliceBlue
****************************
AliceBlue
AntiqueWhite
Aquamarine
Azure
ten
eleven
Brown
BurlyWood
CadetBlue
Chartreuse
****************************
AntiqueWhite
Aquamarine
Azure
ten
eleven
Brown
BurlyWood
CadetBlue
Chartreuse
****************************
AntiqueWhite
Aquamarine
Azure
ten
eleven
****************************
AntiqueWhite
Aquamarine
Azure
ten
eleven
Brown
BurlyWood
CadetBlue
Chartreuse
****************************
c.contains(Azure) = true
c.containsAll(c2) = true
c2.isEmpty() = true
****************************
PapayaWhip
PeachPuff
Peru
Pink
Plum
PowderBlue
Purple
after c.clear():[]
*/
```

（转右栏）

为了演示除了 Collection 接口之外我们没用到其他东西，这里创建了几个包含不同数据集的 ArrayList，并将其向上转型为 Collection 对象。

3.9 可选的操作

Collection 接口中用来执行各种添加和移除操作的方法是**可选的操作**。这意味着实现类并不一定要提供这些方法的功能定义。

这种定义接口的方式并不常见。正如我们所看到的，一个接口就是一个契约。这相当于说："无论你选择如何实现这个接口，我都保证你可以向这个对象发送这些消息。"（这里的接口既可以指正式的 `interface` 关键字，也可以更具一般性地指"任何类或子类支持的方法"。）但是"可选"的操作违反了这个非常基本的原则。在这种情况下，调用某些方法不会执行有意义的行为，而是会抛出异常！编译时的类型安全似乎被丢弃了。

情况并没有那么糟糕。如果一个操作是可选的，编译器仍然会限制我们只能调用该接口中的方法。它不像动态语言那样，可以在任何对象上调用任何方法，但在运行时才能发现某个特定的调用能否工作。[1] 此外，大多数以 Collection 为参数的方法只从该 Collection 中读取，而且 Collection 的所有"读取方法"都不是可选的。

为什么我们要把方法定义为"可选"的呢？这样做可以防止在设计中出现大量的接口。集合类库的其他设计往往会产生大量接口来描述主题的每一种变化，令人困惑。甚至不可能通过接口捕获所有的特殊情况，因为总会有人发明各种新的接口。这种"不支持的操作"方法实现了 Java 集合类库的一个重要目标：这些集合类很容易学习和使用。不支持的操作是一种特殊情况，可以推迟到必要时再进行。然而，为了使这种方法发挥作用，需要注意以下两点。

1. `UnsupportedOperationException` 必须是小概率事件。也就是说，在大部分情况下，所有的操作都能工作，而某个操作只在特殊情况下不被支持。在 Java 集合类库中确实是这样，因为我们在 99% 的时间中使用的 ArrayList、LinkedList、HashSet 和 HashMap，以及其他的具体实现，都支持所有的操作。这种设计确实提供了一个"后门"，可以创建一个新的 Collection，不用为 Collection 接口中的所有方法都提供有意义的定义，而它仍然可以与现有的库融为一体。

2. 当某个操作不被支持时，`UnsupportedOperationException` 很可能在实现时就会出现，而不是在将产品交付给客户之后。这是合理的。毕竟，它指示的是一个编程错误：我们错误地使用了某个实现。

值得注意的是，不支持的操作只在运行时才能被发现，所以这代表了动态类型检查。如果你之前用过 C++ 这样的静态类型语言，那么 Java 可能看起来就是另一种静态类型语言。Java 当然有**静态**类型检查，但是也有大量的**动态**类型，所以很难说它绝对属于某种单一类型的语言。一旦注意到这一点，你就会开始看到 Java 中动态类型检查的其他

---

[1] 虽然这种描述听起来很奇怪，而且可能没什么用，但是我们已经看到了，这种动态行为非常强大，特别是在基础卷第 19 章中。

例子了。

不支持的操作

不支持的操作常常来源于底层由固定大小的数据结构支撑的集合。当使用 Arrays.asList() 方法将一个数组转变为一个 List 时，我们就会得到这样的集合。通过 Collections 类中的"**不可修改**"（unmodifiable）的方法，我们可以**选择**让任何集合（包括 Map）抛出 UnsupportedOperationException。下面的示例演示了这两种情况：

```java
// collectiontopics/Unsupported.java
// Java 集合类中不支持的操作
import java.util.*;

public class Unsupported {
  static void
  check(String description, Runnable tst) {
    try {
      tst.run();
    } catch(Exception e) {
      System.out.println(description + "(): " + e);
    }
  }
  static void test(String msg, List<String> list) {
    System.out.println("--- " + msg + " ---");
    Collection<String> c = list;
    Collection<String> subList = list.subList(1,8);
    // sublist 的副本：
    Collection<String> c2 = new ArrayList<>(subList);
    check("retainAll", () -> c.retainAll(c2));
    check("removeAll", () -> c.removeAll(c2));
    check("clear", () -> c.clear());
    check("add", () -> c.add("X"));
    check("addAll", () -> c.addAll(c2));
    check("remove", () -> c.remove("C"));
    // List.set() 方法修改了值，但是
    // 没有改变该数据结构的大小：
    check("List.set", () -> list.set(0, "X"));
  }
  public static void main(String[] args) {
    List<String> list = Arrays.asList(
      "A B C D E F G H I J K L".split(" "));
    test("Modifiable Copy", new ArrayList<>(list));
    test("Arrays.asList()", list);
    test("unmodifiableList()",
      Collections.unmodifiableList(
        new ArrayList<>(list)));
  }
}
```

```
/* 输出:
--- Modifiable Copy ---
--- Arrays.asList() ---
retainAll(): java.lang.UnsupportedOperationException
removeAll(): java.lang.UnsupportedOperationException
clear(): java.lang.UnsupportedOperationException
add(): java.lang.UnsupportedOperationException
addAll(): java.lang.UnsupportedOperationException
remove(): java.lang.UnsupportedOperationException
--- unmodifiableList() ---
retainAll(): java.lang.UnsupportedOperationException
removeAll(): java.lang.UnsupportedOperationException
clear(): java.lang.UnsupportedOperationException
add(): java.lang.UnsupportedOperationException
addAll(): java.lang.UnsupportedOperationException
remove(): java.lang.UnsupportedOperationException
List.set(): java.lang.UnsupportedOperationException
*/
```

因为 Arrays.asList() 会产生一个底层由固定大小的数组支撑的 List，所以只支持那些不会改变数组大小的操作是合理的。任何会改变底层数据结构大小的方法都会导致 UnsupportedOperationException，用以说明调用了某个不支持的方法（这是一个编程错误）。

注意，总是可以将 Arrays.asList() 的结果当作一个构造器参数传递给任何 Collection（也可以使用 addAll() 方法或 static Collections.addAll() 方法）来创建一个支持各种操作的普通集合，我们在 main() 中第一次调用 test() 时对此进行了展示。这样的调用会生成一个新的、可以调整大小的底层数据结构。

Collections 类中的"不可修改"的方法将该集合包到一个代理中。如果执行了会以任何方式修改该集合的任何操作，这个代理就会抛出 UnsupportedOperationException。使用这些方法的目的是产生一个"常量化"的集合对象。稍后将展示这些"不可修改"的 Collections 方法的完整列表。

test() 中的最后一个 check() 检查了 set() 方法，该方法是 List 的一部分。这里，"不支持的操作"技术的粒度就派上了用场——对于 Arrays.asList() 返回的对象和 Collections.unmodifiableList() 返回的对象，所生成的"接口"可能只有一个方法的差别。Arrays.asList() 返回的是一个固定大小的 List，但是 Collections.unmodifiableList() 生成的是一个无法修改的列表。在输出中可以看到，可以修改 Arrays.asList() 所返回 List 中的元素，因为这不会破坏该 List 的"固定大小"本质。但是显然，unmodifiableList() 的结果应该是不能以任何方式修改的。如果使用接口，还需要两个额外的接口：一个带有可以工作的

set() 方法，一个不带。对于 Collection 的各种不可修改的子类型，都需要额外的接口。

对于以集合作为参数的方法，其文档应该说明哪些可选的方法是必须实现的。

3.10 Set 与存储顺序

基础卷第 12 章中的 Set 示例对基本 Set 上的操作提供了很好的介绍。然而，为方便起见，那些示例使用了诸如 Integer 和 String 等预定义的 Java 类型，它们被设计为可以在集合中使用。在创建我们自己的类型时，请注意 Set 需要某种方式来维护存储顺序（Map 也是，我们很快就会看到），而 Set 的不同实现之间会有差别。因此，不同的 Set 不仅有不同的行为，还对我们要放到某个特定 Set 中的对象的类型有不同的要求（见表 3-2）。

表 3-2

集 合	行 为
Set（interface）	向 Set 中添加的每个元素都必须是唯一的，Set 不会添加重复的元素。被添加到 Set 中的元素必须至少定义了 equals() 方法，用以确定对象的唯一性。Set 继承了 Collection，而且没有添加任何东西。Set 接口不保证以任何特定的顺序维护元素
HashSet*	用于对快速查找时间要求较高的 Set。要添加的元素必须定义 hashCode() 和 equals() 方法
TreeSet	有序的 Set，底层是一个树结构。通过这种方式，我们可以从某个 Set 提取出一个有序的序列。要添加的元素必须实现 Comparable 接口
LinkedHashSet	拥有 HashSet 的查找速度，但是内部使用了一个链表维护着我们添加元素的顺序（即插入顺序）。因此，当我们在这个 Set 上迭代时，结果是以插入顺序出现的。要加入的元素必须定义 hashCode() 和 equals() 方法

HashSet 上的星号表示，在没有其他约束的情况下，它应该是我们的默认选择，因为它对速度做了优化。

关于如何定义 hashCode()，我们会在附录 C 中描述。在定义自己的类时，如果其对象会以哈希方式或树方式存储，则必须为其创建 equals() 方法。但是只有在其对象会被放入 HashSet 或 LinkedHashSet 中时（HashSet 的情况更常见，因为它应该是我们用作 Set 实现的第一选择），才必须定义 hashCode() 方法。然而，为了确保良好的编程风格，在重写 equals() 时也要重写 hashCode()。

这个示例演示了要配合某个特定的 Set 实现来使用时，一个类型需要实现的方法：

```
// collectiontopics/TypesForSets.java
// 要将自己的类型放到 Set 中，需要实现如下必要的方法
import java.util.*;
import java.util.function.*;
```

```java
import java.util.Objects;

class SetType {
  protected int i;
  SetType(int n) { i = n; }
  @Override public boolean equals(Object o) {
    return o instanceof SetType &&
      Objects.equals(i, ((SetType)o).i);
  }
  @Override public String toString() {
    return Integer.toString(i);
  }
}

class HashType extends SetType {
  HashType(int n) { super(n); }
  @Override public int hashCode() {
    return Objects.hashCode(i);
  }
}

class TreeType extends SetType
implements Comparable<TreeType> {
  TreeType(int n) { super(n); }
  @Override public int compareTo(TreeType arg) {
    return Integer.compare(arg.i, i);
  }
  // 等价于:
  // return arg.i < i ? -1 : (arg.i == i ? 0 : 1);
  }
}

public class TypesForSets {
  static <T> void
  fill(Set<T> set, Function<Integer, T> type) {
    for(int i = 10; i >= 5; i--) // 降序
      set.add(type.apply(i));
    for(int i = 0; i < 5; i++) // 升序
      set.add(type.apply(i));
  }
  static <T> void
  test(Set<T> set, Function<Integer, T> type) {
    fill(set, type);
    fill(set, type); // 尝试添加重复的对象
    fill(set, type);
    System.out.println(set);
  }
  public static void main(String[] args) {
    test(new HashSet<>(), HashType::new);
    test(new LinkedHashSet<>(), HashType::new);
    test(new TreeSet<>(), TreeType::new);
    // 无法工作:
    test(new HashSet<>(), SetType::new);
    test(new HashSet<>(), TreeType::new);
```

```
      test(new LinkedHashSet<>(), SetType::new);
      test(new LinkedHashSet<>(), TreeType::new);
      try {
        test(new TreeSet<>(), SetType::new);
      } catch(Exception e) {
        System.out.println(e.getMessage());
      }
      try {
        test(new TreeSet<>(), HashType::new);
      } catch(Exception e) {
        System.out.println(e.getMessage());
      }
    }
  }
  /* 输出：
  [0, 1, 2, 3, 4, 5, 6, 7, 8, 9, 10]
  [10, 9, 8, 7, 6, 5, 0, 1, 2, 3, 4]
  [10, 9, 8, 7, 6, 5, 4, 3, 2, 1, 0]
  [4, 1, 5, 1, 0, 5, 6, 8, 7, 0, 2, 8, 4, 9, 6, 10, 6, 7,
  2, 9, 10, 3, 8, 4, 10, 3, 9, 5, 3, 7, 1, 2, 0]
  [9, 8, 4, 8, 5, 0, 1, 6, 2, 9, 3, 7, 2, 2, 7, 0, 9, 5,
  5, 6, 4, 7, 10, 1, 6, 4, 1, 10, 3, 3, 0, 10, 8]
  [10, 9, 8, 7, 6, 5, 0, 1, 2, 3, 4, 10, 9, 8, 7, 6, 5,
  0, 1, 2, 3, 4, 10, 9, 8, 7, 6, 5, 0, 1, 2, 3, 4]
  [10, 9, 8, 7, 6, 5, 0, 1, 2, 3, 4, 10, 9, 8, 7, 6, 5,
  0, 1, 2, 3, 4, 10, 9, 8, 7, 6, 5, 0, 1, 2, 3, 4]
  SetType cannot be cast to java.lang.Comparable
  HashType cannot be cast to java.lang.Comparable
  */
```

为了证明哪些方法对于某个特定的 Set 是必要的，同时为了避免代码重复，我们创建了 3 个类。

第一个是基类 SetType，它保存了一个 int，并且可以通过 toString() 输出这个值。保存在 Set 中的所有类都必须有一个 equals() 方法，因此这个方法也被放到了基类中。相等性是基于 int i 的值来判断的。

第二个是继承自 SetType 的 HashType，为了让该类型的对象可以放入 Set 的某个哈希实现中，加入了必要的 hashCode() 方法。

第三个是 TreeType。要在某种有序集合，比如在 SortedSet（TreeSet 是其唯一实现）中使用对象，TreeType 所实现的 Comparable 接口是必要的。注意，在 compareTo() 中**没有**使用 return i-i2 这种"简单且明显"的形式。因为这是一种常见的编程错误，所以只有 i 和 i2 都是"无符号"（unsigned）的 int 时才能正常工作（前提是 Java **有** unsigned 关键字，然而它并没有）。对于 Java 的带符号的 int，它就会有问题了，因为这样的值没有大到可以表示两个带符号 int 数的差。如果 i 是一个很大的正整型数，而 j 是一个绝对值很大的负整型数，i-j 将溢出并返回一个负值，这就出错了。

我们通常希望 `compareTo()` 方法能产生与 `equals()` 方法一致的自然排序。如果对于某个特定的比较，`equals()` 的结果为 `true`，那么 `compareTo()` 的结果应该为零；如果 `equals()` 的结果为 `false`，那么 `compareTo()` 的结果应该是一个非零值。

在 `TypesForSets` 中，为避免代码重复，`fill()` 和 `test()` 都是使用泛型定义的。为了验证 `Set` 的行为，`test()` 在用于测试的 `set` 上调用了 3 次 `fill()`，尝试引入重复的对象。`fill()` 可以接受一个任意类型的 `Set`，而 `Function` 对象负责产生这个类型。因为这个示例中用到的所有对象都有一个接受单个 `int` 参数的构造器，所以我们可以将这个构造器当作 `Function` 来传递，它负责提供填充 `Set` 的对象。

注意，为了看出所得的存储顺序，`fill()` 方法按照降序添加了第一批元素，按照升序添加了第二批元素。输出表明，`HashSet` 是按照升序保存元素的。然而在附录 C 中，我们会看到这是偶然情况，因为哈希会创建自己的存储顺序。只是因为我们的值是一个简单的 `int`，所以它才是按照升序排列的。`LinkedHashSet` 会按照元素插入的顺序来保存它们，而 `TreeSet` 则按照排序结果来维护元素（在这个例子中是降序，这是因为受到了 `compareTo()` 的实现方式的影响）。

对于 `Set` 要求的那些操作，如果我们尝试使用的类型并没有正确地提供支持，就会出问题。`SetType` 或 `TreeType` 对象没有包含重新定义的 `hashCode()` 方法，把这样的对象放入任何哈希实现中时，都会出现重复的值，这就破坏了 `Set` 的主要约定。这让我们相当不安，因为它甚至不会抛出运行时错误。然而，默认的 `hashCode()` 是合法的，所以这虽然并不正确，却是合法的行为。要确保这种程序的正确性，唯一可靠的方法是将单元测试加入我们的构建系统中。

如果尝试在 `TreeSet` 中使用一个没有实现 `Comparable` 的类型，我们会得到一个更明确的结果：当 `TreeSet` 尝试将该对象用作 `Comparable` 时，会抛出异常。

SortedSet

`SortedSet` 中的元素必然是有序排列的。`SortedSet` 接口中的下列方法提供了更多功能。

- `Comparator comparator()`：生成用于该 `Set` 的比较器对象；如果选择自然排序，则返回 `null`。
- `Object first()`：生成最小的元素。
- `Object last()`：生成最大的元素。
- `SortedSet subSet(fromElement, toElement)`：生成该 `Set` 的一个视图，包含从 `fromElement`（包含）到 `toElement`（不包含）的元素。

- SortedSet headSet(toElement)：生成该 Set 的一个视图，包含小于 toElement 的所有元素。
- SortedSet tailSet(fromElement)：生成该 Set 的一个视图，包含大于或等于 fromElement 的所有元素。

下面是一个简单的演示：

```java
// collectiontopics/SortedSetDemo.java
import java.util.*;
import static java.util.stream.Collectors.*;

public class SortedSetDemo {
  public static void main(String[] args) {
    SortedSet<String> sortedSet =
      Arrays.stream(
        "one two three four five six seven eight"
        .split(" "))
        .collect(toCollection(TreeSet::new));
    System.out.println(sortedSet);
    String low = sortedSet.first();
    String high = sortedSet.last();
    System.out.println(low);
    System.out.println(high);
    Iterator<String> it = sortedSet.iterator();
    for(int i = 0; i <= 6; i++) {
      if(i == 3) low = it.next();
      if(i == 6) high = it.next();
      else it.next();
    }
    System.out.println(low);
    System.out.println(high);
    System.out.println(sortedSet.subSet(low, high));
    System.out.println(sortedSet.headSet(high));
    System.out.println(sortedSet.tailSet(low));
  }
}
/* 输出：
[eight, five, four, one, seven, six, three, two]
eight
two
one
two
[one, seven, six, three]
[eight, five, four, one, seven, six, three]
[one, seven, six, three, two]
*/
```

注意，SortedSet 意味着"根据该对象的比较函数来排序"，而不是根据"插入顺序"。要保留插入顺序的话，可以使用 LinkedHashSet。

3.11 Queue

Queue 的实现有很多种，其中大部分是为并发应用设计的。有些实现的差别在于排序行为，而不是性能。下面是一个基本的示例，涵盖了大多数 Queue 实现，包括基于并发的那些。我们在一端放入元素，从另一端取出：

```java
// collectiontopics/QueueBehavior.java
// 对比基本行为
import java.util.*;
import java.util.stream.*;
import java.util.concurrent.*;

public class QueueBehavior {
  static Stream<String> strings() {
    return Arrays.stream(
      ("one two three four five six seven " +
      "eight nine ten").split(" "));
  }
  static void test(int id, Queue<String> queue) {
    System.out.print(id + ": ");
    strings().forEach(queue::offer);
    while(queue.peek() != null)
      System.out.print(queue.remove() + " ");
    System.out.println();
  }
  public static void main(String[] args) {
    int count = 10;
    test(1, new LinkedList<>());
    test(2, new PriorityQueue<>());
    test(3, new ArrayBlockingQueue<>(count));
    test(4, new ConcurrentLinkedQueue<>());
    test(5, new LinkedBlockingQueue<>());
    test(6, new PriorityBlockingQueue<>());
    test(7, new ArrayDeque<>());
    test(8, new ConcurrentLinkedDeque<>());
    test(9, new LinkedBlockingDeque<>());
    test(10, new LinkedTransferQueue<>());
    test(11, new SynchronousQueue<>());
  }
}
/* 输出：
1: one two three four five six seven eight nine ten
2: eight five four nine one seven six ten three two
3: one two three four five six seven eight nine ten
4: one two three four five six seven eight nine ten
5: one two three four five six seven eight nine ten
6: eight five four nine one seven six ten three two
7: one two three four five six seven eight nine ten
8: one two three four five six seven eight nine ten
9: one two three four five six seven eight nine ten
10: one two three four five six seven eight nine ten
11:
*/
```

Deque 接口也继承自 Queue。除了优先级队列之外，Queue 都是按照元素的放入顺序来输出元素的。在这个示例中，SynchronousQueue 没有输出任何结果，因为它是一个阻塞队列，每一个插入操作都必须等待另一个线程进行的相应移除操作，反之亦然。

3.11.1 优先级队列

考虑一个待办清单（to-do list），其中的每个对象都包含一个 String，以及分别表示主要优先级和次要优先级的值。清单的顺序通过实现 Comparable 来控制：

```java
// collectiontopics/ToDoList.java
// PriorityQueue 的更复杂应用
import java.util.*;

class ToDoItem implements Comparable<ToDoItem> {
  private char primary;
  private int secondary;
  private String item;
  ToDoItem(String td, char pri, int sec) {
    primary = pri;
    secondary = sec;
    item = td;
  }
  @Override public int compareTo(ToDoItem arg) {
    if(primary > arg.primary)
      return +1;
    if(primary == arg.primary)
      if(secondary > arg.secondary)
        return +1;
      else if(secondary == arg.secondary)
        return 0;
    return -1;
  }
  @Override public String toString() {
    return Character.toString(primary) +
      secondary + ": " + item;
  }
}

class ToDoList {
  public static void main(String[] args) {
    PriorityQueue<ToDoItem> toDo =
      new PriorityQueue<>();
    toDo.add(new ToDoItem("Empty trash", 'C', 4));
    toDo.add(new ToDoItem("Feed dog", 'A', 2));
    toDo.add(new ToDoItem("Feed bird", 'B', 7));
    toDo.add(new ToDoItem("Mow lawn", 'C', 3));
    toDo.add(new ToDoItem("Water lawn", 'A', 1));
    toDo.add(new ToDoItem("Feed cat", 'B', 1));
    while(!toDo.isEmpty())
      System.out.println(toDo.remove());
  }
}
/* 输出：
A1: Water lawn
A2: Feed dog
B1: Feed cat
B7: Feed bird
C3: Mow lawn
C4: Empty trash
*/
```

这里演示了如何通过优先级队列实现对列表条目的自动排序。

3.11.2 Deque

Deque（双端队列）像一个队列，但是可以在任意一端添加和移除元素。Java 6 为 Deque 添加了一个显式的接口。下面通过实现了该接口的类来测试最基本的 Deque 方法：

```java
// collectiontopics/SimpleDeques.java
// 对 Deque 的基本测试
import java.util.*;
import java.util.concurrent.*;
import java.util.function.*;

class CountString implements Supplier<String> {
  private int n = 0;
  CountString() {}
  CountString(int start) { n = start; }
  @Override public String get() {
    return Integer.toString(n++);
  }
}

public class SimpleDeques {
  static void test(Deque<String> deque) {
    CountString s1 = new CountString(),
                s2 = new CountString(20);
    for(int n = 0; n < 8; n++) {
      deque.offerFirst(s1.get());
      deque.offerLast(s2.get()); // 和 offer() 一样
    }
    System.out.println(deque);
    String result = "";
    while(deque.size() > 0) {
      System.out.print(deque.peekFirst() + " ");
      result += deque.pollFirst() + " ";
      System.out.print(deque.peekLast() + " ");
      result += deque.pollLast() + " ";
    }
    System.out.println("\n" + result);
  }
  public static void main(String[] args) {
    int count = 10;
    System.out.println("LinkedList");
    test(new LinkedList<>());
    System.out.println("ArrayDeque");
    test(new ArrayDeque<>());
    System.out.println("LinkedBlockingDeque");
    test(new LinkedBlockingDeque<>(count));
    System.out.println("ConcurrentLinkedDeque");
    test(new ConcurrentLinkedDeque<>());
  }
}
```

```
      }
```
```
/* 输出:
LinkedList
[7, 6, 5, 4, 3, 2, 1, 0, 20, 21, 22, 23, 24, 25, 26,
27]
7 27 6 26 5 25 4 24 3 23 2 22 1 21 0 20
7 27 6 26 5 25 4 24 3 23 2 22 1 21 0 20
ArrayDeque
[7, 6, 5, 4, 3, 2, 1, 0, 20, 21, 22, 23, 24, 25, 26,
27]
7 27 6 26 5 25 4 24 3 23 2 22 1 21 0 20
7 27 6 26 5 25 4 24 3 23 2 22 1 21 0 20
LinkedBlockingDeque
[4, 3, 2, 1, 0, 20, 21, 22, 23, 24]
4 24 3 23 2 22 1 21 0 20
4 24 3 23 2 22 1 21 0 20
ConcurrentLinkedDeque
[7, 6, 5, 4, 3, 2, 1, 0, 20, 21, 22, 23, 24, 25, 26,
27]
7 27 6 26 5 25 4 24 3 23 2 22 1 21 0 20
7 27 6 26 5 25 4 24 3 23 2 22 1 21 0 20
*/
```

这里只使用了 Deque 方法的 offer 和 poll 版本，因为当 LinkedBlockingDeque 的大小有限时，它们不会抛出异常。注意，当 LinkedBlockingDeque 添加元素时，它会在达到其大小限制时停止，然后忽略后续的 offer。

3.12　理解 Map

我们在基础卷第 12 章已经学过，Map（也叫**关联数组**）维护着键 - 值关联（键值对），所以可以使用键来查找值。标准 Java 库包含多个 Map 的基本实现，如 HashMap、TreeMap、LinkedHashMap、WeakHashMap、ConcurrentHashMap 和 IdentityHashMap。它们有相同的基本 Map 接口，但是在行为上有所区别，包括效率、键值对的保存和演示顺序、保存对象的时间长短、在多线程程序中的表现，以及键的相等性如何确定等。从 Map 接口的实现数量可以看出这个工具的重要性。

要更深入地了解 Map，学习如何构建关联数组是很有帮助的。下面是一个极其简单的实现：

```
// collectiontopics/AssociativeArray.java
// 将键和值关联起来

public class AssociativeArray<K, V> {
  private Object[][] pairs;
```

```java
  private int index;
  public AssociativeArray(int length) {
    pairs = new Object[length][2];
  }
  public void put(K key, V value) {
    if(index >= pairs.length)
      throw new ArrayIndexOutOfBoundsException();
    pairs[index++] = new Object[]{ key, value };
  }
  @SuppressWarnings("unchecked")
  public V get(K key) {
    for(int i = 0; i < index; i++)
      if(key.equals(pairs[i][0]))
        return (V)pairs[i][1];
    return null; // 没有找到这个键
  }
  @Override public String toString() {
    StringBuilder result = new StringBuilder();
    for(int i = 0; i < index; i++) {
      result.append(pairs[i][0].toString());
      result.append(" : ");
      result.append(pairs[i][1].toString());
      if(i < index - 1)
        result.append("\n");
    }
    return result.toString();
  }
  public static void main(String[] args) {
    AssociativeArray<String,String> map =
      new AssociativeArray<>(6);
    map.put("sky", "blue");
    map.put("grass", "green");
    map.put("ocean", "dancing");
    map.put("tree", "tall");
    map.put("earth", "brown");
    map.put("sun", "warm");
    try {
      map.put("extra", "object"); // 超过了末尾
    } catch(ArrayIndexOutOfBoundsException e) {
      System.out.println("Too many objects!");
    }
    System.out.println(map);
    System.out.println(map.get("ocean"));
  }
}
/* 输出：
Too many objects!
sky : blue
grass : green
ocean : dancing
tree : tall
earth : brown
sun : warm
dancing
*/
```

关联数组中的基本方法是 put() 和 get()，但是为了便于显示，我们重写了 toString() 来打印键值对。为了演示其可以工作，我们在 main() 中创建了一个由 String 对组成的 AssociativeArray，并打印了生成的映射，之后通过 get() 获得了其中的一个值。

要使用 get() 方法，需要传入我们想查找的 key。它会返回关联的值作为结果；如果

找不到，就返回 null。get() 方法所使用的定位这个值的方法，可能是我们能想象到的方法中效率最差的：从数组最上面开始，使用 equals() 依次比较每个键。但是这里的重点是简洁明了，而非效率。

这个版本是有启发性的，但是性能不高，而且大小是固定的，因而不够灵活。幸运的是，java.util 中的 Map 没有这些问题。

3.12.1 性能

性能是映射的一个基本问题。当搜寻某个键时，在 get() 中使用线性搜索非常慢。HashMap 正是在这一点上提升速度的。它不是缓慢地查找这个键，而是使用了一个叫**哈希码**的特殊值。哈希码是这样一种方法：它从所处理的对象中提取一些信息，并将这些信息转变为该对象的一个"相对唯一"的 int。hashCode() 是根类 Object 中的一个方法，因此所有的 Java 对象都能生成哈希码。HashMap 会调用对象的 hashCode()，并使用这个哈希码来快速查找键。此举可以显著提升性能。[①]

表 3-3 展示了基本的 Map 实现。HashMap 上的星号表示，如果没有其他约束，这应该是我们的默认选择，因为它为速度做了优化。其他实现重视别的特性，因此不像 HashMap 那么快。

表 3-3

集合	行为
HashMap*	基于哈希表实现。（使用这个类代替 Hashtable。）提供了常数时间的键值对插入和定位性能。可以通过构造器来调整其性能，因为构造器支持设置这个哈希表的**容量**（capacity）和**负载因子**（load factor）
LinkedHashMap	类似于 HashMap，但是在遍历时会以插入的顺序或最近最少使用（least-recently-used，LRU）的顺序获得键值对。除遍历之外，性能比 HashMap 稍低。对于遍历的情况，因为它用了链表来维护内部顺序，所以更快一些
TreeMap	基于红黑树实现。当我们查看键或者键值对时，会发现它们是有序的（顺序通过 Comparable 或 Comparator 来确定）。TreeMap 的要点是，我们会以有序的方式得到结果。TreeMap 是唯一提供了 subMap() 方法的 Map，它会返回树的一部分
WeakHashMap	由弱键组成的 Map，该映射所引用的对象可以被释放。它是为解决特定类型的问题而设计的。如果在这个映射之外，已经没有指向某个特定键的引用，那么可以对这些键进行垃圾收集
ConcurrentHashMap	线程安全的 Map，没有使用同步锁。在本书第 5 章中讨论
IdentityHashMap	哈希映射，它使用 == 而不是 equals() 来比较键。仅用于解决特殊类型的问题，并非通用

[①] 如果这些加速措施仍然不能满足性能要求，可以通过编写自己的 Map 进一步加速表的查找，并根据自己的特定类型来定制，以避免因为与 Object 类型来回转换而造成延迟。如果对性能有更高的追求，速度爱好者们可以阅读高德纳的《计算机程序设计艺术 卷 3：排序与查找（第 2 版）》（此书中文版已由人民邮电出版社出版，详见：https://www.ituring.com.cn/book/926。——编者注），用数组来代替溢出桶列表（overflow bucket list）。这会带来两个额外的优势：一是可以优化磁盘存储特性，二是可以节省单条记录的创建时间和垃圾收集时间。

哈希是用来在映射中保存元素的最常用方式。

对 Map 中键的要求和对 Set 中元素的要求是一样的，我们在 TypesForSets.java 中已经看到了。任何键都必须有一个 equals() 方法。如果这个键被用在了使用哈希的 Map 中，它必须还有一个适当的 hashCode()。如果这个键被用在了 TreeMap 中，它必须实现 Comparable。

下列示例使用之前定义的 CountMap 测试数据集，演示了可以在 Map 接口上执行的操作：

```java
// collectiontopics/MapOps.java
// 可以在 Map 上执行的操作
import java.util.concurrent.*;
import java.util.*;
import onjava.*;

public class MapOps {
  public static
  void printKeys(Map<Integer,String> map) {
    System.out.print("Size = " + map.size() + ", ");
    System.out.print("Keys: ");
    // 生成由键组成的 Set :
    System.out.println(map.keySet());
  }
  public static
  void test(Map<Integer,String> map) {
    System.out.println(
      map.getClass().getSimpleName());
    map.putAll(new CountMap(25));
    // Map 有用于键的 Set 行为：
    map.putAll(new CountMap(25));
    printKeys(map);
    // 生成由值组成的 Collection :
    System.out.print("Values: ");
    System.out.println(map.values());
    System.out.println(map);
    System.out.println("map.containsKey(11): " +
      map.containsKey(11));
    System.out.println(
      "map.get(11): " + map.get(11));
    System.out.println("map.containsValue(\"F0\"): "
      + map.containsValue("F0"));
    Integer key = map.keySet().iterator().next();
    System.out.println("First key in map: " + key);
    map.remove(key);
    printKeys(map);
    map.clear();
    System.out.println(
      "map.isEmpty(): " + map.isEmpty());
    map.putAll(new CountMap(25));
    // 可以通过在这个 Set 上的操作来修改 Map :
```

```
      map.keySet().removeAll(map.keySet());
      System.out.println(
        "map.isEmpty(): " + map.isEmpty());
  }
  public static void main(String[] args) {
    test(new HashMap<>());
    test(new TreeMap<>());
    test(new LinkedHashMap<>());
    test(new IdentityHashMap<>());
    test(new ConcurrentHashMap<>());
    test(new WeakHashMap<>());
  }
}
/* 输出:（只列出了前 11 行）
HashMap
Size = 25, Keys: [0, 1, 2, 3, 4, 5, 6, 7, 8, 9, 10, 11,
12, 13, 14, 15, 16, 17, 18, 19, 20, 21, 22, 23, 24]
Values: [A0, B0, C0, D0, E0, F0, G0, H0, I0, J0, K0,
L0, M0, N0, O0, P0, Q0, R0, S0, T0, U0, V0, W0, X0, Y0]
{0=A0, 1=B0, 2=C0, 3=D0, 4=E0, 5=F0, 6=G0, 7=H0, 8=I0,
9=J0, 10=K0, 11=L0, 12=M0, 13=N0, 14=O0, 15=P0, 16=Q0,
17=R0, 18=S0, 19=T0, 20=U0, 21=V0, 22=W0, 23=X0, 24=Y0}
map.containsKey(11): true
map.get(11): L0
map.containsValue("F0"): true
First key in map: 0
Size = 24, Keys: [1, 2, 3, 4, 5, 6, 7, 8, 9, 10, 11,
12, 13, 14, 15, 16, 17, 18, 19, 20, 21, 22, 23, 24]
map.isEmpty(): true
map.isEmpty(): true
                           ...
*/
```

printKeys() 方法演示了如何生成 Map 的一个 Collection 视图。keySet() 方法会生成一个由 Map 中的键组成的 Set。打印 values() 方法的结果，得到一个包含了 Map 中所有值的 Collection。（注意，键必须是唯一的，但是值可以重复。）这些 Collection 的底层由 Map 中的结构支撑，所以对 Collection 的任何修改都会在与其关联的 Map 中体现出来。

这个程序的其余部分为每个 Map 操作提供了简单的示例，并测试了每种基本的 Map 类型。

3.12.2 SortedMap

利用 SortedMap（TreeMap 或 ConcurrentSkipListMap 实现了该接口），可以确保键是有序的，从而可以使用 SortedMap 接口中的方法提供的功能。

- Comparator comparator()：生成用于该 Map 的比较器对象；如果使用自然排序，则返回 null。
- T firstKey()：生成最小的键。

- `T lastKey()`：生成最大的键。
- `SortedMap subMap(fromKey, toKey)`：生成该 Map 的一个视图，包含从 fromKey（包含）到 toKey（不包含）的键。
- `SortedMap headMap(toKey)`：生成该 Map 的一个视图，包含小于 toKey 的所有键。
- `SortedMap tailMap(fromKey)`：生成该 Map 的一个视图，包含大于或等于 fromKey 的所有键。

下面是一个与 SortedSetDemo.java 类似的示例，演示了 TreeMap 的其他行为：

```java
// collectiontopics/SortedMapDemo.java
// 可以在 TreeMap 上执行的操作
import java.util.*;
import onjava.*;

public class SortedMapDemo {
  public static void main(String[] args) {
    TreeMap<Integer,String> sortedMap =
      new TreeMap<>(new CountMap(10));
    System.out.println(sortedMap);
    Integer low = sortedMap.firstKey();
    Integer high = sortedMap.lastKey();
    System.out.println(low);
    System.out.println(high);
    Iterator<Integer> it =
      sortedMap.keySet().iterator();
    for(int i = 0; i <= 6; i++) {
      if(i == 3) low = it.next();
      if(i == 6) high = it.next();
      else it.next();
    }
    System.out.println(low);
    System.out.println(high);
    System.out.println(sortedMap.subMap(low, high));
    System.out.println(sortedMap.headMap(high));
    System.out.println(sortedMap.tailMap(low));
  }
}
/* 输出：
{0=A0, 1=B0, 2=C0, 3=D0, 4=E0, 5=F0, 6=G0, 7=H0, 8=I0, 9=J0}
0
9
3
7
{3=D0, 4=E0, 5=F0, 6=G0}
{0=A0, 1=B0, 2=C0, 3=D0, 4=E0, 5=F0, 6=G0}
{3=D0, 4=E0, 5=F0, 6=G0, 7=H0, 8=I0, 9=J0}
*/
```

这里，键值对是按照键的顺序有序存储的。因为 TreeMap 是有序的，"定位"的概念也就有意义了，所以我们可以获得第一个和最后一个元素以及子映射。

3.12.3　LinkedHashMap

为了提高速度，LinkedHashMap 对所有的东西都进行了哈希处理，但是在遍历过程中也会按照插入顺序输出键值对（System.out.println() 在该映射上执行了迭代，所以我们能看到遍历的结果）。此外，可以在 LinkedHashMap 的构造器中进行配置，让它使用基于访问量的最近最少使用（LRU）算法。这样的话，尚未访问的元素就会出现在列表的前面（而且因此成了移除操作的候选）。 这样可以轻松创建为节省空间而进行定期清理的程序。下面这个简单的例子演示了这两个特点：

```java
// collectiontopics/LinkedHashMapDemo.java
// 可以在 LinkedHashMap 上执行的操作
import java.util.*;
import onjava.*;

public class LinkedHashMapDemo {
  public static void main(String[] args) {
    LinkedHashMap<Integer,String> linkedMap =
      new LinkedHashMap<>(new CountMap(9));
    System.out.println(linkedMap);
    // 按照最近最少使用的顺序
    linkedMap =
      new LinkedHashMap<>(16, 0.75f, true);
    linkedMap.putAll(new CountMap(9));
    System.out.println(linkedMap);
    for(int i = 0; i < 6; i++)
      linkedMap.get(i);
    System.out.println(linkedMap);
    linkedMap.get(0);
    System.out.println(linkedMap);
  }
}
/* 输出：
{0=A0, 1=B0, 2=C0, 3=D0, 4=E0, 5=F0, 6=G0, 7=H0, 8=I0}
{0=A0, 1=B0, 2=C0, 3=D0, 4=E0, 5=F0, 6=G0, 7=H0, 8=I0}
{6=G0, 7=H0, 8=I0, 0=A0, 1=B0, 2=C0, 3=D0, 4=E0, 5=F0}
{6=G0, 7=H0, 8=I0, 1=B0, 2=C0, 3=D0, 4=E0, 5=F0, 0=A0}
*/
```

键值对确实是按插入顺序遍历的，即使在 LRU 版本中也是一样。然而，在 LRU 版本中，在（仅有的）6 个条目被访问过之后，最后 3 个条目会被放到列表的前面。然后，当 0 再次被访问之后，它会被移到列表的尾部。

3.13 工具函数

有一些用于集合的独立工具，它们用 `java.util.Collections` 中的 `static` 方法表示。我们已经见过一些，比如 `addAll()`、`reverseOrder()` 和 `binarySearch()`。这里介绍另一些（`synchronized` 和 `unmodifiable` 工具会在后面介绍）。在表 3-4 中，我们看到在相关的情况下使用了泛型。

表 3-4

工具函数	作用
`checkedCollection(Collection<T>, Class<T> type)`、`checkedList(List<T>, Class<T> type)`、`checkedMap(Map<K, V>, Class<K> keyType, Class<V> valueType)`、`checkedSet(Set<T>, Class<T> type)`、`checkedSortedMap(SortedMap<K, V>, Class<K> keyType, Class<V> valueType)`、`checkedSortedSet(SortedSet<T>, Class<T> type)`	从 Collection 或 Collection 的某个具体子类型生成一个动态、类型安全的视图。当无法使用静态检查的版本时，可以使用它们。我们在基础卷 20.12 节演示过
`max(Collection)`、`min(Collection)`	使用参数指向的 Collection 中的对象自然比较方法，计算其中最大或最小的元素
`max(Collection, Comparator)`、`min(Collection, Comparator)`	使用 Comparator 来计算 Collection 中最大或最小的元素
`indexOfSubList(List source, List target)`	计算 target 在 source 中**第一次**出现的位置的起始索引；如果没有出现过，则返回 -1
`lastIndexOfSubList(List source, List target)`	计算 target 在 source 中**最后一次**出现的位置的起始索引；如果没有出现过，则返回 -1
`replaceAll(List<T>, T oldVal, T newVal)`	用 newVal 替换所有的 oldVal
`reverse(List)`	就地将所有的元素变为逆向
`reverseOrder()`、`reverseOrder(Comparator<T>)`	返回一个 Comparator，对于实现了 Comparable<T> 接口的对象组成的集合，它可以逆转其自然排序。第二个版本可以使用所提供的 Comparator 来逆转顺序
`rotate(List, int distance)`	将所有元素向前移动 distance 的距离，对于超出末尾的元素，将其放到列表的开始处
`shuffle(List)`、`shuffle(List, Random)`	随机变换指定的列表。第一种形式有自己的随机源，也可以用第二种形式提供我们的随机源
`sort(List<T>)`、`sort(List<T>, Comparator<? super T> c)`	使用自然顺序对参数指定的 List<T> 排序。第二种形式接受一个 Comparator 来排序
`copy(List<? super T> dest, List<? extends T> src)`	将 src 中的元素复制到 dest 中
`swap(List, int i, int j)`	交换这个 List 中位于位置 i 和 j 处的元素。执行速度可能会比我们手写的代码更快
`fill(List<? super T>, T x)`	将 List 中的所有元素都替换为 x
`nCopies(int n, T x)`	返回一个大小为 n 的不可变 List<T>，其中的所有引用都指向 x
`disjoint(Collection, Collection)`	如果参数中的两个 Collection 没有共同元素，则返回 true
`frequency(Collection, Object x)`	返回这个 Collection 中和 x 等价的元素的数量
`emptyList()`、`emptyMap()`、`emptySet()`	返回一个不可变的空 List、Map 或 Set。它们都是用泛型定义的，所以生成的 Collection 都会被参数化为指定类型

（续）

工具函数	作用
singleton(T x)、singletonList(T x)、singletonMap(K key, V value)	生成一个不可变的 Set<T>、List<T> 或 Map<K, V>，其中包含基于给定参数创建的一个元素
list(Enumeration<T> e)	生成一个 ArrayList<T>，包含按照顺序从（老式的）Enumeration（Iterator 的前身）返回的元素。用于转换遗留代码
enumeration(Collection<T>)	生成一个老式的 Enumeration<T>

注意 min() 和 max() 是配合 Collection 对象使用的，而非 List 对象，所以我们不用担心是否应该对这个 Collection 排序。（前面也提到过，在执行 binarySearch() 之前必须对 List 或数组执行 sort()。）

下面的示例演示了表 3-4 中大部分工具函数的基本用法：

```java
// collectiontopics/Utilities.java
// Collections 工具的简单演示
import java.util.*;

public class Utilities {
  static List<String> list = Arrays.asList(
    "one Two three Four five six one".split(" "));
  public static void main(String[] args) {
    System.out.println(list);
    System.out.println("'list' disjoint (Four)?: " +
      Collections.disjoint(list,
        Collections.singletonList("Four")));
    System.out.println(
      "max: " + Collections.max(list));
    System.out.println(
      "min: " + Collections.min(list));
    System.out.println(
      "max w/ comparator: " + Collections.max(list,
      String.CASE_INSENSITIVE_ORDER));
    System.out.println(
      "min w/ comparator: " + Collections.min(list,
      String.CASE_INSENSITIVE_ORDER));
    List<String> sublist =
      Arrays.asList("Four five six".split(" "));
    System.out.println("indexOfSubList: " +
      Collections.indexOfSubList(list, sublist));
    System.out.println("lastIndexOfSubList: " +
      Collections.lastIndexOfSubList(list, sublist));
    Collections.replaceAll(list, "one", "Yo");
    System.out.println("replaceAll: " + list);
    Collections.reverse(list);
    System.out.println("reverse: " + list);
    Collections.rotate(list, 3);
    System.out.println("rotate: " + list);
```

```java
      List<String> source =
        Arrays.asList("in the matrix".split(" "));
      Collections.copy(list, source);
      System.out.println("copy: " + list);
      Collections.swap(list, 0, list.size() - 1);
      System.out.println("swap: " + list);
      Collections.shuffle(list, new Random(47));
      System.out.println("shuffled: " + list);
      Collections.fill(list, "pop");
      System.out.println("fill: " + list);
      System.out.println("frequency of 'pop': " +
        Collections.frequency(list, "pop"));
      List<String> dups =
        Collections.nCopies(3, "snap");
      System.out.println("dups: " + dups);
      System.out.println("'list' disjoint 'dups'?: " +
        Collections.disjoint(list, dups));
      // 获得一个老式的 Enumeration:
      Enumeration<String> e =
        Collections.enumeration(dups);
      Vector<String> v = new Vector<>();
      while(e.hasMoreElements())
        v.addElement(e.nextElement());
      // 通过 Enumeration 将一个老式的 Vector
      // 转换为一个 List:
      ArrayList<String> arrayList =
        Collections.list(v.elements());
      System.out.println("arrayList: " + arrayList);
  }
}
/* 输出:
[one, Two, three, Four, five, six, one]
'list' disjoint (Four)?: false
max: three
min: Four
max w/ comparator: Two
min w/ comparator: five
indexOfSubList: 3
lastIndexOfSubList: 3
replaceAll: [Yo, Two, three, Four, five, six, Yo]
reverse: [Yo, six, five, Four, three, Two, Yo]
rotate: [three, Two, Yo, Yo, six, five, Four]
copy: [in, the, matrix, Yo, six, five, Four]
swap: [Four, the, matrix, Yo, six, five, in]
shuffled: [six, matrix, the, Four, Yo, five, in]
fill: [pop, pop, pop, pop, pop, pop, pop]
frequency of 'pop': 7
dups: [snap, snap, snap]
'list' disjoint 'dups'?: true
arrayList: [snap, snap, snap]
*/
```

输出说明了每个工具方法的行为。请注意当 min() 和 max() 使用 String.CASE_

INSENSITIVE_ORDER Comparator 时，大小写带来的差别。

3.13.1　List 上的排序和查找

在 List 上执行排序和查找的工具函数虽然与用于对象数组的那些拥有同样的名字和签名，但它们是 Collections 而非 Arrays 的 static 方法。下面的示例使用了来自 Utilities.java 的 list 数据：

```java
// collectiontopics/ListSortSearch.java
// 利用 Collections 工具函数对 List 执行排序和查找
import java.util.*;

public class ListSortSearch {
  public static void main(String[] args) {
    List<String> list =
      new ArrayList<>(Utilities.list);
    list.addAll(Utilities.list);
    System.out.println(list);
    Collections.shuffle(list, new Random(47));
    System.out.println("Shuffled: " + list);
    ListIterator<String> it =
      list.listIterator(10);           // [1]
    while(it.hasNext()) {              // [2]
      it.next();
      it.remove();
    }
    System.out.println("Trimmed: " + list);
    Collections.sort(list);
    System.out.println("Sorted: " + list);
    String key = list.get(7);
    int index = Collections.binarySearch(list, key);
    System.out.println(
      "Location of " + key + " is " + index +
      ", list.get(" + index + ") = " +
      list.get(index));
    Collections.sort(list,
      String.CASE_INSENSITIVE_ORDER);
    System.out.println(
      "Case-insensitive sorted: " + list);
    key = list.get(7);
    index = Collections.binarySearch(list, key,
      String.CASE_INSENSITIVE_ORDER);
    System.out.println(
      "Location of " + key + " is " + index +
      ", list.get(" + index + ") = " +
      list.get(index));
  }
}
```

```
/* 输出:
[one, Two, three, Four, five, six, one, one, Two,
three, Four, five, six, one]
Shuffled: [Four, five, one, one, Two, six, six, three,
three, five, Four, Two, one, one]
Trimmed: [Four, five, one, one, Two, six, six, three,
three, five]
Sorted: [Four, Two, five, five, one, one, six, six,
three, three]
Location of six is 7, list.get(7) = six
Case-insensitive sorted: [five, five, Four, one, one,
six, six, three, three, Two]
Location of three is 7, list.get(7) = three
*/
```

同对数组执行查找和排序一样，如果使用一个 Comparator 来排序，也必须使用同样的 Comparator 来执行 binarySearch()。

Collections.shuffle() 会将 List 的顺序变成随机的。

[1] list.listIterator(10) 在 list 的位置 10 处创建了一个 ListIterator。

[2] 只要在这个 list 的位置 10 及其后面仍有元素，while 循环就会删除它们。

3.13.2 创建不可修改的 Collection 或 Map

通常，创建 Collection 或 Map 的只读版本是非常方便的。Collections 类是这么做的：让某个方法接受一个原始的集合，并返回一个只读的版本。这个方法还有多个变种：用于 Collection 的（在无法把一个 Collection 当作更具体的类型时使用）、用于 List 的、用于 Set 的，以及用于 Map 的。下面这个示例演示了构建每种集合的只读版本的正确方式：

```java
// collectiontopics/ReadOnly.java
// 使用 Collections.unmodifiable 方法
import java.util.*;
import onjava.*;

public class ReadOnly {
  static Collection<String> data =
    new ArrayList<>(Countries.names(6));
  public static void main(String[] args) {
    Collection<String> c =
      Collections.unmodifiableCollection(
        new ArrayList<>(data));
    System.out.println(c); // 读是可以的
    //- c.add("one"); // 不能修改它

    List<String> a = Collections.unmodifiableList(
```

```
      new ArrayList<>(data));
    ListIterator<String> lit = a.listIterator();
    System.out.println(lit.next()); // 读是可以的
    //- lit.add("one"); // 不能修改它

    Set<String> s = Collections.unmodifiableSet(
      new HashSet<>(data));
    System.out.println(s); // 读是可以的
    //- s.add("one"); // 不能修改它

    // 对于 SortedSet :
    Set<String> ss =
      Collections.unmodifiableSortedSet(
        new TreeSet<>(data));

    Map<String,String> m =
      Collections.unmodifiableMap(
        new HashMap<>(Countries.capitals(6)));
    System.out.println(m); // 读是可以的
    //- m.put("Ralph", "Howdy!");

    // 对于 SortedMap :
    Map<String,String> sm =
      Collections.unmodifiableSortedMap(
        new TreeMap<>(Countries.capitals(6)));
  }
}
/* 输出:
[ALGERIA, ANGOLA, BENIN, BOTSWANA, BURKINA FASO,
BURUNDI]
ALGERIA
[BENIN, BOTSWANA, ANGOLA, BURKINA FASO, ALGERIA,
BURUNDI]
{BENIN=Porto-Novo, BOTSWANA=Gaberone, ANGOLA=Luanda,
BURKINA FASO=Ouagadougou, ALGERIA=Algiers,
BURUNDI=Bujumbura}
*/
```

调用某个特定类型的"不可修改"的方法不会触发编译时检查，但是一旦发生转换，只要调用了修改集合内容的方法，就会产生 UnsupportedOperationException。

不管是哪一种情况，在将集合变为只读之前，都必须用有意义的数据填充这个集合。一旦它被加载，最好的方法就是用"不可修改"调用所产生的引用来替换现有的引用。一旦使其成为只读的，我们就不用担心有不小心修改集合内容的风险了。此外，这个工具也允许我们把一个可修改的集合当成某个类中的 private 成员，然后从方法调用返回一个指向该集合的只读引用。因此，我们可以在这个类内修改它，但是其他人只能读取。

3.13.3 同步 Collection 或 Map

synchronized 关键字是**多线程**主题的一个重要部分。多线程主题更为复杂，本书第 5 章会介绍。这里只是要指出，Collections 包含一种自动同步整个集合的方式，其语法与不可修改的方法类似：

```java
// collectiontopics/Synchronization.java
// 使用 Collections.synchronized 方法
import java.util.*;

public class Synchronization {
  public static void main(String[] args) {
    Collection<String> c =
      Collections.synchronizedCollection(
        new ArrayList<>());
    List<String> list = Collections
      .synchronizedList(new ArrayList<>());
    Set<String> s = Collections
      .synchronizedSet(new HashSet<>());
    Set<String> ss = Collections
      .synchronizedSortedSet(new TreeSet<>());
    Map<String,String> m = Collections
      .synchronizedMap(new HashMap<>());
    Map<String,String> sm = Collections
      .synchronizedSortedMap(new TreeMap<>());
  }
}
```

最好是直接通过恰当的"同步"（synchronized）方法来传递新创建的集合，如以上代码所示。这样一来，我们就不会意外地将非同步的版本暴露出去了。

快速失败

Java 集合类也有一种防止多个任务（task）同时修改集合内容的机制。这里使用了更通用的术语"任务"来表示可以独立运行的事物，而没有使用特定于实现的"线程"，因为 Java 即将带来的变化可能会从线程转向更好的抽象。

如果在我们遍历集合的过程中有其他任务介入，比如在这个集合中插入、移除或修改了一个对象，那么问题就出现了：或许我们在集合中已经遍历了这个元素，或许它还未被遍历到，或许集合在我们调用 size() 之后变小了……会出错的场景有很多。Java 集合类库使用了一种**快速失败**（fail-fast）机制，它会寻找任何并非由我们的任务引发的对该集合的修改。如果检测到其他人在修改这个集合，它会立即生成一个 ConcurrentModificationException。这就是所谓的"快速失败"：它不会尝试之后再使用一个更复杂的算法来检测问题。

下面是一个非常基础的对快速失败机制的演示。我们创建了一个迭代器，并在这个迭代器所指向的集合中添加一个元素，就像这样：

```java
// collectiontopics/FailFast.java
// 说明"快速失败"行为
import java.util.*;

public class FailFast {
  public static void main(String[] args) {
    Collection<String> c = new ArrayList<>();
    Iterator<String> it = c.iterator();
    c.add("An object");
    try {
      String s = it.next();
    } catch(ConcurrentModificationException e) {
      System.out.println(e);
    }
  }
}
/* 输出：
java.util.ConcurrentModificationException
*/
```

之所以出现这个异常，是因为在获取了迭代器之后，又尝试向其中放入一个元素。尽管这个程序中没有并发，但是它演示了程序的两个部分可能会如何修改同一集合。这会导致一种不确定的状态，所以这个异常就是通知我们修改自己的代码：在这种情况下，要先把所有的元素都添加到集合中，然后再获取迭代器。

ConcurrentHashMap、CopyOnWriteArrayList 和 CopyOnWriteArraySet 都使用了可以避免 ConcurrentModificationException 的技术。

3.14 持有引用

java.lang.ref 库包含一组类，给垃圾收集提供了更大的灵活性。当存在可能耗尽内存的大对象时，这些类特别有用。有 3 个继承自抽象类 Reference 的类：SoftReference、WeakReference 和 PhantomReference。[1] 如果讨论中的对象只能通过这些 Reference 对象中的某一个来访问，那么它们可以为垃圾收集器提供不同层次的间接控制。

如果某个对象是可达的，就意味着可以在程序中的某个地方找到它。这可能意味着：在栈上有一个普通的引用，直接指向这个对象；我们有一个指向某个对象的引用，而被指向的对象中又有一个引用，指向我们讨论的对象；也许存在多个中间链接。如果某个对象

---

[1] Java 中的引用可以分为 4 种，分别是强引用（strong reference，即下文中提到的普通引用）、软引用（soft reference）、弱引用（weak reference）和虚引用（phantom reference）。这 4 种引用的强度依次减弱。——译者注

是可达的，垃圾收集器就不能释放它，因为它还在被我们的程序所使用。如果某个对象不可达，我们的程序没有办法使用它，那么对这个对象执行垃圾收集就是安全的。

因为我们可以使用 Reference 对象继续持有一个指向那个对象的引用，所以那个对象就是可达的，不过我们也允许垃圾收集器释放该对象。因此，我们有办法使用这个对象，但是如果内存即将耗尽，也可以释放它。

我们是这么做的：使用一个 Reference 对象作为我们和普通引用之间的中介（一个代理）。另外，不能存在指向该对象的其他普通引用（就是没有包在 Reference 对象中的那种）。如果垃圾收集器发现某个对象可以通过一个普通引用访问到，它就不会释放这个对象。

SoftReference、WeakReference 和 PhantomReference 的顺序对应不同级别的可达性，后面的比前面的"更弱"。软引用用于实现对内存敏感的缓存。弱引用用于实现规范映射（canonicalizing mapping）——为节省存储空间，对象的实例可以同时在程序中的多个位置使用——这不会妨碍它们的键（或值）被回收。与 Java 终结机制相比，虚引用用于以更灵活的方式安排事后（post-mortem）清理动作。而且请注意，从 Java 9 开始，Object.finalize() 方法已被废弃。事实证明，从 Java 诞生之初，它就是一个糟糕的、容易引起误解的想法。

在使用 SoftReference 和 WeakReference 时，我们可以选择是否将其放到一个 ReferenceQueue 中（该队列用于事后清理动作），但是 PhantomReference 只能放到 ReferenceQueue 中。下面是一个简单的演示：

```java
// collectiontopics/References.java
// Reference 对象的演示
import java.lang.ref.*;
import java.util.*;

class VeryBig {
  private static final int SIZE = 10000;
  private long[] la = new long[SIZE];
  private String ident;
  VeryBig(String id) { ident = id; }
  @Override public String toString() { return ident; }
  @SuppressWarnings("deprecation")
  @Override protected void finalize() {
    System.out.println("Finalizing " + ident);
  }
}

public class References {
```

```java
    private static ReferenceQueue<VeryBig> rq =
      new ReferenceQueue<>();
    public static void checkQueue() {
      Reference<? extends VeryBig> inq = rq.poll();
      if(inq != null)
        System.out.println("In queue: " + inq.get());
    }
    public static void main(String[] args) {
      int size = 10;
      // 或者通过命令行选择大小:
      if(args.length > 0)
        size = Integer.valueOf(args[0]);
      LinkedList<SoftReference<VeryBig>> sa =
        new LinkedList<>();
      for(int i = 0; i < size; i++) {
        sa.add(new SoftReference<>(
          new VeryBig("Soft " + i), rq));
        System.out.println(
          "Just created: " + sa.getLast());
        checkQueue();
      }
      LinkedList<WeakReference<VeryBig>> wa =
        new LinkedList<>();
      for(int i = 0; i < size; i++) {
        wa.add(new WeakReference<>(
          new VeryBig("Weak " + i), rq));
        System.out.println(
          "Just created: " + wa.getLast());
        checkQueue();
      }
      SoftReference<VeryBig> s =
        new SoftReference<>(new VeryBig("Soft"));
      WeakReference<VeryBig> w =
        new WeakReference<>(new VeryBig("Weak"));
      System.gc();
      LinkedList<PhantomReference<VeryBig>> pa =
        new LinkedList<>();
      for(int i = 0; i < size; i++) {
        pa.add(new PhantomReference<>(
          new VeryBig("Phantom " + i), rq));
        System.out.println(
          "Just created: " + pa.getLast());
        checkQueue();
      }
    }
}
/* 输出（前10行和最后10行）：
Just created: java.lang.ref.SoftReference@19e0bfd
Just created: java.lang.ref.SoftReference@139a55
Just created: java.lang.ref.SoftReference@1db9742
Just created: java.lang.ref.SoftReference@106d69c
```

```
Just created: java.lang.ref.SoftReference@52e922
Just created: java.lang.ref.SoftReference@25154f
Just created: java.lang.ref.SoftReference@10dea4e
Just created: java.lang.ref.SoftReference@647e05
Just created: java.lang.ref.SoftReference@1909752
Just created: java.lang.ref.SoftReference@1f96302
..._____..._____..._____..._____...
Just created: java.lang.ref.PhantomReference@16f6e28
In queue: null
Just created: java.lang.ref.PhantomReference@15fbaa4
In queue: null
Just created: java.lang.ref.PhantomReference@1ee12a7
In queue: null
Just created: java.lang.ref.PhantomReference@10bedb4
In queue: null
Just created: java.lang.ref.PhantomReference@103dbd3
In queue: null
*/
```

运行这个程序，并将输出重定向到一个文本文件中，以便分页查看。我们会看到这些对象被垃圾收集器回收了，但是仍然可以通过 Reference 对象访问它们（可以使用 get() 得到实际的对象引用）。我们还会看到，ReferenceQueue 总是给出一个包含 null 对象的 Reference。

WeakHashMap

Java 集合类库中的 WeakHashMap 持有的是弱引用。这个类使得创建规范映射更容易了。在这样的映射中，对于某个特定的值，我们只创建一个实例，以便节省存储空间。当程序需要这个值时，它就在该映射中查找现有的对象并使用它（而不是从头创建）。该映射可以在其初始化过程中构造这些值，但更有可能按需创建。

这是一种节省存储空间的技术，因此 WeakHashMap 允许垃圾收集器自动清理不用的值，这一点非常方便。放在 WeakHashMap 中的键和值不需要任何特殊处理，它们会自动被该映射包在 WeakReference 中。当这个键不再使用时，它就可以被垃圾收集器清理了。下面是一个简单的演示：

```
// collectiontopics/CanonicalMapping.java
// WeakHashMap 的演示
import java.util.*;
import java.util.stream.*;

public class CanonicalMapping {
  static void showKeys(Map<String, String> m) {
    // 显示排序后的键
```

```
      List<String> keys = new ArrayList<>(m.keySet());
      Collections.sort(keys);
      System.out.println(keys);
    }
    public static void main(String[] args) {
      int size = 100;
      String[] savedKeys = new String[size];
      WeakHashMap<String,String> map =
        new WeakHashMap<>();
      for(int i = 0; i < size; i++) {
        String key = String.format("%03d", i);
        String value = Integer.toString(i);
        if(i % 3 == 0)
          savedKeys[i] = key; // 当作"真正"的引用来保存
        map.put(key, value);
      }
      showKeys(map);
      System.gc();
      showKeys(map);
    }
}
```

通过将 String 用作这个映射的键，我们就自动获得了 String 的 hashCode() 和 equals()，所以它作为哈希数据结构中的键是没问题的（参见附录 C）。

键每隔两个就有一个被放在 savedKeys 数组中，这就告诉垃圾收集器，这个键还处于活跃使用中。所有的键值对都放在名为 map 的 WeakHashMap 中。

在通过调用 System.gc() 强制让垃圾收集器运行之后，你将看到，只有那些键被保存到 savedKeys 中的元素才会继续存在。

3.15　Java 1.0/1.1 的集合类

不幸的是，很多代码是用 Java 1.0/1.1 的集合类编写的，有时连新代码也是这样。在编写新代码时，千万不要使用旧的集合类。旧的集合类很有限，所以没有那么多可说的。它们已经过时了，所以我会尽量避免过分强调它们的丑陋设计。

3.15.1　Vector 和 Enumeration

在 Java 1.0/1.1 中，唯一可以自动扩展的序列就是 Vector，所以它被大量使用。它的缺点过多，无法在这里详述（参见 *Thinking in Java*）。基本上，我们可以把它看成方法名又怪又长的 ArrayList。在修订后的 Java 集合类库中，Vector 被调整了，所以可以作为 Collection 和 List 使用。结果表明，这么做并不合适，因为它会让人误以为 Vector 变得更好了。然而实际上，它之所以被包含进来，只是为了支持旧的 Java 代码。

Java 1.0/1.1 版本的迭代器选择了一个新名字：enumeration，而不是使用大家耳熟能详的术语 iterator。Enumeration 接口比 Iterator 小，只有两个方法，而且使用了更长的方法名：boolean hasMoreElements() 在枚举包含更多元素时产生 true，而 Object nextElement() 在有更多元素时返回这个枚举的下一个元素（否则抛出异常）。

Enumeration 只是一个接口，而非实现。新的库有时仍在使用旧的 Enumeration，这很遗憾，但通常是无害的。如果可能，一定要在自己的代码中使用 Iterator，但是也要做好准备应对某些库交给你的 Enumeration。

此外，通过 Collections.enumeration() 方法，可以为任何 Collection 生成一个 Enumeration，如以下示例所示：

```java
// collectiontopics/Enumerations.java
// Java 1.0/1.1 中的 Vector 和 Enumeration
import java.util.*;
import onjava.*;

public class Enumerations {
  public static void main(String[] args) {
    Vector<String> v =
      new Vector<>(Countries.names(10));
    Enumeration<String> e = v.elements();
    while(e.hasMoreElements())
      System.out.print(e.nextElement() + ", ");
    // 为任何 Collection 生成一个 Enumeration：
    e = Collections.enumeration(new ArrayList<>());
  }
}
/* 输出：
ALGERIA, ANGOLA, BENIN, BOTSWANA, BURKINA FASO,
BURUNDI, CAMEROON, CAPE VERDE, CENTRAL AFRICAN
REPUBLIC, CHAD,
*/
```

要生成 Enumeration，可以调用 elements()，然后使用它来执行前向迭代。

最后一行创建了一个 ArrayList，并使用 enumeration() 方法将其从 ArrayList 的 Iterator 适配为 Enumeration。因此，如果我们有想用 Enumeration 的旧代码，仍然可以使用新的集合类。

3.15.2 Hashtable

正如我们在本章的性能对比中看到的，基本的 Hashtable 与 HashMap 非常相似，甚至包括方法名。在新的代码中，没有理由使用 Hashtable 而不用 HashMap。

3.15.3 Stack

栈很早就和 LinkedList 一起引入了。Java 1.0/1.1 中的 Stack 比较奇怪：它没有以组合方式使用 Vector，而是**继承**了 Vector。因此，它有 Vector 的所有特性和行为，外加 Stack 的一些行为。我们很难知道，这是设计者认真思考过、认为有价值的做法，还是没有经过深思熟虑的幼稚做法。无论如何，我们清楚的是，它在匆忙发布之前没有经过评估，所以这个糟糕的设计仍然如幽灵一般存在（不要使用它）。

下面是对 Stack 的简单演示，它将 enum 的 String 表示压到栈中。它还演示了我们可以非常轻松地把 LinkedList 当作栈来使用，也可以使用基础卷第 12 章创建的 Stack 类：

```java
// collectiontopics/Stacks.java
// Stack 类的演示
import java.util.*;

enum Month { JANUARY, FEBRUARY, MARCH, APRIL,
  MAY, JUNE, JULY, AUGUST, SEPTEMBER,
  OCTOBER, NOVEMBER }

public class Stacks {
  public static void main(String[] args) {
    Stack<String> stack = new Stack<>();
    for(Month m : Month.values())
      stack.push(m.toString());
    System.out.println("stack = " + stack);
    // 把栈当作一个 Vector：
    stack.addElement("The last line");
    System.out.println(
      "element 5 = " + stack.elementAt(5));
    System.out.println("popping elements:");
    while(!stack.empty())
      System.out.print(stack.pop() + " ");

    // 把 LinkedList 当作一个栈：
    LinkedList<String> lstack = new LinkedList<>();
    for(Month m : Month.values())
      lstack.addFirst(m.toString());
    System.out.println("lstack = " + lstack);
    while(!lstack.isEmpty())
      System.out.print(lstack.removeFirst() + " ");

    // 使用基础卷第 12 章创建的 Stack 类：
    onjava.Stack<String> stack2 =
      new onjava.Stack<>();
    for(Month m : Month.values())
      stack2.push(m.toString());
    System.out.println("stack2 = " + stack2);
    while(!stack2.isEmpty())
```

```
            System.out.print(stack2.pop() + " ");
  }
}
        /* 输出：
        stack = [JANUARY, FEBRUARY, MARCH, APRIL, MAY, JUNE,
        JULY, AUGUST, SEPTEMBER, OCTOBER, NOVEMBER]
        element 5 = JUNE
        popping elements:
        The last line NOVEMBER OCTOBER SEPTEMBER AUGUST JULY
        JUNE MAY APRIL MARCH FEBRUARY JANUARY lstack =
        [NOVEMBER, OCTOBER, SEPTEMBER, AUGUST, JULY, JUNE, MAY,
        APRIL, MARCH, FEBRUARY, JANUARY]
        NOVEMBER OCTOBER SEPTEMBER AUGUST JULY JUNE MAY APRIL
        MARCH FEBRUARY JANUARY stack2 = [NOVEMBER, OCTOBER,
        SEPTEMBER, AUGUST, JULY, JUNE, MAY, APRIL, MARCH,
        FEBRUARY, JANUARY]
        NOVEMBER OCTOBER SEPTEMBER AUGUST JULY JUNE MAY APRIL
        MARCH FEBRUARY JANUARY
        */
```

String 表示是从 Month 常量生成的，用 push() 方法插入 Stack 中，之后再用 pop() 方法从栈顶取出。为了说明问题，我们也在 Stack 对象上执行了 Vector 操作。之所以可以这么做，是因为存在继承，一个 Stack 就是一个 Vector（is-a 关系）。因此，所有可以在 Vector 上执行的操作都可以在 Stack 上执行，比如 elementAt()。

如前所述，当需要栈的行为时，请使用 LinkedList 或者从 LinkedList 类创建而来的 onjava.Stack 类。

3.15.4 BitSet

BitSet 可以高效地存储一组开关数据。这里的高效是从存储空间大小的角度来看的。如果希望高效访问，它比使用原生的数组稍慢一些。

此外，BitSet 的最小长度相当于一个 long 值：64 位。这意味着，如果要保存更小的东西（比如 8 位的数据），BitSet 会浪费空间。如果大小是一个问题，那么创建自己的类或者使用一个数组来保存自己的标志信息是更明智的选择。（只有当要创建很多包含开关信息列表的对象时才需要这么做，而且只应该根据性能剖析等指标来决定。如果只是因为自己觉得某个东西太大而做出这个决定，最终会造成不必要的复杂性，并浪费大量时间。）

普通的集合通常会随着我们添加更多的元素而扩展，BitSet 也是如此。下面的示例演示了 BitSet 是如何工作的：

```
// collectiontopics/Bits.java
// BitSet 的演示
import java.util.*;
```

```java
public class Bits {
  public static void printBitSet(BitSet b) {
    System.out.println("bits: " + b);
    StringBuilder bbits = new StringBuilder();
    for(int j = 0; j < b.size() ; j++)
      bbits.append(b.get(j) ? "1" : "0");
    System.out.println("bit pattern: " + bbits);
  }
  public static void main(String[] args) {
    Random rand = new Random(47);
    // 得到 nextInt() 的最低有效位:
    byte bt = (byte)rand.nextInt();
    BitSet bb = new BitSet();
    for(int i = 7; i >= 0; i--)
      if(((1 << i) &  bt) != 0)
        bb.set(i);
      else
        bb.clear(i);
    System.out.println("byte value: " + bt);
    printBitSet(bb);

    short st = (short)rand.nextInt();
    BitSet bs = new BitSet();
    for(int i = 15; i >= 0; i--)
      if(((1 << i) &  st) != 0)
        bs.set(i);
      else
        bs.clear(i);
    System.out.println("short value: " + st);
    printBitSet(bs);

    int it = rand.nextInt();
    BitSet bi = new BitSet();
    for(int i = 31; i >= 0; i--)
      if(((1 << i) &  it) != 0)
        bi.set(i);
      else
        bi.clear(i);
    System.out.println("int value: " + it);
    printBitSet(bi);

    // 测试 BitSet 大于 64 位的情况:
    BitSet b127 = new BitSet();
    b127.set(127);
    System.out.println("set bit 127: " + b127);
    BitSet b255 = new BitSet(65);
    b255.set(255);
    System.out.println("set bit 255: " + b255);
    BitSet b1023 = new BitSet(512);
    b1023.set(1023);
    b1023.set(1024);
    System.out.println("set bit 1023: " + b1023);
  }
```

```
}
/* 输出:
byte value: -107
bits: {0, 2, 4, 7}
bit pattern: 10101001000000000000000000000000
0000000000000000000000
short value: 1302
bits: {1, 2, 4, 8, 10}
bit pattern: 01101000101000000000000000000000
0000000000000000000000
int value: -2014573909
bits: {0, 1, 3, 5, 7, 9, 11, 18, 19, 21, 22, 23, 24,
25, 26, 31}
bit pattern: 11010101010100000011011111100001000000000
0000000000000000000000
set bit 127: {127}
set bit 255: {255}
set bit 1023: {1023, 1024}
*/
```

随机数生成器创建了随机的 byte、short 和 int。每一个都被转换为 BitSet 中相应的位模式。因为一个 BitSet 是 64 位的，所以这样做很好，不会导致其大小增加。然后我们创建了更大的 BitSet。注意，BitSet 会在必要时扩展。

对于数量固定、可以命名的一组标志，EnumSet（参见本书第 1 章）通常是比 BitSet 更好的选择，因为 EnumSet 允许我们操作名字，而不是数字化的位位置（bit location），从而减少了错误。EnumSet 还可以防止意外添加新的标志位置，误添加可能会导致一些难以发现的严重错误。使用 BitSet 而不是 EnumSet 的理由很少：在运行之前不知道到底需要多少个标志，给标志指派名字不切实际，或者我们需要 BitSet 提供的特殊操作（参见 BitSet 和 EnumSet 的 JDK 文档）。

3.16 总结

可以说，集合是编程语言中最常用的工具。有些语言（如 Python）甚至内置了基本的集合组件（列表、映射和集）。

正如我们在基础卷第 12 章中看到的，借助集合，我们可以轻松地实现一些很有用的功能。某些时候，为了正确地使用集合，我们必须加深对它的了解。特别是，要编写自己的 hashCode() 方法，必须足够了解哈希操作（还必须知道何时需要编写）。而且，我们必须对各种集合实现有足够的了解，以便根据需要选择适合的那个。本章介绍了这些概念，还讨论了关于集合类库的其他一些有用的细节。现在，你已经为在日常编程中使用 Java

集合类做好了充足的准备。

  设计集合类库是非常困难的（对于大部分库设计问题而言也是如此）。在 C++ 中，集合类用很多不同的类构建了基础。这比之前的 C++ 集合类要好（之前几乎什么都没有），但是它并不能被照搬到 Java 中。这是一种极端；另一种极端是，我见过一个只由一个类（collection）组成的集合类库，它同时担当着线性序列和关联数组的角色。Java 集合类库尝试在能力和复杂性之间取得平衡，结果就是，有些地方看起来有点奇怪。与早期 Java 库中的一些设计决策不同的是，这些奇怪的地方并非偶然，而是仔细权衡了复杂性之后的结果。

04 注解

> 注解（又叫作**元数据**）使我们可以用正式的方式为代码添加信息，这样就可以在将来方便地使用这些数据。[①]

注解的出现，部分原因是为了迎合将元数据绑定到源代码文件（而非保存在额外的文档中）的趋势。同时 Java 也受到了来自其他语言（如 C#）特性的压力，这也是对此的一个回应。

注解是 Java 5 的一项重要语言更新。它提供了用 Java 无法表达、却是完整表述程序所需的信息。因此，注解使你可以用某种格式来保存和程序有关的额外信息，编译器会验证该格式的正确性。注解可以生成描述符文件，甚至还可以生成新的类定义，并帮助你减轻编写"样板"代码的负担。通过注解，可以将这些元数据保存在 Java 源代码中，并拥有以下优势：更整洁的代码；编译时的类型检查；为注解构建处理工具的注解 API。虽然 Java 中预定义了几种类型的元数据，

---

① Jeremy Meyer 专门来到 Crested Butte，并花了两周和我一起编写本章，他的帮助是无价的。

但通常来说，要添加什么样的注解类型，以及用它们来做什么，完全由你决定。

注解的语法十分简单，主要是在语言中添加 @ 符号。Java 5 引入了第一批定义在 java.lang 中的 3 个通用内建注解。

- @Override：用来声明该方法的定义会重写基类中的某个方法。如果不小心拼错了方法名，或者使用了不恰当的签名，该注解会使编译器报错。①
- @Deprecated：如果该元素被使用了，则编译器会发出警告。
- @SuppressWarnings：关闭不当的编译警告。

以下是 Java 7 和 Java 8 新增的注解。

- @SafeVarargs：Java 7 引入，用于在使用泛型作为可变参数的方法或构造器中关闭对调用者的警告。
- @FunctionalInterface：Java 8 引入，用于表明类型声明是函数式接口。

另外还有 5 个注解类型用于创建新注解，你将在本章学习它们。

每当你创建涉及重复工作的类或接口时，通常都可以用注解来自动化及简化该过程。例如 Enterprise JavaBeans（EJB）中的许多额外工作已被 EJB3 中的注解替代。

注解可以替代一些已有系统，如 XDoclet（一个创建注解风格文档的独立文档工具）。对比来看，注解是真正的语言组件，因此是结构化的，并可接受编译时类型检查。将所有信息都保存在真正的代码中而不是注释中，会使代码更整洁，且更便于维护。通过直接使用或扩展注解 API 和工具，或者使用外部的字节码处理库（如本章后面所述），可以对源代码以及字节码执行强大的检查和操作。

4.1 基本语法

在下面的示例中，testExecute() 方法添加了 @Test 注解。该注解本身并不会做任何事，只是编译器会确保在 CLASSPATH 中存在 @Test 注解的定义。本章稍后会创建一个通过反射来运行该方法的工具。

```
// annotations/Testable.java
package annotations;
```

---

① 毫无疑问，这受到了 C# 中一个相似功能的启发。C# 的该功能是一个关键字，而并非注解，并且是由编译器强制执行的。也就是说，当你在 C# 中重载某个方法时，必须使用 override 关键字，而在 Java 中，对应的 @Override 注解是可选的。

```
import onjava.atunit.*;

public class Testable {
  public void execute() {
    System.out.println("Executing..");
  }
  @Test
  void testExecute() { execute(); }
}
```

增加了注解的方法和其他方法并无区别，本例中的 @Test 注解可以和任何修饰符一起配合使用，如 public、static 和 void。从语法角度看，注解的使用方式和修饰符基本相同。

4.1.1 定义注解

以下代码是对前文中注解的定义。注解的定义看起来非常像接口的定义。实际上，注解和任何其他 Java 接口一样，也会编译成类文件。

```
// onjava/atunit/Test.java
// @Test 标签
package onjava.atunit;
import java.lang.annotation.*;

@Target(ElementType.METHOD)
@Retention(RetentionPolicy.RUNTIME)
public @interface Test {}
```

除了 @ 符号外，@Test 的定义非常像一个空接口。注解的定义也要求必须有**元注解** @Target 和 @Retention。@Target 定义了你可以在何处应用该注解（例如方法或字段）。@Retention 定义了该注解在源代码（SOURCE）、类文件（CLASS）或运行时（RUNTIME）中是否可用。

注解通常包含一些可以设定值的**元素**。程序或工具在处理注解时可以使用这些参数。元素看起来比较像接口方法，只不过你可以为其指定默认值。

没有任何元素的注解（如上面的 @Test）称为**标记注解**。

下面是一个用于跟踪某项目中用例的简单注解，程序员会给某个特定用例所需的所有方法或方法集都加上注解。项目经理可以通过计算已实现的用例数来了解项目的进度，维护项目的开发人员可以轻松地找到需要更新的用例，或者在系统内调试业务规则。

```
// annotations/UseCase.java
import java.lang.annotation.*;

@Target(ElementType.METHOD)
```

```
@Retention(RetentionPolicy.RUNTIME)
public @interface UseCase {
  int id();
  String description() default "no description";
}
```

注意，id 和 description 与方法声明很相似。因为 id 会受到编译器的类型检查，所以可以放心地用它将跟踪数据库链接到用例文档和源代码。description 元素有个默认值，如果在方法被注解时未指定该元素的值，则注解处理器会使用该默认值。

看一下以下这个类，其中的 3 个方法被注解为用例：

```
// annotations/PasswordUtils.java
import java.util.*;

public class PasswordUtils {
  @UseCase(id = 47, description =
  "Passwords must contain at least one numeric")
  public boolean validatePassword(String passwd) {
    return (passwd.matches("\\w*\\d\\w*"));
  }
  @UseCase(id = 48)
  public String encryptPassword(String passwd) {
   return new StringBuilder(passwd)
    .reverse().toString();
  }
  @UseCase(id = 49, description =
  "New passwords can't equal previously used ones")
  public boolean checkForNewPassword(
    List<String> prevPasswords, String passwd) {
    return !prevPasswords.contains(passwd);
  }
}
```

注解元素定义值的方式是，在 @UseCase 声明后的圆括号中，用"名 - 值"对形式来表示。此处 encryptPassword() 方法的注解并未给 description 元素传入值，因此当类经过注解处理器的处理后，@interface UseCase 中定义的默认值便会出现在此处。

想象一下，你可以先用这种方法来"勾勒"出你的系统，然后在构建时逐渐完善其功能。

4.1.2 元注解

Java 语言中目前只定义了 5 个标准注解（前面已介绍）和 5 个元注解（见表 4-1）。元注解是为了对注解进行注解。

表 4-1

注解	效果
@Target	该注解可应用的地方。可能的 ElementType 参数包括： - CONSTRUCTOR——构造器声明 - FIELD——字段声明（包括枚举常量） - LOCAL_VARIABLE——本地变量声明 - METHOD——方法声明 - PACKAGE——包声明 - PARAMETRE——参数声明 - TYPE——类、接口（包括注解类型）或枚举的声明
@Retention	注解信息可以保存多久。可能的 RetentionPolicy 参数包括： - SOURCE——注解会被编译器丢弃 - CLASS——注解在类文件中可被编译器使用，但会被虚拟机丢弃 - RUNTIME——注解在运行时仍被虚拟机保留，因此可以通过反射读取到注解信息
@Documented	在 Javadoc 中引入该注解
@Inherited	允许子类继承父注解
@Repeatable	可以多次应用于同一个声明（Java 8）

大多数时候，你可以定义自己的注解，然后自行编写处理器来处理它们。

4.2 编写注解处理器

如果没有工具来读取注解，那它实际并不会比注释带来更多帮助。使用注解的过程中，很重要的一点是创建并使用**注解处理器**。Java 为反射 API 提供了扩展，以帮助创建这些工具。Java 同时还提供了一个 javac 编译器钩子，用来在编译时使用注解。

以下示例是一个非常简单的注解处理器，它读取被注解的 PasswordUtils（密码工具）类，然后利用反射来查找 @UseCase 标签。通过给定的 id 值列表，该注解列出了它找到的所有用例，并报告所有丢失的用例。

```java
// annotations/UseCaseTracker.java
import java.util.*;
import java.util.stream.*;
import java.lang.reflect.*;

public class UseCaseTracker {
  public static void
  trackUseCases(List<Integer> useCases, Class<?> cl) {
    for(Method m : cl.getDeclaredMethods()) {
      UseCase uc = m.getAnnotation(UseCase.class);
      if(uc != null) {
        System.out.println("Found Use Case " +
          uc.id() + "\n  " + uc.description());
        useCases.remove(Integer.valueOf(uc.id()));
      }
    }
```

```
    useCases.forEach(i ->
      System.out.println("Missing use case " + i));
  }
  public static void main(String[] args) {
    List<Integer> useCases = IntStream.range(47, 51)
      .boxed().collect(Collectors.toList());
    trackUseCases(useCases, PasswordUtils.class);
  }
}
```

```
/* 输出:
Found Use Case 49
  New passwords can't equal previously used ones
Found Use Case 48
  no description
Found Use Case 47
  Passwords must contain at least one numeric
Missing use case 50
*/
```

此处同时使用了反射方法 getDeclaredMethods() 和从 AnnotatedElement 接口（诸如 Class、Method 以及 Field 等这样的类都会实现该接口）中继承实现的 getAnnotation() 方法，该方法返回指定类型的注解对象，在本例中即 UseCase。如果在此注解方法上没有该指定类型的注解，将会返回 null。元素的值通过调用 id() 和 description() 方法提取出来。注意，encryptPassword() 方法的注解中并未指定 description 描述，因此当在该注解上调用 description() 方法时，上述处理器会调取默认值 no description。

4.2.1 注解元素

在之前的 UseCase.java 中定义了 @UseCase 标签，其中包含 int 元素 id 和 String 元素 description。以下列出了注解所允许的所有元素类型：

- 所有的基本类型（int、float、boolean 等）
- String（字符串）
- Class（类）
- enum（枚举）
- Annotation（注解）
- 以上任何类型的数组

如果尝试使用任何其他类型，编译器都会报错。注意，任何包装类都是不允许使用的，但由于有自动装箱机制，因此这实际上并不会造成限制。注解也可以作为元素的类型，正如你稍后会看到的，内嵌注解是非常有用的。

4.2.2 默认值的限制

编译器对元素的默认值要求非常苛刻。所有元素都需要有确定的值，这意味着元素要么有默认值，要么由使用该注解的类来设定值。

还有另一个限制：不论是在源代码中声明时，还是在注解中定义默认值时，任何非基本类型元素都不能赋值为 null。这导致很难让处理器去判断某个元素存在与否，因为所有元素在所有注解声明中都是有效存在的。但可以通过检查该元素是否为特殊值（如空字符串或负值）来绕过这个限制：

```
// annotations/SimulatingNull.java
import java.lang.annotation.*;

@Target(ElementType.METHOD)
@Retention(RetentionPolicy.RUNTIME)
public @interface SimulatingNull {
  int id() default -1;
  String description() default "";
}
```

这是定义注解时的一个经典技巧。

4.2.3 生成外部文件

有些框架要求一些额外信息来配合源代码共同工作，在使用这种框架时，注解特别有用。诸如 Enterprise JavaBeans（即 EJB）这样的技术（在 EJB3 出现之前）需要大量的接口和部署描述文件作为"样板"代码，它们以相同的方式为每个 bean 进行定义。Web 服务、自定义标签库，以及 Toplink、Hibernate 等对象/关系映射工具（ORM）通常也需要代码之外的 XML 描述文件。每定义一个 Java 类，程序员都必须经过一个乏味的配置信息的过程，比如配置类名、包名等——这些都是类中本来就有的信息。无论你什么时候使用外部描述符文件，最终都会得到关于一个类的两个独立的信息源，这常常导致代码的信息同步问题。同时这也要求该项目的程序员除了写 Java 程序外，还必须知道如何编写这些描述符文件。

假如你想实现一套基本的 ORM 功能来自动化数据库表的创建，你便可以通过 XML 描述符文件来指定类名、类中的每一个成员，以及数据库映射信息。而如果使用注解，你可以将所有的信息都维护在单个源代码文件中。要实现此功能，你需要注解来定义数据库表名、字段信息，以及要映射到属性的 SQL 类型。

以下示例是一个注解，它会让注解处理器创建一个数据库表：

```java
// annotations/database/DBTable.java
package annotations.database;
import java.lang.annotation.*;

@Target(ElementType.TYPE) // 只适用于类
@Retention(RetentionPolicy.RUNTIME)
public @interface DBTable {
  String name() default "";
}
```

在 @Target 注解中指定的每个 ElementType（元素类型）都是一条约束，它告诉编译器注解只能被应用于该特定类型。你可以指定一个单值的 enum 元素类型，也可以指定一个用逗号分隔的任意值组成的列表。如果想将注解应用于所有的 ElementType，则可以将 @Target 注解全部去掉，虽然这么做不太常见。

注意，@DBTable 中有个 name() 元素，该注解可以通过它为处理器要创建的数据库表指定表名。

以下示例是表字段的注解：

```java
// annotations/database/Constraints.java
package annotations.database;
import java.lang.annotation.*;

@Target(ElementType.FIELD)
@Retention(RetentionPolicy.RUNTIME)
public @interface Constraints {
  boolean primaryKey() default false;
  boolean allowNull() default true;
  boolean unique() default false;
}
```

```java
// annotations/database/SQLString.java
package annotations.database;
import java.lang.annotation.*;

@Target(ElementType.FIELD)
@Retention(RetentionPolicy.RUNTIME)
public @interface SQLString {
  int value() default 0;
  String name() default "";
  Constraints constraints() default @Constraints;
}
```

```java
// annotations/database/SQLInteger.java
package annotations.database;
import java.lang.annotation.*;

@Target(ElementType.FIELD)
```

```java
@Retention(RetentionPolicy.RUNTIME)
public @interface SQLInteger {
  String name() default "";
  Constraints constraints() default @Constraints;
}
```

@Constraints 注解使得处理器可以提取出数据库表的元数据，这相当于数据库提供的一个小的约束子集，不过它可以帮助你形成一个整体的概念。通过为 primaryKey()、allowNull() 和 unique() 元素设置合理的默认值，可以帮使用者减少很多编码工作。

另外两个 @interface 用于定义 SQL 的类型。同样，为了使该框架更好用，可以为每个额外的 SQL 类型都定义一个注解。在本例中，两个注解就足够了。

这些类型都有一个 name() 元素和一个 constraints() 元素，后者利用嵌套注解的特性嵌入字段类型的数据库约束信息。注意 constraints() 元素的默认值是 @Constraints。该注解类型后面的圆括号中没有指定元素值，因此 constraints() 的默认值实际上是一个自身带有一套默认值的 @Constraints 注解。如果想将内嵌的 @Constraints 注解的唯一性默认设置为 true，可以像下面这样定义它的元素：

```java
// annotations/database/Uniqueness.java
// 嵌套注解示例：
package annotations.database;

public @interface Uniqueness {
  Constraints constraints()
  default @Constraints(unique = true);
}
```

以下示例是一个使用了该注解的简单类：

```java
// annotations/database/Member.java
package annotations.database;

@DBTable(name = "MEMBER")
public class Member {
  @SQLString(30) String firstName;
  @SQLString(50) String lastName;
  @SQLInteger Integer age;
  @SQLString(value = 30,
    constraints = @Constraints(primaryKey = true))
  String reference;
  static int memberCount;
  public String getReference() { return reference; }
  public String getFirstName() { return firstName; }
  public String getLastName() { return lastName; }
  @Override public String toString() {
    return reference;
```

```
    }
    public Integer getAge() { return age; }
}
```

类注解 @DBTable 被赋值为 MEMBER，以用作表名。属性 firstName 和 lastName 都被注解为 @SQLString，并且分别被赋值为 30 和 50。这些注解很有意思，原因有二：首先，它们都用到了内嵌注解 @Constraints 中的默认值；其次，它们使用了快捷格式——如果将注解中的元素名定义为 value，那么只要它是唯一指定的元素类型，就无须使用"名-值"对的语法，只需要在圆括号内直接指定该值即可。这种方式适用于任何合法的元素类型，该方法限制你必须将元素命名为"value"，不过在如前所述的情况下，这确实促成了有意义且易读的注解规范。

```
@SQLString(30)
```

处理器会用该值来设定待创建的 SQL 字段的长度。

默认值的语法虽然简洁，但很快就会变得复杂起来。来看看字段 reference 上的注解，其中有一个 @SQLString 注解，但它同时又必须是数据库的主键，因此在内嵌注解 @Constraint 中必须设置元素类型 primaryKey。麻烦就在这里，现在你不得不在内嵌注解中使用相当冗长的"名-值"对格式，重新指定元素名和 @interface 的名称。但是因为被特殊命名的元素 value 不再是唯一被指定的元素值，所以你无法继续使用快捷格式。正如你所看到的，最终结果并不优雅。

替换方案

对于这个问题，还有其他方法来创建注解。举例来说，可以写一个叫 @TableColumn 的注解类，其中包含一个 enum 元素，来定义诸如 STRING、INTEGER、FLOAT 这样的值。这样就不再需要为每个 SQL 类型都写一个 @interface 了，但也使你无法再用 size（长度）或 precision（精度）等额外的元素来进一步修饰类型，而这些可能会更有用。

你也可以使用 String 元素来描述实际的 SQL 类型（比如 VARCHAR(30) 或 INTEGER）。这使得你可以修饰类型，却将 Java 类型和 SQL 类型的映射关系在代码中绑定了，这并非好的设计。你肯定不想每当数据库有变化时就重新编译一遍代码。更优雅的方法是，告诉注解处理器你需要什么"口味"（flavor）的 SQL，然后处理器在执行时再来处理这些细节。

第三种可行的方法是同时使用两个注解类型来注解目标字段——@Constraints 和相应的 SQL 类型（比如 @SQLInteger）。这不太优雅，但是只要你需要，编译器就允许对目标增加任意个注解。在 Java 8 中使用多注解时，同一个注解可以重复使用。

4.2.4 注解不支持继承

我们无法对 @interface 使用 extends 关键字。这很可惜，一套优雅的方案应该像之前所建议的一样，定义一个注解 @TableColumn，该注解内部包含一个内嵌注解 @SQLType，由此就可以从 @SQLType 继承所有的 SQL 类型，如 @SQLInteger 和 @SQLString。这样可以减少编码工作，并使语法更简洁。目前看不到 Java 未来版本中要支持注解继承的迹象，在这种情况下，上面这个例子应该是你目前的最佳选择了。

4.2.5 实现处理器

下面这个示例演示了注解处理器如何读取类文件，检查其数据库注解，并生成 SQL 命令来创建数据库：

```java
// annotations/database/TableCreator.java
// 基于反射的注解处理器
// {java annotations.database.TableCreator
// annotations.database.Member}
package annotations.database;
import java.lang.annotation.*;
import java.lang.reflect.*;
import java.util.*;

public class TableCreator {
  public static void
  main(String[] args) throws Exception {
    if(args.length < 1) {
      System.out.println(
        "arguments: annotated classes");
      System.exit(0);
    }
    for(String className : args) {
      Class<?> cl = Class.forName(className);
      DBTable dbTable = cl.getAnnotation(DBTable.class);
      if(dbTable == null) {
        System.out.println(
          "No DBTable annotations in class " +
          className);
        continue;
      }
      String tableName = dbTable.name();
      // 如果 name 为空，则使用 Class name：
      if(tableName.length() < 1)
        tableName = cl.getName().toUpperCase();
      List<String> columnDefs = new ArrayList<>();
      for(Field field : cl.getDeclaredFields()) {
        String columnName = null;
        Annotation[] anns =
          field.getDeclaredAnnotations();
```

```java
      if(anns.length < 1)
        continue; // 不是数据库表字段
      if(anns[0] instanceof SQLInteger) {
        SQLInteger sInt = (SQLInteger) anns[0];
        // 如果 name 未指定，使用字段名
        if(sInt.name().length() < 1)
          columnName = field.getName().toUpperCase();
        else
          columnName = sInt.name();
        columnDefs.add(columnName + " INT" +
          getConstraints(sInt.constraints()));
      }
      if(anns[0] instanceof SQLString) {
        SQLString sString = (SQLString) anns[0];
        // 如果 name 未指定，使用字段名
        if(sString.name().length() < 1)
          columnName = field.getName().toUpperCase();
        else
          columnName = sString.name();
        columnDefs.add(columnName + " VARCHAR(" +
          sString.value() + ")" +
          getConstraints(sString.constraints()));
      }
      StringBuilder createCommand = new StringBuilder(
        "CREATE TABLE " + tableName + "(");
      for(String columnDef : columnDefs)
        createCommand.append(
          "\n    " + columnDef + ",");
      // 移除尾部的逗号
      String tableCreate = createCommand.substring(
        0, createCommand.length() - 1) + ");";
      System.out.println("Table Creation SQL for " +
        className + " is:\n" + tableCreate);
    }
  }
}
private static
String getConstraints(Constraints con) {
  String constraints = "";
  if(!con.allowNull())
    constraints += " NOT NULL";
  if(con.primaryKey())
    constraints += " PRIMARY KEY";
  if(con.unique())
    constraints += " UNIQUE";
  return constraints;
}
}
/* 输出：
Table Creation SQL for annotations.database.Member is:
CREATE TABLE MEMBER(
```

```
    FIRSTNAME VARCHAR(30));
Table Creation SQL for annotations.database.Member is:
CREATE TABLE MEMBER(
    FIRSTNAME VARCHAR(30),
    LASTNAME VARCHAR(50));
Table Creation SQL for annotations.database.Member is:
CREATE TABLE MEMBER(
    FIRSTNAME VARCHAR(30),
    LASTNAME VARCHAR(50),
    AGE INT);
Table Creation SQL for annotations.database.Member is:
CREATE TABLE MEMBER(
    FIRSTNAME VARCHAR(30),
    LASTNAME VARCHAR(50),
    AGE INT,
    REFERENCE VARCHAR(30) PRIMARY KEY);
*/
```

main() 方法会遍历命令行中的所有类名，forName() 方法负责加载所有类，而 getAnnotation(DBTable.class) 则检查类上是否有 @DBTable 注解。如果有，则会找到表名并保存下来。然后通过 getDeclaredAnnotations() 加载和校验类中所有的字段。该方法返回定义在某个方法上的所有注解。instanceof 操作符用来确定这些注解是否是 @SQLInteger 和 @SQLString 类型，不论是哪种，都会用表字段名来创建相关的 String 片段。注意，因为无法继承注解接口，所以使用 getDeclaredAnnotations() 是唯一一种能实现近似多态行为的方式。

内嵌的 @Constraint 注解会被传入 getConstraints() 方法，该方法用于创建包含 SQL 约束的字符串。

值得一提的是，用上述技巧来定义一套 ORM 是个略不成熟的方案。如果使用将表名作为参数的 @DBTable 类型，那么只要表名有变更，你就得重新编译 Java 代码，但你可能并不希望如此。有许多可用框架可以实现关系型数据库的 ORM，并且越来越多的框架开始使用注解。

4.3 用 javac 处理注解

通过 javac，你可以创建编译时注解处理器，并将注解应用于 Java 源文件，而不是编译后的类文件。不过这里有个重要的限制：无法通过注解处理器来修改源代码。唯一能影响结果的方法是创建新的文件。

如果注解处理器创建了一个新的源文件,则在新一轮处理中会检查该文件自身的注解。该工具会一轮接着一轮地持续处理,直到不再有新的源文件被创建,然后就编译所有的源文件。

你编写的每个注解都需要自己的处理器,但是 javac 可以轻松地将若干注解处理器进行组合。你可以指定多个要处理的类,并且还可以添加监听器来接收一轮处理完成的通知。

本节中的示例可带你入门,但是如果你需要深入了解,就要做好刻苦钻研的准备,多从 Google 和 Stack Overflow 上查找资料。

4.3.1 最简单的处理器

让我们从定义一个能想到的最简单的处理器(只是编译和测试一点东西)开始。下面是该注解的定义:

```java
// annotations/simplest/Simple.java
// 一个非常简单的注解
package annotations.simplest;
import java.lang.annotation.Retention;
import java.lang.annotation.RetentionPolicy;
import java.lang.annotation.Target;
import java.lang.annotation.ElementType;

@Retention(RetentionPolicy.SOURCE)
@Target({ElementType.TYPE, ElementType.METHOD,
        ElementType.CONSTRUCTOR,
        ElementType.ANNOTATION_TYPE,
        ElementType.PACKAGE, ElementType.FIELD,
        ElementType.LOCAL_VARIABLE})
public @interface Simple {
    String value() default "-default-";
}
```

@Retention 现在成了 SOURCE,这意味着该注解不会存活到编译后的代码中。对于编译期的注解操作,并不需要这么做——这只是为了表明此时 javac 是唯一有机会处理注解的代理。

@Target 声明列举了几乎所有可能的目标类型(除了 PACKAGE),这里同样也只是为了演示。

以下是用来测试的示例:

```java
// annotations/simplest/SimpleTest.java
// 测试"Simple"注解
// {java annotations.simplest.SimpleTest}
```

```java
package annotations.simplest;

@Simple
public class SimpleTest {
  @Simple
  int i;
  @Simple
  public SimpleTest() {}
  @Simple
  public void foo() {
    System.out.println("SimpleTest.foo()");
  }
  @Simple
  public void bar(String s, int i, float f) {
    System.out.println("SimpleTest.bar()");
  }
  @Simple
  public static void main(String[] args) {
    @Simple
    SimpleTest st = new SimpleTest();
    st.foo();
  }
}
/* 输出：
SimpleTest.foo()
*/
```

此处，我们用 `@Simple` 注解了所有 `@Target` 声明所允许的内容。

SimpleTest.java 只要求 Simple.java 能成功编译，虽然编译的过程中什么都没有发生。javac 允许使用 `@Simple` 注解（只要它还存在），但是并不会对它做任何事，直到我们创建了一个注解处理器，并将其绑定到编译器中。

以下示例是个非常简单的处理器，它所做的只是打印注解的信息：

```java
// annotations/simplest/SimpleProcessor.java
// 一个非常简单的注解处理器
package annotations.simplest;
import javax.annotation.processing.*;
import javax.lang.model.SourceVersion;
import javax.lang.model.element.*;
import java.util.*;

@SupportedAnnotationTypes(
  "annotations.simplest.Simple")
@SupportedSourceVersion(SourceVersion.RELEASE_8)
public class SimpleProcessor
extends AbstractProcessor {
  @Override public boolean process(
    Set<? extends TypeElement> annotations,
    RoundEnvironment env) {
    for(TypeElement t : annotations)
      System.out.println(t);
```

```
      for(Element el :
        env.getElementsAnnotatedWith(Simple.class))
        display(el);
      return false;
    }
    private void display(Element el) {
      System.out.println("==== " + el + " ====");
      System.out.println(el.getKind() +
        " : " + el.getModifiers() +
        " : " + el.getSimpleName() +
        " : " + el.asType());
      if(el.getKind().equals(ElementKind.CLASS)) {
        TypeElement te = (TypeElement)el;
        System.out.println(te.getQualifiedName());
        System.out.println(te.getSuperclass());
        System.out.println(te.getEnclosedElements());
      }
      if(el.getKind().equals(ElementKind.METHOD)) {
        ExecutableElement ex = (ExecutableElement)el;
        System.out.print(ex.getReturnType() + " ");
        System.out.print(ex.getSimpleName() + "(");
        System.out.println(ex.getParameters() + ")");
      }
    }
  }
```

已被废弃的旧 apt（即 Annotation Processing Tool，编译时注解处理器）版本的注解处理器需要额外的方法来确定哪些注解和 Java 版本可以被支持。而现在你可以简单地通过 @SupportedAnnotationTypes 和 @SupportedSourceVersion 注解来达到相同的目的（这个示例很好地诠释了注解如何做到简化代码）。

此处唯一需要实现的方法是 process()，其中包含了所有的逻辑。第一个参数会告诉你有哪些注解，第二个参数则包含余下的所有信息。此处我们做的只是把注解（只有一个）都打印了出来，要了解其他功能，请参考 TypeElement 文档。

通过 process() 方法的第二个参数，我们遍历所有被 @Simple 注解的元素，并且对每个元素都调用了 display() 方法。每个 Element 都可以携带自身的基本信息，例如 getModifiers() 能够告诉我们它是否是 public 和 static 的。

Element 只能执行编译器解析过的所有基本对象共有的操作，而类和方法等则需要提取出额外的信息。因此（如果你能找到相关的文档，可能发现这是显而易见的，但是我能找到的所有文档都没提到这一点，因此我只能在 Stack Overflow 上寻找线索）你需要先检查它是哪种 ElementKind，然后向下转型为更具体的元素类型——此处指针对 CLASS 的 TypeElement，以及针对 METHOD 的 ExecutableElement。然后，你就可以对这些 Element 类型调用额外的方法了。

动态向下转型（不会在编译期被检查）是一种"很不 Java"的处理方式，因此看起来非常不直观，这也可能是我从来不想这么做的原因。相反，我花了好几天来研究应该如何读取信息，至少用已废弃的 apt 方式来实现都会多少更直观一些。目前为止，我仍然没有找到任何证据表明上述形式是规范，但在我看来它就是规范了。

如果你只是正常地编译 SimpleTest.java，不会得到任何结果。想要得到注解的输出，就需要加上 -processor 标识和注解处理器类：

```
javac -processor annotations.simplest.SimpleProcessor SimpleTest.java
```

然后编译器会输出如下的结果：

```
annotations.simplest.Simple
==== annotations.simplest.SimpleTest ====
CLASS : [public] : SimpleTest : annotations.simplest.SimpleTest
annotations.simplest.SimpleTest
java.lang.Object
i,SimpleTest(),foo(),bar(java.lang.String,int,float),main(java.lang.String[])
==== i ====
FIELD : [] : i : int
==== SimpleTest() ====
CONSTRUCTOR : [public] : <init> : ()void
==== foo() ====
METHOD : [public] : foo : ()void
void foo()
==== bar(java.lang.String,int,float) ====
METHOD : [public] : bar : (java.lang.String,int,float)void
void bar(s,i,f)
==== main(java.lang.String[]) ====
METHOD : [public, static] : main : (java.lang.String[])void
void main(args)
```

这可以让你初步了解各种你日后可以探索的内容，包括参数名、类型、返回值等。

4.3.2 更复杂的处理器

当你创建了一个配合 javac 使用的注解处理器后，便无法使用 Java 的反射功能，因为此时操作的是源代码，而不是编译后的类。各种 mirror（镜子）[1] 可以解决该问题，方法是让你在未编译的源代码中查看方法、字段、类型。

以下示例是一个注解，它从一个类中提取 public 方法，以将它们转换为接口：

---

[1] Java 的设计者腼腆地提示，镜子就是指你发现反射的地方。

```
// annotations/ifx/ExtractInterface.java
// 基于javac的注解处理
package annotations.ifx;
import java.lang.annotation.*;

@Target(ElementType.TYPE)
@Retention(RetentionPolicy.SOURCE)
public @interface ExtractInterface {
  String interfaceName() default "-!!-";
}
```

其中 RetentionPolicy 是 SOURCE，这是因为从类中提取接口后，就没有必要在类文件中继续保留该注解了。下面的测试类提供了一些可组成接口的 public 方法：

```
// annotations/ifx/Multiplier.java
// 基于javac的注解处理
// {java annotations.ifx.Multiplier}
package annotations.ifx;

@ExtractInterface(interfaceName="IMultiplier")
public class Multiplier {
  public boolean flag = false;
  private int n = 0;
  public int multiply(int x, int y) {
    int total = 0;
    for(int i = 0; i < x; i++)
      total = add(total, y);
    return total;
  }
  public int fortySeven() { return 47; }
  private int add(int x, int y) {
    return x + y;
  }
  public double timesTen(double arg) {
    return arg * 10;
  }
  public static void main(String[] args) {
    Multiplier m = new Multiplier();
    System.out.println(
      "11 * 16 = " + m.multiply(11, 16));
  }
}
```

```
/* 输出:
11 * 16 = 176
*/
```

Multiplier 类（只能用于正整型）中有个 multiply() 方法，它多次调用私有的 add() 方法，以实现相乘操作。add() 方法不是 public 的，因此并不属于接口。其他方法则提供了一些语法的变体。该注解的 interfaceName 被赋值为 IMultiplier，以作为要创建的接口名。

下面的示例是一个编译期处理器，它会提取出感兴趣的方法，并创建新接口的源代码文件（该源文件之后会作为"编译阶段"的一部分，被自动编译）：

```java
// annotations/ifx/IfaceExtractorProcessor.java
// 基于javac的注解处理
package annotations.ifx;
import javax.annotation.processing.*;
import javax.lang.model.SourceVersion;
import javax.lang.model.element.*;
import javax.lang.model.util.*;
import java.util.*;
import java.util.stream.*;
import java.io.*;

@SupportedAnnotationTypes(
  "annotations.ifx.ExtractInterface")
@SupportedSourceVersion(SourceVersion.RELEASE_8)
public class IfaceExtractorProcessor
extends AbstractProcessor {
  private ArrayList<Element>
    interfaceMethods = new ArrayList<>();
  Elements elementUtils;
  private ProcessingEnvironment processingEnv;
  @Override public void init(
    ProcessingEnvironment processingEnv) {
    this.processingEnv = processingEnv;
    elementUtils = processingEnv.getElementUtils();
  }
  @Override public boolean process(
    Set<? extends TypeElement> annotations,
    RoundEnvironment env) {
    for(Element elem:env.getElementsAnnotatedWith(
        ExtractInterface.class)) {
      String interfaceName = elem.getAnnotation(
        ExtractInterface.class).interfaceName();
      for(Element enclosed :
          elem.getEnclosedElements()) {
        if(enclosed.getKind()
           .equals(ElementKind.METHOD) &&
           enclosed.getModifiers()
           .contains(Modifier.PUBLIC) &&
           !enclosed.getModifiers()
           .contains(Modifier.STATIC)) {
          interfaceMethods.add(enclosed);
        }
      }
      if(interfaceMethods.size() > 0)
        writeInterfaceFile(interfaceName);
    }
    return false;
  }
  private void
  writeInterfaceFile(String interfaceName) {
    try(
      Writer writer = processingEnv.getFiler()
```

```
          .createSourceFile(interfaceName)
          .openWriter()
    ) {
      String packageName = elementUtils
        .getPackageOf(interfaceMethods
                      .get(0)).toString();
      writer.write(
        "package " + packageName + ";\n");
      writer.write("public interface " +
        interfaceName + " {\n");
      for(Element elem : interfaceMethods) {
        ExecutableElement method =
          (ExecutableElement)elem;
        String signature = "  public ";
        signature += method.getReturnType() + " ";
        signature += method.getSimpleName();
        signature += createArgList(
          method.getParameters());
        System.out.println(signature);
        writer.write(signature + ";\n");
      }
      writer.write("}");
    } catch(Exception e) {
      throw new RuntimeException(e);
    }
  }
  private String createArgList(
    List<? extends VariableElement> parameters) {
    String args = parameters.stream()
      .map(p -> p.asType() + " " + p.getSimpleName())
      .collect(Collectors.joining(", "));
    return "(" + args + ")";
  }
}
```

Elements 对象 elementUtils 是个 static 工具的集合，我们通过它来在 writeInterfaceFile() 中找到包名。

getEnclosedElements() 方法生成被某个特定元素"围住"的所有元素。此处，该类围住了其所有的组件。通过 getKind()，我们可以找到所有的 public 和 static 的方法，并将它们添加到 interfaceMethods 列表中。然后 writeInterfaceFile() 通过该列表来生成新的接口定义。注意在 writeInterfaceFile() 中对 ExecutableElement 的向下转型使得我们可以提取所有的方法信息。createArgList() 则是一个生成参数列表的辅助方法。

Filer（由 getFiler() 生成）是一种创建新文件的 PrintWriter。之所以使用 Filer 对象而非某个普通的 PrintWriter，是因为 Filer 对象允许 javac 持续跟踪你创建的所有新文件，从而可以检查它们的注解，并在额外的"编译阶段"中编译它们。

如下代码是使用处理器来编译的命令行指令：

```
javac -processor annotations.ifx.IfaceExtractorProcessor Multiplier.java
```

它所生成的 IMultiplier.java 文件，看起来会像下面这样（通过上面的处理器中的 println() 语句，你也许能够猜到）：

```
package annotations.ifx;
public interface IMultiplier {
  public int multiply(int x, int y);
  public int fortySeven();
  public double timesTen(double arg);
}
```

该文件同样也会被 javac 所编译（作为"编译阶段"的一部分），因此你可以在同一个目录中看到 IMultiplier.class 文件。

4.4　基于注解的单元测试

单元测试是一种常见做法——通过为类中的每个方法都创建一个或多个测试，以定期检测类中各部分的行为是否正确。目前在 Java 中最流行的单元测试工具是 JUnit（参见基础卷第 16 章）。JUnit 4 引入了注解。[1] 在加入注解之前，JUnit 的一个主要问题是需要大量的额外工作来设置和运行 JUnit 单元测试。随着时间的推移，这种情况已有所好转，但是注解则使测试更接近"你可能拥有的最简单的单元测试系统"这个目标。

在 JUnit 4 之前的版本中，你需要创建一个单独的类来持有你的单元测试代码。而在 JUnit 4 中，你可以将单元测试集成到要测试的类中，从而将各种耗时和故障降到最低值。这种方法还有额外的好处——它测试 private 方法就和测试 public 方法一样简单。

下面这个测试框架的示例是基于注解的，因此叫作 @Unit。只用一个 @Test 注解来标识出需要测试的方法，是最基础也可能是你最常用的测试形式。也可以选择让测试方法不接收参数，仅返回一个 boolean 值来表示测试是成功还是失败。你可以给测试方法起任何你喜欢的名字。同样，@Unit 注解的测试方法可以支持任何你想要的访问权限，包括 private。

要使用 @Unit，你需要引入 onjava.atunit，用 @Unit 注解标签标记出适当的方法和字段（你会在后面的示例中学到），然后让构建系统在结果类上运行 @Unit。下面是个简单的

---

[1] 我原本想基于这里所述的设计思路实现一个"更好的 JUnit"。然而，看起来 JUnit 4 也同样引入了很多相同的设计思路，因此还是跟随 JUnit 的原生版本比较简单。

示例：

```java
// annotations/AtUnitExample1.java
// {java onjava.atunit.AtUnit
// build/classes/java/main/annotations/AtUnitExample1.class}
package annotations;
import onjava.atunit.*;
import onjava.*;

public class AtUnitExample1 {
  public String methodOne() {
    return "This is methodOne";
  }
  public int methodTwo() {
    System.out.println("This is methodTwo");
    return 2;
  }
  @Test
  boolean methodOneTest() {
    return methodOne().equals("This is methodOne");
  }
  @Test
  boolean m2() { return methodTwo() == 2; }
  @Test
  private boolean m3() { return true; }
  // 错误输出展示:
  @Test
  boolean failureTest() { return false; }
  @Test
  boolean anotherDisappointment() {
    return false;
  }
}
/* 输出:
annotations.AtUnitExample1
  . anotherDisappointment (failed)
  . methodOneTest
  . failureTest (failed)
  . m2 This is methodTwo

  . m3
(5 tests)

>>> 2 FAILURES <<<
  annotations.AtUnitExample1: anotherDisappointment
  annotations.AtUnitExample1: failureTest
*/
```

要用 @Unit 测试的类必须放在包中。

methodOneTest()、m2()、m3()、failureTest() 和 anotherDisappointment() 等方法前面的 @Test 注解告诉 @Unit 要将这些方法作为单元测试来运行，它同样也会确保这些方法不接收任何参数，并且返回值为 boolean 或 void。你编写单元测试时，只需要确定测试是成功还是失败，并分别返回 true 或者 false（对于返回 boolean 值的方法）。

如果你对 JUnit 很熟悉，同样会注意到 @Unit 具有更丰富的信息输出。你可以看到正在运行的测试，因此测试产生的输出会更有用，并且最后它会告诉你导致失败的类和测试用例。

如果将测试方法嵌入类中对你来说并不适用，那么你就无须这么做。要创建非嵌入式的测试，最简单的方法是使用继承：

```java
// annotations/AUExternalTest.java
// 创建非嵌入的测试
// {java onjava.atunit.AtUnit
// build/classes/java/main/annotations/AUExternalTest.class}
package annotations;
import onjava.atunit.*;
import onjava.*;

public class
AUExternalTest extends AtUnitExample1 {
  @Test
  boolean tMethodOne() {
    return methodOne().equals("This is methodOne");
  }
  @Test
  boolean tMethodTwo() {
    return methodTwo() == 2;
  }
}
/* 输出:
annotations.AUExternalTest
  . tMethodOne
  . tMethodTwo This is methodTwo

OK (2 tests)
*/
```

以上示例同样展示了灵活命名的好处。这里，直接用于测试某方法的 @Test 方法以该方法的方法名前面加 "t" 来命名（我并不是推荐这种命名方式，只是举了个可能的例子）。

你也可以用组合的方式来创建非嵌入的测试：

```java
// annotations/AUComposition.java
// 创建非嵌入的测试
// {java onjava.atunit.AtUnit
// build/classes/java/main/annotations/AUComposition.class}
package annotations;
import onjava.atunit.*;
import onjava.*;
```

```
public class AUComposition {
  AtUnitExample1 testObject = new AtUnitExample1();
  @Test
  boolean tMethodOne() {
    return testObject.methodOne()
      .equals("This is methodOne");
  }
  @Test
  boolean tMethodTwo() {
    return testObject.methodTwo() == 2;
  }
}
```

```
/* 输出：
annotations.AUComposition
  . tMethodOne
  . tMethodTwo This is
methodTwo

OK (2 tests)
*/
```

这里给每个测试都创建了一个 AUComposition 对象，因此也为每个测试都创建了一个新的 testObject 成员。

和 JUnit 不同，这里并没有专门的"assert"（断言）方法，而是使用了 @Test 方法的第二种形式，返回了 void（或者 boolean，如果你仍然希望在这里返回 true 或 false）。如果要验证成功，你可以使用 Java 的断言语句。Java 断言一般在 java 命令行指令中由 -ea 标签来启用，但是 @Unit 会自动启用断言。如果要表示失败，你甚至可以使用异常。@Unit 的设计目标之一是尽可能不增加语法复杂度，而 Java 断言和异常则是报告错误所必需的。如果测试方法引发了失败断言或者异常，则会被视为测试失败，但是 @Unit 并不会因此阻塞——它会持续运行，直到所有的测试都运行完毕。如下例所示：

```
// annotations/AtUnitExample2.java
// 断言和异常可以在 @Tests 中使用
// {java onjava.atunit.AtUnit
// build/classes/java/main/annotations/AtUnitExample2.class}
package annotations;
import java.io.*;
import onjava.atunit.*;
import onjava.*;

public class AtUnitExample2 {
  public String methodOne() {
    return "This is methodOne";
  }
  public int methodTwo() {
    System.out.println("This is methodTwo");
    return 2;
  }
  @Test
  void assertExample() {
    assert methodOne().equals("This is methodOne");
  }
  @Test
```

```
  void assertFailureExample() {
    assert 1 == 2: "What a surprise!";
  }
  @Test
  void exceptionExample() throws IOException {
    try(FileInputStream fis =
        new FileInputStream("nofile.txt")) {} // 抛出
  }
  @Test
  boolean assertAndReturn() {
    // 附带消息的断言：
    assert methodTwo() == 2: "methodTwo must equal 2";
    return methodOne().equals("This is methodOne");
  }
}
/* 输出：
annotations.AtUnitExample2
  . assertFailureExample java.lang.AssertionError: What
a surprise!
(failed)
  . assertExample
  . exceptionExample java.io.FileNotFoundException:
nofile.txt (The system cannot find the file specified)
(failed)
  . assertAndReturn This is methodTwo

(4 tests)

>>> 2 FAILURES <<<
  annotations.AtUnitExample2: assertFailureExample
  annotations.AtUnitExample2: exceptionExample
*/
```

下面是用断言实现的非嵌入式测试，它对 java.util.HashSet 做了一些简单的测试：

```
// annotations/HashSetTest.java
// {java onjava.atunit.AtUnit
// build/classes/java/main/annotations/HashSetTest.class}
package annotations;
import java.util.*;
import onjava.atunit.*;
import onjava.*;

public class HashSetTest {
  HashSet<String> testObject = new HashSet<>();
  @Test
  void initialization() {
    assert testObject.isEmpty();
  }
  @Test
  void tContains() {
    testObject.add("one");
```

```
/* 输出：
annotations.HashSetTest
  . tContains
  . initialization
  . tRemove
OK (3 tests)
*/
```

```
    assert testObject.contains("one");
  }
  @Test
  void tRemove() {
    testObject.add("one");
    testObject.remove("one");
    assert testObject.isEmpty();
  }
}
```

在没有其他约束的情况下，继承的方式看起来似乎更简单。

在每个单元测试中，@Unit 都通过无参数的构造方法，为每个要测试的类创建了一个对象。测试会在该对象上进行，然后该对象会被丢弃，以防止各种副作用渗透到其他单元测试中。这里依赖无参数的构造方法来创建对象。如果没有无参数的构造函数，或者需要更复杂的构造函数，你需要创建一个静态方法来构建对象，并添加 @TestObjectCreate 注解，就像下面这样：

```
// annotations/AtUnitExample3.java
// {java onjava.atunit.AtUnit
// build/classes/java/main/annotations/AtUnitExample3.class}
package annotations;
import onjava.atunit.*;
import onjava.*;

public class AtUnitExample3 {
  private int n;
  public AtUnitExample3(int n) { this.n = n; }
  public int getN() { return n; }
  public String methodOne() {
    return "This is methodOne";
  }
  public int methodTwo() {
    System.out.println("This is methodTwo");
    return 2;
  }
  @TestObjectCreate
  static AtUnitExample3 create() {
    return new AtUnitExample3(47);
  }
  @Test
  boolean initialization() { return n == 47; }
  @Test
  boolean methodOneTest() {
    return methodOne().equals("This is methodOne");
  }
  @Test
  boolean m2() { return methodTwo() == 2; }
}
/* 输出：
annotations.AtUnitExample3
  . initialization
  . methodOneTest
  . m2 This is methodTwo

OK (3 tests)
*/
```

@TestObjectCreate 方法必须是静态的，并且必须返回你测试的类型的对象。@Unit 程序会确保这些。

有时你需要额外的字段来支持单元测试。@TestProperty 注解可以标识仅用于单元测试的字段（这样在交付给客户前便可以随意移除这些字段）。下面是一个示例，它读取一个被 String.split() 方法切割后的字符串的值，该值会作为输入来生成测试对象：

```java
// annotations/AtUnitExample4.java
// {java onjava.atunit.AtUnit
// build/classes/java/main/annotations/AtUnitExample4.class}
// {VisuallyInspectOutput}
package annotations;
import java.util.*;
import onjava.atunit.*;
import onjava.*;

public class AtUnitExample4 {
  static String theory = "All brontosauruses " +
    "are thin at one end, much MUCH thicker in the " +
    "middle, and then thin again at the far end.";
  private String word;
  private Random rand = new Random(); // 基于时间因素的随机种子
  public AtUnitExample4(String word) {
    this.word = word;
  }
  public String getWord() { return word; }
  public String scrambleWord() {
    List<Character> chars = Arrays.asList(
      ConvertTo.boxed(word.toCharArray()));
    Collections.shuffle(chars, rand);
    StringBuilder result = new StringBuilder();
    for(char ch : chars)
      result.append(ch);
    return result.toString();
  }
  @TestProperty
  static List<String> input =
    Arrays.asList(theory.split(" "));
  @TestProperty
  static Iterator<String> words = input.iterator();
  @TestObjectCreate
  static AtUnitExample4 create() {
    if(words.hasNext())
      return new AtUnitExample4(words.next());
    else
      return null;
  }
  @Test
  boolean words() {
    System.out.println("'" + getWord() + "'");
```

```
    return getWord().equals("are");
  }
  @Test
  boolean scramble1() {
    // 用指定的种子得到可验证的结果：
    rand = new Random(47);
    System.out.println("'" + getWord() + "'");
    String scrambled = scrambleWord();
    System.out.println(scrambled);
    return scrambled.equals("lAl");
  }
  @Test
  boolean scramble2() {
    rand = new Random(74);
    System.out.println("'" + getWord() + "'");
    String scrambled = scrambleWord();
    System.out.println(scrambled);
    return scrambled.equals("tsaeborornussu");
  }
}
/* 输出：
annotations.AtUnitExample4
  . words 'All'
(failed)
  . scramble1 'brontosauruses'
ntsaueorosurbs
(failed)
  . scramble2 'are'
are
(failed)
(3 tests)

>>> 3 FAILURES <<<
  annotations.AtUnitExample4: words
  annotations.AtUnitExample4: scramble1
  annotations.AtUnitExample4: scramble2
*/
```

@TestProperty 同样可以用来标识在测试期间可用，但自身并不是测试的方法。

这个程序依赖于测试执行的顺序，通常来说这并不是一种好的实现方式。

如果测试对象的创建过程需要执行初始化，而且需要在稍后清理对象，你可以选择添加一个静态的 @TestObjectCleanup 方法，以在使用完测试对象后执行清理工作。在下一个示例中，@TestObjectCreate 通过打开一个文件来创建各个测试对象，因此必须在丢弃测试对象前关闭该文件。

```
// annotations/AtUnitExample5.java
// {java onjava.atunit.AtUnit}
// build/classes/java/main/annotations/AtUnitExample5.class}
```

```java
package annotations;
import java.io.*;
import onjava.atunit.*;
import onjava.*;

public class AtUnitExample5 {
  private String text;
  public AtUnitExample5(String text) {
    this.text = text;
  }
  @Override public String toString() { return text; }
  @TestProperty
  static PrintWriter output;
  @TestProperty
  static int counter;
  @TestObjectCreate
  static AtUnitExample5 create() {
    String id = Integer.toString(counter++);
    try {
      output = new PrintWriter("Test" + id + ".txt");
    } catch(IOException e) {
      throw new RuntimeException(e);
    }
    return new AtUnitExample5(id);
  }
  @TestObjectCleanup
  static void cleanup(AtUnitExample5 tobj) {
    System.out.println("Running cleanup");
    output.close();
  }
  @Test
  boolean test1() {
    output.print("test1");
    return true;
  }
  @Test
  boolean test2() {
    output.print("test2");
    return true;
  }
  @Test
  boolean test3() {
    output.print("test3");
    return true;
  }
}
/* 输出：
annotations.AtUnitExample5
  . test1
Running cleanup
  . test3
Running cleanup
  . test2
Running cleanup
OK (3 tests)
*/
```

从以上输出可以看出，在每项测试之后，清理方法都被自动执行了。

4.4.1 在 @Unit 中使用泛型

泛型会带来一个特别的问题，因为你无法"笼统地测试"，而只能对特定的类型参数或参数集合进行测试。解决这个问题的办法很简单：从泛型类的某个具体版本继承一个测试类。

下面的示例简单地实现了一个栈：

```java
// annotations/StackL.java
// 用 LinkedList 构建的栈
package annotations;
import java.util.*;

public class StackL<T> {
  private LinkedList<T> list = new LinkedList<>();
  public void push(T v) { list.addFirst(v); }
  public T top() { return list.getFirst(); }
  public T pop() { return list.removeFirst(); }
}
```

如果要测试字符串的版本，则从 StackL<String> 继承一个测试类：

```java
// annotations/StackLStringTst.java
// 将 @Unit 应用于泛型
// {java onjava.atunit.AtUnit
// build/classes/java/main/annotations/StackLStringTst.class}
package annotations;
import onjava.atunit.*;
import onjava.*;

public class
StackLStringTst extends StackL<String> {
  @Test
  void tPush() {
    push("one");
    assert top().equals("one");
    push("two");
    assert top().equals("two");
  }
  @Test
  void tPop() {
    push("one");
    push("two");
    assert pop().equals("two");
    assert pop().equals("one");
  }
  @Test
  void tTop() {
    push("A");
    push("B");
    assert top().equals("B");
    assert top().equals("B");
  }
}
/* 输出：
annotations.
StackLStringTst
  . tPop
  . tTop
  . tPush
OK (3 tests)
*/
```

继承的唯一潜在缺点是，你会失去访问被测试类中 private 方法的能力。如果你不希望这样，则可以将该方法设为 `protected`，或者添加一个非私有的 `@TestProperty` 方法来调用该私有方法（本章稍后介绍的 `AtUnitRemover` 工具会从产品代码中剥离 `@TestProperty` 方法）。

`@Unit` 会查找包含适当注解的类文件，然后执行 `@Test` 方法。对于 `@Unit` 测试系统，我的主要目标是使它极度透明，仅需要添加 `@Test` 方法即可上手使用，而不需要其他的特殊代码或知识（现代版本的 JUnit 遵从了这个思路）。要想不添加任何新的障碍就实现测试的编写是基本不可能的，因此 `@Unit` 会尽力使过程变得更轻松。只有这样，你才会更愿意编写测试。

4.4.2　实现 @Unit

首先，我们来定义所有的注解类型。这些都是很简单的标签，并没有任何字段。在本章开始已经介绍了 `@Test` 标签的定义，接下来是其余的注解：

```java
// onjava/atunit/TestObjectCreate.java
// @Unit @TestObjectCreate 标签
package onjava.atunit;
import java.lang.annotation.*;

@Target(ElementType.METHOD)
@Retention(RetentionPolicy.RUNTIME)
public @interface TestObjectCreate {}
```

```java
// onjava/atunit/TestObjectCleanup.java
// @Unit @TestObjectCleanup 标签
package onjava.atunit;
import java.lang.annotation.*;

@Target(ElementType.METHOD)
@Retention(RetentionPolicy.RUNTIME)
public @interface TestObjectCleanup {}
```

```java
// onjava/atunit/TestProperty.java
// @Unit @TestProperty 标签
package onjava.atunit;
import java.lang.annotation.*;

// 字段（Field）和方法（Method）都可以被标记为属性（property）：
@Target({ElementType.FIELD, ElementType.METHOD})
@Retention(RetentionPolicy.RUNTIME)
public @interface TestProperty {}
```

所有测试的生命周期都限定在 `RUNTIME` 内，这是因为 `@Unit` 系统必须在编译后的代码中检测到这些测试。

为了实现运行这些测试的系统，我们利用反射来提取注解。程序会利用这些信息来决定如何构建测试对象，并在其上运行测试。注解使得结果出乎意料地简单易懂：

```java
// onjava/atunit/AtUnit.java
// 一个基于注解的单元测试框架
// {java onjava.atunit.AtUnit}
package onjava.atunit;
import java.lang.reflect.*;
import java.io.*;
import java.util.*;
import java.nio.file.*;
import java.util.stream.*;
import onjava.*;

public class AtUnit implements ProcessFiles.Strategy {
  static Class<?> testClass;
  static List<String> failedTests= new ArrayList<>();
  static long testsRun = 0;
  static long failures = 0;
  public static void
  main(String[] args) throws Exception {
    ClassLoader.getSystemClassLoader()
      .setDefaultAssertionStatus(true); // 启用断言
    new ProcessFiles(new AtUnit(), "class").start(args);
    if(failures == 0)
      System.out.println("OK (" + testsRun + " tests)");
    else {
      System.out.println("(" + testsRun + " tests)");
      System.out.println(
        "\n>>> " + failures + " FAILURE" +
        (failures > 1 ? "S" : "") + " <<<");
      for(String failed : failedTests)
        System.out.println("  " + failed);
    }
  }
  @Override public void process(File cFile) {
    try {
      String cName = ClassNameFinder.thisClass(
        Files.readAllBytes(cFile.toPath()));
      if(!cName.startsWith("public:"))
        return;
      cName = cName.split(":")[1];
      if(!cName.contains("."))
        return; // 忽略未包装的类
      testClass = Class.forName(cName);
    } catch(IOException | ClassNotFoundException e) {
      throw new RuntimeException(e);
    }
    TestMethods testMethods = new TestMethods();
    Method creator = null;
    Method cleanup = null;
    for(Method m : testClass.getDeclaredMethods()) {
      testMethods.addIfTestMethod(m);
      if(creator == null)
        creator = checkForCreatorMethod(m);
      if(cleanup == null
```

```
        cleanup = checkForCleanupMethod(m);
    }
    if(testMethods.size() > 0) {
      if(creator == null)
        try {
          if(!Modifier.isPublic(testClass
              .getDeclaredConstructor()
              .getModifiers())) {
            System.out.println("Error: " + testClass +
              " zero-argument constructor must be public");
            System.exit(1);
          }
        } catch(NoSuchMethodException e) {
          // 同步的无参构造器，没有问题
        }
      System.out.println(testClass.getName());
    }
    for(Method m : testMethods) {
      System.out.print("  . " + m.getName() + " ");
      try {
        Object testObject = createTestObject(creator);
        boolean success = false;
        try {
          if(m.getReturnType().equals(boolean.class))
            success = (Boolean)m.invoke(testObject);
          else {
            m.invoke(testObject);
            success = true; // 如果没有断言失败
          }
        } catch(InvocationTargetException e) {
          // 实际的异常在 e 中：
          System.out.println(e.getCause());
        }
        System.out.println(success ? "" : "(failed)");
        testsRun++;
        if(!success) {
          failures++;
          failedTests.add(testClass.getName() +
            ": " + m.getName());
        }
        if(cleanup != null)
          cleanup.invoke(testObject, testObject);
      } catch(IllegalAccessException |
              IllegalArgumentException |
              InvocationTargetException e) {
        throw new RuntimeException(e);
      }
    }
  }
}
public static
class TestMethods extends ArrayList<Method> {
  void addIfTestMethod(Method m) {
    if(m.getAnnotation(Test.class) == null)
```

```
      return;
    if(!(m.getReturnType().equals(boolean.class) ||
        m.getReturnType().equals(void.class)))
      throw new RuntimeException("@Test method" +
        " must return boolean or void");
    m.setAccessible(true); // 如果是private的，等等
    add(m);
  }
}
private static
Method checkForCreatorMethod(Method m) {
  if(m.getAnnotation(TestObjectCreate.class) == null)
    return null;
  if(!m.getReturnType().equals(testClass))
    throw new RuntimeException("@TestObjectCreate " +
      "must return instance of Class to be tested");
  if((m.getModifiers() &
      java.lang.reflect.Modifier.STATIC) < 1)
    throw new RuntimeException("@TestObjectCreate " +
      "must be static.");
  m.setAccessible(true);
  return m;
}
private static
Method checkForCleanupMethod(Method m) {
  if(m.getAnnotation(TestObjectCleanup.class) == null)
    return null;
  if(!m.getReturnType().equals(void.class))
    throw new RuntimeException("@TestObjectCleanup " +
      "must return void");
  if((m.getModifiers() &
      java.lang.reflect.Modifier.STATIC) < 1)
    throw new RuntimeException("@TestObjectCleanup " +
      "must be static.");
  if(m.getParameterTypes().length == 0 ||
      m.getParameterTypes()[0] != testClass)
    throw new RuntimeException("@TestObjectCleanup " +
      "must take an argument of the tested type.");
  m.setAccessible(true);
  return m;
}
private static Object
createTestObject(Method creator) {
  if(creator != null) {
    try {
      return creator.invoke(testClass);
    } catch(IllegalAccessException |
            IllegalArgumentException |
            InvocationTargetException e) {
      throw new RuntimeException("Couldn't run " +
        "@TestObject (creator) method.");
    }
  } else { // 使用无参数的构造器:
```

```
    try {
      return testClass
        .getConstructor().newInstance();
    } catch(InstantiationException |
            NoSuchMethodException |
            InvocationTargetException |
            IllegalAccessException e) {
      throw new RuntimeException(
        "Couldn't create a test object. " +
        "Try using a @TestObject method.");
    }
  }
}
```

尽管这可能能算是"过早重构"（因为本书中只用过一次），AtUnit.java 使用了另一个叫作 ProcessFiles 的工具来单步遍历命令行中的每个参数，以及确定它是目录还是文件，并进行相应的处理。其中包含了一个可定制的 Strategy（策略）接口，因此可应用于多种方案实现。

```
// onjava/ProcessFiles.java
package onjava;
import java.io.*;
import java.nio.file.*;

public class ProcessFiles {
  public interface Strategy {
    void process(File file);
  }
  private Strategy strategy;
  private String ext;
  public ProcessFiles(Strategy strategy, String ext) {
    this.strategy = strategy;
    this.ext = ext;
  }
  public void start(String[] args) {
    try {
      if(args.length == 0)
        processDirectoryTree(new File("."));
      else
        for(String arg : args) {
          File fileArg = new File(arg);
          if(fileArg.isDirectory())
            processDirectoryTree(fileArg);
          else {
            // 用户可以去掉后缀名：
            if(!arg.endsWith("." + ext))
              arg += "." + ext;
            strategy.process(
              new File(arg).getCanonicalFile());
          }
        }
```

```
    } catch(IOException e) {
      throw new RuntimeException(e);
    }
  }
  public void
  processDirectoryTree(File root) throws IOException {
    PathMatcher matcher = FileSystems.getDefault()
      .getPathMatcher("glob:**/*.{" + ext + "}");
    Files.walk(root.toPath())
      .filter(matcher::matches)
      .forEach(p -> strategy.process(p.toFile()));
  }
}
```

AtUnit 类实现了 ProcessFiles.Strategy，其中包含 process() 方法。由此，AtUnit 的实例可以传递给 ProcessFiles 构造器。构造器的第二个参数告诉 ProcessFiles 去查找所有文件名后缀为 .class 的文件。

以下是简单的用法示例：

```
// annotations/DemoProcessFiles.java
import onjava.ProcessFiles;

public class DemoProcessFiles {
  public static void main(String[] args) {
    new ProcessFiles(file -> System.out.println(file),
      "java").start(args);
  }
}
/* 输出：
.\AtUnitExample1.java
.\AtUnitExample2.java
.\AtUnitExample3.java
.\AtUnitExample4.java
.\AtUnitExample5.java
.\AUComposition.java
.\AUExternalTest.java
.\database\Constraints.java
.\database\DBTable.java
.\database\Member.java
.\database\SQLInteger.java
.\database\SQLString.java
.\database\TableCreator.java
.\database\Uniqueness.java
.\DemoProcessFiles.java
```
（转右栏）
```
.\HashSetTest.java
.\ifx\ExtractInterface.java
.\ifx\IfaceExtractorProcessor.java
.\ifx\Multiplier.java
.\PasswordUtils.java
.\simplest\Simple.java
.\simplest\SimpleProcessor.java
.\simplest\SimpleTest.java
.\SimulatingNull.java
.\StackL.java
.\StackLStringTst.java
.\Testable.java
.\UseCase.java
.\UseCaseTracker.java
*/
```

在没有命令行参数的情况下，程序会遍历当前的目录树。你还可以提供多个参数，可以是类文件（不论文件名是否带有 .class 后缀）或目录。

回到我们关于 AtUnit.java 的讨论，@Unit 会自动找到可测试的类和方法，因此并不需要"套件"机制。

AtUnit.java 在寻找类文件时有个必须解决的问题：从类文件名无法确切地得知限定的类名（包括包名）。要获取这个信息，就必须分析类文件。这并非易事，但也并非做不到。[①] 当找到一个 .class 文件时，程序会打开该文件，读取它的二进制数据，并传给 ClassNameFinder.thisClass()。此处，我们将进入"字节码工程"的领域，因为我们实际上已经在分析类文件的内容了。

```java
// onjava/atunit/ClassNameFinder.java
// {java onjava.atunit.ClassNameFinder}
package onjava.atunit;
import java.io.*;
import java.nio.file.*;
import java.util.*;
import onjava.*;

public class ClassNameFinder {
  public static String thisClass(byte[] classBytes) {
    Map<Integer,Integer> offsetTable = new HashMap<>();
    Map<Integer,String> classNameTable =
      new HashMap<>();
    try {
      DataInputStream data = new DataInputStream(
        new ByteArrayInputStream(classBytes));
      int magic = data.readInt();    // 0xcafebabe
      int minorVersion = data.readShort();
      int majorVersion = data.readShort();
      int constantPoolCount = data.readShort();
      int[] constantPool = new int[constantPoolCount];
      for(int i = 1; i < constantPoolCount; i++) {
        int tag = data.read();
        // int tableSize;
        switch(tag) {
          case 1: // UTF
            int length = data.readShort();
            char[] bytes = new char[length];
            for(int k = 0; k < bytes.length; k++)
              bytes[k] = (char)data.read();
            String className = new String(bytes);
            classNameTable.put(i, className);
            break;
          case 5: // LONG
          case 6: // DOUBLE
            data.readLong(); // 丢弃 8 字节
```

[①] 我和 Jeremy Meyer 一起花了几乎一整天才搞清楚这件事。

```
              i++; // 必要的特殊处理，跳过此处
              break;
            case 7: // CLASS
              int offset = data.readShort();
              offsetTable.put(i, offset);
              break;
            case 8: // STRING
              data.readShort(); // 抛弃2字节
              break;
            case 3:  // INTEGER
            case 4:  // FLOAT
            case 9:  // FIELD_REF
            case 10: // METHOD_REF
            case 11: // INTERFACE_METHOD_REF
            case 12: // NAME_AND_TYPE
            case 18: // Invoke Dynamic (动态调用指令)
              data.readInt(); // 抛弃4字节
              break;
            case 15: // Method Handle (方法句柄)
              data.readByte();
              data.readShort();
              break;
            case 16: // Method Type (方法类型)
              data.readShort();
              break;
            default:
              throw
                new RuntimeException("Bad tag " + tag);
          }
        }
        short accessFlags = data.readShort();
        String access = (accessFlags & 0x0001) == 0 ?
          "nonpublic:" : "public:";
        int thisClass = data.readShort();
        int superClass = data.readShort();
        return access + classNameTable.get(
          offsetTable.get(thisClass)).replace('/', '.');
      } catch(IOException | RuntimeException e) {
        throw new RuntimeException(e);
      }
    }
  }
  // 示范：
  public static void
  main(String[] args) throws Exception {
    PathMatcher matcher = FileSystems.getDefault()
      .getPathMatcher("glob:**/*.class");
    // 遍历整个树：
    Files.walk(Paths.get("."))
      .filter(matcher::matches)
      .map(p -> {
          try {
            return thisClass(Files.readAllBytes(p));
          } catch(Exception e) {
```

```
            throw new RuntimeException(e);
          }
        })
      .filter(s -> s.startsWith("public:"))
      // .filter(s -> s.indexOf('$') >= 0)
      .map(s -> s.split(":")[1])
      .filter(s -> !s.startsWith("enums."))
      .filter(s -> s.contains("."))
      .forEach(System.out::println);
  }
}
/* 输出:
onjava.ArrayShow
onjava.atunit.AtUnit$TestMethods
onjava.atunit.AtUnit
onjava.atunit.ClassNameFinder
onjava.atunit.Test
onjava.atunit.TestObjectCleanup
onjava.atunit.TestObjectCreate
onjava.atunit.TestProperty
onjava.BasicSupplier
onjava.CollectionMethodDifferences
onjava.ConvertTo
onjava.Count$Boolean
onjava.Count$Byte
onjava.Count$Character
onjava.Count$Double
onjava.Count$Float
onjava.Count$Integer
onjava.Count$Long
onjava.Count$Pboolean
onjava.Count$Pbyte
onjava.Count$Pchar
onjava.Count$Pdouble
onjava.Count$Pfloat
onjava.Count$Pint
onjava.Count$Plong
onjava.Count$Pshort
onjava.Count$Short
onjava.Count
onjava.CountingIntegerList
onjava.CountMap
onjava.Countries
onjava.Enums
onjava.FillMap
onjava.HTMLColors
onjava.MouseClick
onjava.Nap
onjava.Null
onjava.Operations
onjava.OSExecute
onjava.OSExecuteException
onjava.Pair
onjava.ProcessFiles$Strategy
onjava.ProcessFiles
onjava.Rand$Boolean
onjava.Rand$Byte
onjava.Rand$Character
onjava.Rand$Double
onjava.Rand$Float
onjava.Rand$Integer
onjava.Rand$Long
onjava.Rand$Pboolean
onjava.Rand$Pbyte
onjava.Rand$Pchar
onjava.Rand$Pdouble
onjava.Rand$Pfloat
onjava.Rand$Pint
onjava.Rand$Plong
onjava.Rand$Pshort
onjava.Rand$Short
onjava.Rand$String
onjava.Rand
onjava.Range
onjava.Repeat
onjava.RmDir
onjava.Sets
onjava.Stack
onjava.Suppliers
onjava.TimedAbort
onjava.Timer
onjava.Tuple
onjava.Tuple2
onjava.Tuple3
onjava.Tuple4
onjava.Tuple5
onjava.TypeCounter
*/
```

（转右栏）

虽然这里不可能深入挖掘所有细节，但每个类文件都已遵从了一种特定的格式，我也已经尽量使用了有意义的字段名来表示从 `ByteArrayInputStream` 中取出的数据片段。还可以通过对输入流执行的读取的长度，来得知每个数据片段的大小。举例来说，任何类文件的头 32 位永远是十六进制的"魔术数字"：0xcafebabe，[①] 并且之后的两个短整型位是版本信息。常量池保存了程序所需的常量，因此大小不固定。接下来的短整型则告知了常量池的大小，由此可以分配一个大小合适的数组。常量池中的每个项都可以是固定长度或长度可变的值，因此我们必须检查每个项开头的标签，由此来决定该如何处理它，即例子中的 `switch` 语句。此处，我们并不去试图精确地分析类文件中的所有数据，而仅仅是逐步遍历数据，然后保存其中感兴趣的部分，因此你会发现大量的数据被丢弃了。类的信息被保存在 `classNameTable` 和 `offsetTable` 中，在读取完常量池后，程序会找到 `thisClass` 的信息，它是 `offsetTable` 的索引，`offsetTable` 则生成 `classNameTable` 的索引，而 `classNameTable` 则生成类名。

回到 AtUnit.java，`process()` 方法现在得到了类名，我们进而可以查看它是否包含 `.`，这代表它是否在包中，不在包中的类会被忽略。如果类在包中，则会由标准的类加载器通过 `Class.forName()` 方法来加载该类。现在可以来分析该类中的 `@Unit` 注解了。

我们只需要找到三样东西：`@Test` 方法（保存在 `TestMethods` 列表中）、`@TestObjectCreate` 方法和 `@TestObjectCleanup` 方法。正如你在代码中所见，这几样东西是通过调用相关方法（用于查找注解）找到的。

如果找到了任何 `@Test` 方法，便会显示出类名，由此可以看到当前正在发生些什么，接下来便会执行各项测试。这意味着会打印方法名，然后调用 `createTestObject()` 方法，后者会使用 `@TestObjectCreate` 方法（如果该方法存在；否则会回退到无参数的构造器）。一旦创建了测试对象，便会对该对象执行测试方法。如果测试返回 `boolean`，结果便会被捕获；如果没有，且没有抛出异常（异常会在断言失败或任何其他异常发生时抛出），我们便认为测试成功了。如果抛出了异常，便会打印出异常信息以告知细节。如果发生任何失败，失败数会累加，并且类名和方法名会被追加到 `failedTests`，由此可以在测试结束后报告所有错误信息。

[①] 关于该魔术数字的意义，衍生了很多版本的传说。

4.5 总结

注解是一个很受欢迎的 Java 新特性。它是一种结构化且接受类型检查的向代码中添加元数据的方法，并且不会导致代码被渲染得混乱和不可读。它可以帮助我们免除部署描述文件和其他生成文件的编写工作。Javadoc 中的 `@deprecated` 标签被 `@Deprecated` 注解所取代，仅这一点便印证了由合适的注解来描述代码组件的信息优于用注释来做同样的事。

Java 中仅有少量的注解，这意味着如果你没有在别处找到相关的库，便需要自行创建注解以及相关逻辑。利用 javac 附带的注解处理器，只需一步就可以编译新创建的文件，以简化构建过程。

各种 API 和框架的开发者将逐渐引入注解，使其成为工具包的一部分。通过 @Unit 系统，你可以想象，注解很可能会给 Java 的编程体验带来巨大的改变。

05

并发编程

> '但是我不想成为那种疯子。'爱丽丝说。
> '嗯……这由不得你,'猫说,'这里的人都是疯子。我是,你也是。'
> '你怎么知道我疯了?'爱丽丝说。
> '你肯定是疯了,'猫说,'否则你不会来到这里。'
> ——《爱丽丝梦游仙境》,第 6 章

到目前为止,我们一直在以一种类似于文学创作中"意识流"(剧情按照人物的思路顺序发展)的叙事手法在编程:剧情按部就班地发生着,每一步的发展都尽在掌握。比如,明明给一个变量赋值为 5,过一会儿却发现它变成了 47 这种事是绝对不会发生的。

现在我们即将进入并发的世界,你会发现 5 变成 47 这种事真的会发生。你的经验将不再可靠,代码时常不按预期运行。大多数时候,它只在一定条件下才会正常工作,而你需要知晓并谙熟其中之道。

打个比方,我们日常生活在牛顿经典力学的世界里:物质都有质量,并在下落过程中进行势能和动量的转换;电线都有电阻;光按直线传播。而一旦我们进入一个非常小、非常热、非常冷或非常巨大的世界(我们无法生存),则一切都会改变。我们无法确定一种物质是粒子还是波,光是否会受到重力的影响，一些物质是否变成了超导体。

相比之前的单线意识流创作手法，我们现在更像是进入了多角色多线叙事的特工小说世界。特工 A 在一块特殊的石头下面藏了一个微缩胶卷，而当特工 B 来取的时候，它很可能已经被特工 C 拿走了。这类小说通常不会展现所有细节，你很可能看到最后都没搞明白到底发生了些什么。

构建并发程序的过程非常像"叠叠乐"（Jenga）这个游戏，每次从积木塔中抽出一块积木放到塔顶时，都可能使整座塔倒塌。每座积木塔以及每个程序都有各自的特性，其搭建方法往往不能简单地照搬。

本章旨在对并发做一个最基础的介绍。尽管我使用了主流的 Java 8 来演示相关原理，但对这一主题的探讨远没有达到面面俱到。我的目的是，通过讲解必要的基本知识，让你能够快速明白并发的复杂和危险，并在日后的艰难探索中保持足够的敬畏心。

更多复杂的底层细节，请参见本书第 6 章。如果想深入研究该领域，强烈推荐 Brian Goetz 等人编著的《Java 并发编程实战》。虽然该书比本书早诞生 10 余年，但它仍然涵盖了必须掌握的知识要点。恰好本章及本书第 6 章也非常适合作为该书的预习材料。另一本值得推荐的书是 Bill Venners 所著的《深入 Java 虚拟机》，该书深入阐述了 JVM 的内部机制，包括线程。

5.1 令人迷惑的术语

并发（concurrent）、并行（parallel）、多任务（multitasking）、多进程（multiprocessing）、多线程（multithreading）、分布式系统（distributed systems）等——这些术语在大量编程书中常常被滥用甚至混淆。Brian Goetz 在其 2016 年的演讲 "From Concurrent to Parallel" 中首次指出该问题，并给出了一个较为合理的区分方法：

- 并发是指如何正确、高效地控制共享资源；
- 并行是指如何利用更多的资源来产生更快速的响应。

这样定义相当不错，但数十年的混用历史不是一朝可以改变的。通常来说，人们习惯于简单地只用"并发"一词来指代相关领域的一切，很多时候连我自己都形成了这种思维——甚至包括 Brian Goetz 的《Java 并发编程实战》在内的大部分书也在书名中使用了这个词。

并发通常是指"多个任务在进行"，而并行则更多是指"多个任务在同时执行"。你立刻就会发现这样定义有问题：并行说的同样也是多个任务在"进行"。区别在于细节，在于如何"进行"的。当然，定义也的确有重合之处：并行类的程序有时在单 CPU 上也可以运行，而有些并发编程系统也能利用多处理器的优势。

还有另外一种方式，定义的重点在于"性能究竟是在何处慢下来的"。

并发

同时处理多个任务，即不必等待一个任务完成就能开始处理其他任务。并发解决的是阻塞问题，即一个任务必须要等待非其可控的外部条件满足后才能继续执行，最常见的例子是 I/O，一个任务必须要等待输入才能执行（即被阻塞），类似的场景称为 I/O 密集型问题。

并行

同时在多处执行多个任务。并行解决的是所谓的计算密集型问题，即通过把任务分成多个部分，并在多个处理器上执行，从而提升程序运行的速度。

从上述定义可以看出术语易混淆的原因：两者的关键都是"同时处理多个任务"，而并行则额外包括了多处理器分布式处理的概念。更重要的是，这两者解决的是不同类型的问题：对于 I/O 密集型问题，并行可能起不到什么明显的作用，因为瓶颈不在于速度，而在于阻塞；而对于计算密集型问题，如果想用并发在单处理器上解决，则多半会徒劳无功。这两种思路都是想在更短的时间内做更多的事，但是它们实现加速的方法是不同的，这取决于不同问题所带来的不同约束的核心矛盾不同。

这两个概念混淆的另一个主要原因是，很多编程语言（包括 Java）使用了相同的机制——**线程**——来实现并发和并行。

我们甚至可以用更细分的定义（当然这些并不是标准术语）。

- **纯并发**（purely concurrent）：多个任务在单 CPU 上运行。纯粹并发系统会比时序系统更快地生成结果，但是无法利用多处理器进一步提升性能。
- **并发式并行**（concurrent-parallel）：应用并发技术，使程序能利用多处理器并更快地产生结果。
- **并行式并发**（parallel-concurrent）：使用并行编程技术编写的程序，而且即便只有一个处理器也能运行（Java 8 的 Stream 就是很好的例子）。
- **纯并行**（purely parallel）：只能在多处理器上运行。

在某些情况下，这可能是一个便于理解的分类方法。

支持并发的语言和库似乎是解决"抽象泄露"（leaky abstraction）问题的完美可选方案。抽象的目的是把不影响核心思路的具体实现"抽象掉"，让你不受细枝末节的困扰。如果抽象泄露了（内部实现细节），这些细枝末节会不断制造麻烦，变得喧宾夺主，不管你多么努力地隐藏它们。

我很想知道究竟是否存在真正意义上的抽象。在编写这类程序时，你永远无法忽略底层的系统和工具，甚至还要关注 CPU 缓存执行的细节。最终，如果你足够谨慎的话，程序终于跑起来了，结果换个环境它又不行了。这有时是因为两台机器的配置不同，有时是因为对负载预估不同。这和 Java 本身无关，而是由并发和并行的本质决定的。

你可能会反驳说，纯函数式（pure functional）语言就不会有这些局限。确实，纯函数式语言能解决很多并发问题，所以如果你面临复杂的高并发场景，你可能会考虑用纯函数式语言来实现该部分功能。但即便如此，假如你所写的系统里用到了队列，而程序没有调试妥当，并且请求流量估算不准，或者被打满（流量打满意味着各种难以预料的状况都可能发生），那么队列会被填满阻塞，或者溢出。总之，你必须清楚所有细节，而任何小问题都可能击溃你的系统。并发真的是一个非常特殊的编程领域。

并发的新定义

数十年来，我不断和并发作各种斗争，而其中最大的挑战绝对是如何给它一个简单的定义。在编写本章时，我终于觉得自己想通了：

并发是一系列聚焦于如何减少等待并提升性能的技术。

这样表述实在很模糊，下面我再具体细化一下。

- 它是一个**集合**：有多种方法可以解决问题，这是导致并发很难定义的困难之一，因为不同技术间的差异非常大。
- 它是**关于性能的技术**：没错，并发归根结底就是要使你的程序运行得更快。在 Java 中，并发技巧性很强，但也很难，所以切记，只有遇到重大的性能问题时才能使用。即便如此，也要尽量用最简单的方法来实现需求，因为并发很容易就会变得失控。
- 关于"减少等待"，这一点非常重要，也很微妙。不论运行在多少处理器上（假设），只有某种等待发生了，优化才有收益。比如请求 I/O 资源并且立刻就成功了，那么就没有延迟，也就不需要优化。再比如在多处理器上运行多任务程序，如果每台处理器都在满负荷运行，任务间没有相互等待，尝试提升吞吐量也就毫无意义。唯一需要应用并发的时机是程序中某些地方被迫等待。等待会以多种形式呈现,这也恰巧印证了为什么并发领域有如此多的解决方法。

有必要再次强调,本定义的关键意义在**等待**上。如果没有等待发生,也就没有提速的可能。如果有等待发生,则有很多方法可以优化,具体取决于很多因素,包括系统配置、要解决的问题类型,以及很多其他方面。

5.2 并发的超能力

想象你是一部科幻电影中的角色。有一栋拥有 1000 万个房间的高楼,你的目标巧妙地藏身于其中一个房间,而你需要找到它。你进入大楼,沿着走廊往里走,走廊尽头是岔路口。

如果仅靠你一个人,你得花 100 辈子才能完成这个任务。

这时假设你拥有一种超能力,可以克隆一个自己的分身,然后让分身沿着岔路继续走下去,你自己则继续往前走。每次遇到岔路或者上楼的楼梯,你就重复克隆的操作。最终整个大楼的每一条走廊尽头都会有你的一个分身。

每条走廊有 1000 个房间,你的超能力不太够,同时最多只能克隆出 50 个自己。

一旦一个分身进入一个房间,他需要寻遍每一个角落。这时他启动了第二个超能力,将自己分裂成 100 万个纳米机器人,每个机器人都飞入或爬入某个隐蔽的角落。你不用明白这是怎么一回事儿,只要知道可以这么干就行。这些纳米机器人自行搜寻房间,完成之后就重新组装成你的分身。你忽然间就知道目标是否在这个房间里了。

我很想说:"科幻电影里的超能力?那不就是并发嘛!"每来一个新任务,就克隆一个分身,就这么简单。但问题是不论我们用哪种模型来描述这个奇迹,最终都会陷入抽象泄露的困境中。

其中一种泄露是这样的:在理想世界中,每次克隆你都需要复制一套硬件处理器来运行新的分身。但这显然不可能做到,实际上你的机器只会有 4 个或 8 个处理器(常见情况)。你可能有更多的处理器,但在很多情况下也很可能只有 1 个。在抽象的相关探讨中,物理处理器的分配方式不仅会被泄露,而且甚至还能左右你的决策。

我们再来修改一下电影设定。现在,当每个分身到达一个房间门口时,他都需要敲门,并且等待开门。如果每个分身都能分配一个处理器,那就不会有什么问题。处理器挂起等待就行,直到有人开门。但是,如果只有 8 个处理器,却有成千上万个分身,我们并不希望处理器被闲置,仅仅因为一个分身等待开门,并由此导致阻塞。我们希望能将处理器分配给当前能干活的分身,所以需要一套将处理器在任务间切换分配的机制。

很多模型可以将处理器的实际数量对外屏蔽,并允许你伪装成有很多处理器的样子。但在某些场合下,该机制可能失效,这时就必须清楚地知道处理器的实际数量,由此才能围绕实际数量做出更优决策。

一个关键的决定因素是处理器是否多于一个。如果只有一个处理器,那么它要承受额外的任务切换带来的性能损耗,这时并发反而会使系统变得**更慢**。

这可能会使你认定,在单处理器场景下,永远不该使用并发。然而在某些情况下,并发模型能在开发上带来更高的便利性,这时即使慢一点也是值得的。

在刚才说的分身敲门并等待的场景下,即使是单处理器系统,并发也能带来好处,它可以从一个等待中(**被阻塞**)的任务切换到另一个已经准备好的任务。但是如果所有的任务永远都能保持运行不被阻塞,那么切换任务的损耗反而会使整个系统都慢下来。这时,并发只能在确实拥有多处理器时才有意义。

假设你要破解某种加密,同时在破解中的 worker 越多,破解速度就越快。假定每个 worker 都能尽量充分地使用你分配给它的 CPU 时间,那么最好的情况就是每个 worker 都能独享一个处理器,在这种情况下(计算密集场景),你的程序应该将 worker 的数量设置为和处理器的数量完全相等。

假设有一个负责电话应答的客服部门,客服团队人数有限,但是可能会有很多电话打进来,客服人员(相当于处理器)必须在一部电话上完成整个通话才能挂断,其他打进的电话则必须等待。

在童话故事《鞋匠和精灵》里,鞋匠要做的鞋太多了,当他睡觉时,一群精灵会来帮他做鞋,此时就变成了分布式处理。但即使有很多物理处理器(精灵),性能还是会在某些部件生产上遇到瓶颈,比如做鞋底需要的时间最长,那么这个环节就会限制鞋的整体生产速度,相应地也会迫使你优化解决方案。

因此,不同的问题会促成不同的方案设计。理想很美好,我们可以把任务分解成很多子任务并让它们"各自独立运行",但现实则往往是另一回事,物理现实会不断打击美好的理想。

麻烦还不止这些。假设有个生产蛋糕的工厂,我们已经在某种程度上实现了将蛋糕生产的任务按人拆分，但问题还是来了。工人 A 即将执行将蛋糕放入包装盒的步骤，盒子也已经就绪，结果当工人 A 开始放入蛋糕的操作时，工人 B 抢先一步将另一个蛋糕放进去了，而工人 A 并未停止手头的动作，结果"吧唧"一下，两个蛋糕碰到了一起，形状也都破坏了。这就是导致常见的**竞态条件**（race condition）的"共享内存"问题，其结果取决于哪位工人会先将他的蛋糕放入盒子（典型的解决方案是使用锁机制，总有一位工人能先锁定盒子，从而避免别人的蛋糕放入）。

这种问题通常发生于同时执行的任务互相干扰时，而且常常表现得复杂微妙、难以预测，因此可以说并发是"理论上充满确定性，实际上充满不确定性"。意思是说，你总是感觉你的并发程序经过了周全的考虑，代码检测没问题，运行也正确。但现实中更常见的情况是，写出来的并发程序只是看起来运行正常，但是在某些条件下会出问题。这些条件可能永远不会真的形成，或者形成的概率低到在测试过程中根本不会发现。实际上，一般不太可能为并发程序编写生成失败条件的测试用例，这些错误通常又非常偶发，因此最终只会默默转化为客户的抱怨。这就是学习并发最有说服力的理由之一：一旦忽略，便可能遭到反噬。

并发因此显得充满危险，而如果你因而心生畏惧，那这很大程度上是件好事。虽然 Java 8 在并发上做了很大的改进，但仍旧未提供诸如编译时校验或异常检查那样可以明确告知错误的安全防范手段。在并发层面上，你只能靠你自己。只有抱着求知若渴、敢于质疑和积极向上的态度，你才能编写出可靠的 Java 并发程序。

5.3　并发为速度而生

了解了并发编程的各种风险之后，你可能会想，是否值得为了并发而承担这些风险。答案是"不值得，除非你的程序跑得不够快"。这样你便会经过慎重的考虑后再判断程序是否不够快。不要轻易打开并发编程这个"潘多拉魔盒"，如果你能找到其他方法提升程序性能，比如换一台更高配置的机器，或者通过分析找到性能瓶颈，然后换一个更快的算法等，那么就先这么做。只有确定别无选择时，你才能开始使用并发，并且只能在隔离的环境中使用。

速度问题乍一看很简单：如果你希望程序能跑快点儿，那就将程序拆分成片段，然后将片段分别放到独立的处理器上运行。随着处理器时钟频率的提升变得越来越困难（至少对常规芯片来说如此），相较于单纯提升芯片速度，速度的提升越来越倾向于多处理器并行的方向。要想使程序运行得更快，你需要学习如何利用更多的处理器，而这正是并发能够带给你的。

利用多处理器机器，可以将多任务分发到这些处理器上，这将显著提升吞吐量。这是那些强大的多核 Web 服务器常用的方案，它们可以通过程序为每个请求分配一个线程，从而将大量的用户请求分发到多个 CPU 上。

不过，并发常常也能提升运行在**单处理器**上的程序性能。这听起来有点反常。你可能会想，将并发程序运行在单处理器上，实际应该会比让程序全部按顺序执行带来更多开销，因为**上下文切换**（context switch，在任务之间切换）会导致额外的开销。表面看来，让程序的所有片段按顺序执行（就像单任务一样）可以节省上下文切换的开销，看起来会是更经济可行的方案。

给情况带来变化的是**阻塞**。如果程序中的某个任务由于程序无法控制的某些外部条件（典型的如 I/O）导致无法继续执行，我们便认为该任务或者该线程被阻塞了（在之前的科幻故事中，克隆的分身敲门后便一直在等待开门）。如果不使用并发，整个程序都会停下，直到外部条件改变。而如果使用并发，即使一个任务阻塞了，程序中的其他任务仍可继续执行，由此程序持续向前运转。事实上，从性能的角度来看，除非有任务可能阻塞，否则没有理由在单处理器上使用并发。

有个提升单处理器系统性能的常见例子：**事件驱动编程**（event-driven programming）。在用户界面编程领域中的应用设想一个程序执行某项耗时较长的操作，最终导致无法处理用户的输入，从而进入无响应状态。如果程序里有个"退出"按钮，你可能不想在每段代码里都轮询它。如果无法确保程序员会记得执行检查操作，最终便会催生出难用的代码。如果没有并发，要想实现一个高响应的用户界面，唯一的方法是对所有的任务都定期检查用户输入。而通过创建一个独立的线程负责响应用户的输入，程序便可以保证一定程度的响应性。

有个简单的实现并发的方法是在操作系统层面使用**多进程**，这和多线程有所区别。进程是在自有地址空间中运行的自包含程序。多进程听起来很不错，因为操作系统通常会将各个进程相互隔离，因此不会相互干扰，这使得用多进程的方式编程会很容易。对比来看，多线程会共享诸如内存和 I/O 等资源，因此编写多线程程序的基本难点之一便是在各个线程驱动的任务间调度这些资源，以使其同一时刻只能被一个任务访问。

有些人甚至主张多进程是唯一合理的并发方法[①]，但不幸的是，多进程通常存在数量和

[①] 例如，Eric Raymond 在《UNIX 编程艺术》一书中进行了有力的论证。

开销限制，这影响了多进程在并发领域的适用性。(最终你会习惯于抱怨并发的标准："这种方法在某些情况下有效，但在其他情况下则不行。")

有些编程语言的设计是将各个并发任务隔离起来，这通常叫作**函数式语言**（functional language），其中的每个函数调用都不会产生副作用（所以也无法干预其他函数），因此可以视为独立的任务来驱动。Erlang 便是这种语言之一，它引入了保障任务间通信的安全机制。如果你发现程序中有某部分需要大量使用并发，并且在构建这部分程序时遇到了大量问题，你可能需要考虑用专门用于并发的语言来实现这部分程序。

Java 则采用了一种更传统的方法，在其按顺序执行的语言特性基础上增加了多线程的支持。[①] 多线程并不是在多任务操作系统中 fork[②] 出额外的进程，而是在执行程序所维护的单个进程内部创建多任务。

并发会带来额外开销，并引入更多复杂性，但也能在程序设计、资源协调、用户便利性等方面带来很大的提升。总体来说，并发可以帮你创造出更松耦合的设计；否则，部分代码会迫使你在一些本可以用并发来正常处理的操作上付出更多的精力。

5.4 Java 并发四定律

经过了多年和 Java 并发的缠斗，我研究出了以下四条定律。

1. 不要使用并发。
2. 一切都不可信，一切都很重要。
3. 能运行并不代表没有问题。
4. 你终究要理解并发。

这四条定律专门代表 Java 设计中容易出现的问题，尽管它们也可以引申到某些其他语言。不过，确实有语言是专门针对这类问题来设计的。

5.4.1 不要使用并发

（并且不要自己动手来做。）

想要避免陷入并发带来的各种麻烦，最简单的方法就是不要使用并发。虽然并发很诱人，看起来在处理某些简单问题时也足够安全，但使用它仍然会有无数的陷阱和隐患。如

---

① 有人可能会说，试图强行在顺序语言上捆绑并发性是注定会失败的方法，但你必须得出自己的结论。
② fork() 是类 UNIX 系统中用于创建进程的函数，因此经常用 "fork" 来表示创建进程/线程的动作。——译者注

果可以避免使用并发，你的日子会好过很多。

并发**唯一**的价值在于速度。如果你的程序跑得不够快（这里要小心，如果只是你想要它跑得更快，这并不算合理的理由），那就先尝试一下分析工具（参见基础卷第 16 章 16.7 节），看看是否能发现一些其他可以优化的地方。

如果你必须使用并发，那就尽量用最简单、最安全的方法来解决问题，尽量使用知名的库，少写自己的代码。在并发中，不存在"太简单"这种东西，小聪明是你的敌人。

5.4.2　一切都不可信，一切都很重要

在不使用并发的情况下编程时，你已习惯于在你的世界中，期待着稳定的秩序和一致性。就像给一个变量赋值一样简单，很显然这一切都会永远正确地运行着。

而在并发的世界中，有些事情可能是真的，有些可能不是，因此你必须假定一切都不可信。你需要永远保持质疑，甚至给一个变量赋值这种小事，都不一定按照你的预期运行，并变得不可捉摸。我已经很习惯那种本来觉得理所当然行得通，但发现实际并非如此的感觉。

所有在非并发编程中可以忽略的事，在并发中都突然变得重要了。举例来说，你现在必须了解处理器缓存，以及如何保持本地缓存和主存中数据的一致性。你必须掌握对象构建的深层复杂性，这样你的构造方法才不会意外暴露数据，导致被其他的线程修改等。要注意的事还有很多。

虽然这些话题对你来说过于复杂，本章无法提供足够的专业知识（再次推荐《Java 并发编程实战》），但你还是必须警惕它们。

5.4.3　能运行并不代表没有问题

你很容易写出一个看起来运行正常，但实际上有问题的并发程序。而这个问题只有在满足最罕见的条件时，才会将自己暴露出来。它必然会在你完成部署后出现，成为用户反馈的一个问题。

- 你无法证明并发程序是正确的，你只能（有时候）证明它是不正确的。
- 大多数时候，你甚至都不应该使用多线程：如果出问题了，你很可能都无法发现问题在哪里。
- 你通常无法编写有效的测试，所以要想发现 bug，就必须依靠代码检查，这要求你对并发有着深度的理解。

- 即使是能正常运行的程序也只能在其设计参数下才能正常运行。当超出预先设计的参数范围时，大部分并发程序都会以某种方式运行失败。

在其他的 Java 主题中，我们逐渐发展出一种决定论的感觉。所有的事都按照语言所承诺（或暗许）的那样发展，这很舒服，也是我们想要的——毕竟，编程语言的重点就是要让机器按照我们所想的运行。从具有确定性的编程世界跨越到并发编程的领域，我们遇到了叫作"邓宁 - 克鲁格效应"（Dunning-Kruger Effect）的认知偏差，其内容可以简单总结为"你懂得越少，你以为自己懂得就越多"。意思是"……相对不熟练的人会有虚幻的优越感，错误地认为他们的能力比实际要强很多"。

我自己的经验是，不论你多么确信你的代码是线程安全的，它都可能会出现问题。你会非常容易轻信你已经了解了所有的隐患，然后几个月或几年后，你会接触到一些新的概念，使你意识到几乎所有你写出的代码都会受到并发方面的 bug 影响。编译器也无法帮你发现所有错误。要想不出问题，就必须好好分析你的代码，并时刻在脑中记住所有并发方面的隐患。

在 Java 中，对于所有不涉及并发的领域来说，"没有明显的 bug，也没有编译器报警"基本就可以证明一切 OK 了。但是对于并发，这什么都证明不了。在这里，你最需要警惕的就是你的"自信"。

5.4.4　你终究要理解并发

看完前面三条定律后，你可能开始有点被并发吓到，并会想："我已经逃避并发这么久了，没准儿我可以一直躲下去。"

这个反应很理性。你可能知道其他一些更适合编写并发程序的语言——有些甚至是在 JVM 中运行的（因此可以更好地和 Java 配合），比如 Clojure 或者 Kotlin。为什么不用这些语言来实现需要并发的部分，然后用 Java 来实现其他的部分呢？

唉，你不可能这么容易就躲掉的。

- 即使你并未显式地创建线程，你使用的框架也可能（隐式地）创建——比如 Swing 的 GUI（graphical user interface，图形用户界面）库，或者像 Timer 这么简单的类。
- 最坏的情况是，当你编写一些组件时，你必须假定这些组件会拿到多线程环境中使用。即使你可以放弃兼容多线程，并声明你的组件是"非线程安全"的，你也必须足够清楚地认识到这样的声明的重要性和意义。

人们有时会建议不要在一本介绍编程语言的书中讲解并发，这过于进阶了。他们认为并发是一个独立的主题，应该被单独对待，而那些日常开发中的少见情况（比如 GUI）则可以通过专门的套路来处理。所以如果可以避免使用，为什么还要讲解并发这么复杂的主题呢？

唉，如果真是这样就好了。不幸的是，你无法选择线程何时在 Java 程序中出现，不能仅仅因为你并未手动开启一个线程，就认为你可以避免编写线程代码。举例来说，Web 系统是最常见的 Java 应用之一，并且也是基于多线程的——Web 服务器通常是多处理器的，而且并行是利用这些处理器的理想方式。虽然这样的系统看似简单，但想要写出正确的代码，就必须了解并发。

Java 是一种支持多线程的语言，因此并发问题也会时刻相随，不论你是否注意到它们。因此有许多运行中的 Java 系统，要么靠运气来正常运行，要么大部分时间运行正常，但时不时由于一些未知的 bug 而神秘地出了问题。有时这种不稳定性影响不大，但有时会导致重要数据的丢失。如果你没有对并发问题保持最基本的警惕性，你最后可能还以为问题是出在其他地方，而不是代码中。这些类型的问题也会在将程序迁移到多处理器环境的过程中被暴露或放大。基本而言，理解并发可以使你意识到明明看起来正确的程序，也可能会表现出错误的行为。

5.5　残酷的事实

当人类开始烹饪食物时，他们大大减少了身体分解和消化食物所需的能量。烹饪创造了一种"外化的胃"，并由此将这部分能量节省下来做其他的事。对火的使用促进了人类文明的发展。

利用计算机和网络技术，我们创造了"外化的脑"，由此开始了第二轮基础技术的升级。我们只触及了技术的一些表层皮毛，便已经引发了各种变化，比如设计生物机制的能力，并已经能够看到对文化发展方面的巨大推进（在过去，人们必须通过实地旅行来进行文化的交流融合；而现在，人们已经开始通过互联网来进行跨文化交流了）。这些变化所带来的影响和裨益已经远远超出了科幻作家的预测能力（他们尤其难以预测文化和个人的改变，甚至是技术革新带来的次生效应）。

由于这些技术给人类社会带来了如此大的变化，因此看到其中不计其数的挫折和失败的试验也毫不意外。事实上，技术演进依赖于大量的试验，其中大部分注定会失败。这些试验在前进的道路上是必不可少的。

Java 是在自信、热情和紧迫的氛围中创造出来的。在创造一门编程语言时，非常容易让人觉得语言永远可以像最初时那么可塑。你可以不断地尝试，如果发现出了问题，修复就好。编程语言独有的特性是这样的——它们会经历和水一样的形态变化阶段：气态、液态，最终变成固态。在气态阶段，语言看起来有着无限的灵活性，并且很容易让人们觉得一直都会这样下去。但是，一旦人们开始使用你的语言，变更带来的影响便会扩大，也更容易受到环境的制约。语言设计的过程本身就是一门艺术。

紧迫则源于互联网最初的崛起。这看起来就像一场比赛，谁最先通过开始那扇门，谁就能"胜出"（确实，像 Java、JavaScript 和 PHP 这样的热门语言证明了这一点）。唉，匆忙设计语言所产生的认知负担和技术债务最终会追上我们的。

对于语言来说，仅具备图灵完备性（Turing-completeness）还不够，还需要更多：语言必须具备富有创造性的表达能力，而不是用不必要的细节来拖累我们。本来要解放我们的精神能力，结果却反过来再次困住了它，这毫无意义。我承认，尽管存在这些问题，我们还是取得了惊人的成就；但我也看到，如果没有这些问题，我们的成就还能更大。

热情促使最初的 Java 设计者不断加入新特性，因为看起来这些特性都是必需的。信心（以及语言最初的气态阶段）使得他们认为任何问题都可以稍后再解决。在历史发展的某个时间点，有人决定将任何添加进 Java 的东西都永远固定下来——这是信心促成的决定，相信第一个决定永远都是那个正确的决定，所以我们可以看到各种不明智的决定给 Java 带来的混乱景象。其中的一些决定最终并未造成很大的影响。举例来说，你可以告诉人们不要使用 Vector，但它会一直保留在语言中以兼容老代码。

Java 1.0 中就引入了多线程。毫无疑问，支持并发是语言设计的一项基本决策，会影响到该语言的方方面面，所以很难想象到后面再添加该特性（会有多么麻烦）。公平地说，当时并不清楚并发究竟有多么重要。C 等其他语言将多线程视为一种扩展特性，所以 Java 的设计者也照搬了这个设计，引入了一个 Thread 类，以及必要的 JVM 支持（这比你想象的更复杂）。

C 语言是低级语言，这会限制你的野心。这些限制使得扩展多线程库看起来合情合理。然而 Java 更宏大的雄心壮志，使得在将低级语言模型直接照搬到高级语言的过程中，很快就暴露出了根基上的问题。这些错配问题体现于 Thread 类中大量方法的弃用，在随后的为并发提供更好抽象的高级库的浪潮中也暴露了出来。

不幸的是，要想在高级语言中较好地支持并发，所有的语言特性都会受到影响，包括那些最基本的特性，比如标识符是否代表可变值。使一切**都不可变**，以在函数和方法中防止副作用的发生，这样的思路对并发编程的简化产生了深远的影响（这也是纯函数式编程语言的基础）。但在当时，对于主流语言的设计者来说，这种想法看上去非常奇怪。最初的 Java 设计者要么还没有意识到这些想法，要么认为它们太特别了，会让许多潜在的语言使用者望而却步。我们可以宽宏大量地说，当时的语言设计社区还没有足够的经验来理解在线程库中打补丁的影响。

Java 的经历向我们表明，结果是灾难性的。程序员很容易陷入陷阱，以为 Java 多线程并不复杂。复杂微妙的并发 bug 会使看似能正常运行的程序实际上千疮百孔。

为了获得正确的并发性，语言特性的设计必须从最开始就考虑到并发。箭已离弦，无法回头，Java 永远不可能成为一个专为并发设计的语言，而只是一种可以支持并发的语言。

令人惊叹的是，尽管有这么多无法弥补的重大问题，Java 还是发展到了现在的程度。Java 的后续版本已经增加了库以提升并发的抽象水平。事实上，我之前绝没有想到 Java 8 会带来如此大的提升：并行流和 CompletableFuture——这是史诗般的魔法，我还将非常惊讶地看到更多魔法的诞生。

这些改进非常有用，我们也会在本章中聚焦于并行流和 CompletableFuture。虽然它们可以很大程度上简化你对并发的理解，并让之后的代码更简洁，但基本的问题还是存在：由于 Java 最初的设计，你的任何代码都是脆弱的，因此你必须理解其复杂性和隐患。Java 中的多线程永远都不简单，也并不安全。要想改变这种体验，只能再指望一门新的语言了。

5.6 本章剩余部分

本章剩余部分将涵盖的内容如下。请记住，本章的重点是使用最新的高级 Java 并发框架。相比旧版本，新版本会使你轻松不少。不过你还是会在历史代码中遇到一些低级的并发工具。你自己偶尔可能也会被迫使用这些工具。本书第 6 章中包含了对部分更原始的 Java 并发元素的介绍。

- 5.7 节

目前为止，我在本书中已经强调过由 Java 8 的 Stream 所带来的升级过的语法。现在你可以更舒心地使用（并且我希望你会喜欢上）这些语法了，而且还能得到额外的好处：可以通过简单地在表达式中增加 parallel() 来并行化一个流。这是一种更为简单、强大、甚至令人惊叹的利用多处理器的方式。

通过增加 parallel() 来提升速度看起来易如反掌，但可惜的是，正如你刚才在本章的 5.5 节中所学到的，这永远不会那么容易。我会演示并讲解盲目地将 parallel() 添加到

Stream 表达式中会带来的问题。

● 5.8 节

任务（task）是可以独立运行的代码片段。为了阐明创建和运行任务的基本原理，本节中会介绍一种比并行流或 CompletableFuture 更简单的机制：Executor。Executor 管理着一个低级 Thread 的对象池（Java 中并发的最原始形态）。你可以创建一个任务，然后将其传递给一个 Executor 来执行。

Executor 有若干种类型，分别用于不同的目的。此处，我们会演示最简洁的方式，这也是最简单和最好的创建及运行任务的方法。

● 5.9 节

任务是独立运行的，因此需要一套结束运行的机制。这通常会用一个标识来实现，并且会带来共享内存方面的问题。我们会用 Java 的"原子"（Atomic）库来规避相关问题。

● 5.10 节

假设你把衣服交给干洗店，店家会先给你一张收据，然后你去忙其他的事情，最后等衣服洗干净后，你再来取走。这张收据就是你和由干洗机在后台执行的干洗任务之间的关联。这便是 Java 5 中引入的 Future 所采用的方式。

Future 比之前的方式要稍微简单一些，但你还是得露面，并用收据取走你干洗的衣服。如果衣服没洗完，你还得等着。对于管道方式的操作来说，Future 真的没帮上多少忙。

Java 8 中的 CompletableFuture 是个好得多的方案：它允许你将多种操作连接到一起，这样就不必手动编写代码来连接一系列操作了。有了 CompletableFuture，想要将"采购食材、混合食材、烹饪食物、上菜、清理菜肴、储存菜肴"等操作作为一系列相关联的任务来执行，就会变得很简单了。

● 5.11 节、5.12 节

有些任务必须要等待（即**阻塞**（block））其他任务的执行结果。一个阻塞中的任务很可能在等待另一个阻塞的任务，而后者可能又在等待另一个阻塞的任务，以此类推。如果这条阻塞任务链最后指向了第一个阻塞任务，形成了循环，那么谁都无法继续下去，此时就形成了**死锁**（deadlock）。

如果死锁并没有在程序刚开始运行时就立刻暴露出来，那么最可怕的问题就发生了。

你的系统变成了**易死锁**（deadlock-prone）的系统，并且只会在一定条件下才发生死锁。程序可能在特定的平台（比如你的开发机）上运行良好，但一旦部署到别的硬件上，就开始发生死锁了。

死锁通常由很隐蔽的编程错误引发，一系列无意间的决策最终凑巧形成了依赖循环。本节会以一个经典的例子来诠释死锁那难以捉摸的天性。

- 5.13 节

本章最后，我们会模拟创建比萨的过程，先用并行流的方式来实现，然后再用 CompletableFuture 实现。这不只是为了比较两种方法，更重要的是为了诠释：要想提升程序的速度，你需要付出多大的努力。

5.7 并行流

Java 8 的流有个很大的好处：在某些时候，流很容易并行化。这来自库的精心设计，特别是流使用的**内部迭代**（internal iteration）的方式——也就是说，流会控制它们自身的迭代器。特别是流会使用一种称为**分流器**（spliterator）的特殊迭代器，其设计要求是要易于自动分割。这就产生了一个相当神奇的结果，即通过简单地直接使用 .parallel()，流中的一切就突然都可以作为一组并行的任务来运行了。如果代码使用了 Stream，那么就可以轻而易举通过并行化来提升速度。

举例来说，看一下基础卷第 14 章的 Prime.java。寻找素数是个很耗时的过程，我们可以通过给程序增加计时来证明：

```
// concurrent/ParallelPrime.java
import java.util.*;
import java.util.stream.*;
import static java.util.stream.LongStream.*;
import java.io.*;
import java.nio.file.*;
import onjava.Timer;

public class ParallelPrime {
  static final int COUNT = 100_000;
  public static boolean isPrime(long n) {
    return rangeClosed(2, (long)Math.sqrt(n))
      .noneMatch(i -> n % i == 0);
  }
  public static void main(String[] args)
    throws IOException {
    Timer timer = new Timer();
```

```
/* 输出：
1635
*/
```

```
    List<String> primes =
      iterate(2, i -> i + 1)
        .parallel()                          // [1]
        .filter(ParallelPrime::isPrime)
        .limit(COUNT)
        .mapToObj(Long::toString)
        .collect(Collectors.toList());
    System.out.println(timer.duration());
    Files.write(Paths.get("primes.txt"), primes,
      StandardOpenOption.CREATE);
  }
}
```

注意，我们给整个程序计时，这并不是为了做微型基准测试。我们将数据保存到磁盘上，是为了保护程序不受过度优化的影响，如果我们对结果什么都不做，狡猾的编译器可能会发现程序毫无意义，然后便终止了计算（这不大可能发生，但也不绝对如此）。注意用 nio2 库来写文件是多么简单（参见基础卷第 17 章）。

当我注释掉 [1] parallel() 这一行后，耗时大概是使用 parallel() 时的 3 倍。

并行流看起来十分诱人。你所需要的只是将程序要解决的问题转换为流，然后插入 parallel() 来提升速度。事实上，有时就是那么简单，但不幸的是，这样做会有很多隐患。

5.7.1　parallel() 并非灵丹妙药

为了探寻流和并行流的不确定性，我们来看一个貌似很简单的问题：对一系列递增的数字求和。事实证明，有很多方法可以实现它，而我会冒着一定风险，通过计时来比较它们——我会尽量小心，但也要承认，在计时代码执行的时候，我可能会掉入某个基本的（并发）陷阱。结果可能不会完全精确（比如 JVM 可能会没有"预热"），但是我觉得这段代码最终多少能给出一些有用的结论。

我首先会实现一个 timeTest() 方法，它接受 LongSupplier 为参数，并测量调用 getAsLong() 的执行耗时，然后将结果和 checkValue 做比较，最后显示结果。

注意，这里必须全部严格使用 long 类型，我一开始花了些时间追查那些隐蔽的溢出问题，后来才意识到是自己忽略了使用 long 类型的重要性。

这里所有关于时间和内存的数值和讨论，都是基于"我的"机器。你在自己的机器上运行这段代码，情况可能会有所不同。

```
// concurrent/Summing.java
import java.util.stream.*;
import java.util.function.*;
```

```java
import onjava.Timer;

public class Summing {
  static void timeTest(String id, long checkValue,
    LongSupplier operation) {
    System.out.print(id + ": ");
    Timer timer = new Timer();
    long result = operation.getAsLong();
    if(result == checkValue)
      System.out.println(timer.duration() + "ms");
    else
      System.out.format("result: %d%ncheckValue: %d%n",
        result, checkValue);
  }
  public static final int SZ = 100_000_000;
  // 甚至连这样都可以运行：
  // public static final int SZ = 1_000_000_000;
  public static final long CHECK =
    (long)SZ * ((long)SZ + 1)/2; // 高斯公式
  public static void main(String[] args) {
    System.out.println(CHECK);
    timeTest("Sum Stream", CHECK, () ->
      LongStream.rangeClosed(0, SZ).sum());
    timeTest("Sum Stream Parallel", CHECK, () ->
      LongStream.rangeClosed(0, SZ).parallel().sum());
    timeTest("Sum Iterated", CHECK, () ->
      LongStream.iterate(0, i -> i + 1)
        .limit(SZ+1).sum());
    // 超过 1 000 000 后会开始变慢或者内存溢出
    // timeTest("Sum Iterated Parallel", CHECK, () ->
    //   LongStream.iterate(0, i -> i + 1)
    //     .parallel()
    //     .limit(SZ+1).sum());
  }
}
/* 输出：
5000000050000000
Sum Stream: 841ms
Sum Stream Parallel: 179ms
Sum Iterated: 4051ms
*/
```

CHECK 值是由数学家卡尔·弗里德里希·高斯在 18 世纪晚期还在读小学时发明的公式计算得出的。

main() 方法的第一个版本用了最简单的方法，生成 Stream，并调用 sum()。我们可以看到流带来的好处，可以在保证内存不溢出的情况下，处理 10 亿大小的 SZ（我用了较小一些的数字，这样程序不会运行那么久）。用 parallel() 来进行基本的 range 操作则会显著提升程序的运行速度。

如果使用 iterate() 方法来生成序列，则速度会大为减慢，大概是因为每生成一个数字，都会调用一次 lambda 表达式。但是如果我们试着并行化该操作，结果不仅会比非并行的版本慢，而且当 SZ 超出 100 万后，还会导致内存溢出（在某些机器上）。当然，既然有了 range() 方法,你就肯定不会用 iterate(),但如果你要生成简单序列之外的东西,你就肯定得用 iterate()。使用 parallel() 是相当合理的想法,但会生成这些意外的结果。我们会在下一节探讨内存限制的原因,但是可以先对流并行算法做一些初步的观察总结。

- 流的并行化将输入的数据拆分成多个片段,这样就可以针对这些独立的数据片段应用各种算法。
- 数组的切分非常轻量、均匀,并且可以完全掌握切分的大小。
- 链表则完全没有这些属性,对链表"切分"仅仅意味着会将其拆分成"第一个元素"和"其余的部分",这并没有什么实际用处。
- 无状态生成器的表现很像数组,以上对 range 的使用就是无状态的。
- 迭代式生成器的表现很像链表,iterate() 就是一个迭代式生成器。

现在试着来实现这个需求:先给一个数组填充值,然后再对其求和。由于数组只分配了一次,看起来我们似乎不大可能遇到垃圾收集的时机问题。

首先试着将一个数组用基本类型 long 来填充:

```
// concurrent/Summing2.java
// {ExcludeFromTravisCI}
import java.util.*;

public class Summing2 {
  static long basicSum(long[] ia) {
    long sum = 0;
    int size = ia.length;
    for(int i = 0; i < size; i++)
      sum += ia[i];
    return sum;
  }
  // 在我的机器内存溢出前,SZ 的近似最大值:
  public static final int SZ = 20_000_000;
  public static final long CHECK =
    (long)SZ * ((long)SZ + 1)/2;
  public static void main(String[] args) {
    System.out.println(CHECK);
    long[] la = new long[SZ+1];
    Arrays.parallelSetAll(la, i -> i);
    Summing.timeTest("Array Stream Sum", CHECK, () ->
      Arrays.stream(la).sum());
    Summing.timeTest("Parallel", CHECK, () ->
      Arrays.stream(la).parallel().sum());
    Summing.timeTest("Basic Sum", CHECK, () ->
      basicSum(la));
    // 破坏性求和:
    Summing.timeTest("parallelPrefix", CHECK, () -> {
```

```
/* 输出:
200000010000000
Array Stream Sum: 166ms
Parallel: 30ms
Basic Sum: 45ms
parallelPrefix: 53ms
*/
```

```
    Arrays.parallelPrefix(la, Long::sum);
    return la[la.length - 1];
  });
  }
}
```

第一个限制来自内存大小，由于数组是预先分配的，所以我们无法创建任何和之前版本大小相近的东西。并行化可以提速，甚至比只是用 basicSum() 遍历还要快一点儿。有趣的是，Arrays.parallelPrefix() 看起来反而会使程序慢下来。然而所有这些技术，在其他条件下可能会更适用——这就是为什么你无法预先确定该怎么做，而只能"试着来"。

最后，换成装箱后的 Long，再来看看效果：

```
// concurrent/Summing3.java
// {ExcludeFromTravisCI}
import java.util.*;

public class Summing3 {
  static long basicSum(Long[] ia) {
    long sum = 0;
    int size = ia.length;
    for(int i = 0; i < size; i++)
      sum += ia[i];
    return sum;
  }
  // 在我的机器内存溢出前，SZ 的近似最大值：
  public static final int SZ = 10_000_000;
  public static final long CHECK =
    (long)SZ * ((long)SZ + 1)/2;
  public static void main(String[] args) {
    System.out.println(CHECK);
    Long[] aL = new Long[SZ+1];
    Arrays.parallelSetAll(aL, i -> (long)i);
    Summing.timeTest("Long Array Stream Reduce",
      CHECK, () ->
      Arrays.stream(aL).reduce(0L, Long::sum));
    Summing.timeTest("Long Basic Sum", CHECK, () ->
      basicSum(aL));
    // 破坏性求和：
    Summing.timeTest("Long parallelPrefix",CHECK, ()-> {
      Arrays.parallelPrefix(aL, Long::sum);
      return aL[aL.length - 1];
    });
  }
}
/* 输出：
50000005000000
Long Array Stream Reduce: 1510ms
Long Basic Sum: 35ms
Long parallelPrefix: 4306ms
*/
```

现在可用的内存大概减少了一半，各处计算所需的时间都呈爆炸式增长，除了 basicSum()，它只是简单地遍历了一遍数组。意外的是，Arrays.parallelPrefix() 明显比其他所有方法都慢很多。

还有 parallel() 的实现版本，如果将其放在上面的程序中，则会导致漫长的垃圾收集过程，所以我把它单独拿了出来：

```
// concurrent/Summing4.java
// {ExcludeFromTravisCI}
import java.util.*;

public class Summing4 {
  public static void main(String[] args) {
    System.out.println(Summing3.CHECK);
    Long[] aL = new Long[Summing3.SZ+1];
    Arrays.parallelSetAll(aL, i -> (long)i);
    Summing.timeTest("Long Parallel",
      Summing3.CHECK, () ->
      Arrays.stream(aL)
        .parallel()
        .reduce(0L,Long::sum));
  }
}
/* 输出：
50000005000000
Long Parallel: 1147ms
*/
```

该方式比未使用 parallel() 的版本要稍微快一点，但差距不大。

处理器的缓存机制是导致耗时增加的主要原因之一。由于 Summing2.java 中用的是基本类型 long，因此数组 la 是一段连续的内存，处理器会更容易预测到对这个数组的使用情况，从而将数组元素保存在缓存中以备后续所需，而访问缓存远远比跳出去访问主存要快。Long parallelPrefix 的计算看起来似乎受到了影响，因为它每次计算都要读取 2 个数组元素，还要将结果写回数组，每次这样的操作都会对 Long 生成一个缓存外的引用。

在 Summing3.java 和 Summing4.java 中，aL 都是 Long 型的数组，并不是一段连续的数值数组，而是一段连续的 Long 型对象引用的数组。尽管该数组很可能会保存在缓存中，但其指向的那些对象几乎永远在缓存之外。

这些例子都使用了不同的 SZ 值展现了内存的各种限制。为了便于比较不同的耗时，下面是将 SZ 设置为最小值 1000 万时的结果：

```
Sum Stream: 69ms
Sum Stream Parallel: 18ms
Sum Iterated: 277ms
Array Stream Sum: 57ms
Parallel: 14ms
Basic Sum: 16ms
parallelPrefix: 28ms
Long Array Stream Reduce: 1046ms
Long Basic Sum: 21ms
Long parallelPrefix: 3287ms
Long Parallel: 1008ms
```

由于 Java 8 中的各种内建"并行"工具非常好用，我发现它们已经被当成了神奇的灵丹妙药："只要使用 parallel()，就能让程序跑得更快！"我希望能让大家开始意识到，这么想完全是错误的。盲目地应用内建 parallel 操作有时反而能让程序变得特别慢。

5.7.2 `parallel()` 和 `limit()` 的作用

`parallel()` 还能带来更进一步的影响。和其他语言一样，流是围绕无限流的模型设计的。如果要处理有限数量的元素，就需要使用集合，以及专门为有限大小的集合所设计的相关算法。如果使用无限流，则需要使用这些专门为流优化后的算法。

Java 8 合并了以上这两种情况。举例来说，Collection 没有内建的 `map()` 操作，Collection 和 Map 中唯一的流式批处理操作是 `forEach()`。如果你想要执行类似 `map()` 和 `reduce()` 的操作，就需要首先将 Collection 转换为 Stream，这样才能有这些操作可用：

```java
// concurrent/CollectionIntoStream.java
import onjava.*;
import java.util.*;
import java.util.stream.*;

public class CollectionIntoStream {
  public static void main(String[] args) {
    List<String> strings =
      Stream.generate(new Rand.String(5))
        .limit(10)
        .collect(Collectors.toList());
    strings.forEach(System.out::println);
    // 转换为 Stream，以执行更多操作：
    String result = strings.stream()
      .map(String::toUpperCase)
      .map(s -> s.substring(2))
      .reduce(":", (s1, s2) -> s1 + s2);
    System.out.println(result);
  }
}
/* 输出：
btpen
pccux
szgvg
meinn
eeloz
tdvew
cippc
ygpoa
lkljl
bynxt
:PENCUXGVGINNLOZVEWPPCPOALJLNXT
*/
```

Collection 确实支持一些批处理操作，如 `removeAll()`、`removeIf()` 和 `retainAll()`，但这些都是破坏性的操作。ConcurrentHashMap 则对 `forEach` 和 `reduce` 操作有特别全面的支持。

在许多场景中，单纯地对集合调用 `stream()` 或 `parallelStream()` 是没有任何问题的。然而有时，Stream 和 Collection 同时出现则会造成意外，下面是个有趣的奇怪现象：

```java
// concurrent/ParallelStreamPuzzle.java
import java.util.*;
import java.util.function.*;
import java.util.stream.*;
```

```
public class ParallelStreamPuzzle {
  static class IntGenerator
  implements Supplier<Integer> {
    private int current = 0;
    @Override public Integer get() {
      return current++;
    }
  }
  public static void main(String[] args) {
    List<Integer> x =
      Stream.generate(new IntGenerator())
        .limit(10)
        .parallel()                              // [1]
        .collect(Collectors.toList());
    System.out.println(x);
  }
}
/* Output:
[0, 1, 2, 3, 4, 5, 6, 7, 8, 9]
*/
```

如果注释掉 [1]，则每次都会如预期的那样，得到：

```
[0, 1, 2, 3, 4, 5, 6, 7, 8, 9]
```

但是如果取消注释，使用了 .parallel()，则程序似乎变成了一个随机数生成器，输出（每次运行都不一样）类似这样：

```
[0, 3, 6, 8, 11, 14, 17, 20, 23, 26]
```

为何如此简单的程序也会这么不稳定？来想想我们此刻的目的："并行生成"。这意味着什么呢？一堆线程全都运行在一个生成器上，然后以某种方式选择一组有限的结果？代码看起来很简单，但最终造成了特别混乱的状况。

为了研究这个问题，需要添加一些工具。我们处理的是多线程，因此必须捕获所有的追踪信息，并保存到并发数据结构中。此处用到了 ConcurrentLinkedDeque：

```
// concurrent/ParallelStreamPuzzle2.java
import java.util.*;
import java.util.function.*;
import java.util.stream.*;
import java.util.concurrent.*;
import java.util.concurrent.atomic.*;
import java.nio.file.*;

/* 输出：
[1, 5, 7, 8, 9, 11, 13, 15, 18, 21]
*/

public class ParallelStreamPuzzle2 {
  public static final Deque<String> trace =
    new ConcurrentLinkedDeque<>();
```

```
  static class
  IntGenerator implements Supplier<Integer> {
    private AtomicInteger current =
      new AtomicInteger();
    @Override public Integer get() {
      trace.add(current.get() + ": " +
        Thread.currentThread().getName());
      return current.getAndIncrement();
    }
  }
  public static void
  main(String[] args) throws Exception {
    List<Integer> x =
      Stream.generate(new IntGenerator())
        .limit(10)
        .parallel()
        .collect(Collectors.toList());
    System.out.println(x);
    Files.write(Paths.get("PSP2.txt"), trace);
  }
}
```

current 由线程安全的 AtomicInteger 类来定义，以避免竞态条件的发生，parallel() 则允许多线程来调用 get() 方法：

当你看到 PSP2.txt 时，可能会感到吃惊，IntGenerator.get() 竟然被调用了 1024 次！

```
0: main
1: ForkJoinPool.commonPool-worker-1
2: ForkJoinPool.commonPool-worker-2
3: ForkJoinPool.commonPool-worker-2
4: ForkJoinPool.commonPool-worker-1
5: ForkJoinPool.commonPool-worker-1
6: ForkJoinPool.commonPool-worker-1
7: ForkJoinPool.commonPool-worker-1
8: ForkJoinPool.commonPool-worker-4
9: ForkJoinPool.commonPool-worker-4
10: ForkJoinPool.commonPool-worker-4
11: main
12: main
13: main
14: main
15: main
...
1017: ForkJoinPool.commonPool-worker-1
1018: ForkJoinPool.commonPool-worker-6
1019: ForkJoinPool.commonPool-worker-6
1020: ForkJoinPool.commonPool-worker-1
1021: ForkJoinPool.commonPool-worker-1
1022: ForkJoinPool.commonPool-worker-1
1023: ForkJoinPool.commonPool-worker-1
```

这些分块的大小似乎是由内部实现决定的（试试向 limit() 传入不同的参数，可以看到不同的分块大小）。将 parallel() 和 limit() 一起配合使用，可以告诉程序预先选取一组值，以作为流输出。

试着想象一下这里面究竟发生了些什么：流抽象了一个可按需生产的无限序列。当你让它以并行方式生成流时，实际是在让所有的线程都尽可能调用 get() 方法。加上 limit() 意味着你想要的是"只需要**一些**"。基本上，如果你同时使用 parallel() 和 limit()，那就是在请求随机的输出——对当前要解决的需求来说，这可能没什么问题，但你这么做的时候，必须要清楚这一点。此功能只适合高手使用，并不能拿来作为证明"Java 运行有问题"的理由。

对于该问题来说，怎样的实现方式更合理呢？如果你想要生成 int 流，可以使用 IntStream.range()，就像这样：

```java
// concurrent/ParallelStreamPuzzle3.java
// {VisuallyInspectOutput}
import java.util.*;
import java.util.stream.*;

public class ParallelStreamPuzzle3 {
  public static void main(String[] args) {
    List<Integer> x = IntStream.range(0, 30)
      .peek(e -> System.out.println(e + ": " +
        Thread.currentThread().getName()))
      .limit(10)
      .parallel()
      .boxed()
      .collect(Collectors.toList());
    System.out.println(x);
  }
}
/* 输出：
8: main
6: ForkJoinPool.commonPool-worker-5
3: ForkJoinPool.commonPool-worker-7
5: ForkJoinPool.commonPool-worker-5
1: ForkJoinPool.commonPool-worker-3
2: ForkJoinPool.commonPool-worker-6
4: ForkJoinPool.commonPool-worker-1
0: ForkJoinPool.commonPool-worker-4
7: ForkJoinPool.commonPool-worker-1
9: ForkJoinPool.commonPool-worker-2
[0, 1, 2, 3, 4, 5, 6, 7, 8, 9]
*/
```

为了证明 .parallel() 确实有用，我增加了对 peek()（流的一个函数，大部分情况下用来调试）的调用：它会从流中拉取出一个值（来进行想要的操作），但并不会影响在流中传递下去的元素。注意，这会干扰线程的行为，此处我只是为了给大家演示，并不是真的在调试。

你还可以看到增加了对 `boxed()` 的调用，它接收 int 流，并将其转换为 Integer 流。

现在我们得到了生成不同值的多个线程，但也只会根据要求生成 10 个值，而不是 1024 个线程来生成 10 个值。

这会快一些吗？更好的问题是：怎样才能开始变得合理？当然不是指这么小的数据量，上下文切换的开销很可能远远超过任何并行化所带来的速度提升。难以想象，什么时候才有必要用并行生成一个简单的数字序列。如果你要使用的对象的创建开销非常大，这种情况下可能有必要——但这些只是猜测，只有通过测试才能知道答案。记住那条格言："先运行起来，再优化，并且仅在必要时。" `parallel()` 和 `limit()` 的搭配使用仅限高手（特此澄清，我并不认为自己是高手）。

5.7.3 并行流只是看起来很简单

实际上，在很多情况下，并行流确实可以毫不费力地更快生成结果。但是如你所见，简单粗暴地在 Stream 操作上使用 `parallel()` 并不一定是安全的做法。在使用 `parallel()` 之前，你必须了解并行化可能会对你的操作带来的怎样的影响，究竟是有利还是有害。这里面最根本的错误便是将并行化奉为灵丹妙药——它并不是。流意味着你无须完全重写代码，就可以使其并行化工作。但流永远无法取代你的思考，包括对并行化运行原理的理解，以及它是否能帮你达成目标。

5.8 创建和运行任务

如果你无法通过并行流来实现并发，那么你就必须自行创建和运行任务。稍后你会看到，理想的 Java 8 版本的实现方式是 `CompletableFuture`，不过我们会用更多的基本工具来介绍该概念。

Java 并发的历史始于非常原始和简陋的机制，并且遍布各种优化尝试的痕迹。这部分内容主要收录在本书第 6 章中。而在这里，我们会通过一套经典的用法来展现创建并运行任务的最简及最佳方法。和所有并发的事物一样，这里面也会有各种各样的变体，但这些变体要么收录在本书第 6 章中，要么超出了本书的范围。

5.8.1 Task 和 Executor

在 Java 的早期版本中，要使用多线程，就需要直接创建你自己的 `Thread` 对象，甚至是实现它们的子类以创建自定义的特殊"任务 – 线程"对象。你需要手动调用构造器，并且自行启动线程。

创建所有这些线程的开销都非常大，所以现在并不鼓励这些手动的方法。Java 5 专门新增了一些类来处理**线程池**。你不再需要为每个不同的任务类型都创建一个新的 Thread 子类，而只需将任务创建为一个单独的类型，然后传递给某个 ExecutorService 来运行该任务。该 ExecutorService 会为你管理多线程，并且在线程完成任务后不会丢弃，而是会回收它们。

一开始，我们会创建一个几乎什么都不做的任务。它会"睡眠"（挂起执行）100 毫秒，然后显示它的标识符和正在执行任务的 Thread 名称，最后结束：

```java
// concurrent/NapTask.java
import onjava.Nap;

public class NapTask implements Runnable {
  final int id;
  public NapTask(int id) { this.id = id; }
  @Override public void run() {
    new Nap(0.1); // 秒
    System.out.println(this + " " +
      Thread.currentThread().getName());
  }
  @Override public String toString() {
    return "NapTask[" + id + "]";
  }
}
```

这是个简单的 Runnable：一个包含 run() 方法的类。它在实现中并未实际运行任何任务。

我们通过 Nap 类来实现"睡眠"：

```java
// onjava/Nap.java
package onjava;
import java.util.concurrent.*;

public class Nap {
  public Nap(double t) { // 秒
    try {
      TimeUnit.MILLISECONDS.sleep((int)(1000 * t));
    } catch(InterruptedException e) {
      throw new RuntimeException(e);
    }
  }
  public Nap(double t, String msg) {
    this(t);
    System.out.println(msg);
  }
}
```

为了消除异常处理产生的视觉干扰，它被定义为公共工具类。第二个构造器在时间结束后会显示出相关信息。

调用 TimeUnit.MILLISECONDS.sleep()，会获得"当前线程"，并让它按参数中传入的时长睡眠，这意味着该线程将被挂起。但这并不意味着底层的处理器停止了。操作系统会切换到某些其他任务，比如运行计算机上的其他窗口。操作系统的**任务管理器**会定期检查 sleep() 是否到时间了。如果时间到了，线程会被"唤醒"，并继续分配给处理器时间。

你可以看到，sleep() 会抛出 InterruptedException 异常，这是 Java 早期设计的产物，通过立刻跳出任务来终止它们。由于这容易产生不稳定的状态，后续便不再鼓励如此终止任务了。然而我们必须捕获各种情况下的该异常，以应对必要或不可控的任务终止。

要执行任务，我们从最简单的方式开始，即 SingleThreadExecutor：

```
// concurrent/SingleThreadExecutor.java
import java.util.concurrent.*;
import java.util.stream.*;
import onjava.*;

public class SingleThreadExecutor {
  public static void main(String[] args) {
    ExecutorService exec =
      Executors.newSingleThreadExecutor();
    IntStream.range(0, 10)
      .mapToObj(NapTask::new)
      .forEach(exec::execute);
    System.out.println("All tasks submitted");
    exec.shutdown();
    while(!exec.isTerminated()) {
      System.out.println(
        Thread.currentThread().getName() +
        " awaiting termination");
      new Nap(0.1);
    }
  }
}
/* 输出：
All tasks submitted
main awaiting termination
main awaiting termination
NapTask[0] pool-1-thread-1
main awaiting termination
NapTask[1] pool-1-thread-1
NapTask[2] pool-1-thread-1
main awaiting termination
NapTask[3] pool-1-thread-1
main awaiting termination
main awaiting termination
NapTask[4] pool-1-thread-1
NapTask[5] pool-1-thread-1
main awaiting termination
NapTask[6] pool-1-thread-1
main awaiting termination
NapTask[7] pool-1-thread-1
main awaiting termination
NapTask[8] pool-1-thread-1
main awaiting termination
NapTask[9] pool-1-thread-1
*/
```

（转右栏）

首先请注意,并不存在 SingleThreadExecutor 类。newSingleThreadExecutor() 是 Executors 中的工厂方法,用于创建特定类型的 ExecutorService。[①]

我创建了 10 个 NapTask 并将它们提交到了 ExecutorService,这意味着它们会自行启动。然而同时,main() 方法会继续处理其他的事。当我调用 exec.shutdown() 时,会告诉 ExecutorService 完成所有已提交的任务,但不再接收任何新任务。不过此时那些任务仍在运行,所以我们在退出 main() 前必须等待这些任务完成。这是通过检查 exec.isTerminated() 的结果来实现的,所有的任务都完成后,该方法会返回 true。

注意 main() 中的线程的名字是 main,除此以外只有唯一的一个线程 pool-1-thread-1。同样,从交错的输出也可以看出,这两个线程确实在并发地运行着。

如果你仅仅调用 exec.shutdown(),程序会在所有任务完成后立即结束。也就是说,while(!exec.isTerminated()) 并不是必需的:

```java
// concurrent/SingleThreadExecutor2.java
import java.util.concurrent.*;
import java.util.stream.*;

public class SingleThreadExecutor2 {
  public static void main(String[] args)
    throws InterruptedException {
    ExecutorService exec =
      Executors.newSingleThreadExecutor();
    IntStream.range(0, 10)
      .mapToObj(NapTask::new)
      .forEach(exec::execute);
    exec.shutdown();
  }
}
/* 输出:
NapTask[0] pool-1-thread-1
NapTask[1] pool-1-thread-1
NapTask[2] pool-1-thread-1
NapTask[3] pool-1-thread-1
NapTask[4] pool-1-thread-1
NapTask[5] pool-1-thread-1
NapTask[6] pool-1-thread-1
NapTask[7] pool-1-thread-1
NapTask[8] pool-1-thread-1
NapTask[9] pool-1-thread-1
*/
```

一旦调用了 exec.shutdown(),此后再想提交新任务,就会抛出 RejectedExecutionException 异常:

```java
// concurrent/MoreTasksAfterShutdown.java
import java.util.concurrent.*;

public class MoreTasksAfterShutdown {
  public static void main(String[] args) {
    ExecutorService exec =
      Executors.newSingleThreadExecutor();
    exec.execute(new NapTask(1));
    exec.shutdown();
```

[①] 尽管有些错乱,但这确是个有趣的方法。我们一般会期望用显式的类来表达公有接口上的不同行为。

```
        try {
          exec.execute(new NapTask(99));
        } catch(RejectedExecutionException e) {
          System.out.println(e);
        }
      }
    }
```

```
/* 输出:
java.util.concurrent.RejectedExecutionException: Task
NapTask[99] rejected from java.util.concurrent.ThreadPo
olExecutor@106d69c[Shutting down, pool size = 1, active
threads = 1, queued tasks = 0, completed tasks = 0]
NapTask[1] pool-1-thread-1
*/
```

exec.shutdown() 的兄弟方法是 exec.shutdownNow()，其作用是不再接受新任务，同时还会尝试通过中断来停止所有正在运行的任务。再次提醒，中断线程容易引发混乱和错误，并不鼓励这么做。

5.8.2 使用更多的线程

使用多线程的主要目的（几乎）总是使任务完成得更快一些，所以我们为何要将自己限制在 SingleThreadExecutor 中呢？看看 Javadoc 中的 Executors 部分，你会发现更多的选择。比如，CachedThreadPool：

```java
// concurrent/CachedThreadPool.java
import java.util.concurrent.*;
import java.util.stream.*;

public class CachedThreadPool {
  public static void main(String[] args) {
    ExecutorService exec =
      Executors.newCachedThreadPool();
    IntStream.range(0, 10)
      .mapToObj(NapTask::new)
      .forEach(exec::execute);
    exec.shutdown();
  }
}
```

```
/* 输出:
NapTask[1] pool-1-thread-2
NapTask[5] pool-1-thread-6
NapTask[4] pool-1-thread-5
NapTask[8] pool-1-thread-9
NapTask[0] pool-1-thread-1
NapTask[9] pool-1-thread-10
NapTask[2] pool-1-thread-3
NapTask[3] pool-1-thread-4
NapTask[6] pool-1-thread-7
NapTask[7] pool-1-thread-8
*/
```

当你运行该程序时，你会注意到它完成得更快。这很合理，因为不再用同一个线程来按顺序运行所有任务，而是每个任务都能获得属于自己的线程，所以它们都是并行的了。这看起来似乎并没有缺点，由此也很难想通还会有谁会去用 SingleThreadExecutor。

要理解这个问题，我们需要一个更复杂的任务：

```
// concurrent/InterferingTask.java
```

```
public class InterferingTask implements Runnable {
  final int id;
  private static Integer val = 0;
  public InterferingTask(int id) { this.id = id; }
  @Override public void run() {
    for(int i = 0; i < 100; i++)
      val++;
    System.out.println(id + " " +
      Thread.currentThread().getName() + " " + val);
  }
}
```

每个任务都会使 val 自增 100 次。这看起来够简单了，让我们用 CachedThreadPool 来试试：

```
// concurrent/CachedThreadPool2.java
import java.util.concurrent.*;
import java.util.stream.*;

public class CachedThreadPool2 {
  public static void main(String[] args) {
    ExecutorService exec =
      Executors.newCachedThreadPool();
    IntStream.range(0, 10)
      .mapToObj(InterferingTask::new)
      .forEach(exec::execute);
    exec.shutdown();
  }
}
/* 输出：
3 pool-1-thread-4 100
2 pool-1-thread-3 200
6 pool-1-thread-7 300
7 pool-1-thread-8 400
0 pool-1-thread-1 514
1 pool-1-thread-2 527
4 pool-1-thread-5 627
5 pool-1-thread-6 727
9 pool-1-thread-10 827
8 pool-1-thread-9 927
*/
```

输出结果并不如我们所料，并且每次运行结果都不一样。问题在于所有的任务都在试图对单例的 val 进行写操作，它们在相互打架。我们认为这样的类是**非线程安全**（not thread-safe）的。再来看看用 SingleThreadExecutor 会是怎样的情形：

```
// concurrent/SingleThreadExecutor3.java
import java.util.concurrent.*;
import java.util.stream.*;

public class SingleThreadExecutor3 {
  public static void main(String[] args)
    throws InterruptedException {
    ExecutorService exec =
      Executors.newSingleThreadExecutor();
    IntStream.range(0, 10)
      .mapToObj(InterferingTask::new)
      .forEach(exec::execute);
    exec.shutdown();
  }
}
/* 输出：
0 pool-1-thread-1 100
1 pool-1-thread-1 200
2 pool-1-thread-1 300
3 pool-1-thread-1 400
4 pool-1-thread-1 500
5 pool-1-thread-1 600
6 pool-1-thread-1 700
7 pool-1-thread-1 800
8 pool-1-thread-1 900
9 pool-1-thread-1 1000
*/
```

现在我们每次都能得到一致的结果了，尽管 InterferingTask 缺乏线程安全性。这是 SingleThreadExecutor 最大的好处——由于其同时只会运行一项任务，这些任务永远不会互相影响，因此保证了线程安全性。这样的现象称为**线程封闭**（thread confinement），因为将多个任务运行在单线程上可以限制它们之间的影响。线程封闭限制了提速，但也节省了很多困难的调试和重写工作。

5.8.3 生成结果

由于 InterferingTask 是一个 Runnable 的实现，它并没有返回值，因此只能通过副作用生成结果，即通过控制环境来生成（而不是直接返回）结果。副作用是并发编程的主要问题之一，原因正如我们在 CachedThreadPool2.java 中所见。InterferingTask 中的 val 称作**可变共享状态**（mutable shared state），正是它带来了问题：多个任务同时修改同一个变量会导致所谓的竞态条件。结果由哪个任务抢先得到终点并修改了变量（以及其他各种可能性）而决定。

避免竞态条件的最好方法是避免使用可变共享状态。我们可以称其为**自私儿童原则**（selfish child principle）：什么都不共享。

对于 InterferingTask，如果能消除副作用，并且只返回任务结果就好了。要达到这个目的，我们需要创建一个 Callable，而不是 Runnable：

```java
// concurrent/CountingTask.java
import java.util.concurrent.*;

public class CountingTask implements Callable<Integer> {
  final int id;
  public CountingTask(int id) { this.id = id; }
  @Override public Integer call() {
    Integer val = 0;
    for(int i = 0; i < 100; i++)
      val++;
    System.out.println(id + " " +
      Thread.currentThread().getName() + " " + val);
    return val;
  }
}
```

call() 完全独立地生成结果，独立于任何其他的 CountingTask，这意味着并不会存在可变共享状态。

ExecutorService 允许在集合中通过 invokeAll() 来启动所有的 Callable：

```
// concurrent/CachedThreadPool3.java
```

```
import java.util.*;
import java.util.concurrent.*;
import java.util.stream.*;

public class CachedThreadPool3 {
  public static Integer
  extractResult(Future<Integer> f) {
    try {
      return f.get();
    } catch(Exception e) {
      throw new RuntimeException(e);
    }
  }
  public static void main(String[] args)
    throws InterruptedException {
    ExecutorService exec =
      Executors.newCachedThreadPool();
    List<CountingTask> tasks =
      IntStream.range(0, 10)
        .mapToObj(CountingTask::new)
        .collect(Collectors.toList());
    List<Future<Integer>> futures =
      exec.invokeAll(tasks);
    Integer sum = futures.stream()
      .map(CachedThreadPool3::extractResult)
      .reduce(0, Integer::sum);
    System.out.println("sum = " + sum);
    exec.shutdown();
  }
}
/* 输出：
0 pool-1-thread-1 100
9 pool-1-thread-1 100
1 pool-1-thread-2 100
2 pool-1-thread-3 100
3 pool-1-thread-4 100
6 pool-1-thread-7 100
7 pool-1-thread-8 100
4 pool-1-thread-5 100
5 pool-1-thread-6 100
8 pool-1-thread-9 100
sum = 1000
*/
```

只有当所有的任务都完成时，invokeAll() 才会返回由 Future 组成的 List，每个 Future 都对应一个任务。Future 是 Java 5 引入的机制，它允许你提交一个任务，并且不需要等待它完成。此处，我们使用了 ExecutorService.submit()：

```
// concurrent/Futures.java
import java.util.*;
import java.util.concurrent.*;
import java.util.stream.*;

public class Futures {
  public static void main(String[] args)
    throws InterruptedException, ExecutionException {
    ExecutorService exec =
      Executors.newSingleThreadExecutor();
    Future<Integer> f =
      exec.submit(new CountingTask(99));
    System.out.println(f.get());            // [1]
    exec.shutdown();
  }
}
/* 输出：
99 pool-1-thread-1 100
100
*/
```

[1] 当你对尚未完成任务的 Future 调用 get() 方法时，调用会持续阻塞（等待），直到结果可用。

但这意味着，在 CachedThreadPool3.java 中，该 Future 似乎有些冗余，因为在所有任务都完成之前，invokeAll() 甚至不会返回。然而，此处 Future 并不是用于延迟得到结果，而是为了捕获任何可能发生的异常。

还要注意 CachedThreadPool3.java 中提取结果部分的混乱状况。get() 会抛出异常，因此 extractResult() 是在 Stream 内部完成提取的。

由于在调用 get() 时 Future 会阻塞，因此它只是将等待任务完成的问题推迟了。最终，Future 被认为是一个无效的解决办法，现在人们并不鼓励使用它，而是更推荐 Java 8 的 CompletableFuture，本章稍后会介绍它。当然，在各种遗留的库中，你还是会遇到 Future。

我们可以用简单得多也更优雅的方式——并行 Stream——来解决这个问题：

```java
// concurrent/CountingStream.java
// {VisuallyInspectOutput}
import java.util.*;
import java.util.concurrent.*;
import java.util.stream.*;

public class CountingStream {
    public static void main(String[] args) {
        System.out.println(
            IntStream.range(0, 10)
                .parallel()
                .mapToObj(CountingTask::new)
                .map(ct -> ct.call())
                .reduce(0, Integer::sum));
    }
}
/* 输出：
1 ForkJoinPool.commonPool-worker-3 100
8 ForkJoinPool.commonPool-worker-2 100
0 ForkJoinPool.commonPool-worker-6 100
2 ForkJoinPool.commonPool-worker-1 100
4 ForkJoinPool.commonPool-worker-5 100
9 ForkJoinPool.commonPool-worker-7 100
6 main 100
7 ForkJoinPool.commonPool-worker-4 100
5 ForkJoinPool.commonPool-worker-2 100
3 ForkJoinPool.commonPool-worker-3 100
1000
*/
```

这样不仅更容易理解，而且我们所需做的只是将 parallel() 插入一个顺序操作中，然后一切就突然都以并发的方式运行了。

5.8.4　作为任务的 lambda 与方法引用

有了 lambda 和方法引用，你将不再仅限于使用 Runnable 和 Callable。由于 Java 8 通过匹配签名的方式支持 lambda 和方法引用 [也就是支持**结构一致性**（structural conformance）]，我们可以将**非** Runnable 或 Callable 类型的参数传递给 ExecutorService：

```java
// concurrent/LambdasAndMethodReferences.java
import java.util.concurrent.*;

class NotRunnable {
  public void go() {
    System.out.println("NotRunnable");
  }
}

class NotCallable {
  public Integer get() {
    System.out.println("NotCallable");
    return 1;
  }
}

public class LambdasAndMethodReferences {
  public static void main(String[] args)
    throws InterruptedException {
    ExecutorService exec =
      Executors.newCachedThreadPool();
    exec.submit(() -> System.out.println("Lambda1"));
    exec.submit(new NotRunnable()::go);
    exec.submit(() -> {
      System.out.println("Lambda2");
      return 1;
    });
    exec.submit(new NotCallable()::get);
    exec.shutdown();
  }
}
/* 输出：
Lambda1
NotRunnable
Lambda2
NotCallable
*/
```

此处，前两次 submit() 调用可以替代为 'execute()' 调用。所有的 submit() 调用都会返回 Future，你可以在第二次的两个调用中用它们来提取结果。

5.9　终止长时间运行的任务

并发程序通常会运行耗时较长的任务。Callable 的任务在完成时会返回值，虽然这给了它有限的生命周期，但仍然还是可以运行很久。Runnable 任务有时会被设置为永久运行的后台进程。你时常会需要某种方式来在 Runnable/Callable 任务正常结束前提前终止它们，比如要关闭某个程序的时候。

Java 最初的设计提供了某种机制（为了向后兼容，该机制仍然存在）来**中断**（interrupt）

正在运行的任务，中断机制在阻塞方面存在一些问题。中断任务是混乱复杂的，因为你必须理解所有可能导致中断发生的状态，以及可能导致的数据丢失。中断被认为是一种反模式，但由于向后兼容设计的残留，我们仍然不得不捕获 InterruptedException 异常。

终止任务最好的方法是设置一个任务会定期检查的标识，由此任务可以通过自己的关闭流程来优雅地终止。你可以请求任务在合适的时候自行终止，而不是在某一时刻突然拔掉任务的插头。这样的结果永远都比中断更好，而且代码也更清晰、更好理解。

像这样终止任务听起来足够简单：设置一个任务可见的 boolean 标识。修改任务，让其定期检查该标识，并优雅地终止。你确实只需要这么做，但还是有麻烦的地方——我们的老敌人——共享可变状态。如果该标识可以被其他任务操作，那么就有可能产生冲突。

如果你研究 Java 的相关文献，会发现很多解决该问题的方法，最常见的是通过 volatile 关键字。我们应该使用更简单的技术，以避免 volatile 所带来的所有不确定性，具体详见本书第 6 章。

Java 5 引入了 Atomic 类，它提供了一组类型，让你可以无须担心并发问题，并放心使用。下面我们引入 AtomicBoolean 标识来告诉任务自行清理并退出：

```java
// concurrent/QuittableTask.java
import java.util.concurrent.atomic.AtomicBoolean;
import onjava.Nap;

public class QuittableTask implements Runnable {
  final int id;
  public QuittableTask(int id) { this.id = id; }
  private AtomicBoolean running =
    new AtomicBoolean(true);
  public void quit() { running.set(false); }
  @Override public void run() {
    while(running.get())                 // [1]
      new Nap(0.1);
    System.out.print(id + " ");          // [2]
  }
}
```

虽然多个任务可以成功地在同一个实例中调用 quit()，但 AtomicBoolean 阻止了多个任务同时修改 running，由此保证 quit() 方法是线程安全的。

[1] 只要 running 标识还是 true，该任务的 run() 方法就会持续执行。

[2] 在任务退出后才会执行本行输出。

running 需要定义为 AtomicBoolean，这一点体现出了 Java 并发程序编写的最大难点之一。如果将 running 设为一个普通的 boolean，你可能永远不会看到程序在执行中出现问题。的确，在本例中，可能永远不会出现任何问题——但代码仍然是不安全的。要编写测试代码来暴露问题，这很难甚至不可能做到。因此，并没有实时的反馈来告诉你：你做错了。一般来说，唯一让你可以写出线程安全代码的方法，就是掌握所有可能导致问题的隐患。

为了测试，我们启动了很多个 QuittableTask 任务，然后关闭它们。试试将 COUNT 值设置得更大：

```java
// concurrent/QuittingTasks.java
import java.util.*;
import java.util.stream.*;
import java.util.concurrent.*;
import onjava.Nap;

public class QuittingTasks {
  public static final int COUNT = 150;
  public static void main(String[] args) {
    ExecutorService es =
      Executors.newCachedThreadPool();
    List<QuittableTask> tasks =
      IntStream.range(1, COUNT)
        .mapToObj(QuittableTask::new)
        .peek(qt -> es.execute(qt))
        .collect(Collectors.toList());
    new Nap(1);
    tasks.forEach(QuittableTask::quit);
    es.shutdown();
  }
}
/* 输出：
11 23 20 12 24 16 19 15 35 147 32 27 7 4 28 31 8 83 3 1
13 9 5 2 6 18 14 17 21 25 22 26 29 30 33 34 37 41 38 46
45 49 50 53 57 58 54 69 104 112 40 73 74 115 116 70 119
77 81 56 85 78 82 111 86 90 48 89 36 108 107 44 55 52
43 60 63 59 64 71 68 67 75 76 72 79 80 84 87 88 66 91
10 95 65 96 94 92 62 100 61 93 47 39 51 99 103 128 123
127 124 140 120 139 136 135 143 148 144 105 102 131 101
132 98 97 149 137 134 42 106 110 109 114 133 113 117
118 130 129 126 121 125 122 138 141 145 142 146
*/
```

在将任务收录到 List 中之前，我通过 peek() 将 QuittableTask 传给 ExecutorService。

main() 保证只要还有任务在运行，程序就不会退出。任务关闭的顺序和创建的顺序并不一致，即使每个任务是按顺序调用 quit() 方法的。这些独立运行的任务对信号的响应是不可控的。

5.10 CompletableFuture

为了让你有个初步印象，下面先将 QuittingTasks.java 改为用 CompletableFuture 来实现：

```java
// concurrent/QuittingCompletable.java
import java.util.*;
import java.util.stream.*;
import java.util.concurrent.*;
import onjava.Nap;

public class QuittingCompletable {
  public static void main(String[] args) {
    List<QuittableTask> tasks =
      IntStream.range(1, QuittingTasks.COUNT)
        .mapToObj(QuittableTask::new)
        .collect(Collectors.toList());
    List<CompletableFuture<Void>> cfutures =
      tasks.stream()
        .map(CompletableFuture::runAsync)
        .collect(Collectors.toList());
    new Nap(1);
    tasks.forEach(QuittableTask::quit);
    cfutures.forEach(CompletableFuture::join);
  }
}
/* 输出：
6 7 5 9 11 12 13 14 15 16 17 18 19 20 21 22 23 10 24 26
27 28 29 30 31 32 25 33 8 36 37 38 39 40 41 42 43 44 45
46 47 48 49 50 51 52 53 54 55 56 57 58 59 60 61 62 63
64 65 66 67 68 69 70 71 72 73 74 75 76 77 78 79 80 81
82 83 84 85 86 87 88 89 90 91 92 93 94 95 96 97 98 99
100 101 102 103 104 105 106 107 108 34 110 111 112 113
114 115 116 117 118 119 120 121 109 123 124 125 126 127
128 129 130 131 35 132 122 134 133 136 135 138 140 141
142 143 144 145 146 147 148 149 137 139 3 1 2 4
*/
```

和在 QuittingTasks.java 中一样，tasks 是 List<QuittableTask> 类型，但是在本例中，并没有使用 peek() 来将 QuittableTask 逐个提交给 ExecutorService，而是在 cfutures 的创建过程中将任务传给了 CompletableFuture::runAsync，这样就会执行 QuittableTask.run()，并返回 CompletableFuture<Void>。由于 run() 并不会返回任何东西，因此在这里我只用了 CompletableFuture 来调用 join()，以等待完成。

本例中需要注意一个重点：并不要求用 ExecutorService 来运行任务。这是由 CompletableFuture 管理的（虽然可以选择实现自定义的 ExecutorService）。你也无须调用 shutdown()，事实上除非你和我一样显式地调用 join()，否则程序会在第一时间退出，而不会等待任务的完成。

本例只是一个起点，你很快就会看到 CompletableFuture 的能力比这要强得多。

5.10.1 基本用法

下面这个类通过静态的 work() 方法对该类对象执行了某些操作：

```java
// concurrent/Machina.java
import onjava.Nap;

public class Machina {
  public enum State {
    START, ONE, TWO, THREE, END;
    State step() {
      if(equals(END)) return END;
      return values()[ordinal() + 1];
    }
  }
  private State state = State.START;
  private final int id;
  public Machina(int id) { this.id = id; }
  public static Machina work(Machina m) {
    if(!m.state.equals(State.END)){
      new Nap(0.1);
      m.state = m.state.step();
    }
    System.out.println(m);
    return m;
  }
  @Override public String toString() {
    return "Machina" + id + ": " +
      (state.equals(State.END)? "complete" : state);
  }
}
```

这其实是个**有限状态机**，不过它是随便写的，因为并没有分支……它只是从一条路径的头部移动到尾部。work() 方法使状态机从一个状态移动到下一个状态，并请求了 100 毫秒来执行该 "work"。

还可以利用 CompletableFuture，通过 completedFuture() 方法来包装一个对象：

```java
// concurrent/CompletedMachina.java
import java.util.concurrent.*;

public class CompletedMachina {
  public static void main(String[] args) {
    CompletableFuture<Machina> cf =
      CompletableFuture.completedFuture(
```

```
      new Machina(0));
    try {
      Machina m = cf.get();   // 不会阻塞
    } catch(InterruptedException |
            ExecutionException e) {
      throw new RuntimeException(e);
    }
  }
}
```

completedFuture() 创建了一个"已完成"的 CompletableFuture。这种 future 唯一能做的事是 get() 内部对象，所以乍一看这样做并没有什么用处。

注意 CompletableFuture 的类型为它所包含的对象，这很重要。

一般来说，get() 会阻塞正在等待结果的被调用线程。该阻塞可以通过 InterruptedException 或 ExecutionException 来退出。在本场景下，由于 CompletableFuture 已经完成，因此永远不会发生阻塞，当时就能得到结果。

有趣的是，一旦将 Machina 用 CompletableFuture 包装起来，我们就会发现，可以通过在 CompletableFuture 上增加操作来控制其包含的对象：

```java
// concurrent/CompletableApply.java
import java.util.concurrent.*;

public class CompletableApply {
  public static void main(String[] args) {
    CompletableFuture<Machina> cf =
      CompletableFuture.completedFuture(
        new Machina(0));
    CompletableFuture<Machina> cf2 =
      cf.thenApply(Machina::work);
    CompletableFuture<Machina> cf3 =
      cf2.thenApply(Machina::work);
    CompletableFuture<Machina> cf4 =
      cf3.thenApply(Machina::work);
    CompletableFuture<Machina> cf5 =
      cf4.thenApply(Machina::work);
  }
}
/* 输出：
Machina0: ONE
Machina0: TWO
Machina0: THREE
Machina0: complete
*/
```

thenApply() 用到了一个接收输入并生成输出的 Function。在本例中，work() 这个 Function 返回和输入相同的类型，由此每个得出的 CompletableFuture 都仍旧是 Machina 类型，但是（类似于 Stream 中的 map()）Function 也可以返回不同的类型，这可以从返回类型看出来。

可以从中看出 CompletableFuture 的一些本质：当执行某个操作时，它们会自动对其所携带的对象拆开包装，再重新包装。这样你就不会陷入混乱的细节，从而可以大幅简化代码的编写和理解工作。

我们可以消除中间变量，将多个操作串联起来，就像我们使用 Stream 时那样：

```java
// concurrent/CompletableApplyChained.java
import java.util.concurrent.*;
import onjava.Timer;

public class CompletableApplyChained {
  public static void main(String[] args) {
    Timer timer = new Timer();
    CompletableFuture<Machina> cf =
      CompletableFuture.completedFuture(
        new Machina(0))
      .thenApply(Machina::work)
      .thenApply(Machina::work)
      .thenApply(Machina::work)
      .thenApply(Machina::work);
    System.out.println(timer.duration());
  }
}
/* 输出：
Machina0: ONE
Machina0: TWO
Machina0: THREE
Machina0: complete
521
*/
```

此处我们还增加了一个 Timer，通过它可以看到每一步都增加了 100 毫秒，并且还有一些额外的开销。

使用 CompletableFuture 有个重要的好处——会促使我们应用**自私儿童原则**（什么都不共享）。默认情况下，通过 thenApply() 来应用函数并不会产生任何通信，它只是接收参数并返回结果。这是函数式编程的基础之一，也是它如此适合并发的原因之一[①]。并行流和 CompletableFuture 便是基于这些原则而设计的。只要你决定怎样都不分享任何数据（分享很容易发生，甚至会意外发生），就可以写出相当安全的并发程序。

操作是通过调用 thenApply() 开始的。本例中，CompletableFuture 的创建过程会等到所有任务都完成后才会完成。虽然这有时很有用，但更多的价值还是在于可以开启所有的任务，然后你就可以在任务运行时继续做其他的事情。我们通过在操作最后增加 Async 来实现该效果：

```java
// concurrent/CompletableApplyAsync.java
import java.util.concurrent.*;
import onjava.*;

public class CompletableApplyAsync {
  public static void main(String[] args) {
    Timer timer = new Timer();
    CompletableFuture<Machina> cf =
```

[①] 不，永远都不会有纯粹的函数式 Java。最切合实际的期望是能有一个全新的、基于 JVM 运行的语言。

```
    CompletableFuture.completedFuture(            /* 输出:
      new Machina(0))                             103
    .thenApplyAsync(Machina::work)                Machina0: ONE
    .thenApplyAsync(Machina::work)                Machina0: TWO
    .thenApplyAsync(Machina::work)                Machina0: THREE
    .thenApplyAsync(Machina::work);               Machina0: complete
  System.out.println(timer.duration());           Machina0: complete
  System.out.println(cf.join());                  545
  System.out.println(timer.duration());           */
  }
}
```

同步调用（即我们平时使用的那种）意味着"完成工作后返回"，而**异步**调用则意味着"立即返回，同时在后台继续工作"。正如你所见，现在 cf 的创建过程变快了很多。对 thenApplyAsync() 的每次调用都会立刻返回，这样就可以立即执行下一个调用，整个链式调用序列就会比之前快得多了。

执行速度确实就是那么快，在没有调用 cf.join() 的情况下，程序在任务完成前就退出了（试试去掉该行代码）。对 join() 的调用会一直阻塞 main() 线程的执行，直到 cf 操作完成，我们可以看到大部分时间确实花在了这个地方。

这种可以"立刻返回"的**异步**能力依赖于 CompletableFuture 库的某些背后操作。通常来说，该库需要将你请求的操作链保存为一组**回调**（callback）。当第一个后台操作完成并返回后，第二个后台操作必须接收相应的 Machina 并开始工作，然后当该操作完成后，下一个操作继续，以此类推。但是由于这里并非由程序调用栈控制的普通函数调用序列，其调用顺序会丢失，因此改用回调来存储，即一个记录了函数地址的表格。

幸运的是，关于回调，你只需要知道这么多。程序员将手动操作带来的混乱称为"回调地狱"。通过 Async 调用，CompletableFuture 会为你管理所有的回调。除非你知道系统里有特别的影响因素，否则你通常会使用 Async 调用。

5.10.2 其他操作

如果你查看 Javadoc 中 CompletableFuture 的相关内容，可以看到它有很多方法，但其中大部分都是各种不同操作的变种。举例来说，其中有 thenApply() 和它的变种 thenApplyAsync()，以及 thenApplyAsync() 的另一种形式，它接收参数 Executor 来运行任务（本书中不会涉及带 Executor 的版本）。

以下示例演示了所有的"基本"操作，这些操作既不会涉及两个 CompletableFuture 的合并，也不会涉及异常（这些稍后会讨论）。首先，我们应该复用下面这两个工具，以简化代码并增加便利性：

```java
// concurrent/CompletableUtilities.java
package onjava;
import java.util.concurrent.*;

public class CompletableUtilities {
  // 获取并展示 CF 中存储的值:
  public static void showr(CompletableFuture<?> c) {
    try {
      System.out.println(c.get());
    } catch(InterruptedException
            | ExecutionException e) {
      throw new RuntimeException(e);
    }
  }
  // 针对无值的 CF 操作:
  public static void voidr(CompletableFuture<Void> c) {
    try {
      c.get(); // 返回 void
    } catch(InterruptedException
            | ExecutionException e) {
      throw new RuntimeException(e);
    }
  }
}
```

showr() 调用了 CompletableFuture<Integer> 的 get()，并显示了结果，同时对两个可能的异常进行了捕获。voidr() 是 showr() 针对 CompletableFuture<Void> 的实现版本，也就是说，即只会在任务完成或失败时用来显示（show）的那类 CompletableFuture。

简单起见，下面的 CompletableFuture 仅仅包装了 Integer 类型。cfi() 则是一个简化的方法，其在一个完整的 CompletableFuture<Integer> 内部包装了一个 int 类型。

```java
// concurrent/CompletableOperations.java
import java.util.concurrent.*;
import static onjava.CompletableUtilities.*;

public class CompletableOperations {
  static CompletableFuture<Integer> cfi(int i) {
    return
      CompletableFuture.completedFuture(
        Integer.valueOf(i));
  }
  public static void main(String[] args) {
    showr(cfi(1)); // 基本测试
    voidr(cfi(2).runAsync(() ->
      System.out.println("runAsync")));
    voidr(cfi(3).thenRunAsync(() ->
      System.out.println("thenRunAsync")));
    voidr(CompletableFuture.runAsync(() ->
      System.out.println("runAsync is static")));
```

```java
      showr(CompletableFuture.supplyAsync(() -> 99));
      voidr(cfi(4).thenAcceptAsync(i ->
        System.out.println("thenAcceptAsync: " + i)));
      showr(cfi(5).thenApplyAsync(i -> i + 42));
      showr(cfi(6).thenComposeAsync(i -> cfi(i + 99)));
      CompletableFuture<Integer> c = cfi(7);
      c.obtrudeValue(111);
      showr(c);
      showr(cfi(8).toCompletableFuture());
      c = new CompletableFuture<>();
      c.complete(9);
      showr(c);
      c = new CompletableFuture<>();
      c.cancel(true);
      System.out.println("cancelled: " +
        c.isCancelled());
      System.out.println("completed exceptionally: " +
        c.isCompletedExceptionally());
      System.out.println("done: " + c.isDone());
      System.out.println(c);
      c = new CompletableFuture<>();
      System.out.println(c.getNow(777));
      c = new CompletableFuture<>();
      c.thenApplyAsync(i -> i + 42)
        .thenApplyAsync(i -> i * 12);
      System.out.println("dependents: " +
        c.getNumberOfDependents());
      c.thenApplyAsync(i -> i / 2);
      System.out.println("dependents: " +
        c.getNumberOfDependents());
  }
}
/* 输出：
1
runAsync
thenRunAsync
runAsync is static
99
thenAcceptAsync: 4
47
105
111
8
9
cancelled: true
completed exceptionally: true
done: true
java.util.concurrent.CompletableFuture@1629346[Complete
d exceptionally]
777
dependents: 1
dependents: 2
*/
```

main() 中包含了一系列测试，这些测试可由它们的 int 值引用。cfi(1) 证明了 showr() 运行正确。cfi(2) 是个调用 runAsync() 的例子。Runnable 不会返回任何值，因此结果是个 CompletableFuture<Void>，并且用到了 voidr()。

注意，由于 cfi(3) 的缘故，thenRunAsync() 和 runAsync() 似乎完全一样。区别体现在之后的测试中：runAsync() 是静态方法，所以你通常不会像在 cfi(2) 中一样调用它，而是会像在 QuittingCompletable.java 中一样调用它。再往后的测试演示了 supplyAsync() 同样也是静态的，但它依赖 Supplier 而不是 Runnable，并且会生成 CompletableFuture<Integer>，而不是 CompletableFuture<Void>。

then 系列方法针对已有的 CompletableFuture<Integer> 进行操作。不同于 thenRunAsync()，用于 cfi(4)、cfi(5) 和 cfi(6) 的系列 then 方法接收未包装的 Integer 类型作为参数。正如你在 voidr() 的用法中可以看到的，thenAcceptAsync() 接收 Consumer 作为参数，所以不会返回结果。thenApplyAsync() 接收 Function 作为参数，因此会返回结果（可以是和参数不同的类型）。thenComposeAsync() 和 thenApplyAsync() 非常像，只是它的 Function 必须返回已在 CompletableFuture 中被包装后的结果。

示例 cfi(7) 演示了 obtrudeValue() 方法，它强制输入一个值作为结果。cfi(8) 使用了 toCompletableFuture() 以从当前的 CompletionStage 生成 CompletableFuture。c.complete(9) 演示了你可以如何通过传入结果来让一个 future 完成执行（而 obtrudeValue() 则可以强制用自己的结果来替换这个结果）。

如果你 cancel()（取消）了 CompletableFuture，它同样会变成"已完成"（done），并且是异常情况下的完成（completed exceptionally）。

getNow() 方法要么返回 CompletableFuture 的完整值，要么返回 getNow() 的替代参数（如果该 future 尚未完成）。

最后来看看 dependent[①] 的概念。如果我们将两次对 CompletableFuture 的 thenApplyAsync() 调用连在一起，dependent 的数量仍然还是一个。但是如果我们直接将另一个 thenApplyAsync() 添加到 c，那么就有了两个 dependent：两个连续的调用和一个额外的调用。这说明了一个单独的 CompletionStage 可以在其完成后，基于它的结果 fork 出多个新任务。

5.10.3　合并多个 CompletableFuture

CompletableFuture 中的第二类方法接收两个 CompletableFuture 作为参数，并以多种

---

① 依赖项，即正在等待该 CompletableFuture 完成的 CompletableFuture 的预估数量。——译者注

方式将其合并。一般来说一个 CompletableFuture 会先于另一个执行完成，两者看起来就像在彼此竞争。这些方法使你可以用不同的方式处理结果。

为了测试上述方法，我们会创建一个任务，该任务的参数之一是完成该任务所需的时长，由此我们可以控制首先完成哪个 CompletableFuture：

```java
// concurrent/Workable.java
import java.util.concurrent.*;
import onjava.Nap;

public class Workable {
  String id;
  final double duration;
  public Workable(String id, double duration) {
    this.id = id;
    this.duration = duration;
  }
  @Override public String toString() {
    return "Workable[" + id + "]";
  }
  public static Workable work(Workable tt) {
    new Nap(tt.duration); // 秒
    tt.id = tt.id + "W";
    System.out.println(tt);
    return tt;
  }
  public static CompletableFuture<Workable>
  make(String id, double duration) {
    return
      CompletableFuture.completedFuture(
        new Workable(id, duration))
        .thenApplyAsync(Workable::work);
  }
}
```

在 make() 中，work() 方法被用于 CompletableFuture。work() 花了 duration 中的时间完成执行，然后将字母 W 附加到 id 的后面，以标识该"work"已完成。

现在我们可以创建多个互相竞争的 CompletableFuture，并通过 CompletableFuture 库中的多种方法将它们互相关联起来：

```java
// concurrent/DualCompletableOperations.java
import java.util.concurrent.*;
import static onjava.CompletableUtilities.*;

public class DualCompletableOperations {
  static CompletableFuture<Workable> cfA, cfB;
  static void init() {
```

```java
    cfA = Workable.make("A", 0.15);
    cfB = Workable.make("B", 0.10); // 总是胜出
}
static void join() {
    cfA.join();
    cfB.join();
    System.out.println("*****************");
}
public static void main(String[] args) {
    init();
    voidr(cfA.runAfterEitherAsync(cfB, () ->
        System.out.println("runAfterEither")));
    join();

    init();
    voidr(cfA.runAfterBothAsync(cfB, () ->
        System.out.println("runAfterBoth")));
    join();

    init();
    showr(cfA.applyToEitherAsync(cfB, w -> {
        System.out.println("applyToEither: " + w);
        return w;
    }));
    join();

    init();
    voidr(cfA.acceptEitherAsync(cfB, w -> {
        System.out.println("acceptEither: " + w);
    }));
    join();

    init();
    voidr(cfA.thenAcceptBothAsync(cfB, (w1, w2) -> {
        System.out.println("thenAcceptBoth: "
            + w1 + ", " + w2);
    }));
    join();

    init();
    showr(cfA.thenCombineAsync(cfB, (w1, w2) -> {
        System.out.println("thenCombine: "
            + w1 + ", " + w2);
        return w1;
    }));
    join();

    init();
    CompletableFuture<Workable>
        cfC = Workable.make("C", 0.08),
        cfD = Workable.make("D", 0.09);
    CompletableFuture.anyOf(cfA, cfB, cfC, cfD)
        .thenRunAsync(() ->
```

```
      System.out.println("anyOf"));
  join();
  
  init();
  cfC = Workable.make("C", 0.08);
  cfD = Workable.make("D", 0.09);
  CompletableFuture.allOf(cfA, cfB, cfC, cfD)
    .thenRunAsync(() ->
      System.out.println("allOf"));
  join();
  }
}
/* 输出：
Workable[BW]
runAfterEither
Workable[AW]
*****************
Workable[BW]
Workable[AW]
runAfterBoth
*****************
Workable[BW]
applyToEither: Workable[BW]
Workable[BW]
Workable[AW]
*****************
Workable[BW]
acceptEither: Workable[BW]
Workable[AW]
*****************
Workable[BW]
Workable[AW]
thenAcceptBoth: Workable[AW], Workable[BW]
*****************
Workable[BW]
Workable[AW]
thenCombine: Workable[AW], Workable[BW]
Workable[AW]
*****************
Workable[CW]
anyOf
Workable[DW]
Workable[BW]
Workable[AW]
*****************
Workable[CW]
Workable[DW]
Workable[BW]
Workable[AW]
*****************
allOf
*/
```

为了便于访问，cfA 和 cfB 都是静态的。init() 方法对这两者进行初始化，其中使用 "B" 初始化的 cfB 被赋予了更短的延时，因此 cfB 总是会"胜出"。join() 是另一个便利的方法，cfA 和 cfB 分别都调用了 join() 方法，并显示出分界线。

所有这些"双重"（dual）的方法都使用了一个 CompletableFuture 作为调用方法的对象（并在其上调用方法），以及另一个 CompletableFuture 作为第一个参数，后面再跟上要执行的操作。

通过 showr() 和 voidr() 的用法，可以看到 "run" 和 "accept" 是终结操作，而 "apply" 和 "combine" 则生成新的承担负载（payload-bearing）的 CompletableFuture。

这些方法名都是自释性良好（self-explanatory）的，并且你可以通过输出验证这一点。其中 combineAsync() 是个特别有趣的方法，它会等待两个 CompletableFuture 完成，再将两者传给 BiFunction，然后 BiFunction 将结果合并（join）到最终 CompletableFuture 的荷载中。

5.10.4 模拟场景应用

为了示范如何通过 CompletableFuture 来将一系列的操作捆绑到一起，我们来模拟制作蛋糕的过程。第一步是准备配料（ingredient），并将它们混入面糊（batter）中：

```java
// concurrent/Batter.java
import java.util.concurrent.*;
import onjava.Nap;

public class Batter {
  static class Eggs {}
  static class Milk {}
  static class Sugar {}
  static class Flour {}
  static <T> T prepare(T ingredient) {
    new Nap(0.1);
    return ingredient;
  }
  static <T> CompletableFuture<T> prep(T ingredient) {
    return CompletableFuture
      .completedFuture(ingredient)
      .thenApplyAsync(Batter::prepare);
  }
  public static CompletableFuture<Batter> mix() {
    CompletableFuture<Eggs> eggs = prep(new Eggs());
    CompletableFuture<Milk> milk = prep(new Milk());
    CompletableFuture<Sugar> sugar = prep(new Sugar());
    CompletableFuture<Flour> flour = prep(new Flour());
    CompletableFuture
```

```
      .allOf(eggs, milk, sugar, flour)
      .join();
    new Nap(0.1); // 混合时间
    return
      CompletableFuture.completedFuture(new Batter());
  }
}
```

每种配料都需要一些时间来准备。allOf() 会等待所有的配料准备完毕，然后再花些时间将它们混合并放入面糊中。

下一步是将一份面糊分摊到 4 个平底锅中，然后开始烘焙。成品会以 CompletableFuture 类型的 Stream 形式返回：

```
// concurrent/Baked.java
import java.util.concurrent.*;
import java.util.stream.*;
import onjava.Nap;

public class Baked {
  static class Pan {}
  static Pan pan(Batter b) {
    new Nap(0.1);
    return new Pan();
  }
  static Baked heat(Pan p) {
    new Nap(0.1);
    return new Baked();
  }
  static CompletableFuture<Baked>
  bake(CompletableFuture<Batter> cfb) {
    return cfb
      .thenApplyAsync(Baked::pan)
      .thenApplyAsync(Baked::heat);
  }
  public static
  Stream<CompletableFuture<Baked>> batch() {
    CompletableFuture<Batter> batter = Batter.mix();
    return Stream.of(bake(batter), bake(batter),
                     bake(batter), bake(batter));
  }
}
```

最后，我们创建了一份 Frosting（糖霜），并将其洒在蛋糕上：

```
// concurrent/FrostedCake.java
import java.util.concurrent.*;
import java.util.stream.*;
import onjava.Nap;
```

```
final class Frosting {
  private Frosting() {}
  static CompletableFuture<Frosting> make() {
    new Nap(0.1);
    return CompletableFuture
      .completedFuture(new Frosting());
  }
}

public class FrostedCake {
  public FrostedCake(Baked baked, Frosting frosting) {
    new Nap(0.1);
  }
  @Override public String toString() {
    return "FrostedCake";
  }
  public static void main(String[] args) {
    Baked.batch().forEach(baked -> baked
      .thenCombineAsync(Frosting.make(),
        (cake, frosting) ->
          new FrostedCake(cake, frosting))
      .thenAcceptAsync(System.out::println)
      .join());
  }
}
```

一旦你适应了 CompletableFuture 背后的设计理念，它们就会变得相当好用。

5.10.5 异常

和 CompletableFuture 对处理链中的对象加以包装的方式一样，它还可以缓存异常。在处理过程中调用者并不会对此有所感知，这种效果只会在尝试提取结果时体现出来。为了演示其运作机制，我们先来创建一个会在特定条件下抛出异常的类。

```
// concurrent/Breakable.java
import java.util.concurrent.*;

public class Breakable {
  String id;
  private int failcount;
  public Breakable(String id, int failcount) {
    this.id = id;
    this.failcount = failcount;
  }
  @Override public String toString() {
    return "Breakable_" + id +
      " [" + failcount + "]";
  }
  public static Breakable work(Breakable b) {
    if(--b.failcount == 0) {
```

```
      System.out.println(
        "Throwing Exception for " + b.id + "");
      throw new RuntimeException(
        "Breakable_" + b.id + " failed");
    }
    System.out.println(b);
    return b;
  }
}
```

通过正整型的 `failcount`（失败数），每次向 `work()` 方法传递对象，`failcount` 都会递减。当它等于 0 的时候，`work()` 会抛出异常。如果直接传入值为 0 的 `failcount`，则永远不会抛出异常。

注意，该程序在异常抛出时会发出异常报告。

在随后的 `test()` 方法中，`work()` 被多次应用于 `Breakable`，所以如果 `failcount` 在范围内，则会抛出异常。不过，在从 A 到 E 的测试中，你可以从输出看到异常被抛出，但并不会显露出来：

```
// concurrent/CompletableExceptions.java
import java.util.concurrent.*;

public class CompletableExceptions {
  static CompletableFuture<Breakable>
  test(String id, int failcount) {
    return
      CompletableFuture.completedFuture(
        new Breakable(id, failcount))
          .thenApply(Breakable::work)
          .thenApply(Breakable::work)
          .thenApply(Breakable::work)
          .thenApply(Breakable::work);
  }
  public static void main(String[] args) {
    // 异常不会显露出来……
    test("A", 1);
    test("B", 2);
    test("C", 3);
    test("D", 4);
    test("E", 5);
    // ……直到你尝试获取结果：
    try {
      test("F", 2).get(); // 或者 join()
    } catch(Exception e) {
      System.out.println(e.getMessage());
    }
    // 测试异常
    System.out.println(
```

```
    test("G", 2).isCompletedExceptionally());
// 算作"完成":
System.out.println(test("H", 2).isDone());
// 强制产生异常:
CompletableFuture<Integer> cfi =
  new CompletableFuture<>();
System.out.println("done? " + cfi.isDone());
cfi.completeExceptionally(
  new RuntimeException("forced"));
try {
  cfi.get();
} catch(Exception e) {
  System.out.println(e.getMessage());
}
  }
}
/* 输出:
Throwing Exception for A
Breakable_B [1]
Throwing Exception for B
Breakable_C [2]
Breakable_C [1]
Throwing Exception for C
Breakable_D [3]
Breakable_D [2]
Breakable_D [1]
Throwing Exception for D
Breakable_E [4]
Breakable_E [3]
Breakable_E [2]
Breakable_E [1]
Breakable_F [1]
Throwing Exception for F
java.lang.RuntimeException: Breakable_F failed
Breakable_G [1]
Throwing Exception for G
true
Breakable_H [1]
Throwing Exception for H
true
done? false
java.lang.RuntimeException: forced
*/
```

从 A 到 E 的测试执行到抛出异常的节点时，什么都没有发生。只有在测试 F 中调用 get() 时，才会看到抛出的异常。

从测试 G 可以看出，你可以先检查处理过程中是否有异常抛出，而不必真的抛出该异常。然而，测试 H 告诉我们该异常仍符合"完成"的条件，不论其是否真的成功。

代码的最后一部分演示了你可以如何向 CompletableFuture 插入异常，不论是否出现任何失败。

相比于在合并或获取结果时粗暴地使用 try-catch，我们更倾向于利用 CompletableFuture 所带来的更为先进的机制来自动地响应异常。你只需照搬在所有 CompletableFuture 中看到的方式即可：在调用链中插入 CompletableFuture 调用。一共有 3 个选项：exceptionally()、handle() 以及 whenComplete()：

```java
// concurrent/CatchCompletableExceptions.java
import java.util.concurrent.*;

public class CatchCompletableExceptions {
  static void handleException(int failcount) {
    // 只在有异常时才调用该函数，必须生成和输入相同的类型：
    CompletableExceptions
      .test("exceptionally", failcount)
      .exceptionally((ex) -> { // Function
        if(ex == null)
          System.out.println("I don't get it yet");
        return new Breakable(ex.getMessage(), 0);
      })
      .thenAccept(str ->
        System.out.println("result: " + str));

    // 创建新结果（恢复）：
    CompletableExceptions
      .test("handle", failcount)
      .handle((result, fail) -> { // BiFunction
        if(fail != null)
          return "Failure recovery object";
        else
          return result + " is good";
      })
      .thenAccept(str ->
        System.out.println("result: " + str));

    // 做了一些逻辑处理，但仍然向下传递相同的结果：
    CompletableExceptions
      .test("whenComplete", failcount)
      .whenComplete((result, fail) -> { // BiConsumer
        if(fail != null)
          System.out.println("It failed");
        else
          System.out.println(result + " OK");
      })
      .thenAccept(r ->
        System.out.println("result: " + r));
  }
  public static void main(String[] args) {
    System.out.println("**** Failure Mode ****");
    handleException(2);
```

```
        System.out.println("**** Success Mode ****");
        handleException(0);
    }
}
/* 输出：
**** Failure Mode ****
Breakable_exceptionally [1]
Throwing Exception for exceptionally
result: Breakable_java.lang.RuntimeException:
Breakable_exceptionally failed [0]
Breakable_handle [1]
Throwing Exception for handle
result: Failure recovery object
Breakable_whenComplete [1]
Throwing Exception for whenComplete
It failed
**** Success Mode ****
Breakable_exceptionally [-1]
Breakable_exceptionally [-2]
Breakable_exceptionally [-3]
Breakable_exceptionally [-4]
result: Breakable_exceptionally [-4]
Breakable_handle [-1]
Breakable_handle [-2]
Breakable_handle [-3]
Breakable_handle [-4]
result: Breakable_handle [-4] is good
Breakable_whenComplete [-1]
Breakable_whenComplete [-2]
Breakable_whenComplete [-3]
Breakable_whenComplete [-4]
Breakable_whenComplete [-4] OK
result: Breakable_whenComplete [-4]
*/
```

只有在出现异常时，exceptionally() 参数才会运行。exceptionally() 的限制在于 Function 返回值的类型必须和输入相同。将一个正确的对象插回到流，可使 exceptionally() 恢复到工作状态。

handle() 总是会被调用的，而且你必须检查 fail 是否为 true，以确定是否有异常发生。但是 handle() 可生成任意新类型，因此它允许你执行处理，而不是像 exceptionally() 那样只是恢复。

whenComplete() 和 handle() 类似，都必须测试失败情况，但是参数是消费者，只会使用而不会修改传递中的 result 对象。

1. 流异常

我们对 CompletableExceptions.java 做些改动，来看看 CompletableFuture 的异常和

Stream 的异常有何区别：

```
// concurrent/StreamExceptions.java
import java.util.concurrent.*;
import java.util.stream.*;

public class StreamExceptions {
  static Stream<Breakable>
  test(String id, int failcount) {
    return
      Stream.of(new Breakable(id, failcount))
        .map(Breakable::work)
        .map(Breakable::work)
        .map(Breakable::work)
        .map(Breakable::work);
  }
  public static void main(String[] args) {
    // 甚至没有进行任何操作……
    test("A", 1);
    test("B", 2);
    Stream<Breakable> c = test("C", 3);
    test("D", 4);
    test("E", 5);
    // ……直到出现终结操作
    System.out.println("Entering try");
    try {
      c.forEach(System.out::println);   // [1]
    } catch(Exception e) {
      System.out.println(e.getMessage());
    }
  }
}
/* 输出：
Entering try
Breakable_C [2]
Breakable_C [1]
Throwing Exception for C
Breakable_C failed
*/
```

使用 CompletableFuture 时，我们看到了从测试 A 到测试 E 的执行过程。但是使用 Stream 时，甚至直到你应用到终结操作（如 [1] 处的 forEach()）之前，什么都不会发生。CompletableFuture 会执行任务，并捕捉任何异常，以备后续的结果取回。因为 Stream 的机制是在终结操作前不做任何事，所以这两者并不好直接比较。不过 Stream 肯定不会保存异常。

2. 检查型异常

CompletableFuture 和并行 Stream 都不支持包含检查型异常的操作。因此，你必须在调用操作的时候处理检查型异常，而这会明显降低代码的优雅性：

```
// concurrent/ThrowsChecked.java
import java.util.stream.*;
import java.util.concurrent.*;
```

```java
public class ThrowsChecked {
  class Checked extends Exception {}
  static ThrowsChecked nochecked(ThrowsChecked tc) {
    return tc;
  }
  static ThrowsChecked
  withchecked(ThrowsChecked tc) throws Checked {
    return tc;
  }
  static void testStream() {
    Stream.of(new ThrowsChecked())
      .map(ThrowsChecked::nochecked)
      // .map(ThrowsChecked::withchecked);      // [1]
      .map(tc -> {
        try {
          return withchecked(tc);
        } catch(Checked e) {
          throw new RuntimeException(e);
        }
      });
  }
  static void testCompletableFuture() {
    CompletableFuture
      .completedFuture(new ThrowsChecked())
      .thenApply(ThrowsChecked::nochecked)
      // .thenApply(ThrowsChecked::withchecked);  // [2]
      .thenApply(tc -> {
        try {
          return withchecked(tc);
        } catch(Checked e) {
          throw new RuntimeException(e);
        }
      });
  }
}
```

如果你想像使用 nochecked() 的引用一样，使用 withchecked() 的方法引用，编译器就会在 [1] 和 [2] 处报错。所以，你必须写出 lambda 表达式（或者写一个不会抛出异常的包装方法）。

5.11 死锁

因为任务会被阻塞，所以一项任务 A 很可能因等待另一项任务 B 而卡住，而任务 B 又在等待另一项任务 C，以此类推，直到整个链条指回到一项正在等待任务 A 的任务。此时形成了一个互相等待的无限循环，谁都动不了。这称为**死锁**。[1]

---

[1] 当两个任务有能力改变它们的状态，也就是说不会被阻塞，但永远不会取得任何有效进展时，也会产生活锁。

如果你刚开始运行程序时就死锁了，那么你马上就可以去追查 bug。真正的问题在于，你的程序看起来运行正常，但是存在潜在的死锁风险。这时，你可能无法发现会导致死锁的迹象，这样隐患就潜伏在程序中，直到它突然发生——一般会暴露给使用者（以一种极难重现的方式）。因此，依靠在程序设计中的谨慎小心来防止死锁发生，是并发系统开发极为重要的一部分。

Edsger Dijkstra 提出的**哲学家用餐问题**（Dining Philosophers problem），是对死锁的经典诠释。这个问题可大致描述为，有 5 位哲学家（此处描述的场景对数量没有限制），他们一部分时间在思考，另一部分时间在吃饭。当他们思考时，并不需要任何共享资源，但是吃饭时只能使用有限数量的餐具。在最初的故事版本中，餐具是叉子。每位哲学家要使用两个叉子，从桌子中间的碗里取意大利面吃。在更有说服力的版本中，餐具则是筷子。显然，每位哲学家需要使用两根筷子。

困难出现了：作为哲学家，他们没什么钱，所以只能负担得起 5 根筷子（可泛化为：筷子数和人数相同）。哲学家们围着一张圆桌就座，每两位哲学家中间放着一根筷子。当其中一位哲学家想要吃东西时，他必须从左右手边各拿起一根筷子。如果坐在旁边的另一位哲学家正在使用他想要的筷子，那么这位哲学家就必须等待，直到那根筷子处于可用状态。

下面这个 StickHolder（筷子持有者）类将一个 Chopstick（筷子）类保存在一个长度为 1 的 BlockingQueue 中，以进行管理。BlockingQueue 是一种线程安全的集合，专门用于并发程序，如果调用 take() 且队列为空，它就会阻塞（等待）。一旦新的元素被放入队列，阻塞就会被解除，并会返回该元素值：

```java
// concurrent/StickHolder.java
import java.util.concurrent.*;

public class StickHolder {
  private static class Chopstick {}
  private Chopstick stick = new Chopstick();
  private BlockingQueue<Chopstick> holder =
    new ArrayBlockingQueue<>(1);
  public StickHolder() { putDown(); }
  public void pickUp() {
    try {
      holder.take(); // 不可用时会阻塞
    } catch(InterruptedException e) {
      throw new RuntimeException(e);
    }
  }
  public void putDown() {
    try {
      holder.put(stick);
```

```
      } catch(InterruptedException e) {
        throw new RuntimeException(e);
      }
    }
  }
}
```

简单起见，Chopstick 不会由 StickHolder 实际生成，而仅在类中是私有的。如果你调用 pickUp()（拿起）且筷子当前不可用，pickUp() 会一直阻塞，直到筷子被其他 Philosopher 通过调用 putDown()（放下）被返回。注意本类中所有的线程安全性都是由 BlockingQueue 来保证的。

每个 Philosopher 都是一个任务，它尝试从左右两边 pickUp()（拿起）筷子来吃饭，然后通过 putDown() 来释放这些筷子：

```
// concurrent/Philosopher.java

public class Philosopher implements Runnable {
  private final int seat;
  private final StickHolder left, right;
  public Philosopher(int seat,
    StickHolder left, StickHolder right) {
    this.seat = seat;
    this.left = left;
    this.right = right;
  }
  @Override public String toString() {
    return "P" + seat;
  }
  @Override public void run() {
    while(true) {
      // System.out.println("Thinking");    // [1]
      right.pickUp();
      left.pickUp();
      System.out.println(this + " eating");
      right.putDown();
      left.putDown();
    }
  }
}
```

两个 Philosopher 不可能成功地同时 take() 同一根筷子。另外，如果一根筷子已经被一个 Philosopher 拿走，下一个尝试拿走同一根筷子的 Philosopher 会阻塞，等待该筷子被释放。

结果是一个看似没问题的程序死锁了。为了使语法更清晰，我把集合改用数组来实现：

```java
// concurrent/DiningPhilosophers.java
// 隐藏的死锁
// {ExcludeFromGradle} 该代码在 Gradle 中有问题   // 本书作者使用 Gradle 管理所有代码，读者可
忽略该行注释，不影响内容表达。——译者注
import java.util.*;
import java.util.concurrent.*;
import onjava.Nap;

public class DiningPhilosophers {
  private StickHolder[] sticks;
  private Philosopher[] philosophers;
  public DiningPhilosophers(int n) {
    sticks = new StickHolder[n];
    Arrays.setAll(sticks, i -> new StickHolder());
    philosophers = new Philosopher[n];
    Arrays.setAll(philosophers, i ->
      new Philosopher(i,
        sticks[i], sticks[(i + 1) % n]));   // [1]
    // 通过颠倒筷子的顺序来修正死锁：
    // philosophers[1] =                    // [2]
    //   new Philosopher(0, sticks[0], sticks[1]);
    Arrays.stream(philosophers)
      .forEach(CompletableFuture::runAsync); // [3]
  }
  public static void main(String[] args) {
    // 立刻返回
    new DiningPhilosophers(5);              // [4]
    // 保持 main() 不退出：
    new Nap(3, "Shutdown");
  }
}
```

当你发现输出停止时，程序便死锁了。不过基于你的机器配置，死锁也可能不会发生。这似乎依赖于你机器的 CPU 核数[1]。看起来双核的机器并不会发生死锁，但在双核以上的机器上似乎很快就会出现。该行为甚至使本例更好地诠释了死锁，因为你可能在双核机器上编写程序（如果双核确实是蒙蔽了死锁的原因），并且确信程序运行正确，而直到将程序安装到另一台不同的机器上时，才开始出现死锁。而且要注意的是，程序不会轻易发生死锁，并不意味着程序不会在双核机器上发生死锁。程序仍然是有死锁风险的，只是很难出现——这可以说是最糟糕的情况了，毕竟这类问题很难暴露。

在 DiningPhilosophers 的构造器中，每个 Philosopher 都被分配了左右各一个 StickHolder 的引用。除了最后一位 Philosopher，所有的 Philosopher 都通过将自身定位于下一双

---

[1] 这并不是超线程。通常每个核都会有两个超线程，当请求查询内核数量时，本书中使用的 Java 版本会返回超线程的数量（而不是物理核数）。超线程极大地加快了上下文切换的速度，但只有实际的内核会进行这项工作。

筷子中间来进行初始化。最后一位 Philosopher 则被分配了第 0 根筷子，作为右边的筷子的引用，由此完成了圆桌上的闭环。这是因为最后一位 Philosopher 正好坐在第一位 Philosopher 旁边，他们共享第 0 根筷子。代码 [1] 演示了通过模数 n 选择右手边的筷子，并将最后一位 Philosopher 指回到第一位 Philosopher 旁边。

现在所有的 Philosopher 都可以开始吃东西了，每一位 Philosopher 都在等待邻座的 Philosopher 放下手中的筷子。

为了在 [3] 处逐个启动 Philosopher，我调用了 runAsync()，这意味着 DiningPhilosophers 的构造器会在 [4] 处立刻返回。如果不想办法阻止 main() 完成，本程序就会很快退出，也做不了什么事。Nap 对象阻塞了 main() 的退出，然后经过 3 秒后，强制退出该（想必）已死锁的程序。

在当前配置下，Philosopher 实际上并没有时间思考，因此他们全都因吃饭而抢夺筷子，死锁很快就会发生。你可以做以下改变：

1. 在行 [4] 处，通过增大值来添加更多的 Philosopher；
2. 取消 Philosopher.java 中行 [2] 处的注释。

上面的任何一项改变都会降低死锁的可能性，这演示了编写并发程序的危险性，以及不能因为程序看起来"在我的机器上运行良好"所以就相信它是安全的。你很容易说服自己相信程序是不会死锁的，即使事实并非如此。本例之所以有趣，正因为它演示了一个程序可以在存在死锁风险的情况下依然看似运行正确。

为了修复该问题，我们观察到，出现死锁需要同时满足以下 4 个条件。

1. 互斥。这些任务使用的至少一项资源必须不是共享的。此处，一根筷子同时只能被一位 Philosopher 使用。
2. 至少一个任务必须持有一项资源，并且等待正被另一个任务持有的资源。也就是说，如果要出现死锁，一位 Philosopher 必须持有一根筷子，并且正在等待另一根。
3. 不能从一个任务中抢走一项资源。任务只能以正常的事件释放资源。我们的 Philosopher 很有礼貌，他们并不会从其他 Philosopher 手里抢走筷子。
4. 会发生循环等待，其中一个任务等待另一个任务持有的资源，另一个任务又在等待另一个任务持有的资源，以此类推，直到某个任务正在等待第一个任务持有的资源，由此一切都陷入了死循环。在 DiningPhilosophers.java 中，由于每一位 Philosopher 都在试图先获取右边的筷子，再获取左边的，因此发生了循环等待。

由于所有的这些条件必须同时发生才能导致死锁，因此你必须阻止其中一个，以避免死锁。在本程序中，有个避免死锁的简单办法：破坏第四个条件。该条件会发生是因为每一位 Philosopher 都试图按指定顺序拿起筷子：先右边，再左边。因此，每一位 Philosopher 都有可能一边拿着右边的筷子，一边等着左边的筷子，这就导致了循环等待。然而，如果其中一位 Philosopher 尝试先拿起左边的筷子，该 Philosopher 就绝不会阻碍右边的 Philosopher 拿起筷子，这样就阻止了循环等待。

在 DiningPhilosophers.java 中，取消在 [1] 处及下一行的注释。这会将原本的 philosophers[1] 替换为一位颠倒了拿筷子顺序的 Philosopher。通过确保第二个 Philosopher 先拿起和放下左边的筷子，我们消除了死锁的可能性。

这只是解决该问题的办法之一。你还可以通过阻止其他条件来解决。

语言层面上的支持无法帮助你避免死锁，你只能通过谨慎的设计来避免这个问题。对于那些正在试图调试死锁程序的人来说，这句话可能会让他们灰心丧气。当然，要避免并发问题，最简单有效的办法就是**永远不要共享资源**——不幸的是，总有躲不掉的时候。

5.12 构造器并不是线程安全的

想象一下对象的构造过程，很容易认为其是线程安全的。毕竟，在新对象完成初始化之前，甚至都没人能看到这个对象，又怎么可能去竞争该对象呢？的确，Java 语言规范（Java Language Specification, JLS）明确地说过：

将构造器设为同步并没有实际意义，因为这样做对正在构造的对象上锁。在对象的所有构造器完成工作之前，其他线程通常无法使用该对象。

不幸的是，对象的构造过程和其他任何场景一样，面对共享内存的并发问题时都是脆弱的。然而，其机制可能更加微妙。

思考一下用一个静态字段为每个对象自动创建一个唯一标识的过程。为了测试不同的实现，我们从一个接口开始：

```java
// concurrent/HasID.java

public interface HasID {
  int getID();
}
```

然后简单地实现该接口：

```java
// concurrent/StaticIDField.java

public class StaticIDField implements HasID {
  private static int counter = 0;
  private int id = counter++;
  @Override public int getID() { return id; }
}
```

这大概是你能想到的最简单稳妥的类了。它甚至都没有一个可能带来麻烦的显式构造器。为了看看生成多个创建对象的并发任务时会发生什么，接下来写了一个测试工具：

```java
// concurrent/IDChecker.java
import java.util.*;
import java.util.function.*;
import java.util.stream.*;
import java.util.concurrent.*;
import com.google.common.collect.Sets;

public class IDChecker {
  public static final int SIZE = 100_000;
  static class MakeObjects
  implements Supplier<List<Integer>> {
    private Supplier<HasID> gen;
    MakeObjects(Supplier<HasID> gen) {
      this.gen = gen;
    }
    @Override public List<Integer> get() {
      return
        Stream.generate(gen)
          .limit(SIZE)
          .map(HasID::getID)
          .collect(Collectors.toList());
    }
  }
  public static void test(Supplier<HasID> gen) {
    CompletableFuture<List<Integer>>
      groupA = CompletableFuture
        .supplyAsync(new MakeObjects(gen)),
      groupB = CompletableFuture
        .supplyAsync(new MakeObjects(gen));
    groupA.thenAcceptBoth(groupB, (a, b) -> {
      System.out.println(
        Sets.intersection(
          Sets.newHashSet(a),
          Sets.newHashSet(b)).size());
    }).join();
  }
}
```

MakeObjects 类是个 Supplier，通过 get() 方法来生成 List<Integer>。该 List 是通过从每个 HasID 对象中提取 id 而生成的。test() 方法创建了两个并行的 CompletableFuture 来运行 MakeObjects 的 supplier，然后接收两者的结果，并通过 Guava 库 Sets.intersection() 来找出这两个 List<Integer> 中有多少个 id 是相同的（Guava 比 retainAll() 要快很多）。

现在可以测试 StaticIDField 了：

```
// concurrent/TestStaticIDField.java

public class TestStaticIDField {
  public static void main(String[] args) {
    IDChecker.test(StaticIDField::new);
  }
}
```

```
/* 输出：
21397
*/
```

重复 id 的数量相当多。显然，单纯的 static int 对于构造过程来说并不安全。下面我们通过 AtomicInteger 来让该过程变为线程安全的：

```
// concurrent/GuardedIDField.java
import java.util.concurrent.atomic.*;

public class GuardedIDField implements HasID {
  private static AtomicInteger counter =
    new AtomicInteger();
  private int id = counter.getAndIncrement();
  @Override public int getID() { return id; }
  public static void main(String[] args) {
    IDChecker.test(GuardedIDField::new);
  }
}
```

```
/* 输出：
0
*/
```

构造器甚至有种更巧妙的办法来共享状态，即通过构造器参数：

```
// concurrent/SharedConstructorArgument.java
import java.util.concurrent.atomic.*;

interface SharedArg {
  int get();
}

class Unsafe implements SharedArg {
  private int i = 0;
  @Override public int get() { return i++; }
}

class Safe implements SharedArg {
  private static AtomicInteger counter =
    new AtomicInteger();
```

```
/* 输出：
16537
0
*/
```

```
    @Override public int get() {
      return counter.getAndIncrement();
    }
  }

  class SharedUser implements HasID {
    private final int id;
    SharedUser(SharedArg sa) {
      id = sa.get();
    }
    @Override public int getID() { return id; }
  }

  public class SharedConstructorArgument {
    public static void main(String[] args) {
      Unsafe unsafe = new Unsafe();
      IDChecker.test(() -> new SharedUser(unsafe));
      Safe safe = new Safe();
      IDChecker.test(() -> new SharedUser(safe));
    }
  }
```

此处 SharedUser 的构造器共享了相同的参数。尽管 SharedUser 是通过完全无害且合理的方式来使用它的参数的，但**构造器的调用方式**导致了冲突。SharedUser 甚至无法知道它以这种方式被使用了，更不用说去控制它了！

虽然语言层面并不支持 synchronized 修饰的构造器，但是可以通过 synchronized 语句块，来创建自己的（同步）构造器（要了解 synchronized 关键字，请参阅本书第 6 章）。虽然 JLS 声明"……这会阻塞正在创建的对象"，但这并不是真的——构造器事实上是个静态方法，因此 synchronized 的构造器实际上对 Class 对象上锁。我们可以通过创建自己的静态对象并对它上锁，来复现该过程：

```
// concurrent/SynchronizedConstructor.java
import java.util.concurrent.atomic.*;

class SyncConstructor implements HasID {
  private final int id;
  private static Object
    constructorLock = new Object();
  SyncConstructor(SharedArg sa) {
    synchronized(constructorLock) {
      id = sa.get();
    }
  }
  @Override public int getID() { return id; }
}

public class SynchronizedConstructor {
```

```
/* 输出：
0
*/
```

```java
  public static void main(String[] args) {
    Unsafe unsafe = new Unsafe();
    IDChecker.test(() ->
      new SyncConstructor(unsafe));
  }
}
```

对 Unsafe 类的共享现在是安全的了。

另一种方法是将构造器设为私有的（因此会阻止继承），并实现一个静态的**工厂方法**来生成新的对象：

```java
// concurrent/SynchronizedFactory.java
import java.util.concurrent.atomic.*;

final class SyncFactory implements HasID {
  private final int id;
  private SyncFactory(SharedArg sa) {
    id = sa.get();
  }
  @Override public int getID() { return id; }
  public static synchronized
  SyncFactory factory(SharedArg sa) {
    return new SyncFactory(sa);
  }
}

public class SynchronizedFactory {
  public static void main(String[] args) {
    Unsafe unsafe = new Unsafe();
    IDChecker.test(() ->
      SyncFactory.factory(unsafe));
  }
}
/* 输出：
0
*/
```

通过将静态**工厂方法**设为同步的，你在构造过程中对 Class 对象上了锁。

以上这些示例强调了在并发 Java 程序中，发现和管理共享状态是多么难以捉摸。即使你采用了 "什么都不共享" 的策略，意外的共享还是格外容易发生。

5.13 工作量、复杂性、成本

假设你要制作一个比萨。在制作过程中，从当前步骤到下一个步骤所需的工作量在这里用枚举的一部分来表示：

```java
// concurrent/Pizza.java
import java.util.function.*;
import onjava.Nap;
```

```java
public class Pizza {
  public enum Step {
    DOUGH(4), ROLLED(1), SAUCED(1), CHEESED(2),
    TOPPED(5), BAKED(2), SLICED(1), BOXED(0);
    int effort; // 需要用来到达下一步
    Step(int effort) { this.effort = effort; }
    Step forward() {
      if(equals(BOXED)) return BOXED;
      new Nap(effort * 0.1);
      return values()[ordinal() + 1];
    }
  }
  private Step step = Step.DOUGH;
  private final int id;
  public Pizza(int id) { this.id = id; }
  public Pizza next() {
    step = step.forward();
    System.out.println("Pizza " + id + ": " + step);
    return this;
  }
  public Pizza next(Step previousStep) {
    if(!step.equals(previousStep))
      throw new IllegalStateException("Expected " +
        previousStep + " but found " + step);
    return next();
  }
  public Pizza roll() { return next(Step.DOUGH); }
  public Pizza sauce() { return next(Step.ROLLED); }
  public Pizza cheese() { return next(Step.SAUCED); }
  public Pizza toppings() { return next(Step.CHEESED); }
  public Pizza bake() { return next(Step.TOPPED); }
  public Pizza slice() { return next(Step.BAKED); }
  public Pizza box() { return next(Step.SLICED); }
  public boolean complete() {
    return step.equals(Step.BOXED);
  }
  @Override public String toString() {
    return "Pizza" + id + ": " +
      (step.equals(Step.BOXED)? "complete" : step);
  }
}
```

和 Machina.java 一样，这又是一个简单的状态机。当比萨被放到盒子中时，状态机到达终点。

如果是一个人在做一个比萨，那么所有的步骤都是线性的，一个接着一个：

```java
// concurrent/OnePizza.java
import onjava.Timer;
```

```java
public class OnePizza {
  public static void main(String[] args) {
    Pizza za = new Pizza(0);
    System.out.println(
      Timer.duration(() -> {
        while(!za.complete())
          za.next();
      }));
  }
}
```

```
/* 输出：
Pizza 0: ROLLED
Pizza 0: SAUCED
Pizza 0: CHEESED
Pizza 0: TOPPED
Pizza 0: BAKED
Pizza 0: SLICED
Pizza 0: BOXED
1665
*/
```

时间单位是毫秒，并且将所有步骤的工作量相加起来，结果和我们期望的一致。

如果你像这样制作了 5 个比萨，那么你会预料到耗时会变成 5 倍。但是如果你想要更快呢？可以从并行流的方式开始：

```java
// concurrent/PizzaStreams.java
import java.util.*;
import java.util.stream.*;
import onjava.Timer;

public class PizzaStreams {
  static final int QUANTITY = 5;
  public static void main(String[] args) {
    Timer timer = new Timer();
    IntStream.range(0, QUANTITY)
      .mapToObj(Pizza::new)
      .parallel()                    // [1]
      .forEach(za -> {
        while(!za.complete())
          za.next();
      });
    System.out.println(timer.duration());
  }
}
```

```
/* 输出：
Pizza 0: ROLLED
Pizza 1: ROLLED
Pizza 2: ROLLED
Pizza 3: ROLLED
Pizza 4: ROLLED
Pizza 1: SAUCED
Pizza 0: SAUCED
Pizza 4: SAUCED
Pizza 2: SAUCED
Pizza 3: SAUCED
Pizza 3: CHEESED
Pizza 2: CHEESED
Pizza 4: CHEESED
Pizza 0: CHEESED
Pizza 1: CHEESED
Pizza 3: TOPPED
Pizza 2: TOPPED
Pizza 4: TOPPED
Pizza 1: TOPPED
Pizza 0: TOPPED
Pizza 4: BAKED
Pizza 2: BAKED
Pizza 3: BAKED
Pizza 0: BAKED
Pizza 1: BAKED
Pizza 4: SLICED
Pizza 3: SLICED
Pizza 2: SLICED
Pizza 0: SLICED
Pizza 1: SLICED
Pizza 2: BOXED
Pizza 4: BOXED
Pizza 3: BOXED
Pizza 1: BOXED
Pizza 0: BOXED
1797
*/
```

现在我们用差不多制作 1 个比萨的时间，制作了 5 个比萨。试着移除行 [1]，可以验证如果不用并行就又变成了 5 倍耗时。还可以试着将 QUANTITY 改为 4、8、10、16 和 17，再看看区别，猜猜背后的原因是什么。

PizzaStreams.java 是在 forEach() 中完成所有的工作的。如果我们将独立的步骤 map 起来，又会发生什么变化呢？

```java
// concurrent/PizzaParallelSteps.java
import java.util.*;
import java.util.stream.*;
import onjava.Timer;

public class PizzaParallelSteps {
  static final int QUANTITY = 5;
  public static void main(String[] args) {
    Timer timer = new Timer();
    IntStream.range(0, QUANTITY)
      .mapToObj(Pizza::new)
      .parallel()
      .map(Pizza::roll)
      .map(Pizza::sauce)
      .map(Pizza::cheese)
      .map(Pizza::toppings)
      .map(Pizza::bake)
      .map(Pizza::slice)
      .map(Pizza::box)
      .forEach(za -> System.out.println(za));
    System.out.println(timer.duration());
  }
}
/* 输出：
Pizza 1: ROLLED
Pizza 2: ROLLED
Pizza 0: ROLLED
Pizza 3: ROLLED
Pizza 4: ROLLED
Pizza 4: SAUCED
Pizza 2: SAUCED
Pizza 0: SAUCED
Pizza 1: SAUCED
Pizza 3: SAUCED
Pizza 1: CHEESED
Pizza 3: CHEESED
Pizza 4: CHEESED
Pizza 2: CHEESED
Pizza 0: CHEESED
Pizza 3: TOPPED
Pizza 1: TOPPED
Pizza 0: TOPPED
Pizza 2: TOPPED
Pizza 4: TOPPED
Pizza 2: BAKED
Pizza 3: BAKED
Pizza 1: BAKED
Pizza 0: BAKED
Pizza 4: BAKED
Pizza 3: SLICED
Pizza 1: SLICED
Pizza 2: SLICED
Pizza 0: SLICED
Pizza 4: SLICED
Pizza 3: BOXED
Pizza3: complete
Pizza 1: BOXED
Pizza1: complete
Pizza 2: BOXED
Pizza 0: BOXED
Pizza0: complete
Pizza2: complete
Pizza 4: BOXED
Pizza4: complete
1766
*/
```

答案是"没有变化"。想想也并不奇怪，因为每个比萨都需要所有的步骤按顺序执行，所以并没有机会让我们可以像在 PizzaParallelSteps.java 中那样，通过分离各个步骤来进一步提速。

我们可以用 CompletableFuture 来重写该示例：

```java
// concurrent/CompletablePizza.java
import java.util.*;
import java.util.concurrent.*;
import java.util.stream.*;
import onjava.Timer;

public class CompletablePizza {
  static final int QUANTITY = 5;
  public static CompletableFuture<Pizza>
  makeCF(Pizza za) {
    return CompletableFuture
      .completedFuture(za)
      .thenApplyAsync(Pizza::roll)
      .thenApplyAsync(Pizza::sauce)
      .thenApplyAsync(Pizza::cheese)
      .thenApplyAsync(Pizza::toppings)
      .thenApplyAsync(Pizza::bake)
      .thenApplyAsync(Pizza::slice)
      .thenApplyAsync(Pizza::box);
  }
  public static void
  show(CompletableFuture<Pizza> cf) {
    try {
      System.out.println(cf.get());
    } catch(Exception e) {
      throw new RuntimeException(e);
    }
  }
  public static void main(String[] args) {
    Timer timer = new Timer();
    List<CompletableFuture<Pizza>> pizzas =
      IntStream.range(0, QUANTITY)
        .mapToObj(Pizza::new)
        .map(CompletablePizza::makeCF)
        .collect(Collectors.toList());
    System.out.println(timer.duration());
    pizzas.forEach(CompletablePizza::show);
    System.out.println(timer.duration());
  }
}
```

```
/* 输出：
98
Pizza 1: ROLLED
Pizza 3: ROLLED
Pizza 4: ROLLED
Pizza 2: ROLLED
Pizza 0: ROLLED
Pizza 1: SAUCED
Pizza 2: SAUCED
Pizza 3: SAUCED
Pizza 0: SAUCED
Pizza 4: SAUCED
Pizza 2: CHEESED
Pizza 0: CHEESED
Pizza 3: CHEESED
Pizza 1: CHEESED
Pizza 4: CHEESED
Pizza 2: TOPPED
Pizza 4: TOPPED
Pizza 1: TOPPED
Pizza 3: TOPPED
Pizza 0: TOPPED
Pizza 0: BAKED
Pizza 4: BAKED
Pizza 1: BAKED
Pizza 3: BAKED
Pizza 2: BAKED
Pizza 3: SLICED
Pizza 1: SLICED
Pizza 2: SLICED
Pizza 4: SLICED
Pizza 0: SLICED
Pizza 0: BOXED
Pizza 2: BOXED
Pizza 4: BOXED
Pizza 1: BOXED
Pizza0: complete
Pizza1: complete
Pizza2: complete
Pizza 3: BOXED
Pizza3: complete
Pizza4: complete
1762
*/
```

并行流和 CompletableFuture 是 Java 并发工具中最发达的技术。不论何时你都应该优先选择其中之一。并行流方案最适合解决**无脑并行**[1]类型问题，即那种很容易将数据拆分成无差别、易处理的片段来处理的问题（要自己实现这部分的话，你得先撸起袖子好好钻研 Spliterator 的文档）。CompletableFuture 处理的工作片段最好是各不相同的，这样效果最好。CompletableFuture 看起来更像是面向任务的，而不是面向数据的。

从刚才的比萨示例来看，结果看起来并没有什么不同——实际上并行流的解决方案看起来更清晰，仅出于这个原因，我发现并行流作为第一次尝试变得更有吸引力了。

制作一份比萨需要一定的时间。不论你使用哪种并发方案，最好的办法就是在制作一份比萨的时间里，制作出 n 份比萨。在本例中很容易看出这一点，但是当你面对更复杂的问题时，你很可能会迷失其中。一般来说，在项目开始时，通过粗略的估算可以很快得出可能的最大吞吐量，这会阻止你想要让它运行更快的想法。

如果你确实需要使用并发，并行 Stream 和 CompletableFuture 可能很容易带来显著的收益。但是当你想要进一步优化时，请一定要小心，因其耗费的成本和精力可能很快就会远远超过优化的好处。

5.14 总结

应用并发的唯一正当理由是"太多等待"。这理论上也应该包括用户界面的响应性，但因为 Java 实际上并不适合构建用户界面，[2]所以这里仅仅是指"你的程序跑得不够快"。

如果并发很简单，那么就没有理由不去使用它。正因为并发很难，所以你应该考虑清楚是否值得为此付出努力。你能用其他办法提升速度吗？比如，迁移到更快的硬件上（相比于人力成本，硬件成本可能低很多），或者将程序拆分成不同片段，并运行在不同的机器上？

奥卡姆剃刀原理经常被误解。我已经至少在一部电影中看到它被定义为"最简单的方案就是最合适的"，就好像某种法律似的。它其实只是一个准则：当面临多个可选方案时，先尝试需要最少假设的方案。在编程世界中，它已经演化为"尝试可能有效的最简单方案"。当你对一种工具有所了解后——正如你现在对并发一样——这个工具会变得很有诱惑力，让你想去使用它，或者你提前就认定了你的方案肯定能"运行得很快"，以便从一

---

[1] 原文为 embarrassingly parallel，此习语来源于英文世界中的某个典故，用来表达可以不假思索地用并行来解决的问题类型。——译者注
[2] 有现成的库，并且这门语言本来也计划支持实现该目的（即用户界面），但是在实际应用中极少这么做，以至于可以说"永远不会这么做"。

开始就证明并发设计是合理的。但是奥卡姆剃刀原理的编程版本告诉我们，应该首先尝试最简单的方法（也会是开发成本最低的方案），看看是否能满足需求。

由于我的（专业）背景比较偏底层（物理学和计算机工程学），我倾向于想象所有系统底层的执行代价。我已经数不清有多少次非常肯定地认为，最简单的方式绝不可能跑得很快，但在尝试后发现它已经足够快了。

5.14.1 缺点

并发的主要缺点列举如下。

1. 线程等待共享资源时会导致系统变慢。
2. 管理线程需要额外的 CPU 开销。
3. 不合理的设计决策会导致不必要的复杂性。
4. 会带来饥饿、竞争、死锁、活锁（多个线程在各自独立的任务上工作，导致整体无法完成）等病态现象。
5. 平台间不一致。通过几个例子，我发现在某些机器上竞态条件很快就会出现，但在另一些机器上则不会。如果你在后者上开发程序，在发布后你可能会大吃一惊。

另外，并发的应用是一门艺术。Java 的设计理念是，你需要多少对象，就可以创建多少对象——至少理论上如此。[1] 然而，Thread 并不是典型的对象：每个线程都有自己的执行环境，包括一个栈和其他必要的元素，这使得线程比一个普通的对象要大很多。大部分环境下，系统内存只够你创建出几千个 Thread 对象（超出则会导致内存溢出）。但通常只会需要少量线程来解决问题，所以这算不上什么很严重的限制，但是对于某些设计来说，这会成为一种约束，可能会迫使你改用一个完全不同的方案。

共享内存的陷阱

并发伴随的一个主要难题是多个任务可能会共享一个资源，比如一个对象的内存，你必须保证这些任务不会同时读写该资源。

我花了好几年研究并发，并与之斗争。而我现在已经想通了，你永远都不能相信一个使用了共享内存的并发程序能够正确运行。你可以发现程序有问题，但你永远无法证明它没问题。这是众所周知的并发准则之一。[2]

---

[1] 例如，在工程中为有限元分析创建数百万个对象，如果没有**享元**（Flyweight）设计模式，用 Java 来实现可能不现实。

[2] 在科学中，一个理论无法被证明为真，但要想使其成立，它就必须（至少）是**可证伪**的。对于并发来说，我们大部分时间甚至无法证伪。

我遇到过很多人，他们对自己正确编写多线程程序的能力充满信心。我偶尔会想，我也可以。曾经有一次我开发一个程序，起初编写的时候，我仅有一台单 CPU 的机器。我能够说服我自己，因为我觉得我了解 Java 的各种工具，所以程序肯定是正确的，并且在我的单 CPU 机器上，程序没有出问题。

后来到了多 CPU 的机器上，我惊讶地发现程序出问题了，但这仅仅是问题之一。这并不是 Java 的错——"一次编写，到处运行"这个宗旨可能无法延伸到并发程序从单 CPU 机器到多 CPU 机器的迁移上。这是并发自身的根本性问题。你实际上能够在单 CPU 机器上发现一些问题，但还有些其他的问题，直到程序运行到多 CPU 机器上，线程真正开始并行运行时，它们才会显现。

而另外一个例子，也就是"哲学家用餐问题"，它能够被轻易地修正，使得死锁再难发生，这给了你一切都很完美的错觉。

只要涉及共享内存并发，你就永远不能对自己的编程能力太过自信。

5.14.2　Java 核心设计的失败之处

如果你感觉被 Java 并发搞得头昏脑涨，那么事实证明实际上并不是只有你一个人有这种感觉。看一下 Javadoc 中关于 Thread 类的部分，究竟有多少方法被废弃了。这些都是 Java 语言设计者犯过的错，因为他们在设计语言期间，对并发了解得还不够深入。

后来的 Java 版本中增加的一些库方案，最终被证明是无效甚至无用的。幸运的是，Java 8 中的并行 Stream 和 CompletableFuture 非常有价值，不过当你面对遗留代码时，仍然会接触到那些旧方案。

在本书其他地方，我提到过 Java 的重要问题之一：所有失败的尝试都永远嵌入了语言或库中。Java 并发则尤其如此。与其说这里面有很多错误——尽管确实是有很多错误——不如说是有很多为了解决问题而做过的尝试。好的一面是，这些尝试催生了更好、更简单的设计；不好的一面是，在找到合适的方案之前，你很容易迷失在那些旧的设计中。

5.14.3　其他的库

本章聚焦于并行 Stream 和 CompletableFuture 这两个相对安全和简单的工具上，并且只触及部分 Java 标准库中更细粒度的工具。为了避免让你一下子消化太多知识，此处并没有涉及一些你在实际工作中可能会用到的库。我们用到了 Atomic 类、ConcurrentLinkedDeque、ExecutorService 和 ArrayBlockingQueue。本书第 6 章涵盖了一些其他的工具，而 Javadoc 中的 java.util.concurrent 部分也值得你去探索。不过要注意，有些库组件已经被更好的新组件取代了。

5.14.4 设想一种为并发而设计的语言

大体来说，使用并发需要谨慎且节制。如果你需要使用并发，就尽量使用最新的方案：并行 Stream 或 CompletableFuture。设计它们的目的就是（在 Java 世界中）尽可能让你远离麻烦（如果你不会试图共享内存的话）。

如果你遇到的并发问题变得越来越大、越来越复杂，超出了高级 Java 体系的支持能力，就得考虑专门为并发设计的语言了。只在程序中需要并发的部分使用这种语言，这是有可能做到的。在编写本书时，基于 JVM 的语言中，最纯粹的函数式语言是 Clojure（Lisp 的一个版本）和 Frege（Haskell 的一种实现）。你可以用这两种语言编写你程序中的并发部分，然后通过 JVM 轻松地和 Java 主程序代码互动。另外，你还可以选择通过**异构语言函数接口**（foreign function interface, FFI），跳出 JVM 和另一种专用于并发的语言通信。在撰写本书时，gRPC 是最有前景的方法。

你很容易依赖一门语言，并且陷入一种想要用该语言解决一切问题的状态。一个常见的例子是构造 HTML/JavaScript 用户界面的工具，那些工具实际上设计粗糙、非常难用。实际上有很多库可以让你选择用你最喜欢的语言来实现该功能（比如，Scala.js 可以让你用 Scala 来实现）。

考虑心理上的便利性，这十分合理，但我希望你能通过本章（以及本书第 6 章）清楚地认识到 Java 并发是个很难逃离的深渊。要记住所有的陷阱，同时还要能通过目视检查发现代码的问题，这里面所需的知识比 Java 语言的其他任何部分都要难。

不论用某种语言或库实现并发看起来多么简单，你都要视并发为黑魔法。如果不给予其足够的重视，总有一天你会吃亏的。

5.14.5 延伸阅读

《Java 并发编程实战》，作者是 Brian Goetz、Tim Peierl、Joshua Bloch、Joseph Bowbeer、David Holmes 和 Doug Lea——基本上，这些名字就是 Java 并发世界中的 "名人录"。

《Java 并发编程》（第 2 版），Doug Lea 著。尽管该书成书明显早于 Java 5，但 Doug 的大部分工作催生了 java.util.concurrent 库，因此本书对于全面理解并发问题来说至关重要。这本书超越了 Java 的范畴，从跨越语言和技术的层面讨论了并发问题。虽然现在来看，这本书的某些地方有些过时，但还是值得多读几遍（最好间隔几个月以助于理解）。Doug 是世界上少数真正理解并发的人之一，因此此书值得细读。

06

底层并发

> 虽然从来都不建议自行编写 Java 底层并发的代码,但是理解其运行机制通常会很有帮助。

本书第 5 章中从上层应用角度介绍了并发的概念,包括更新、更安全的 Java 并发编程结构(并行流和 CompletableFuture)。本章会介绍 Java 中的底层并发概念,这样你在读到这些代码的时候,就能对其有一定的了解。同时,你也能更为深入地理解并发的通用问题。

在 Java 的早期版本中,底层并发的概念是并发编程中很重要的一部分。我们将探讨这些技术的复杂性,以及为什么应该避免使用它们。本书第 5 章介绍了较新版本的 Java(特别是 Java 8)所带来的改进后的技术,极大降低了并发(原本较高)的实现难度。

6.1 什么是线程?

并发将程序分割成为多个各自独立运行的任务。每个任务都由**执行线程**（thread of execution）所驱动，通常简称为**线程**（thread）。线程是操作系统进程内部按单一顺序执行的控制流。由此一个进程可以包含多个并发执行的任务，而你编写程序时就好像每个任务都有一个自己的处理器一样。多线程模型提升了编程的便利性，可以简化在单个程序中调度多个任务的工作。操作系统会通过处理器为你的各个线程分配执行时间。

Java 并发的核心机制是 Thread 类。在这门语言最初的版本中，Thread 本来是打算让程序员直接创建和管理的。随着语言的演进，人们发现了更好的方法，中间机制——特别是 Executor——被加入进来，以消除自行管理多线程的精神负担（和出现的错误）。最后甚至发展出了更优于 Executor 的机制，如本书第 5 章中所示。

Thread 是一种将任务和处理器关联起来的软件结构。虽然 Thread 的创建和使用看起来和任何其他类没什么不同，但是它们的内在区别其实很大。在你创建 Thread 时，JVM 会在一块专为 Thread 保留的内存区域中分配一大块空间，从而为任务的运行提供所需的一切。

- 一个程序计数器，指示要执行的下一条 JVM 字节码指令。
- 一个支持 Java 代码执行的栈，包含该线程到达当前执行节点前所调用过的方法的相关信息。它同时还包含正在执行的方法的所有本地变量（包括基本类型和堆上对象的引用）。在**每个线程**中，该栈的大小通常在 64KB 和 1MB 之间。[①]
- 一个用于本地代码的栈。
- **本地线程变量**（thread-local variable）存储。
- 用于控制线程的状态维护变量。

包括 main() 在内的所有代码都运行在某个线程之内。只要有方法被调用，当前的程序计数器便会跳向该线程的栈，然后栈指针会向下移动足够的距离，创建出一个**栈帧**（stack frame），其中包含该方法的所有本地变量、参数以及返回值的存储。所有的本地基本类型会直接放在该栈上。方法所创建的对象的引用全部保存在该栈帧内，而对象本身则放到堆上。堆只有一个，被程序中所有的线程所共享。

除此之外，Thread 必须由操作系统（OS）进行注册，这样它才可以在某个时机被真正关联到处理器上。这会作为 Thread 构造过程的一部分来管理。Java 通过底层操作系统的这种机制来管理多线程的执行。

---

① 在某些平台上，尤其是 Windows，可能非常难以找出栈空间的默认值。可以通过 -Xss 标志调整栈空间大小。

6.1.1　最佳线程数

如果你仔细看看本书第 5 章中使用的 CachedThreadPool 的例子，便会发现 ExecutorService 为我们提交的每个任务都分配了一个线程。不过，CountingStream.java 中的并行 Stream 只分配了 8 个线程（工作线程 worker 1~7，以及分配给 main() 的一个线程，该线程被巧妙地用于一个额外的并行流）。如果你试着增加 range() 的上限，则会发现并不会创建出额外的线程。这是为什么呢？

我们可以检测出当前机器上的处理器数量：

```
// lowlevel/NumberOfProcessors.java

public class NumberOfProcessors {
  public static void main(String[] args) {
    System.out.println(
      Runtime.getRuntime().availableProcessors());
  }
}
/* 输出:
8
*/
```

我的机器（使用的是 Intel Core i7 处理器）是 4 核的，每个核心可作为 2 个**超线程**（hyperthread，一项硬件上的技术，可以在单个处理器上实现非常快速的上下文切换，某些情况下可以让该处理器看起来就像是两条硬件级的线程）。虽然在"较新"的机器上这是一项常规配置（在编写本书时），但是你可能会在 CountingStream.java 中看到不同的结果——在默认线程数相同的情况下。

你的操作系统可能有办法得到处理器的更多信息。例如，在 Windows 10 中，单击"开始"按钮，在搜索栏中输入"任务管理器"并按下回车键，选择"性能"标签页，便可以看到关于硬件的所有类型的信息，包括"核心"和"逻辑处理器"。

事实证明，线程"通常"的最佳数量就是可用处理器的数量（对某些特定问题来说可能不是这样）。这其中的缘由来自 Java 线程间的**上下文切换**开销：保存要挂起的线程的当前状态，读取要继续执行的线程在进入挂起状态时（所保存）的实时状态。对于 8 个处理器和 8 个（计算密集型）Java 线程的情况来说，JVM 永远不需要在运行这 8 个任务时进行上下文切换。对于任务数量少于处理器数量的情况，分配更多线程则并无益处。

Intel 的超线程定义了"逻辑处理器"的数量，它并没有增加实际的计算吞吐量——该特性在硬件层维护了额外的线程上下文，可以提升上下文切换的速度，这有助于提升如用户界面的响应速度等方面的性能。对于计算密集型任务来说，可以考虑将线程数设置为和物理核心（而非超线程）的数量一致。虽然 Java 会将每个超线程都视作一个处理器，但这似乎是被 Intel 对超线程的过度营销所误导的结果。尽管如此，为了简化代码编写，我会直接让 JVM 来决定默认线程数，你也需要在你自己的生产环境中进行验证。这并不意味着在所有场景中都适合将线程数和处理器数设为相同的。相反，该方式主要用于计算密集型的问题场景。

6.1.2 我可以创建多少线程

Thread 对象中占据空间最多的部分是储存待执行方法的 Java 栈。注意 Thread 对象的大小会随不同的操作系统而不同。下面这段代码测试了这一点，它会不断地创建 Thread 对象，直到 JVM 内存耗尽。

```java
// lowlevel/ThreadSize.java
// {ExcludeFromGradle} 运行时间过长或者会持续挂起
import java.util.concurrent.*;
import onjava.Nap;

public class ThreadSize {
  static class Dummy extends Thread {
    @Override public void run() { new Nap(1); }
  }
  public static void main(String[] args) {
    ExecutorService exec =
      Executors.newCachedThreadPool();
    int count = 0;
    try {
      while(true) {
        exec.execute(new Dummy());
        count++;
      }
    } catch(Error e) {
      System.out.println(
        e.getClass().getSimpleName() + ": " + count);
      System.exit(0);
    } finally {
      exec.shutdown();
    }
  }
}
```

只要你一直向 CachedThreadPool 传入任务，它就会不断地创建 Thread。向 execute() 传入 Dummy 对象即可开启任务，分配一个新的 Thread（如果没有现成可用的 Thread）。Nap() 的参数必须足够大，这样任务就不会进入完成阶段，使得现有的 Thread 被释放，以执行新任务。只要不断传入任务并且使任务无法完成，CachedThreadPool 就最终会耗尽内存。

我不一定总能在每台机器上都制造出内存溢出的错误。在某台机器上，我看到这样的结果：

```
>java ThreadSize
OutOfMemoryError: 2816
```

可以用 -Xss 标志减少每个 Thread 的栈大小。允许的最小栈空间是 64KB：

```
>java -Xss64K ThreadSize
OutOfMemoryError: 4952
```

如果将栈的大小增大到 2MB，能分配的线程数则会大大减少：

```
>java -Xss2M ThreadSize
OutOfMemoryError: 722
```

Windows 的默认栈大小是 320KB，这一点可以通过以下事实来验证——在完全不设置栈大小的时候，我们在输出中得到了（和将栈大小设置为 320KB 时）相同的线程数量：

```
>java -Xss320K ThreadSize
OutOfMemoryError: 2816
```

也可以用 -Xmx 标志增大 JVM 的最大可分配内存：

```
>java -Xss64K -Xmx5M ThreadSize
OutOfMemoryError: 5703
```

注意，操作系统也可能会对允许的线程数施加限制。

因此，"我可以创建多少线程"这个问题的答案是"几千个"。不过，如果你发现自己真的分配了几千个线程，就可能需要重新考虑你的方法了——真正应该问的问题是"我需要多少线程"。

"工作窃取"线程池（WorkStealingPool）

这是一种能基于所有可用处理器（如 JVM 可检测出的）自动创建线程池的 ExecutorService。

```java
// lowlevel/WorkStealingPool.java
import java.util.stream.*;
import java.util.concurrent.*;

class ShowThread implements Runnable {
  @Override public void run() {
    System.out.println(
      Thread.currentThread().getName());
  }
}
```

```
}
public class WorkStealingPool {
  public static void main(String[] args)
    throws InterruptedException {
    System.out.println(
      Runtime.getRuntime().availableProcessors());
    ExecutorService exec =
      Executors.newWorkStealingPool();
    IntStream.range(0, 10)
      .mapToObj(n -> new ShowThread())
      .forEach(exec::execute);
    exec.awaitTermination(1, TimeUnit.SECONDS);
  }
}
/* 输出：
8
ForkJoinPool-1-worker-1
ForkJoinPool-1-worker-2
ForkJoinPool-1-worker-1
ForkJoinPool-1-worker-3
ForkJoinPool-1-worker-4
ForkJoinPool-1-worker-4
ForkJoinPool-1-worker-2
ForkJoinPool-1-worker-3
ForkJoinPool-1-worker-5
ForkJoinPool-1-worker-1
*/
```

工作窃取（Work Stealing）算法使得已完成自身输入队列中所有工作项的线程可以"窃取"其他队列中的工作项。该算法的目的是在执行计算密集型任务时，能够跨处理器分发工作项，由此最大化所有可用处理器的利用率，Java 的 fork/join 框架中同样也用到了该算法。

6.2 捕获异常

下面这个示例可能会出乎你的意料：

```
// lowlevel/SwallowedException.java
import java.util.concurrent.*;

public class SwallowedException {
  public static void main(String[] args)
    throws InterruptedException {
    ExecutorService exec =
      Executors.newSingleThreadExecutor();
    exec.submit(() -> {
      throw new RuntimeException();
    });
    exec.shutdown();
  }
}
```

这段代码什么都没输出（不过如果你将 submit() 替换为 execute()，便会看到异常）。这说明在线程内部抛出异常很需要技巧，也需要特别仔细的操作。

你无法捕获已逃离线程的异常。一旦异常逃逸到任务的 run() 方法之外，便会扩散到控制台，除非采用专门的步骤来捕获这种不正确的异常。

下面这个任务抛出了传播到其 run() 方法之外的异常，而 main() 则显示了运行该任务时发生的情况：

```
// lowlevel/ExceptionThread.java
// {ThrowsException}
import java.util.concurrent.*;

public class ExceptionThread implements Runnable {
  @Override public void run() {
    throw new RuntimeException();
  }
  public static void main(String[] args) {
    ExecutorService es =
      Executors.newCachedThreadPool();
    es.execute(new ExceptionThread());
    es.shutdown();
  }
}
/* 输出：
___[ Error Output ]___
Exception in thread "pool-1-thread-1"
java.lang.RuntimeException
        at ExceptionThread.run(ExceptionThread.java:7)
        at java.util.concurrent.ThreadPoolExecutor.runW
orker(ThreadPoolExecutor.java:1142)
        at java.util.concurrent.ThreadPoolExecutor$Work
er.run(ThreadPoolExecutor.java:617)
        at java.lang.Thread.run(Thread.java:745)
*/
```

输出如下（调整了一些限定符）：

```
Exception in thread "pool-1-thread-1" RuntimeException
  at ExceptionThread.run(ExceptionThread.java:9)
  at ThreadPoolExecutor.runWorker(...)
  at ThreadPoolExecutor$Worker.run(...)
  at java.lang.Thread.run(Thread.java:745)
```

将 main() 方法体包裹在 try-catch 块中也无法成功运行：

```
// lowlevel/NaiveExceptionHandling.java
// {ThrowsException}
import java.util.concurrent.*;

public class NaiveExceptionHandling {
  public static void main(String[] args) {
    ExecutorService es =
      Executors.newCachedThreadPool();
    try {
      es.execute(new ExceptionThread());
    } catch(RuntimeException ue) {
      // 该语句不会执行!
      System.out.println("Exception was handled!");
```

```
    } finally {
      es.shutdown();
    }
  }
}
/* 输出：
___[ Error Output ]___
Exception in thread "pool-1-thread-1"
java.lang.RuntimeException
        at ExceptionThread.run(ExceptionThread.java:7)
        at java.util.concurrent.ThreadPoolExecutor.runW
orker(ThreadPoolExecutor.java:1142)
        at java.util.concurrent.ThreadPoolExecutor$Work
er.run(ThreadPoolExecutor.java:617)
        at java.lang.Thread.run(Thread.java:745)
*/
```

以上代码生成了和前面的示例相同的结果：一个未捕获的异常。要解决这个问题，就必须改变该 Executor 生成线程的方式。

Thread.UncaughtExceptionHandler 接口会向每个 Thread 对象添加异常处理程序。当线程由于未捕获的异常而即将消亡时，Thread.UncaughtExceptionHandler.uncaughtException() 就会被自动调用。要使用该方法，可以创建一个新的 ThreadFactory 类型，它会为它所创建的每个新 Thread 对象添加一个新的 Thread.UncaughtExceptionHandler。将这个 ThreadFactory 工厂传入 Executors.newCachedThreadPool 方法，该方法会创建新 ExecutorService：

```
// lowlevel/CaptureUncaughtException.java
import java.util.concurrent.*;

class ExceptionThread2 implements Runnable {
  @Override public void run() {
    Thread t = Thread.currentThread();
    System.out.println("run() by " + t.getName());
    System.out.println(
      "eh = " + t.getUncaughtExceptionHandler());
    throw new RuntimeException();
  }
}

class MyUncaughtExceptionHandler implements
Thread.UncaughtExceptionHandler {
  @Override
  public void uncaughtException(Thread t, Throwable e) {
    System.out.println("caught " + e);
  }
}

class HandlerThreadFactory implements ThreadFactory {
  @Override public Thread newThread(Runnable r) {
    System.out.println(this + " creating new Thread");
    Thread t = new Thread(r);
    System.out.println("created " + t);
```

```
      t.setUncaughtExceptionHandler(
        new MyUncaughtExceptionHandler());
      System.out.println(
        "eh = " + t.getUncaughtExceptionHandler());
      return t;
    }
  }

  public class CaptureUncaughtException {
    public static void main(String[] args) {
      ExecutorService exec =
        Executors.newCachedThreadPool(
          new HandlerThreadFactory());
      exec.execute(new ExceptionThread2());
      exec.shutdown();
    }
  }
  /* 输出:
  HandlerThreadFactory@106d69c creating new Thread
  created Thread[Thread-0,5,main]
  eh = MyUncaughtExceptionHandler@52e922
  run() by Thread-0
  eh = MyUncaughtExceptionHandler@52e922
  caught java.lang.RuntimeException
  */
```

从附加的跟踪代码可以验证得知，工厂所创建的线程获得了新的 UncaughtExceptionHandler。未捕获的异常现在被 uncaughtException 捕获了。

上例基于具体问题具体分析的方式设置处理器。如果你确定要在所有地方都应用同一个异常处理程序，更简单的方法是设置默认未捕获异常处理程序，该处理器会在 Thread 类内部设置一个静态字段：

```
// lowlevel/SettingDefaultHandler.java
import java.util.concurrent.*;

public class SettingDefaultHandler {
  public static void main(String[] args) {
    Thread.setDefaultUncaughtExceptionHandler(
      new MyUncaughtExceptionHandler());
    ExecutorService es =
      Executors.newCachedThreadPool();
    es.execute(new ExceptionThread());
    es.shutdown();
  }
}
/* 输出:
caught java.lang.RuntimeException
*/
```

只有在为线程专门配置的未捕获异常处理程序不存在的时候，该处理器才会被调用。系统会检查是否存在为线程专门配置的处理器，如果不存在，则会检查该线程组是否专门实现了其 uncaughtException() 方法，如果没有，则会调用 defaultUncaughtExceptionHandler。

可以将该方法和之前介绍的 CompletableFuture 中改进后的方法进行比较。

6.3 共享资源

可以将单线程程序看作一个在问题空间中独自穿行的实体，它一次只能做一件事。因为只有一个实体，所以你永远不需要考虑两个实体试图在同一时间使用同一个资源的问题：例如，两个人试图在同一个地方停车，或者同时穿过一道门，抑或是同时说话。

有了并发，这些就不再是单项任务，而是变成了可能会互相影响的两个任务（甚至更多个）。如果你不去阻止这种冲突，便会出现两个任务试图同时访问同一个银行账户，同时调用同一个打印机进行打印，或者同时调整同一个阀门等情况。

6.3.1 资源竞争

当你启动一项任务来执行某些操作时，该操作的结果可以用两种不同的方法来捕获：通过副作用或者通过返回值。

副作用的方式看起来会更简单：只需用结果来操纵一下环境中的某个东西即可。例如，你的任务可能会执行某些计算，然后直接将结果写入一个集合。

这种方法存在的问题是，集合通常是**共享资源**。当正在运行的任务不止一个时，任何任务都可能同时对一个共享资源进行读或写操作。这便引出了多任务场景最主要的隐患之一：**资源竞争**（resource contention）问题。

在单线程系统中，你永远无须考虑资源竞争的问题，因为你永远都不会同时做两件或更多件事。而当你处理多任务问题时，则必须时刻防止出现资源竞争问题。

解决该问题的一种途径是使用一个能够处理资源竞争问题的集合。如果同一时间有多个任务试图对这样的一个集合进行写操作，该集合能够妥善处理资源竞争问题。在 Java 并发库中能找到一些试图解决资源竞争问题的类，本章中会介绍其中的一小部分，但是并不全面。

看看下面这个示例，一个任务会生成偶数，而其他任务则会消费这些偶数。此处，消费者任务唯一要做的事就是检查这些偶数的有效性。

这个示例中会定义 EvenChecker，即消费者任务，并使其在后续的示例中也能够复用。为了将 EvenChecker 从我们的各种实验用生成器中分离出来，首先创建一个名为 IntGenerator 的抽象类，它包含了 EvenChecker 需要用到的最小必要方法集——next() 方法，而且它可以被取消。

```
// lowlevel/IntGenerator.java
import java.util.concurrent.atomic.AtomicBoolean;

public abstract class IntGenerator {
  private AtomicBoolean canceled =
    new AtomicBoolean();
  public abstract int next();
  public void cancel() { canceled.set(true); }
  public boolean isCanceled() {
    return canceled.get();
  }
}
```

cancel() 会改变 canceled 标志的状态，而 isCanceled() 则会告诉你该标志是否被设置了。由于 canceled 是个 AtomicBoolean，所以它是**原子**（atomic）的，这意味着诸如赋值和返回值这样的简单操作不会相互冲突，因此你不会看到该字段在进行这些简单操作时的中间状态。在本章稍后，你会学到更多关于原子性和 Atomic 类的知识。

任何 IntGenerator 都可以用下面这个 EvenChecker 类进行测试：

```
// lowlevel/EvenChecker.java
import java.util.*;
import java.util.stream.*;
import java.util.concurrent.*;
import onjava.TimedAbort;

public class EvenChecker implements Runnable {
  private IntGenerator generator;
  private final int id;
  public EvenChecker(IntGenerator generator, int id) {
    this.generator = generator;
    this.id = id;
  }
  @Override public void run() {
    while(!generator.isCanceled()) {
      int val = generator.next();
      if(val % 2 != 0) {
        System.out.println(val + " not even!");
        generator.cancel(); // 取消所有的 EvenChecker
      }
    }
  }
  // 测试任意的 IntGenerator：
  public static void test(IntGenerator gp, int count) {
```

```
    List<CompletableFuture<Void>> checkers =
      IntStream.range(0, count)
        .mapToObj(i -> new EvenChecker(gp, i))
        .map(CompletableFuture::runAsync)
        .collect(Collectors.toList());
    checkers.forEach(CompletableFuture::join);
  }
  // 初始计数值:
  public static void test(IntGenerator gp) {
    new TimedAbort(4, "No odd numbers discovered");
    test(gp, 10);
  }
}
```

test() 方法启动了多个访问同一个 IntGenerator 的 EvenChecker。EvenChecker 任务持续从各自的关联 IntGenerator 中读取值并进行测试。如果 IntGenerator 导致失败，则 test() 会报错并返回。

所有依赖于 IntGenerator 对象的 EvenChecker 任务都会检查 IntGenerator，以确认任务是否已被取消。如果 generator.isCanceled() 的结果是 true，run() 方法就会返回。任何 EvenChecker 任务都可以调用 IntGenerator 上的 cancel() 方法，这会使其他所有使用该 IntGenerator 的 EvenChecker 任务优雅地终止。

在这种设计中，多个任务共享了同一个公共资源（即 IntGenerator），监听该资源的终止信号。这样便消除了所谓的竞态条件，即两个以上的任务竞争响应某个条件，并因此发生冲突，或者生成了不一致的结果。

我们必须谨慎地考虑并避免所有可能导致并发系统失败的因素。举例来说，一个任务不能依赖于另一个任务，因为任务的终止顺序是无法保证的。此处，通过让任务依赖于非任务对象，我们消除了竞态条件的隐患。

通常，我们会假设 test() 最终总会失败，因为 EvenChecker 任务能够在 IntGenerator 处于"不正确"的状态时，访问其中的信息。不过，可能要等到 IntGenerator 完成很多次循环后，才会检测出该问题，这取决于操作系统的特性，以及实现方面的其他细节。要确保本书的自动化构建不会卡死，我们使用了 TimedAbort，定义如下：

```
// onjava/TimedAbort.java
// t 秒后终止程序
package onjava;
import java.util.concurrent.*;

public class TimedAbort {
  private volatile boolean restart = true;
```

```
  public TimedAbort(double t, String msg) {
    CompletableFuture.runAsync(() -> {
      try {
        while(restart) {
          restart = false;
          TimeUnit.MILLISECONDS
            .sleep((int)(1000 * t));
        }
      } catch(InterruptedException e) {
        throw new RuntimeException(e);
      }
      System.out.println(msg);
      System.exit(0);
    });
  }
  public TimedAbort(double t) {
    this(t, "TimedAbort " + t);
  }
  public void restart() { restart = true; }
}
```

我们用 lambda 表达式创建了一个 Runnable，通过 CompletableFuture 的 static runAsync() 方法来执行。使用 runAsync() 的价值是可以立刻返回调用。因此，TimedAbort 不会让任何本来可以完成的任务保持开启状态，不过如果执行时间太长，它仍然会终止该任务 [TimedAbort 有时称为**守护进程**（daemon）]。

TimedAbort 同样允许我们将其 restart()（重启），这是为了在需要持续运行某些必要的操作时，可以让程序一直保持运行。

我们可以通过实操，进一步看看 TimedAbort：

```
// lowlevel/TestAbort.java
import onjava.*;

public class TestAbort {
  public static void main(String[] args) {
    new TimedAbort(1);
    System.out.println("Napping for 4");
    new Nap(4);
  }
}
```

```
/* 输出:
Napping for 4
TimedAbort 1.0
*/
```

如果注释掉 Nap() 这一行，程序就会立刻退出，由此可以看出 TimedAbort 并未让程序保持运行。

下面我们会看到第一种 IntGenerator 的实现（EvenProducer），其中的 next() 方法用来生成一系列偶数值：

```
// lowlevel/EvenProducer.java
// 线程冲突时
// {VisuallyInspectOutput}

public class EvenProducer extends IntGenerator {
  private int currentEvenValue = 0;
  @Override public int next() {
    ++currentEvenValue;                    // [1]
    ++currentEvenValue;
    return currentEvenValue;
  }
  public static void main(String[] args) {
    EvenChecker.test(new EvenProducer());
  }
}
/* 输出:
419 not even!
425 not even!
423 not even!
421 not even!
417 not even!
*/
```

[1] 一个任务能在另一个任务执行了 currentEvenValue 的第一个自增操作后调用 next()，但不能在第二个自增操作后调用。这会让该值处于"不正确"的状态。

为了防止这种情况发生，EvenChecker.test() 创建了一组 EvenChecker 对象，以持续读取 EvenProducer 的输出，并检测各个输出值是否是偶数。如果不是，便会报告错误，程序随即终止。

多线程程序的部分问题在于，如果出错的概率非常低，即使存在 bug 的情况下，这些程序也可以看起来运行正常。

要注意，自增操作本身也包括多个执行步骤，而任务有可能在自增操作的中途被线程机制挂起。也就是说，Java 中的自增操作并不是原子操作。因此即使是简单的自增操作，如果不对任务进行必要的保护，也是不安全的。

该程序并不总是在首次生成非偶数值时就终止。所有的任务都不会立刻终止，这也是并发程序的常态了。

6.3.2 解决资源竞争

前面的示例演示了使用线程时存在的重要问题：你永远无法知道一个线程可能在何时运行。想象你坐在餐桌前，手里拿着一根叉子，你正要叉起盘中的最后一块食物，当叉子正伸向食物的时候，食物突然消失了……这是因为你的线程被挂起了，然后另一位食客进入并吃掉了食物。这就是你在编写并发程序时面临的问题。为了让并发正确运行，你需要用某种方式防止两个任务访问同一项资源，至少在关键时期要避免出现这种情况。

避免这种冲突的关键就是在一个任务使用一项资源的时候，对该资源上锁。第一个访问资源的任务必须锁上该资源，然后其他任务就无法访问该资源了，直到锁被解除，然后另一个任务再次上锁并使用该资源，以此类推。如果一辆车的前排座位是有限的资源，那么大喊"前排座位!"的孩子就（在本次旅途中）获得了锁。

要解决线程冲突的问题，可以使用基本的并发方案**将共享资源的访问操作串行化**。这意味着同一时刻只允许一个任务访问共享资源。这通常是通过用一个子句将一段代码包围起来，使得同一时刻只允许一个任务执行该段代码来实现的。由于该语句会产生**互斥**（mutual exclusion）的效果，因此这样的机制通常被称为**互斥锁**（mutex）。

考虑一下卫生间的使用场景：多个人（即多个线程驱动的多个任务）都想单独使用卫生间（即共享资源）。要进入卫生间，一个人需要先敲门以确认卫生间当前是否可用。如果可用，他就会进入卫生间并锁上门。其他任何想要使用卫生间的人就被"阻塞"而不能使用了，因此这些人需要在门口排队等候，直到卫生间重新可用。

在卫生间被释放，允许下一个人进入的时候，这个类比就有点不成立了。实际上并不会有人排队，而且我们并不确定谁能得到卫生间的下一个使用权，因为线程的调度并不是确定的。实际情况是，就好像有一组阻塞的任务在卫生间门口徘徊，当锁住了卫生间的任务开锁出门时，线程调度器会决定哪一个线程接下来进入卫生间。

为了防止资源使用上的冲突，Java 实现了 synchronized 关键字这种形式上的内建支持。当一个任务想要执行一段由 synchronized 关键字所保护的代码段时，Java 编译器会生成代码以确认锁是否可用。如果可用，该任务便会获得锁，执行代码，然后释放锁。

共享资源一般只是以对象形式存在的一段内存，但它也可以是一个文件、一个 I/O 端口，或者诸如打印机这样的事物。要控制对共享资源的访问，首先要将资源放入一个对象中。然后任何使用该资源的方法都可以是 synchronized 的。如果一个任务位于对其中一个 synchronized 方法的调用中，则所有其他任务都会被阻塞，无法进入该对象的**任何一个** synchronized 方法，直到第一个任务从其调用中返回。

通常需要将字段都设为 private 的，而且只通过方法来访问这些字段。可以通过使用 synchronized 关键字声明方法来防止冲突，就像这样：

```
synchronized void f() { /* ... */ }
synchronized void g() { /* ... */ }
```

所有对象都会自动包含一个锁 [也称为**监视器**（monitor）]。在调用任一 synchronized 方法的时候，该对象都会被锁上，该对象的任何其他 synchronized 方法都无法被调用，直到第一个调用完成并释放锁。如果一个任务调用了某对象的 f() 方法，另一个任务便无法

调用同一对象上的 `f()` 或者 `g()`，直到 `f()` 的调用完成并释放锁。因此，存在一个被某特定对象的所有 `synchronized` 方法所共享的锁，这个锁可以防止多个任务在同一时刻对该对象的内存进行写操作。

尤其重要的是，在使用并发时要将字段都设为 `private` 的，否则 `synchronized` 关键字便无法阻止其他的任务直接访问字段，这样便产生了冲突。

一个线程可以多次获得一个对象的锁。如果一个方法调用同一对象上的第二个方法，而第二个方法又调用了同一对象上的另一个方法，便会发生这种情况，以此类推。JVM 会持续对对象被上锁的次数进行计数。如果该对象的锁被解除，该计数便为 0。一个线程第一次获得锁的时候，该计数便为 1。该线程每次获得该同一对象上的又一个锁时，该计数便会加 1。当然，只有首先获取锁的线程才被允许多次获得锁。线程每跳出一次 `synchronized` 方法，该计数便会减 1，直到减为 0，此时完全释放掉锁，让其他线程使用。

每个类还有一个锁（作为该类的 `Class` 对象的一部分），因此 `synchronized static` 的方法可以在类范围内相互上锁，防止同时访问 `static` 的数据。

应该在何时使用同步（synchronize）？可以应用 **Brian 的同步法则**（Brian's Rule of Synchronization）：[1]

如果你在对一个可能接下来会被另一个线程读取的变量进行写操作，或者读取一个可能刚被另一个线程完成写操作的变量，就必须使用同步，并且读操作和写操作都必须用同一个监视器同步。

当类中的多个方法都要处理关键数据时，必须对所有相关方法都进行同步。如果只对其中一个方法进行同步，那么其他的方法就可以无视该对象锁，并被任意调用。这一点很重要：访问关键共享数据的所有操作都必须是 `synchronized` 的，否则便无法正确运行。

6.3.3 将 EvenProducer 同步化

`SynchronizedEvenProducer` 是一个将 `EvenProducer.java` 中的 `next()` 方法修改为 `synchronized` 版本的 `IntGenerator`，这样可以阻止不符合预期的线程访问：

```
// lowlevel/SynchronizedEvenProducer.java
// 使用 synchronized 简化互斥锁
import onjava.Nap;

public class
```

---

[1] 出自《Java 并发编程实战》的作者之一 Brian Goetz。

```
SynchronizedEvenProducer extends IntGenerator {
  private int currentEvenValue = 0;
  @Override public synchronized int next() {      /* 输出:
    ++currentEvenValue;                           No odd numbers discovered
    new Nap(0.01); // 可以更快导致失败              */
    ++currentEvenValue;
    return currentEvenValue;
  }
  public static void main(String[] args) {
    EvenChecker.test(new SynchronizedEvenProducer());
  }
}
```

此处在两次自增操作中间插入了 Nap() 方法，以提升 currentEvenValue 为奇数状态时，上下文切换发生的可能性。由于互斥锁在临界区一次阻止了多个任务，因此这不会导致失败发生。第一个进入 next() 的任务获取了锁，而任何试图获取锁的后续任务都因被阻塞而无法获取锁，直到第一个任务释放锁。同时，调度机制会选择正在等待该锁的另一个任务。这样，一次只有一个任务可以调用该互斥锁所保护的代码。

6.4 volatile 关键字

volatile 可能是 Java 中最微妙、最难用的关键字了。幸运的是，在现在的 Java 中，你实际上可以永远避开使用这个关键字，并且如果你真的看到有代码用到了它，反而应该怀疑——通常来说，这很可能意味着代码太旧了，或者这段代码的编写者并不理解 volatile 或并发可能产生的后果（或者两者兼而有之）。

使用 volatile 主要有三个原因。

6.4.1 字分裂

字分裂（word tearing）出现在当你的数据类型足够大（如 Java 中的 long 和 double，这两者都是 64 位），对某个变量的写操作过程分为两个步骤的时候。JVM 允许将对 64 位数的读写操作分为两次对 32 位数的独立操作[①]，这增加了在读操作中途发生上下文切换的可能性，这样其他任务就会看到错误的结果。之所以称其为"字分裂"，是因为你可能会看到该数仅完成了部分变更时的值。总体来说，一个任务有时可能会在第一步之后、第二步之前读到该变量，从而读到一个垃圾值（对于 boolean 或 int 这样的小变量来说，这不会成为问题，但 long 和 double 除外）。

在没有任何其他保护的时候，将 long 或 double 定义为 volatile 变量，可以防止字分

---

① 注意，这在 64 位处理器上可能不会发生，也就不会存在这种问题。

裂。不过，如果使用 synchronized 或者某个 java.util.concurrent.atomic 类保护了这些变量，volatile 便可以被取代。同样，volatile 无法改变自增操作并不是一个原子操作这个事实。

6.4.2 可见性

第二个问题属于 Java 并发四定律的第二条"一切都不可信，一切都很重要"的一部分（参见本书第 5 章 5.4.2 节）。必须假定每个任务都有自己的处理器，而且每个处理器都有自己的本地缓存。该缓存可以让处理器运行得更快，这是因为处理器不用每次都要从主存中获取数据了，从缓存中读取值所耗费的时间会少得多。

问题之所以出现是因为 Java 试图尽可能提升效率。缓存的全部意义就在于避免从主存读取数据。但是由于并发的缘故，在 Java 应该将数据从主存刷新到本地缓存时，有时情况会变得不明确[1]——这种问题称为**缓存一致性**（cache coherence）问题。

每个线程都可以在处理器缓存中保存变量的本地副本。将字段定义为 volatile 的，可以阻止这些编译器优化，从而直接从内存进行读取，而不会被缓存。一旦在该字段上发生了写操作，所有任务中的所有读操作都会看到（该写操作产生的）变更。如果 volatile 变量恰好存在于本地缓存中，它就会立刻被写入主存，对该字段的所有读操作都会一直发生在主存上。

在以下情况中，应该将变量定义为 volatile 的：

1. 该变量会同时被多个任务访问；
2. 这些访问中至少有一个是写操作；
3. 试图避免使用同步（在现代 Java 中，可以使用高级工具，以取代对同步的使用）。

举例来说，如果你将一个变量作为终止某个任务的标志，该变量必须至少被声明为 volatile 的（尽管这样的一个标志并不一定能保证线程安全）。否则，在一个任务对标志进行了修改后，该修改会保存到本地处理器缓存中，而不会刷新到主存。当其他任务查看该标志时，便看不到这些修改。（我更喜欢用 AtomicBoolean 来实现这样的标志，如本书第 5 章 5.9 节中所述。）

一个任务对自己的变量所做的任何修改都永远对该任务自身可见，因此如果一个变量只会在任务内部使用，那就无须将该变量设为 volatile 的。

如果某个线程对一个变量进行了写操作，而其他线程只需要读取该变量，那也无须将

---

[1] 即可能因为实际执行顺序错乱而导致最终结果不符合预期。——译者注

该变量设为 volatile 的。总体来说，如果多个线程都会对某个变量进行写操作，volatile 就无法解决你的问题，必须改用 synchronized 来防止竞态条件。但有个特殊的例外情况：可以让多个线程对该变量进行写操作，**只要它们不用先读取该变量，再用读出来的值生成新的值写回该变量**。如果这些线程使用结果中的旧值，便会出现竞态条件，因为某个别的线程可以在你的线程正在执行计算的时候修改该变量。即使你一开始做得没有问题，想象一下，在代码的修改或维护期间，或者是对于不了解该问题的其他程序员来说（对 Java 来说尤其容易出现问题，因为程序员往往严重依赖编译期的检查来告诉他们代码是否正确），他们是多么容易忘掉这一点，并引入破坏性的代码变更。

原子性和不确定性（volatility）并不是相同的概念，理解这一点很重要。对非 volatile 变量进行原子操作并不能保证该操作刷新到了主存中。

同步会触发刷新到主存的操作，所以如果某个变量受 synchronized 方法或语句块所保护（或者是 java.util.concurrent.atomic 类型），那就不需要将它设为 volatile 的。

6.4.3 （指令）重排序和先行发生

Java 可能会通过对指令进行重排序来优化性能，只要结果不会造成程序行为上的改变。不过，这里所说的重排序可能会影响到逻辑处理器缓存和主存的交互方式，造成不易察觉的程序 bug。直到 Java 5，这门语言的设计者们才理解并解决了该问题。现在 volatile 关键字可以防止对 volatile 变量的读写指令进行不正确的重排序。这样的重排序规则称为**先行发生**（happens before）保证。

在对 volatile 变量的读或写操作之前出现的指令，保证会在该读或写操作之前执行。同样，任何在 volatile 变量的读或写操作之后的指令，会保证在读或写操作之后执行。例如：

```
// lowlevel/ReOrdering.java

public class ReOrdering implements Runnable {
  int one, two, three, four, five, six;
  volatile int volaTile;
  @Override public void run() {
    one = 1;
    two = 2;
    three = 3;
    volaTile = 92;
    int x = four;
    int y = five;
    int z = six;
  }
}
```

one、two 和 three 的赋值操作可能会被重排序，只要它们都发生在 volatile 写操作之前。同样，x、y 和 z 语句也可能会被重排序，只要它们都发生在 volatile 写操作之后。volatile 操作通常称为**内存栅栏**（memory barrier）。先行发生保证确保了 volatile 变量的读写指令无法穿过内存栅栏而被重排序。

先行发生保证还有另一种效果：当一个线程对某个 volatile 变量执行写操作时，所有在该写操作之前被该线程修改的其他变量——包括非 volatile 变量——也都会被刷新到主存。当一个线程读取 volatile 变量的时候，也会读取到所有和 volatile 变量一起被刷到主存的变量，包括非 volatile 变量。虽然这个特性很重要，解决了 Java 5 之前非常隐蔽的一些 bug，但你不应该依赖该特性，"自动"将周围的变量隐式地变成了 volatile 的。如果你希望某个变量是 volatile 的，就应该让其他所有的代码维护人员明确地知道。

6.4.4 何时使用 volatile

在 Java 的较早版本中，写出一个展示 volatile 必要性的例子并不是多难的事情。如果你搜索一下，便会找到一些这样的例子，但是如果你在 Java 8 上验证这些例子，却看不出效果（目前我还没找到有效的例子）。我努力想写出一个这样的例子，但是没有成功。这可能是由于 JVM 或硬件上的改进，也可能是两个原因都有。对于一些本该使用 volatile 存储但并未这么做的已有程序来说，（volatile 产生的）效果可能会带来益处，这种程序的失败概率比一般程序要低得多，但同时问题也会更难追踪得多。

如果你想要使用 volatile，原因很可能是你想要使一个变量变得线程安全，而又不带来同步导致的开销。由于 volatile 的使用需要掌握一些复杂的技巧，我建议你完全不要使用它，而应该使用本章稍后会介绍的一种 java.util.concurrent.atomic 类。这些类提供了完整的线程安全性，同时又比同步的开销低得多。

如果你想要调试其他人写的并发代码，首先就要找到用了 volatile 的地方，并将它们替换为 Atomic 变量。除非你确定代码的作者对并发有着深刻的理解，否则他们可能是在滥用 volatile。

6.5 原子性

在 Java 多线程的探讨中，有一个常见但错误的说法："原子操作不需要同步。"**原子操作**指的是不会被线程调度器中断的操作，一旦操作开始，直到完成之前，中途就不可能发生任何上下文切换。依赖**原生**（innate）的原子性（原子性是某些特定数据类型内在的属性）会很复杂，也很危险——只有在你是并发专家，或者有并发专家帮助你的情况下，才应该用原生原子性来代替同步或者某些线程安全的数据结构。如果你认为自己已经聪明到可以玩火，那么看看你能否通过下面这个测试。

戈茨测试（Goetz Test）：如果你可以基于现代化的微处理器编写出高性能的 JVM，那么你才有资格考虑自己是否可以抛开同步。[1]

理解原生原子性，并且知道它有助于实现更智能的 java.util.concurrent 库组件，这一点很有用。但是要强烈抵制住依赖它的冲动。

原子性适用于对基本类型（除了 long 和 double 以外）的"简单操作"。它保证了对基本类型变量（除了 long 和 double 以外）的读写操作是不可分割（原子）的。

因此原子操作不会被线程机制中断。专业的程序员可以利用这一点，编写出不需要使用同步的**无锁代码**（lock-free code）。但即便如此，这也是过度简化了。有时，即使原子操作貌似应该很安全，实际上也可能并不安全。本书的读者大概率通不过前面提到的戈茨测试，因此也就不具备用原子操作替代同步的资格。试图移除同步，这通常是过早优化的信号，而且会给你带来很多麻烦。这样做往往还得不到多少好处，甚至没有任何好处。

相较于单处理器系统，在多核系统中，**可见性**（visibility）是比原子性重要得多的问题。一个任务所做的修改，即使是某种意义上不会被中断的原子操作，也可能对其他任务是不可见的（比如该修改可能临时保存在本地处理器缓存中），因此不同的任务看到的程序状态是不同的。另一方面，在多处理器系统中，同步机制会强制某个任务所做的修改在整个程序中都是可见的。如果没有进行同步，修改何时能变得可见，这一点是不确定的。

怎样的操作才算得上是原子操作？字段的赋值和返回值操作可能算是。在 C++ 中，即使是下面两个操作，也**可能**是原子的：

```
i++;     // 在 C++ 中可能是原子操作
i += 2;  // 在 C++ 中可能是原子操作
```

但是在 C++ 中，这取决于编译器和处理器。你无法用 C++ 编写出依赖原子性的跨平台代码，因为 C++[2] 并不像 Java 那样拥有一致的**内存模型**。

在 Java 中，上述操作一定不是原子操作，如以下方法所生成的 JVM 指令所示：

---

[1] 这项测试的一个推论："如果某人暗示多线程很简单，那就要确保这个人没有对你的工程做重要决策。否则，你可就麻烦了。"

[2] 此处指我正在用的版本。后续的标准中可能已经修复了。

```
// lowlevel/NotAtomic.java
// javap -c NotAtomic
// {ExcludeFromGradle}
// {VisuallyInspectOutput}

public class NotAtomic {
  int i;
  void f1() { i++; }
  void f2() { i += 3; }
}
/* 输出:
Compiled from "NotAtomic.java"
public class NotAtomic {
  int i;

  public NotAtomic();
    Code:
      0: aload_0
      1: invokespecial #1 // Method java/lang/Object."<init>":()V
      4: return

  void f1();
    Code:
      0: aload_0
      1: dup
      2: getfield      #2 // Field i:I
      5: iconst_1
      6: iadd
      7: putfield      #2 // Field i:I
     10: return

  void f2();
    Code:
      0: aload_0
      1: dup
      2: getfield      #2 // Field i:I
      5: iconst_3
      6: iadd
      7: putfield      #2 // Field i:I
     10: return
}
*/
```

每个指令都会生成一个"get"（读取）和一个"put"（写入），中间夹杂着一些其他指令。所以在读取和写入操作中间，另一个任务可能会修改该字段，因此该操作并不是原子的。

我们用下面这段代码来测试原子性的这个概念：定义一个抽象类，在其中写一个方法，将一个整型变量按偶数值自增，然后由 run() 方法持续调用该方法。

```java
// lowlevel/IntTestable.java
import java.util.function.*;

public abstract class
IntTestable implements Runnable, IntSupplier {
  abstract void evenIncrement();
  @Override public void run() {
    while(true)
      evenIncrement();
  }
}
```

IntSupplier 是一个带有 getAsInt() 方法的函数式接口。

现在可以创建测试了，以独立任务的方式启动 run()，然后读取值，检查它是否是偶数：

```java
// lowlevel/Atomicity.java
import java.util.concurrent.*;
import onjava.TimedAbort;

public class Atomicity {
  public static void test(IntTestable it) {
    new TimedAbort(4, "No failures found");
    CompletableFuture.runAsync(it);
    while(true) {
      int val = it.getAsInt();
      if(val % 2 != 0) {
        System.out.println("failed with: " + val);
        System.exit(0);
      }
    }
  }
}
```

我们很容易盲目地应用原子性的概念。此处，getAsInt() 看似是安全的原子操作：

```java
// lowlevel/UnsafeReturn.java
import java.util.function.*;
import java.util.concurrent.*;

public class UnsafeReturn extends IntTestable {
  private int i = 0;
  public int getAsInt() { return i; }
  @Override
  public synchronized void evenIncrement() {
    i++; i++;
  }
}
```

```
/* 输出：
failed with: 39
*/
```

```java
  public static void main(String[] args) {
    Atomicity.test(new UnsafeReturn());
  }
}
```

但是，`Atomicity.test()` 在遇到非偶数值时会失败。虽然 `return i` 确实是原子操作，但这里未加上同步，这会导致该值能在对象处于不稳定的中间状态时被读到。此外，`i` 也不是 `volatile` 的，因此会存在可见性问题。`getValue()` 和 `evenIncrement()` 都必须是 `synchronized` 的（这样可以在 `i` 未被设为 `volatile` 的情况下，依然保证其线程安全性）：

```java
// lowlevel/SafeReturn.java
import java.util.function.*;
import java.util.concurrent.*;

public class SafeReturn extends IntTestable {
  private int i = 0;
  public synchronized int getAsInt() { return i; }
  @Override
  public synchronized void evenIncrement() {
    i++; i++;
  }
  public static void main(String[] args) {
    Atomicity.test(new SafeReturn());
  }
}
/* 输出：
No failures found
*/
```

只有并发方面的专家才有足够的能力在这种情况下进行优化。再次强调，要应用 Brian 的同步法则。

6.5.1 Josh 的序列号

作为第二个示例，我们来考虑一件更简单的事：一个生成序列号的类，该类的灵感来自 Joshua Bloch 的 *Effective Java Programming Language Guide* 一书中的第 190 页[①]。每次调用 `nextSerialNumber()` 都必须返回一个唯一值：

```java
// lowlevel/SerialNumbers.java

public class SerialNumbers {
  private volatile int serialNumber = 0;
  public int nextSerialNumber() {
    return serialNumber++; // 非线程安全
  }
}
```

如果你之前用的是 C++，或者有过一些其他的底层开发背景，可能会认为自增操作

---

① 此处页码指的是英文版。本书中文版目前的最新版本为《Effective Java 中文版》（第 3 版）。——译者注

应该是原子操作，因为 C++ 的自增操作经常可以实现为单处理器指令（虽然并不是以一致、可靠、跨平台的方式实现的）。但是，正如前面所提到的，Java 的自增操作并不是原子性的，它分别涉及读和写，因此即使是如此简单的操作，也有发生线程问题的可能。

这个示例中用到了 volatile，只是为了试验它能否起到作用。然而真正的问题在于 nextSerialNumber() 方法没有通过同步的方式访问共享的可变值。

为了测试 SerialNumbers，我们会创建一个不会耗尽内存（导致溢出）的 set，以防检测出问题的时间过长。此处的 CircularSet 复用了用来存储这些 int 的内存，最后覆盖了旧值（复制操作通常都很快，你基本上也可以用 java.util.Set 来替代）：

```java
// lowlevel/CircularSet.java
// 复用了内存，因此无须担心内存溢出
import java.util.*;

public class CircularSet {
  private int[] array;
  private int size;
  private int index = 0;
  public CircularSet(int size) {
    this.size = size;
    array = new int[size];
    // 初始化为非 SerialNumbers 生成的值：
    Arrays.fill(array, -1);
  }
  public synchronized void add(int i) {
    array[index] = i;
    // 包装索引，并覆写旧的元素：
    index = ++index % size;
  }
  public synchronized boolean contains(int val) {
    for(int i = 0; i < size; i++)
      if(array[i] == val) return true;
    return false;
  }
}
```

add() 方法和 contains() 方法是 synchronized 的，以防止线程冲突。

SerialNumberChecker 中含有一个持有最新一批序列号的 CircularSet，以及用于填充 CircularSet 并确保这些序列号唯一的 run() 方法：

```java
// lowlevel/SerialNumberChecker.java
// 测试 SerialNumbers 实现的线程安全性
import java.util.concurrent.*;
import onjava.Nap;
```

```java
public class SerialNumberChecker implements Runnable {
  private CircularSet serials = new CircularSet(1000);
  private SerialNumbers producer;
  public SerialNumberChecker(SerialNumbers producer) {
    this.producer = producer;
  }
  @Override public void run() {
    while(true) {
      int serial = producer.nextSerialNumber();
      if(serials.contains(serial)) {
        System.out.println("Duplicate: " + serial);
        System.exit(0);
      }
      serials.add(serial);
    }
  }
  static void test(SerialNumbers producer) {
    for(int i = 0; i < 10; i++)
      CompletableFuture.runAsync(
        new SerialNumberChecker(producer));
    new Nap(4, "No duplicates detected");
  }
}
```

test() 方法创建了多个 SerialNumberChecker 任务来争夺一个 SerialNumbers 对象。SerialNumberChecker 任务试图生成一个重复的序列号（该过程在多核机器上会更快）。

测试基础版本的 SerialNumbers 类时，运行失败：

```java
// lowlevel/SerialNumberTest.java

public class SerialNumberTest {
  public static void main(String[] args) {
    SerialNumberChecker.test(new SerialNumbers());
  }
}
```

```
/* 输出：
Duplicate: 33280
Duplicate: 33278
Duplicate: 33290
Duplicate: 33277
*/
```

volatile 在这里并未起到作用。要解决这个问题，就需要为 nextSerialNumber() 增加 synchronized 关键字：

```java
// lowlevel/SynchronizedSerialNumbers.java

public class
SynchronizedSerialNumbers extends SerialNumbers {
  private int serialNumber = 0;
  @Override
  public synchronized int nextSerialNumber() {
    return serialNumber++;
  }
  public static void main(String[] args) {
```

```
/* 输出：
No duplicates detected
*/
```

```
    SerialNumberChecker.test(
      new SynchronizedSerialNumbers());
  }
}
```

不再需要 volatile 了，因为 synchronized 关键字确保实现了 volatile 的行为。

对基本类型的读取和赋值操作应该是安全的原子操作。但是如 UnsafeReturn.java 中所示，用原子操作访问处于不稳定中间状态的对象仍然很容易。要尝试解决该问题，处理起来很复杂，同时也很危险。最合理的做法就是遵循 Brian 的同步法则（并且如果可以的话，首先就不要共享变量）。

6.5.2 原子类

Java 5 引入了特殊的原子变量类，如 AtomicInteger、AtomicLong、AtomicReference 等。这些类提供了原子性的更新能力，充分利用了现代处理器的硬件级原子性，实现了快速、无锁的操作。

可以用 AtomicInteger 重写 UnsafeReturn.java：

```
// lowlevel/AtomicIntegerTest.java
import java.util.concurrent.*;
import java.util.concurrent.atomic.*;
import java.util.*;
import onjava.*;

public class AtomicIntegerTest extends IntTestable {
  private AtomicInteger i = new AtomicInteger(0);
  public int getAsInt() { return i.get(); }
  @Override
  public void evenIncrement() { i.addAndGet(2); }
  public static void main(String[] args) {
    Atomicity.test(new AtomicIntegerTest());
  }
}
```

```
/* 输出：
No failures found
*/
```

示例中通过使用 AtomicInteger，去除了 synchronized 关键字。

现在可以重写 SynchronizedEvenProducer.java 以使用 AtomicInteger 了：

```
// lowlevel/AtomicEvenProducer.jav
// 原子类：偶尔可以在常规代码中起到作用
import java.util.concurrent.atomic.*;

public class AtomicEvenProducer extends IntGenerator {
  private AtomicInteger currentEvenValue =
    new AtomicInteger(0);
```

```
  @Override public int next() {
    return currentEvenValue.addAndGet(2);
  }
  public static void main(String[] args) {
    EvenChecker.test(new AtomicEvenProducer());
  }
}
```
```
/* 输出:
No odd numbers discovered
*/
```

通过 `AtomicInteger`，再一次去除了所有其他形式的同步。

下面是使用了 `AtomicInteger` 的 `SerialNumbers` 实现:

```
// lowlevel/AtomicSerialNumbers.java
import java.util.concurrent.atomic.*;

public class
AtomicSerialNumbers extends SerialNumbers {
  private AtomicInteger serialNumber =
    new AtomicInteger();
  @Override public int nextSerialNumber() {
    return serialNumber.getAndIncrement();
  }
  public static void main(String[] args) {
    SerialNumberChecker.test(
      new AtomicSerialNumbers());
  }
}
```
```
/* 输出:
No duplicates detected
*/
```

这些都是只用到单个字段的简单例子，在创建更复杂的类时，必须确定哪些字段需要受到保护，而且在某些情况下，可能最终还是要在方法上使用 `synchronized` 关键字。

6.6 临界区

有时，我们只是想防止多个线程同时访问方法中的部分代码，而不是（防止同时访问）整个方法。要隔离的代码区域称为**临界区**（critical section），它同样也是通过用来保护整个方法的 `synchronized` 关键字来创建的，但使用的语法不同。此处，`synchronized` 用于指定某个对象，该对象的锁用于对花括号内的代码进行同步控制:

```
synchronized(syncObject) {
  // 这段代码同一时间只能被一个任务访问
}
```

这也称为**同步控制块**（synchronized block）。必须先获得 `syncObject` 的锁，才能进入这段代码。如果其他某个任务已经持有了该锁，则只有在锁被释放后才能进入该临界区。如果发生了这种情况，试图获取锁的任务就会被挂起。调度器会定期回来检查锁是否被释放，如果锁被释放，则会唤醒该任务。

使用 synchronized 块（而不是对整个方法都进行同步）的主要目的是提升性能（有时也是为了更灵活的算法——但是在涉及并发的情况下，对灵活性要特别谨慎）。下面的示例演示了对代码块（而不是整个方法）进行同步可以极大地提升方法对于其他任务的可访问性（即并发处理能力）。该示例会统计成功访问 method() 方法的次数，并启动相互竞争以试图调用 method() 的任务：

```java
// lowlevel/SynchronizedComparison.java
// 使用同步控制块而不是同步方法，可以提升访问速度
import java.util.*;
import java.util.stream.*;
import java.util.concurrent.*;
import java.util.concurrent.atomic.*;
import onjava.Nap;

abstract class Guarded {
  AtomicLong callCount = new AtomicLong();
  public abstract void method();
  @Override public String toString() {
    return getClass().getSimpleName() +
      ": " + callCount.get();
  }
}

class SynchronizedMethod extends Guarded {
  @Override public synchronized void method() {
    new Nap(0.01);
    callCount.incrementAndGet();
  }
}

class CriticalSection extends Guarded {
  @Override public void method() {
    new Nap(0.01);
    synchronized(this) {
      callCount.incrementAndGet();
    }
  }
}

class Caller implements Runnable {
  private Guarded g;
  Caller(Guarded g) { this.g = g; }
  private AtomicLong successfulCalls =
    new AtomicLong();
  private AtomicBoolean stop =
    new AtomicBoolean(false);
  @Override public void run() {
    new Timer().schedule(new TimerTask() {
/* 输出：
-> 159
-> 159
-> 159
-> 159
CriticalSection: 636
-> 65
-> 21
-> 11
-> 68
SynchronizedMethod: 165
*/
```

```
    public void run() { stop.set(true); }
  }, 2500);
  while(!stop.get()) {
    g.method();
    successfulCalls.getAndIncrement();
  }
  System.out.println(
    "-> " + successfulCalls.get());
  }
}

public class SynchronizedComparison {
  static void test(Guarded g) {
    List<CompletableFuture<Void>> callers =
      Stream.of(
        new Caller(g),
        new Caller(g),
        new Caller(g),
        new Caller(g))
        .map(CompletableFuture::runAsync)
        .collect(Collectors.toList());
    callers.forEach(CompletableFuture::join);
    System.out.println(g);
  }
  public static void main(String[] args) {
    test(new CriticalSection());
    test(new SynchronizedMethod());
  }
}
```

Guarded 在 callCount 中跟踪记录成功调用 method() 方法的次数。SynchronizedMethod 对整个 method() 方法进行了同步控制，而 CriticalSection 仅用 synchronized 同步控制块对方法的一部分进行了同步控制。这样，消耗时间的 Nap() 方法就可以被排除在 synchronized 控制块之外。从输出可以看出，CriticalSection 中的 method() 的可访问性要高得多。

记住，使用 synchronized 控制块也有风险：它要求你必须能够确定控制块外部的非 synchronized 代码的确是安全的。

Caller 任务试图在给定的时间段内尽可能多次调用 method()（并公布调用次数）。为了建立该时间段，我们使用了 java.util.Timer，虽然有点过时，但仍能起到很好的作用。它接收 TimerTask 为参数，TimerTask 并不是函数式接口，因此我们无法使用 lambda 表达式，只能显式地创建该类（本例中使用了匿名内部类）。在时间段结束后，TimerTask 将 AtomicBoolean stop 标志设为 true，从而退出循环。

test() 方法接收 Guarded 参数，并创建了 4 个 Caller 任务，将其全部添加到同一个 Guarded 对象中，这样它们就会竞争获取 method() 使用的锁。

通常来说，每次的运行结果看上去都不相同。从结果可以看出，CriticalSection 中 method() 方法的可访问次数要远远多于 SynchronizedMethod 中的 method() 方法。这通常也是要使用 synchronized 控制块而不是对整个方法进行同步控制的原因——使其他任务能够更多地访问（只要能保证安全即可）。

6.6.1 在其他对象上进行同步

synchronized 控制块必须在某个对象上进行同步。通常最合理的选择就是通过 synchronized(this) 使用当前对象，这也是之前的示例中 CriticalSection 所采用的方法。通过这种方式，在获得 synchronized 控制块的锁后，便无法调用同一对象中的其他 synchronized 方法和临界区了。因此，在 this 上进行同步控制时，临界区的作用是减少同步控制的范围。

有时必须在另一个对象上进行同步，但如果要这么做，就必须确保所有的相关任务都要在同一个对象上进行同步。下面这个示例演示了一个对象中的方法在不同的锁上进行同步时，两个任务都可以进入该对象。

```java
// lowlevel/SyncOnObject.java
// 在另一个对象上进行同步
import java.util.*;
import java.util.stream.*;
import java.util.concurrent.*;
import onjava.Nap;

class DualSynch {
  ConcurrentLinkedQueue<String> trace =
    new ConcurrentLinkedQueue<>();
  public synchronized void f(boolean nap) {
    for(int i = 0; i < 5; i++) {
      trace.add(String.format("f() " + i));
      if(nap) new Nap(0.01);
    }
  }
  private Object syncObject = new Object();
  public void g(boolean nap) {
    synchronized(syncObject) {
      for(int i = 0; i < 5; i++) {
        trace.add(String.format("g() " + i));
        if(nap) new Nap(0.01);
      }
    }
  }
}

public class SyncOnObject {
  static void test(boolean fNap, boolean gNap) {
```

```
/* 输出：
f() 0
g() 0
g() 1
g() 2
g() 3
g() 4
f() 1
f() 2
f() 3
f() 4
****
f() 0
g() 0
f() 1
f() 2
f() 3
f() 4
g() 1
g() 2
g() 3
g() 4
*/
```

```java
    DualSynch ds = new DualSynch();
    List<CompletableFuture<Void>> cfs =
      Arrays.stream(new Runnable[] {
        () -> ds.f(fNap), () -> ds.g(gNap) })
        .map(CompletableFuture::runAsync)
        .collect(Collectors.toList());
    cfs.forEach(CompletableFuture::join);
    ds.trace.forEach(System.out::println);
  }
  public static void main(String[] args) {
    test(true, false);
    System.out.println("****");
    test(false, true);
  }
}
```

DualSync.f()（通过对整个方法进行同步）在 this 上进行了同步控制，而 g() 则使用了在 syncObject 上进行同步的 synchronized 控制块。因此，这两个同步控制是相互独立的。在 test() 中，通过运行两个分别调用 f() 和 g 的独立任务演示了这一点。fNap 和 gNap 标志分别指示了在 f() 和 g() 中，哪一个应该在其 for 循环中调用 Nap()。举个例子，在 f() 睡眠（nap）时，它仍然继续持有自己的锁，但你可以看到这并未阻止对 g() 的调用，反之亦然。

6.6.2　使用显式 Lock 对象

java.util.concurrent 库提供了显式的互斥机制，定义在 java.util.concurrent.locks 中。Lock 对象必须显式地创建、加锁以及解锁，因此它的语法不如内建的 synchronized 关键字优雅。不过它在解决某些类型的问题时更灵活。下面用显式的 Lock 重写 SynchronizedEvenProducer.java：

```java
// lowlevel/MutexEvenProducer.java
// 用互斥锁防止线程冲突
import java.util.concurrent.locks.*;
import onjava.Nap;

public class MutexEvenProducer extends IntGenerator {
  private int currentEvenValue = 0;
  private Lock lock = new ReentrantLock();
  @Override public int next() {
    lock.lock();
    try {
      ++currentEvenValue;
      new Nap(0.01); // 可以更快地导致失败
      ++currentEvenValue;
      return currentEvenValue;
    } finally {
      lock.unlock();
    }
  }
```

```
  public static void main(String[] args) {
    EvenChecker.test(new MutexEvenProducer());
  }
}
/*
No odd numbers discovered
*/
```

MutexEvenProducer 中增加了一个名为 lock 的互斥锁,并在 next() 中通过 lock() 和 unlock() 方法创建了一个临界区。在使用 Lock 对象的时候,要注意用这种使用方式:在调用 lock() 后,必须放置一个 try-finally 语句,并在 finally 子句中调用 unlock()——这是唯一能保证锁一定会被释放的方法。注意,return 语句必须放在 try 子句中,以确保 unlock() 不会执行得太早,导致将数据暴露给下一个任务。

虽然 try-finally 比使用 synchronized 关键字所需的代码量更多,但也展现了显式 Lock 对象的一个优点。如果在使用 synchronized 关键字时出现了什么问题,便会抛出异常,但是并没有机会执行任何清理操作,以将系统维持在正常的状态。而如果使用显式 Lock 对象,便可以用 finally 子句来维护

总体来说,在使用 synchronized 的时候,要写的代码会更少,同时用户错误发生的概率也会大大降低,因此通常只需要在解决特殊问题的时候才使用显式 Lock 对象。例如,使用 synchronized 关键字时,你无法在开始尝试获取锁后,(如果没有获取成功)就立刻主动放弃,也无法等待一段时间(仍然没有获取成功)后再放弃——要想这样做,就必须使用 concurrent 库:

```
// lowlevel/AttemptLocking.java
// 并发库中的 Lock 让你可以放弃尝试获取锁
import java.util.concurrent.*;
import java.util.concurrent.locks.*;
import onjava.Nap;

public class AttemptLocking {
  private ReentrantLock lock = new ReentrantLock();
  public void untimed() {
    boolean captured = lock.tryLock();
    try {
      System.out.println("tryLock(): " + captured);
    } finally {
      if(captured)
        lock.unlock();
    }
  }
  public void timed() {
    boolean captured = false;
    try {
```

```
        captured = lock.tryLock(2, TimeUnit.SECONDS);
      } catch(InterruptedException e) {
        throw new RuntimeException(e);
      }
      try {
        System.out.println(
          "tryLock(2, TimeUnit.SECONDS): " + captured);
      } finally {
        if(captured)
          lock.unlock();
      }
    }
    public static void main(String[] args) {
      final AttemptLocking al = new AttemptLocking();
      al.untimed(); // True——锁状态为可用
      al.timed();   // True——锁状态为可用
      // 现在创建第二个任务来获取锁:
      CompletableFuture.runAsync( () -> {
          al.lock.lock();
          System.out.println("acquired");
      });
      new Nap(0.1);   // 给第二个任务一个机会
      al.untimed(); // False——任务获得了锁
      al.timed();   // False——任务获得了锁
    }
}
/* 输出:
tryLock(): true
tryLock(2, TimeUnit.SECONDS): true
acquired
tryLock(): false
tryLock(2, TimeUnit.SECONDS): false
*/
```

ReentrantLock 可以尝试获取锁，然后放弃，因此如果其他人已经持有了锁，你便可以决定转而去做其他事情，而不是像 untimed() 方法中那样一直等到锁释放。在 timed() 中，先尝试获取锁，然后可以在 2 秒后放弃（注意 TimeUnit 类指定了单位）。在 main() 中以匿名类的方式创建了一个独立的 Thread，它获取了锁，因此 untimed() 和 timed() 方法有了可以竞争的对象。

相较于内建的 synchronized 锁，显式 Lock 对象提供了更为细粒度的加锁及释放锁的控制。这对于实现专门的同步结构非常有用，例如**交替锁**（hand-over-hand locking）[也称为**锁耦合**（lock coupling）]，可用于遍历链表节点——执行遍历的代码必须先捕获下一个节点的锁，然后才能释放当前节点的锁。

6.7 库组件

java.util.concurrent 库提供了大量用于解决并发问题的类，可以帮助你实现更简单、更稳健的并发程序。不过要注意，相比并行流或 CompletableFuture，这些工具属于更底层的机制。

本节中会探讨一些使用了不同组件的示例，然后对**无锁**（lock-free）库的运行机制稍作探讨。

6.7.1 延迟队列 DelayQueue

这是一种由实现了 Delayed 接口的对象组成的无边界 BlockingQueue（阻塞队列）。一个对象只有在其延迟时间到期后才能从队列中取出。队列是有序的，因此头部对象的延迟时间最长。如果没有到达延迟时间，那么头部元素就相当于不存在，同时 poll() 会返回 null（因此不能将 null 放入队列）。

下面这个示例中，Delayed 对象自身就是任务，而 DelayedTaskConsumer 将最"紧急"的任务（即过期时间最长的任务）从队列中取出并执行。因此，要注意 DelayQueue 是优先级队列的变体。

```java
// lowlevel/DelayQueueDemo.java
import java.util.*;
import java.util.stream.*;
import java.util.concurrent.*;
import static java.util.concurrent.TimeUnit.*;

class DelayedTask implements Runnable, Delayed {
  private static int counter = 0;
  private final int id = counter++;
  private final int delta;
  private final long trigger;
  protected static List<DelayedTask> sequence =
    new ArrayList<>();
  DelayedTask(int delayInMilliseconds) {
    delta = delayInMilliseconds;
    trigger = System.nanoTime() +
      NANOSECONDS.convert(delta, MILLISECONDS);
    sequence.add(this);
  }
  @Override public long getDelay(TimeUnit unit) {
    return unit.convert(
      trigger - System.nanoTime(), NANOSECONDS);
  }
  @Override public int compareTo(Delayed arg) {
    DelayedTask that = (DelayedTask)arg;
    if(trigger < that.trigger) return -1;
    if(trigger > that.trigger) return 1;
    return 0;
  }
  @Override public void run() {
    System.out.print(this + " ");
  }
  @Override public String toString() {
    return
      String.format("[%d] Task %d", delta, id);
  }
  public String summary() {
    return String.format("(%d:%d)", id, delta);
```

```
    }
    public static class EndTask extends DelayedTask {
      EndTask(int delay) { super(delay); }
      @Override public void run() {
        sequence.forEach(dt ->
          System.out.println(dt.summary()));
      }
    }
}

public class DelayQueueDemo {
  public static void
  main(String[] args) throws Exception {
    DelayQueue<DelayedTask> tasks =
      Stream.concat( // 随机延迟：
        new Random(47).ints(20, 0, 4000)
          .mapToObj(DelayedTask::new),
        //增加总结任务
        Stream.of(new DelayedTask.EndTask(4000)))
        .collect(Collectors
          .toCollection(DelayQueue::new));
    while(tasks.size() > 0)
      tasks.take().run();
  }
}
/* 输出：
[128] Task 12 [429] Task 6 [555] Task 2 [551] Task 13
[693] Task 3 [809] Task 15 [961] Task 5 [1258] Task 1
[1258] Task 20 [1520] Task 19 [1861] Task 4 [1998] Task
17 [2200] Task 8 [2207] Task 10 [2288] Task 11 [2522]
Task 9 [2589] Task 14 [2861] Task 18 [2868] Task 7
[3278] Task 16 (0:4000)
(1:1258)
(2:555)
(3:693)
(4:1861)
(5:961)
(6:429)
(7:2868)
(8:2200)
(9:2522)
(10:2207)
(11:2288)
(12:128)
(13:551)
(14:2589)
(15:809)
(16:3278)
(17:1998)
(18:2861)
(19:1520)
(20:1258)
*/
```

DelayedTask 含有一个称为 sequence 的 List<DelayedTask>，它维护了创建任务的顺序，因此我们可以看到该排序确实发生了。

Delayed 接口中有一个方法 getDelay()，它表明了延迟时间还有多久，或延迟时间已经过期了多久。该方法强制我们必须使用 TimeUnit 类，因为参数类型定义就是如此。这个类非常好用，因为无须进行任何计算就可以转换单位。举例来说，delta 的值保存为毫秒，但是 System.nanoTime() 生成的是纳秒。通过指定 delta 值的当前单位和想要转换成的目标单位，就可以转换它的值了，就像这样：

```
NANOSECONDS.convert(delta, MILLISECONDS);
```

在 getDelay() 中，将目标单位作为 unit 参数传入方法，你可以使用该参数，在甚至不知道它是什么时间单位的情况下，就可以将触发时间的时差转换为调用者所请求的单位（这是个策略设计模式的简单示例，将算法的一部分作为参数传入，从而提升了灵活性）。

为了排序，Delayed 接口同样继承了 Comparable 接口。必须实现 compareTo() 方法，这样它就可以实现合理的排序了。

从输出可以看出，任务创建的顺序对执行顺序没有任何影响。相反，任务按照预期以延迟的顺序执行了。

6.7.2 优先级阻塞队列 PriorityBlockingQueue

这基本上是一种能够阻塞读取操作的优先队列。下面的示例中为 Prioritized 指定了优先级数值。一些 Producer 任务的实例将 Prioritized 对象插入 PriorityBlockingQueue，不过在插入操作之间加入了随机的延迟。然后，单个 Consumer 任务在执行 take() 时会呈现出多个选项，PriorityBlockingQueue 则将此时具有最高优先级的 Prioritized 对象传给它。

Prioritized 中的 static counter 是 AtomicInteger 类型。这是必要的，因为同时有多个 Producer 在并行运行，如果不使用 AtomicInteger，就会看到重复的 id 数值。这个问题在本书第 5 章 5.12 节中介绍过。

```
// lowlevel/PriorityBlockingQueueDemo.java
import java.util.*;
import java.util.stream.*;
import java.util.concurrent.*;
import java.util.concurrent.atomic.*;
import onjava.Nap;
```

```java
class Prioritized implements Comparable<Prioritized>  {
  private static AtomicInteger counter =
    new AtomicInteger();
  private final int id = counter.getAndIncrement();
  private final int priority;
  private static List<Prioritized> sequence =
    new CopyOnWriteArrayList<>();
  Prioritized(int priority) {
    this.priority = priority;
    sequence.add(this);
  }
  @Override public int compareTo(Prioritized arg) {
    return priority < arg.priority ? 1 :
      (priority > arg.priority ? -1 : 0);
  }
  @Override public String toString() {
    return String.format(
      "[%d] Prioritized %d", priority, id);
  }
  public void displaySequence() {
    int count = 0;
    for(Prioritized pt : sequence) {
      System.out.printf("(%d:%d)", pt.id, pt.priority);
      if(++count % 5 == 0)
        System.out.println();
    }
  }
  public static class EndSentinel extends Prioritized {
    EndSentinel() { super(-1); }
  }
}

class Producer implements Runnable {
  private static AtomicInteger seed =
    new AtomicInteger(47);
  private SplittableRandom rand =
    new SplittableRandom(seed.getAndAdd(10));
  private Queue<Prioritized> queue;
  Producer(Queue<Prioritized> q) {
    queue = q;
  }
  @Override public void run() {
    rand.ints(10, 0, 20)
      .mapToObj(Prioritized::new)
      .peek(p -> new Nap(rand.nextDouble() / 10))
      .forEach(p -> queue.add(p));
    queue.add(new Prioritized.EndSentinel());
  }
}

class Consumer implements Runnable {
  private PriorityBlockingQueue<Prioritized> q;
  private SplittableRandom rand =
    new SplittableRandom(47);
```

```java
  Consumer(PriorityBlockingQueue<Prioritized> q) {
    this.q = q;
  }
  @Override public void run() {
    while(true) {
      try {
        Prioritized pt = q.take();
        System.out.println(pt);
        if(pt instanceof Prioritized.EndSentinel) {
          pt.displaySequence();
          break;
        }
        new Nap(rand.nextDouble() / 10);
      } catch(InterruptedException e) {
        throw new RuntimeException(e);
      }
    }
  }
}

public class PriorityBlockingQueueDemo {
  public static void main(String[] args) {
    PriorityBlockingQueue<Prioritized> queue =
      new PriorityBlockingQueue<>();
    CompletableFuture.runAsync(new Producer(queue));
    CompletableFuture.runAsync(new Producer(queue));
    CompletableFuture.runAsync(new Producer(queue));
    CompletableFuture.runAsync(new Consumer(queue))
      .join();
  }
}
/* 输出：
[15] Prioritized 1
[17] Prioritized 0
[17] Prioritized 5
[16] Prioritized 6
[14] Prioritized 9
[12] Prioritized 2
[11] Prioritized 4
[11] Prioritized 12
[13] Prioritized 13
[12] Prioritized 16
[14] Prioritized 17
[15] Prioritized 23
[18] Prioritized 26
[16] Prioritized 29
[12] Prioritized 18
[11] Prioritized 30
[11] Prioritized 24
[10] Prioritized 15
[10] Prioritized 22
[8] Prioritized 25
[8] Prioritized 11
[8] Prioritized 10
[6] Prioritized 31
[3] Prioritized 7
[2] Prioritized 20
[1] Prioritized 3
[0] Prioritized 19
[0] Prioritized 8
[0] Prioritized 14
[0] Prioritized 21
[-1] Prioritized 28
(0:17)(1:15)(2:12)(3:1)(4:11)
(5:17)(6:16)(7:3)(8:0)(9:14)
(10:8)(11:8)(12:11)(13:13)(14:0)
(15:10)(16:12)(17:14)(18:12)(19:0)
(20:2)(21:0)(22:10)(23:15)(24:11)
(25:8)(26:18)(27:-1)(28:-1)(29:16)
(30:11)(31:6)(32:-1)
*/
```

（转右栏）

和上一个示例一样，`List sequence` 记住了 `Prioritized` 对象的创建顺序，以便与实际的执行顺序进行比较。`EndSentinel` 是一种特殊类型，它会告诉 `Consumer` 进行关闭。

`Producer` 用 `AtomicInteger` 来为 `SplittableRandom` 提供种子（`seed`），这样不同的 `Producer` 就可以生成不同的序列。因为多个 `Producer` 是并行创建的，所以这很有必要，否则构造过程就不是线程安全的了。

`Producer` 和 `Consumer` 通过 `PriorityBlockingQueue` 相互关联起来。由于队列的阻塞特性提供了所有必要的同步控制，所以请注意，这里不需要任何显式的同步控制——不用考虑在读取该队列的时候队列中是否含有元素，因为该队列在没有元素的时候会阻塞住读取者。

6.7.3　无锁集合

基础卷第 12 章将集合作为基础编程工具进行了强调，这其中也包括并发。因此，早期的集合（如 `Vector` 和 `Hashtable`）有很多方法用到了 `synchronized` 机制。在非多线程程序中使用的时候，这导致了不可接受的开销。在 Java 1.2 中，新的集合库没有用到同步，`Collections` 类中添加了多种声明为 `static synchronized` 的方法，来对不同的集合类型进行同步控制。这虽然是一项改进，因其给了你对集合使用同步的选择权，但由于 `synchronized` 的锁定，开销依然存在。Java 5 增加了专门用来提升线程安全性能的新集合，使用一些技巧消除了锁定。

无锁集合有个有趣的特性：集合可以在读取的同时进行修改，只要读取方只能看见已完成的修改结果。这是通过多种策略实现的。为了了解它们的实现机制，接下来看看其中的几个策略。

1. 复制策略

利用"复制"策略，修改是在部分数据结构的一个单独副本上进行的（或者有时是整个数据结构的副本），该副本在修改过程中不可见。只有在修改完成后，修改后的结构才会安全地与"主"数据结构进行交换，然后读取方才能看到修改。

在 `CopyOnWriteArrayList` 中，写操作会复制整个底层数组。原始数组被留在原地，因此可以在修改副本数组的同时安全地读取（原始数组）。修改完成后，会通过原子操作将新数组交换进去，因此后续的读取操作可以看到新的信息。`CopyOnWriteArrayList` 的一个好处是，它不会在多个迭代器遍历及修改列表的时候抛出异常，因此你无须像过去那样编写专门的代码来避免发生这类异常。

`CopyOnWriteArraySet` 用 `CopyOnWriteArrayList` 实现了自身的无锁行为。

`ConcurrentHashMap` 和 `ConcurrentLinkedQueue` 使用类似的技巧支持并发读写操作，只对部分集合而不是整个集合进行了复制和修改。不过，读取方仍然不会看到任何未完成的修改。`ConcurrentHashMap` 不会抛出 `ConcurrentModificationException`。

2. CAS 操作

在**比较交换**（Compare-And-Swap，CAS）操作中，你从内存中取出一个值，并在计算新值的同时继续用着原始值。然后通过 CAS 指令，将原始值和当前内存中的值进行比较，如果两者相等，则将旧的值替换为你的计算结果，以上所有操作都在单个原子操作中。如果原始值的比较失败了，替换就不会发生，因为这意味着另一个线程已经在此期间修改了内存。在这种情况下，你的代码必须重试，读取新的原始值并重复以上操作。

如果内存竞争不是很严重，那么 CAS 操作基本上无须重试，因此它非常快。相反，`synchronized` 操作每次都需要获得锁和释放锁，这就导致开销要昂贵得多，而又没有额外的好处。随着内存竞争变得激烈，CAS 操作会变慢，因为它必须更频繁地重试，但这是对更激烈的竞争的一种动态响应。CAS 操作的确是一种优雅的方法。

CAS 操作最好的地方在于，很多现代处理器的汇编语言中带有 CAS 指令，而 JVM 中的 CAS 操作（如 `Atomic` 类中的那些操作）利用了这些指令。CAS 指令是硬件级的原子操作，执行这样的操作就如你期望的那么快。

6.8 总结

本章主要是为了让你在遇到底层并发代码的时候，能够对其有一定的了解，尽管这对该主题来说还远谈不上全面。因此，你需要从《Java 并发编程实战》开始入门。理想情况下，该书会彻底把你从 Java 底层并发的领域吓退。如果你没有被吓退，那么你大概率正在忍受**达克效应**（Dunning-Kruger Effect，一种认知偏差，"你知道得越少，对自己的能力就越有信心"）的痛苦。要时刻记住，现在的语言设计者们仍旧在处理由早期语言设计者们的过度自信带来的麻烦事（比如，可以看看 `Thread` 类中有多少方法被废弃了，而 `volatile` 直到 Java 5 才能正常工作）。

下面是并发编程需要遵循的步骤。

1. 不要使用并发，想办法找别的方法来使程序运行得更快。

2.如果必须使用并发，就使用本书第 5 章中演示过的最新的高级工具——并行流和 `CompletableFuture`。

3.不要在任务间共享变量。任何必须在任务间传递的信息都应该通过 `java.util.concurrent` 库中的并发数据结构来共享。

4.如果必须共享变量，要么用 `java.util.concurrent.atomic` 类型，要么对任何可以直接或间接访问这些变量的方法应用 `synchronized`。你很容易落入误以为一切都考虑周到（而实际并没有）的陷阱。真的，试着改用第 3 步的方法吧。

5.如果第 4 步生成结果的速度太慢，可以尝试用 `volatile` 或其他技巧来调整，但如果你正在阅读本书，并且认为你已经做好了尝试这些方法的准备，你便已经超出了自己的能力范围。回到第 1 步吧。

通常仅用 `java.util.concurrent` 库的组件就可以写出并发程序，从而完全避免挑战 `volatile` 和 `synchronized`。注意，在本书第 5 章的示例中，我已经能够做到这一点了。

Java I/O 系统

> 从历史视角理解 I/O 库的演进过程。

7.1 I/O 流

Java 7 引入了一种简单清晰的方法来读写文件及使用目录路径。大多数时候，基础卷第 17 章中讲解的库及各种技巧已经足够应付日常需求了。但是如果你需要解决某些特殊需求、进行底层操作，或者处理历史代码，那么就需要理解本章的内容。

对于语言设计者来说，创建出优秀的输入 / 输出（input/ output, I/O）系统是一项相对较难的任务。从各种不同实现方式的数量之多便可以看出这一点。其中的挑战似乎在于如何考虑到所有的可能性，这不仅涉及 I/O 有很多不同的来源和去处（文件、控制台、网络连接，等等），而且还涉及需要以很多种方式（顺序读取、随机访问、缓冲、字符、按行读取、按字读取，等等）和这些东西打交道。

Java 库的设计者通过创建大量的类来解决这个问题。事实上 Java I/O 系统中如此多的类足以让人望而生畏。随着面向字符、基于 Unicode 的 I/O 类加入原本面向字节的库，Java 1.0 后的 I/O 库也发生了巨大的变化。为了提升性能和功能，还引入了 nio 类（new I/O，即新 I/O，Java 1.4 引入该特性多年后所使用的名称），这些都将在本章 7.3 节中介绍。

因此，在你足够理解 Java I/O 流库并能正确运用之前，有相当多的类需要学习。理解 I/O 库的演进过程同样很有帮助，其原因在于，如果缺乏历史视角，你很快就会被其中的一些类弄晕，并且搞不清楚何时该用、何时不该用。

编程语言的 I/O 库常常会使用流的抽象，将任意数据源或数据接收端表达为一个具有生成或接收数据片段能力的对象。

你一定要理解，Java 8 函数式流相关的类和 I/O 流之间并无关联。这同时也是一个佐证，证明了如果设计者们可以重来一次，他们肯定会改用不同的术语。

I/O 流隐藏了实际的 I/O 设备中数据情况的下列细节。

1. 字节流用于处理原始的二进制数据。
2. 字符流用于处理字符数据。它会自动处理和本地字符集间的相互转换。
3. 缓冲区流提升了性能。它通过减少调用本地 API 的次数，优化了输入和输出。

从 JDK 文档中类的组织方式可以看出，Java 库将 I/O 相关的类分成了输入和输出两类。在 Java 1.0 中，库的设计者们选择让所有和输入相关的类都继承自 `InputStream`，而所有和输出相关的类都继承自 `OutputStream`。从 `InputStream` 类或 `Reader` 类派生出的所有类都含有根方法 `read()`，用于从字节数组中读取单个字节或字节数组。同样，从 `OutputStream` 类或 `Writer` 类派生出的所有类都含有根方法 `write()`，用于写入单个字节或字节数组。不过通常你不会使用这些方法，它们的存在只是为了供其他类使用——这些其他类提供了更有用的接口。

你很少会通过单个类创建流对象，而一般会将多个类分层放在一起来提供所需的功能（这便是**装饰器**设计模式）。生成一个流需要创建多个对象，这正是 Java 1.0 的 I/O 库容易让人迷惑的主要原因。

接下来我会尽量对这些类做一个总体的概述，但也希望你能通过 JDK 文档厘清所有的细节，例如某个类详尽的方法列表。

7.1.1 各种 InputStream 类型

InputStream 用于表示那些从不同源生成输入的类。可能的源有以下几种。

1. 字节数组。
2. 字符串对象。
3. 文件。
4. "管道"(pipe),其运行机制就像物理管道一样:将物体放入一端,然后它会从管道的另一端出来。
5. 其他流组成的序列,这样就可以将这些流合并成单个流。
6. 其他源,例如外部网络连接。

以上这些源各自都有一个关联的 InputStream 子类,如表 7-1 所示。另外,FilterInputStream 也是一种 InputStream 类型,它为"装饰器"类(用于为输入流增加属性或有用的接口)提供了基类。稍后会对此进行讨论。

表 7-1

类	功能	构造器参数	使用方法
ByteArrayInputStream	使内存中的缓冲区可以充当 InputStream	用于提取出字节的缓冲区	作为一种数据源:通过将其连接到 FilterInputStream 对象来提供有用的接口
StringBufferInputStream	将字符串转换为 InputStream	一个字符串。底层实现实际用的是 StringBuffer	作为一种数据源:通过将其连接到 FilterInputStream 对象来提供有用的接口
FileInputStream	用于从一个文件中读取信息	一个用于表示文件名或 File 对象,还有 FileDescriptor 对象的字符串	作为一种数据源:通过将其连接到 FilterInputStream 对象来提供有用的接口
PipedInputStream	用于生成写入到对应的 PipedOutputStream 中的数据。它实现了"管道传输"的概念	PipedOutputStream	作为一种多线程形式的数据源:通过将其连接到 FilterInputStream 对象来提供有用的接口
SequenceInputStream	将两个以上的 InputStream 转换为单个 InputStream	两个 InputStream 对象,或一个作为 InputStream 对象容器的 Enumeration	作为一种数据源:通过将其连接到 FilterInputStream 对象来提供有用的接口
FilterInputStream	作为装饰器接口的抽象类,装饰器用来为其他 InputStream 类提供有用的功能。参见表 7-3	参见表 7-3	参见表 7-3

7.1.2 各种 OutputStream 类型

这一系列的类用于决定输出的去向:究竟是字节数组(但不能是字符串——一般能用字节数组创建字符串)、文件,还是"管道"。

另外，FilterOutputStream 为"装饰器"类（用于为输出流增加属性或有用的接口）提供了基类，如表 7-2 所示。稍后会对此进行讨论。

表 7-2

类	功　能	构造器参数	使用方法
ByteArray-OutputStream	在内存创建一块缓冲区，所有发送到流中的数据都被放在该缓冲区	缓冲区初始大小，为可选参数	用于指定数据的目的地：通过将其连接到 FilterOutputStream 对象来提供有用的接口
FileOutputStream	用于向文件发送信息	用于表示文件名或 File 对象，还有 FileDescriptor 对象的字符串	用于指定数据的目的地：通过将其连接到 FilterOutputStream 对象来提供有用的接口
PipedOutputStream	向其中写入的任何信息都将自动作为对应的 PipedInputStream 的输入。实现了"管道传输"的概念	PipedInputStream	用于为多线程指定数据的目的地：通过将其连接到 FilterOutputStream 对象来提供有用的接口
FilterOutputStream	作为装饰器接口的抽象类，装饰器用来为其他 OutputStream 类提供有用的功能。参见表 7-4	参见表 7-4	参见表 7-4

7.1.3 添加属性和有用的接口

装饰器曾在基础卷第 20 章中介绍过。Java I/O 需要将各种模块以不同方式组合使用，而这就是为什么要使用装饰器设计模式的原因。[①]Java I/O 库中存在这些"过滤器"（filter）类是因为"过滤器"抽象类是所有装饰器的基类。装饰器必须和其所装饰的对象实现同一个接口，但是装饰器也可以扩展接口，个别"过滤器"类中便出现了这种情况。

装饰器模式有一个缺点。在编写程序时，装饰器可以带来更大的灵活性（让你可以轻松地对属性进行混合和匹配），但也会增加代码的复杂性。Java I/O 库难以使用的原因在于，必须创建很多类——包括"核心"的 I/O 类型，再加上所有的装饰器——才能得到想要的单个 I/O 对象。

FilterInputStream 和 FilterOutputStream 这两个类提供了可控制某个特定 InputStream 或 OutputStream 的装饰器接口，而这两个类的命名并不是很直观。它们派生自 I/O 库的基类：InputStream 和 OutputStream，这两个基类是装饰器的必要组成部分（这样装饰器才能为所有被装饰的对象提供公共接口）。

1. 用 FilterInputStream 从 InputStream 中读取

各种 FilterInputStream 类可以完成两件截然不同的事情。其中 DataInputStream

[①] 这不一定是个好的设计决策，特别是和其他语言中的 I/O 库的易用性相比。但这确实是使用这种设计的理由。

用于读取不同的基本类型数据以及字符串对象。（所有的方法名都以"read"开头，如 readByte()、readFloat() 等。）再加上其伴生类 DataOutputStream，使得你可以通过流将基本类型数据从一个地方移动到另一个地方。具体是哪些"地方"则由表 7-1 中的那些类所决定。

余下的 FilterInputStream 类则在内部修改了 InputStream 的行为方式：它是否使用缓冲，是否会记录其所读取的行（以便查询或者设置行数），以及是否可以将一个字符推回到输入流。最后两个类看起来很像是用于支持编译器的构建（它们很可能用于支持"用 Java 构建 Java 编译器"的实验），因此你在常规的编程中可能并不会用到它们。

各种 FilterInputStream 类型如表 7-3 所示。

表 7-3

类	功　能	构造器参数	使用方法
DataInputStream	与 DataOutputStream 配合使用，以可移植的方式从流中读取基本类型（int、char、long，等等）	InputStream	包含用于读取基本类型的全部接口
Buffered- InputStream	用于防止在每次需要更多数据时都进行物理上的读取。相当于声明"使用缓冲区"	InputStream 以及可选参数：（指定）缓冲区大小	这本质上并未提供接口，只是为进程增加缓冲操作而已，需要与接口对象搭配使用
LineNumber- InputStream	记录输入流中的行号，可以调用 getLineNumber() 和 setLineNumber(int)	InputStream	只是增加了行号而已，因此可能需要与接口对象搭配使用
Pushback- InputStream	包含一个单字节回退缓冲区，用于将最后读取的字符推回输入流	InputStream	通常用于编译器的扫描器，一般不会用到

不论所连接的是什么 I/O 设备，你基本每次都会对输入使用缓冲，因此对于 I/O 库来说，相较于几乎每次都强制添加缓冲操作，将无缓冲输入作为特殊情况（或者只是方法调用）来处理会显得更合理。

2. 用 FilterOutputStream 向 OutputStream 中写入

与 DataInputStream 对应的是 DataOutputStream，它可以将各种基本类型和字符串对象格式化输出到流中，这样任何机器上的任何 DataInputStream 就都可以读取这些信息了。其中所有的方法都以 "write" 开头，如 writeByte()、writeFloat()，等等。

PrintStream 最初的目的是以可视化的格式来打印所有的基本数据类型和字符串对象。这和 DataOutputStream 不同，后者的目的是以某种方式将数据元素放入流中，使得 DataInputStream 可以用可移植的方式重建这些元素。

PrintStream 中有两个重要的方法：print() 和 println()，它们会被重载，然后用于打印各种类型。print() 和 println() 之间的区别在于后者会在执行完成后增加一个换行符（即换行）。

PrintStream 可能会出现问题，因为它会"吞掉"所有的 IOException（因此必须用 checkError() 显式地检查错误状态，如果出现了错误，便会返回 true）。另外 PrintStream 的国际化并不完善。这些问题在 PrintWriter 中得到了解决，稍后会进行介绍。

BufferedOutputStream 是个修饰语，它会告诉流使用缓冲，这样就不会在每次向流中写入时都发生物理写操作了。在进行输出时，很可能会经常用到它。

各种 FilterOutputStream 类型如表 7-4 所示。

表 7-4

类	功 能	构造器参数	使用方法
DataOutputStream	与 DataInputStream 搭配使用，这样就能以可移植的方式向流中写入基本类型（int、char、long，等等）了	OutputStream	包含用于写入基本类型数据的全部接口
PrintStream	用于生成格式化的输出。DataOutputStream 负责处理数据的存储，而 PrintStream 则负责处理数据的显示	OutputStream 以及可选参数：boolean，表示是否在每次换行时都清空缓冲区	应该作为 OutputStream 对象的"最终"包装。可能会经常用到
Buffered-OutputStream	用来防止在每次发送数据的时候都发生物理写操作。相当于声明"使用缓冲"。可以调用 flush() 来清空缓冲区	OutputStream 以及可选参数：（指定）缓冲区大小	这本质上并未提供接口，只是为进程增加缓冲操作而已，需要与接口对象搭配使用

7.1.4 各种 Reader 和 Writer

Java 1.1 对基础流式 I/O 库进行了重大的修改。第一眼看到 Reader 和 Writer 类时，你可能会（像我一样）认为这是要用来替换 InputStream 和 OutputStream 类的，但实际上并非如此。虽然原始的流库某些方面已被弃用（如果使用它们，编译器就会发出警告），但 InputStream 和 OutputStream 类仍然以面向字节的 I/O 的形式提供了有价值的功能，而 Reader 和 Writer 类则提供了兼容 Unicode 并且基于字符的 I/O 能力。此外，还有以下两点需要注意。

1. Java 1.1 在 InputStream 和 OutputStream 的继承层次结构中增加了新的类，因此显然整个继承层次结构并没有被替换。

2. 有时必须将"字节"继承层次结构中的类和"字符"继承层次结构中的类结合起来使用。为了实现这个目的，就要用到"适配器"类：InputStreamReader 可以将 InputStream 转换为 Reader，而 OutputStreamWriter 可以将 OutputStream 转换为 Writer。

设计这一系列的 Reader 和 Writer 的主要是为了国际化。旧的一系列 I/O 流类只能支持 8 位字节流, 无法妥善地处理 16 位 Unicode 字符。由于 Unicode 主要用于国际化（并且 Java 原生的 char 也是 16 位 Unicode）, 因此增加 Reader 和 Writer 就是为了在所有的 I/O 操作中都支持 Unicode。另外, 设计新库也是为了提供比旧库更快的操作性能。

1. 数据的来源和去处

几乎所有原始的 Java I/O 流相关的类都有对应的 Reader 和 Writer 类, 以此来提供原生的 Unicode 操作能力。不过在某些地方, 面向字节的 InputStream 和 OutputStream 才是正确的选择, 特别是 java.util.zip 库, 它就是面向字节（而不是面向字符）的。因此, 最明智的做法是尽量使用 Reader 和 Writer 类。否则, 你会在使用面向字节的库时发现问题, 因为代码无法编译成功。

表 7-5 列出了这两套继承层次结构中信息的来源和去处（也就是说数据在物理上来自哪里, 又去往哪里）, 以及其中的类之间的对应关系。

表 7-5

来源和去处：Java 1.0 中的类	Java 1.1 中对应的类
InputStream	Reader 适配器：InputStreamReader
OutputStream	Writer 适配器：OutputStreamWriter
FileInputStream	FileReader
FileOutputStream	FileWriter
StringBufferInputStream（已弃用）	StringReader
（没有对应的类）	StringWriter
ByteArrayInputStream	CharArrayReader
ByteArrayOutputStream	CharArrayWriter
PipedInputStream	PipedReader
PipedOutputStream	PipedWriter

通常来说, 你会发现这两套继承层次结构中的接口即使不完全相同, 也是十分相似的。

2. 改变流的行为

对于 InputStream 和 OutputStream 来说, 会基于特定的需求, 由 FilterInputStream 和 FilterOutputStream 的"装饰器"子类对流进行适配。Reader 和 Writer 这两套继承层次结构中的类继续沿用了这套思想, 但并不完全相同。

在表 7-6 中, 两列间的对应关系相较于表 7-5 而言并不是那么精确。造成这种区别的原因在于类的组织方式不同——虽然 BufferedOutputStream 是 FilterOutputStream 的子类，但 BufferedWriter 不是 FilterWriter 的子类（尽管 FilterWriter 是抽象类，但它没有任何子类，因此把它放在那里似乎只是当作占位符，或者仅仅是为了告诉你有这么一个类而已）。不过这些类的接口却十分相似。

表 7-6

过滤器：Java 1.0 中的类	Java 1.1 中对应的类
FilterInputStream	FilterReader
FilterOutputStream	FilterWriter（没有任何子类的抽象类）
BufferedInputStream	BufferedReader（同样包含 readLine()）
BufferedOutputStream	BufferedWriter
DataInputStream	DataInputStream（除非必须使用 readLine()，否则应该使用 BufferedReader）
PrintStream	PrintWriter
LineNumberInputStream（已弃用）	LineNumberReader
StreamTokenizer	StreamTokenizer（使用以 Reader 作为参数的构造器）
PushbackInputStream	PushbackReader

有一点很清楚：在任何需要使用 readLine() 的时候，都不要使用 DataInputStream 中的该方法（在编译时会报出弃用消息），而应该使用 BufferedReader 中的版本。除此之外，DataInputStream 仍然是 I/O 库中的"首选"成员。

为了更平缓地过渡到使用 PrintWriter，除了接收 Writer 对象的构造器外，它还提供了接收任意 OutputStream 对象的构造器。PrintWriter 的格式化接口实际上与 PrintStream 相同。

Java 5 中增加了 PrintWriter 构造器，以简化对输出执行写操作时的文件创建过程，稍后会介绍。

有一种 PrintWriter 构造器还支持执行自动清空的可选操作，如果设置了构造器的该选项标志，便会在每次 println() 后执行该清空操作。

3. 未发生变化的类

有一些类则从 Java 1.0 中原样保留到了 Java 1.1，如表 7-7 所示。

表 7-7

以下这些 Java 1.0 中的类在 Java 1.1 中没有对应的类
DataOutputStream
File
RandomAccessFile
SequenceInputStream

特别是 DataOutputStream，在使用上没有任何变化，因此如果想要以可传输的方式存储和读取数据，就应该使用 InputStream 和 OutputStream 这两套继承层次结构中的类。

7.1.5　自成一家的 RandomAccessFile

RandomAccessFile 适合用来处理由大小已知的记录组成的文件，由此可以通过 seek() 在各条记录上来回移动，然后读取或者修改记录。文件中各条记录的大小不必相同，只需确定它们的大小以及在文件中的位置即可。

一开始看到这个类的时候，可能很难相信 RandomAccessFile 并非 InputStream 或 OutputStream 系列继承层次结构中的一员。然而它确实和这两个系列的类没有关系，只是碰巧实现了 DataInput 和 DataOutput 接口（DataInputStream 和 DataOutputStream 同样也分别实现了这两个接口）。RandomAccessFile 甚至没有用到现有的 InputStream 或 OutputStream 类的任何功能。它是个从零实现的、完全独立的类，有着完全属于自己（大部分是原生）的方法。这么设计的原因在于 RandomAccessFile 可以使我们在文件中前后移动，所以其行为和其他的 I/O 类型有着本质上的区别。在任何情况下，它都是独立的，是直接从 Object 派生而来的。

从本质上来说，RandomAccessFile 的作用类似于将 DataInputStream 和 DataOutputStream 组合起来使用，再加上几个方法——getFilePointer() 用来找出在文件中当前所处的位置，seek() 用来移动到文件中某个新的位置，而 length() 则用来判断文件的最大大小。此外构造器还需要第二个参数（和 C 中的 fopen() 完全相同），用来标识我们是只要随机读（"r"）还是既要读又要写（"rw"）。它并不支持只写（write-only）文件，由此可以猜测，RandomAccessFile 如果是从 DataInputStream 继承而来的，可能也运行得不错。

只有 RandomAccessFile 才支持检索相关的方法，并且仅适用于文件。BufferedInputStream 确实可以 mark()（标记）某个位置（其值保存在单个内部变量中），以及 reset()（重置）该位置，但这些功能很有限，并没有多大用处。

从 Java 1.4 开始，RandomAccessFile 的绝大多数功能被 nio 的**内存映射文件**（memory-mapped file）所取代，本章稍后会讲述。

7.1.6　I/O 流的典型用法

虽然可以用多种不同方式组合各种 I/O 流的类，但很可能只会用到其中少数几种组合。下面这些示例可以作为 I/O 典型用法的基本参考（如果你确定无法使用基础卷第 17 章中描述的库实现你的需求）。

在这些示例中，异常处理被简化为将异常传递给控制台，但这种方式仅适用于小型示例或者工具程序。在实际的代码中，需要考虑更加复杂的错误处理方式。

1. 缓冲输入文件

如果要打开一个文件用来输入字符，就需要使用由字符串或 `File` 对象作为文件名参数的 `FileReader`。为了提高速度，需要对文件进行缓冲，因此要将产生的引用传递给 `BufferedReader` 的构造器。`BufferedReader` 提供了可以生成 `Stream<String>` 的 `lines()` 方法：

```java
// iostreams/BufferedInputFile.java
// {VisuallyInspectOutput}
import java.io.*;
import java.util.stream.*;

public class BufferedInputFile {
  public static String read(String filename) {
    try(BufferedReader in = new BufferedReader(
      new FileReader(filename))) {
      return in.lines()
        .collect(Collectors.joining("\n"));
    } catch(IOException e) {
      throw new RuntimeException(e);
    }
  }
  public static void main(String[] args) {
    System.out.print(
      read("BufferedInputFile.java"));
  }
}
```

`Collectors.joining()` 在内部通过 `StringBuilder` 来累加出结果，然后通过包含资源的 `try` 子句将文件自动关闭。

2. 从内存输入

下面的示例使用从 `BufferedInputFile.read()` 得到的字符串类型数据创建了 `StringReader`。然后 `read()` 会将字符逐个生成，并显示到控制台：

```java
// iostreams/MemoryInput.java
// {VisuallyInspectOutput}
import java.io.*;

public class MemoryInput {
  public static void
  main(String[] args) throws IOException {
    StringReader in = new StringReader(
      BufferedInputFile.read("MemoryInput.java"));
    int c;
    while((c = in.read()) != -1)
      System.out.print((char)c);
  }
}
```

read() 会以 int 形式返回下一个字符，因此必须将返回值转型为 char 类型才能正确显示。

3. 格式化的内存输入

如果要读取"格式化后的"数据，就需要使用面向字节（而不是面向 char）的 I/O 类 DataInputStream。因此必须使用 InputStream 类（而不是 Reader 类）。InputStream 类可以以字节形式读取任何类型的数据（例如一个文件），不过此处使用的是字符串：

```java
// iostreams/FormattedMemoryInput.java
// {VisuallyInspectOutput}
import java.io.*;

public class FormattedMemoryInput {
  public static void main(String[] args) {
    try(
      DataInputStream in = new DataInputStream(
        new ByteArrayInputStream(
          BufferedInputFile.read(
            "FormattedMemoryInput.java")
              .getBytes()))
    ) {
      while(true)
        System.out.write((char)in.readByte());
    } catch(EOFException e) {
      System.out.println("\nEnd of stream");
    } catch(IOException e) {
      throw new RuntimeException(e);
    }
  }
}
```

ByteArrayInputStream 只能接收字节数组（此处由 String.getBytes() 生成）。得到的 ByteArrayInputStream 便可以作为匹配的 InputStream 类型参数传给 DataInputStream。

如果用 `readByte()` 从 `DataInputStream` 以一次读取一个字节的方式读取字符，那么返回任何字节的值都是合理的结果，所以无法用返回值检测输入是否结束。因此需要使用 `available()` 方法来找出还剩余多少可读取的字符。下面这个示例演示了如何以一次读取一个字节的方式来读取文件：

```java
// iostreams/TestEOF.java
// 测试文件是否结束
// {VisuallyInspectOutput}
import java.io.*;

public class TestEOF {
  public static void main(String[] args) {
    try(
      DataInputStream in = new DataInputStream(
        new BufferedInputStream(
          new FileInputStream("TestEOF.java")))
    ) {
      while(in.available() != 0)
        System.out.write(in.readByte());
    } catch(IOException e) {
      throw new RuntimeException(e);
    }
  }
}
```

注意，`available()` 的工作方式会随着所读取媒介类型的不同而有所不同——该方法的字面意思是"**在没有阻塞的情况下**所能读取的字节数量"。对于文件来说，这意味着整个文件，但是对于不同类型的流来说，则可能并不是这样，因此该方法要谨慎使用。

你也可以在类似的情况中通过捕获异常来检测输入是否结束。不过用异常来控制程序流程一般被认为是错误的异常使用方式。

4. 基本的文件输出

`FileWriter` 对象用于向文件中写入数据。实际上一般会通过将输出包装在 `BufferedWriter` 中来对输出进行缓冲（可以试着移除该包装，看看会对性能产生什么影响——缓冲往往会显著提升 I/O 操作的性能）。此处，`FileWriter` 被装饰为 `PrintWriter` 以提供格式化的能力。以这种方式创建的数据文件可以作为普通的文本文件来读取：

```java
// iostreams/BasicFileOutput.java
// {VisuallyInspectOutput}
import java.io.*;

public class BasicFileOutput {
  static String file = "BasicFileOutput.dat";
```

```
  public static void main(String[] args) {
    try(
      BufferedReader in = new BufferedReader(
        new StringReader(
          BufferedInputFile.read(
            "BasicFileOutput.java")));
      PrintWriter out = new PrintWriter(
        new BufferedWriter(new FileWriter(file)))
    ) {
      in.lines().forEach(out::println);
    } catch(IOException e) {
      throw new RuntimeException(e);
    }
    //
    System.out.println(BufferedInputFile.read(file));
  }
}
```

包含资源的 try 子句清空了缓冲区并关闭了文件。

文本文件输出的快捷方式

Java 5 为 PrintWriter 增加了一个辅助构造器，因此我们无须在每次创建文本文件并向其中写入时都去手动执行装饰工作。下面用该快捷方式对 BasicFileOutput.java 进行了重写：

```
// iostreams/FileOutputShortcut.java
// {VisuallyInspectOutput}
import java.io.*;

public class FileOutputShortcut {
  static String file = "FileOutputShortcut.dat";
  public static void main(String[] args) {
    try(
      BufferedReader in = new BufferedReader(
        new StringReader(BufferedInputFile.read(
          "FileOutputShortcut.java")));
      // 下面便是该快捷方式:
      PrintWriter out = new PrintWriter(file)
    ) {
      in.lines().forEach(out::println);
    } catch(IOException e) {
      throw new RuntimeException(e);
    }
    System.out.println(BufferedInputFile.read(file));
  }
}
```

你仍然实现了缓冲的效果，只是不必自行实现。遗憾的是，其他常见的写入任务并没有提供快捷方式，因此典型的 I/O 流操作仍然包含大量的冗余代码。基础卷第 17 章讲述了如何通过一种不同的方式来极大地简化其他任务。

5. 存储和恢复数据

PrintWriter 可以将数据格式化为人类可读的格式。如果要输出可供另一个流恢复的数据，可以使用 DataOutputStream 来写入数据，并用 DataInputStream 来恢复数据。这里说的流可以是任何形式，不过下面的示例中用的是文件，并为读写进行了缓冲处理。DataOutputStream 和 DataInputStream 都是面向字节的，因此需要使用 InputStream 和 OutputStream。

```java
// iostreams/StoringAndRecoveringData.java
import java.io.*;

public class StoringAndRecoveringData {
  public static void main(String[] args) {
    try(
      DataOutputStream out = new DataOutputStream(
        new BufferedOutputStream(
          new FileOutputStream("Data.txt")))
    ) {
      out.writeDouble(3.14159);
      out.writeUTF("That was pi");
      out.writeDouble(1.41413);
      out.writeUTF("Square root of 2");
    } catch(IOException e) {
      throw new RuntimeException(e);
    }
    try(
      DataInputStream in = new DataInputStream(
        new BufferedInputStream(
          new FileInputStream("Data.txt")))
    ) {
      System.out.println(in.readDouble());
      // 只有 readUTF() 可以正确地恢复 Java 的 UTF 字符串
      System.out.println(in.readUTF());
      System.out.println(in.readDouble());
      System.out.println(in.readUTF());
    } catch(IOException e) {
      throw new RuntimeException(e);
    }
  }
}
/* 输出：
3.14159
That was pi
1.41413
Square root of 2
*/
```

如果使用 DataOutputStream 来写入数据，那么 Java 会确保你可以通过 DataInputStream 精确地恢复数据——不论是在什么平台上写入和读取数据。任何在数据的跨平台问题上花费过时间和精力的人都知道，这非常有价值。如果两个平台上都安装了 Java，这种问题要想使用 DataOutputStream 写入字符串，以便可以让 DataInputStream 来恢复，唯一可靠的方法是使用 UTF-8 编码（本例中由 writeUTF() 和 readUTF() 实现）。UTF-8 是一种多字节的格式，其编码长度会根据实际使用的字符集而有所不同。如果使用的（几乎）全都是 ASCII 字符（仅占用 7 位），Unicode 便会浪费空间和/或带宽，因此 UTF-8 会将 ASCII 字符编码为 1 个字节，非 ASCII 字符则会编码为 2 个或 3 个字节。此外，UTF-8 字符串会将字符串的长度保存在头两个字节。不过 writeUTF() 和 readUTF() 使用的是一种适用于 Java 的特殊 UTF-8 变体（JDK 文档中有关于这些方法的描述），因此如果你用非 Java 程序来读取用 writeUTF() 写入的字符串，就必须编写特殊的代码来妥当地读取该字符串。

有了 writeUTF() 和 readUTF()，便可以在 DataOutputStream 中将字符串和其他类型的数据进行混合，因为我们知道字符串会被妥当地存储为 Unicode，而且可以轻松地用 DataInputStream 来恢复。

writeDouble() 方法会将 double 类型的数值存储到流中，并且与之互补的 readDouble() 方法会对数值进行恢复（还有其他类似的方法用于其他类型的读写操作）。但是为了让这些用于读取的方法都可以正确地工作，就必须知道数据项在流中的确切位置，因为同样可以将存储的 double 类型作为简单的字节（或 char 等）序列进行读取。因此你必须要么对文件中的数据使用固定的格式，要么在你要解析的文件中加入额外的信息来确定数据所在的位置。注意，对象序列化或者 XML（这两种方式均在附录 E 中描述过）可能是更容易存储及读取复杂数据结构的方式。

6. 读写随机访问文件

使用 RandomAccessFile 就类似于组合使用了 DataInputStream 和 DataOutputStream（因为 RandomAccessFile 实现了相同的接口：DataInput 和 DataOutput）。此外可以使用 seek() 来在文件中来回移动，并修改其中的值。

如果使用了 RandomAccessFile，就必须知道文件的布局，以实现对文件的正确操作。RandomAccessFile 含有专门用于读写基本类型和 UTF-8 字符串的方法：

```
// iostreams/UsingRandomAccessFile.java
import java.io.*;
```

---

① XML 是另一种解决在不同计算平台间传输数据的问题的途径，而且不依赖于这些平台上是否安装了 Java。附录 E 对 XML 进行了介绍。

```java
public class UsingRandomAccessFile {
  static String file = "rtest.dat";
  public static void display() {
    try(
      RandomAccessFile rf =
        new RandomAccessFile(file, "r")
    ) {
      for(int i = 0; i < 7; i++)
        System.out.println(
          "Value " + i + ": " + rf.readDouble());
      System.out.println(rf.readUTF());
    } catch(IOException e) {
      throw new RuntimeException(e);
    }
  }
  public static void main(String[] args) {
    try(
      RandomAccessFile rf =
        new RandomAccessFile(file, "rw")
    ) {
      for(int i = 0; i < 7; i++)
        rf.writeDouble(i*1.414);
      rf.writeUTF("The end of the file");
      rf.close();
      display();
    } catch(IOException e) {
      throw new RuntimeException(e);
    }
    try(
      RandomAccessFile rf =
        new RandomAccessFile(file, "rw")
    ) {
      rf.seek(5*8);
      rf.writeDouble(47.0001);
      rf.close();
      display();
    } catch(IOException e) {
      throw new RuntimeException(e);
    }
  }
}
/* 输出:
Value 0: 0.0
Value 1: 1.414
Value 2: 2.828
Value 3: 4.242
Value 4: 5.656
Value 5: 7.069999999999999
Value 6: 8.484
The end of the file
Value 0: 0.0
Value 1: 1.414
Value 2: 2.828
Value 3: 4.242
Value 4: 5.656
Value 5: 47.0001
Value 6: 8.484
The end of the file
*/
```

display() 方法打开了一个文件，并将其中的元素显示为 double 值。在 main() 中，首先创建了文件，然后打开文件并进行了修改。double 的长度永远是个 8 字节，因此如果要 seek()（查找）到第 5 个 double 元素，就必须进行 5*8 的乘法运算来得到要查找的位置。

如之前所提到的，RandomAccessFile 实际上独立于其余的 I/O 类系列，它只不过实现了 DataInput 和 DataOutput 接口。它并不支持装饰器，因此与 InputStream 和 OutputStream 的子类在任何方面都无法结合起来使用。必须假定 RandomAccessFile 已经妥当地进行了缓冲，因为无法为其增加缓冲处理。

构造器的第二个参数提供了唯一的可选项：可以打开一个 RandomAccessFile 来读取（"r"）或者读取并写入（"rw"）。

可以考虑用 nio 的内存映射文件来取代 RandomAccessFile。本章 7.3 节会对此进行描述。

7.1.7 小结

Java I/O 流库确实满足了基本的需求：可以对控制台、文件、内存甚至跨网络进行读写操作。通过继承，可以创建新的输入和输出对象类型。甚至还可以通过重新定义 toString() 方法（在向预期参数类型为 String 的方法传入对象时，会自动调用该对象的 toString() 方法。这是 Java 有限的"自动类型转换"）为流所能接收的对象种类添加简单的扩展性。

不过还有一些问题，I/O 流库的设计和文档并没有给出解答。例如，在打开某个文件并向其输出时，如果会覆盖其原有的内容，则最好抛出异常——某些编程系统允许你打开一个文件用于输出，但前提是该文件尚不存在。而在 Java 中，似乎应该（首先）用 File 对象确定一个文件是否存在，这是因为如果将它作为 FileOutputStream 或 FileWriter 打开，它的内容就肯定会被覆盖。

I/O 流库给我们带来了复杂的使用感受。它的功能很多，并且很轻量。但是如果你尚未理解装饰器设计模式，这种设计便会显得很不直观，因此学习和教授它都需要付出额外的成本。它的完成度也不高——比如说，以前我需要专门编写一些工具，以合理控制读取文本文件所需的代码量——所幸 Java 7 的 nio 使我不必再这么做了。

一旦你真的理解了装饰器模式，并开始在依赖其灵活性的场景中使用该库，你就可以开始享受这种设计模式的好处，这时那些额外的代码成本便不会再那么困扰你了。不过，一定要记得要先确认基础卷第 17 章中介绍的库和技巧肯定解决不了你的问题。

7.2 标准 I/O

术语标准 I/O（standard I/O）指的是 UNIX 中"程序所使用的单一信息流"这个概念（大部分操作系统中以某种形式再现了这种理念）。程序的所有输入都可以来自**标准输入**，所有输出都可以发送到**标准输出**，所有错误都可以发送到**标准错误**。标准 I/O 的价值在于可以很容易地使多个程序串联起来，一个程序的标准输出可以成为另一个程序的标准输入。这是个非常强大的工具。

7.2.1 从标准输入中读取

遵循标准 I/O 模型，Java 实现了 `System.in`、`System.out`，以及 `System.err`。你已经在本书各处看到了如何用 `System.out` 向标准输出写入数据，其中 `System.out` 已经被预先包装为 `PrintStream` 对象。`System.err` 也同样是一个 `PrintStream`，但是 `System.in` 是未经包装的原生 `InputStream`。这意味着虽然可以随时使用 `System.out` 和 `System.err`，但是 `System.in` 则必须在读取前先进行包装。

从输入中读取时，一般会一次读取一行。如果要这么做，就需要将 `System.in` 包装在 `BufferedReader` 中，这就需要用 `InputStreamReader` 将 `System.in` 转换为 `Reader`。下面这个示例会回显出你输入的每一行内容：

```java
// standardio/Echo.java
// 怎样从标准输入中读取
import java.io.*;
import onjava.TimedAbort;

public class Echo {
  public static void main(String[] args) {
    TimedAbort abort = new TimedAbort(2);
    new BufferedReader(
      new InputStreamReader(System.in))
      .lines()
      .peek(ln -> abort.restart())
      .forEach(System.out::println);
    // 使用 Ctrl-Z，或者 2 秒内不进行任何操作，便可终止程序
  }
}
```

`BufferedReader` 中的 `lines()` 方法会返回 `Stream<String>`，这体现出了流模型的灵活性：可以良好地处理标准输入。`peek()` 方法会重启 `TimedAbort` 以保持程序的开启状态（只要输入间断不超过 2 秒）。

7.2.2 将 `System.out` 转换为 `PrintWriter`

`System.out` 是一个 `PrintStream`，而 `PrintStream` 是一个 `OutputStream`。`PrintWriter` 内有一个以 `OutputStream` 为参数的构造器。因此如果需要，就可以通过该构造器将 `System.out` 转换为 `PrintWriter`：

```java
// standardio/ChangeSystemOut.java
// 将 `System.out` 转换为 `PrintWriter`
import java.io.*;

public class ChangeSystemOut {
  public static void main(String[] args) {
```

```
/* 输出：
Hello, world
*/
```

```
    PrintWriter out =
      new PrintWriter(System.out, true);
    out.println("Hello, world");
  }
}
```

重点是要使用带有两个参数的 PrintWriter 构造器，并将第二个参数设为 true 来开启自动清空的功能。否则可能就看不到输出。

7.2.3　标准 I/O 重定向

Java 的 System 类可以通过简单的静态方法调用对标准输入、标准输出以及标准错误的 I/O 标准流进行重定向：

- setIn(InputStream)
- setOut(PrintStream)
- setErr(PrintStream)

如果你突然开始在屏幕上制造大量输出，而且输出滚动的速度快到让你看不清，这时重定向输出就会特别有用。而如果要用命令行程序反复测试特定的用户输入序列，这时重定向输入就很有用。下面这个简单的示例演示了相关方法：

```
// standardio/Redirecting.java
// 演示标准 I/O 重定向
import java.io.*;

public class Redirecting {
  public static void main(String[] args) {
    PrintStream console = System.out;
    try(
      BufferedInputStream in = new BufferedInputStream(
        new FileInputStream("Redirecting.java"));
      PrintStream out = new PrintStream(
        new BufferedOutputStream(
          new FileOutputStream("Redirecting.txt")))
    ) {
      System.setIn(in);
      System.setOut(out);
      System.setErr(out);
      new BufferedReader(
        new InputStreamReader(System.in))
          .lines()
          .forEach(System.out::println);
    } catch(IOException e) {
      throw new RuntimeException(e);
    } finally {
      System.setOut(console);
```

```
      }
    }
}
```

该程序将标准输入附加到一个文件上，并将标准输出和标准错误重定向到另一个文件。它在程序的开头存放了一个引用，指向原始的 `System.out` 对象，并在结尾将系统输出恢复到该对象上。

I/O 重定向操纵的是字节流，而不是字符流，因此这里实际使用的是 `InputStream` 和 `OutputStream`，而不是 `Reader` 和 `Writer`。

7.2.4 进程控制

Java 库提供了一些类，用来从 Java 内部执行操作系统程序，并控制这些程序的输入和输出。

有项任务很常见，即运行某个程序，并将产生的输出发送到控制台。本节中有个实用工具可以简化这项任务。

该工具可能会出现两种类型的错误。第一种是导致异常的普通错误，对于这种错误，只需要重新抛出 `RuntimeException` 异常即可。第二种是进程自身的执行过程中产生的错误。接下来用一个单独的异常来报告这些错误：

```java
// onjava/OSExecuteException.java
package onjava;

public class
OSExecuteException extends RuntimeException {
  public OSExecuteException(String why) {
    super(why);
  }
}
```

如果要运行某个程序，就需要向 `OSExecute.command()` 中传入 `String command`（命令），也就是你要在控制台中运行程序时所输入的那条命令。将这条命令传给 `java.lang.ProcessBuilder` 构造器（其参数为一个 `String` 对象序列），然后生成的 `ProcessBuilder` 对象就启动了：

```java
// onjava/OSExecute.java
// 运行操作系统命令，并将输出发送到控制台
package onjava;
import java.io.*;

public class OSExecute {
```

```
  public static void command(String command) {
    boolean err = false;
    try {
      Process process = new ProcessBuilder(
        command.split(" ")).start();
      try(
        BufferedReader results = new BufferedReader(
          new InputStreamReader(
            process.getInputStream()));
        BufferedReader errors = new BufferedReader(
          new InputStreamReader(
            process.getErrorStream()))
      ) {
        results.lines()
          .forEach(System.out::println);
        err = errors.lines()
          .peek(System.err::println)
          .count() > 0;
      }
    } catch(IOException e) {
      throw new RuntimeException(e);
    }
    if(err)
      throw new OSExecuteException(
        "Errors executing " + command);
  }
}
```

如果要在该程序执行时捕获标准输出流，就需要调用 getInputStream()。这是因为 InputStream 就是用来从中读取数据的。

此处这些行只是直接显示出来了，但你可能还想在 command() 中将它们捕获并返回。

该程序的错误被发送到了标准错误流，并通过调用 getErrorStream() 得以捕获。如果出现任何错误，这些错误便会显示出来，并抛出 OSExecuteException 异常，这样调用方程序便可以处理该问题了。

下面这个示例演示了 OSExecute 的用法：

```
// standardio/OSExecuteDemo.java
// 演示标准 I/O 重定向
// javap -cp build/classes/java/main OSExecuteDemo
// {ExcludeFromGradle}
import onjava.*;

public class OSExecuteDemo {}            /* 输出：
                                         Compiled from "OSExecuteDemo.java"
                                         public class OSExecuteDemo {
                                           public OSExecuteDemo();
                                         }
                                         */
```

这里用到了 javap 反编译器（JDK 中提供的工具）来对程序进行反编译。

7.3　新 I/O 系统

Java 1.4 的 `java.nio.*` 包中所引入的"新" I/O 库，它的目标只有一个：速度。

实际上，"旧" I/O 包已经用 nio 重新实现过了，以利用其带来的速度优势，因此即使没有显式地用 nio 来写代码，也已经享受到了 nio 带来的好处。速度的提升在文件 I/O（稍后会进行探讨）及网络 I/O（比如可用于互联网编程）这两个方面均有体现。

速度的提升来自于其所使用的更接近于操作系统的 I/O 实现方式的结构：**通道**（channel）和**缓冲区**（buffer）。可以将其想象为一座煤矿——通道是含有煤层（即数据）的矿井，而缓冲区则是送入矿井中的手推车。手推车将满车的煤运了回来，你便从手推车中得到了煤。也就是说，你不会直接和通道交互，而是和缓冲区交互，将缓冲区送入到通道中。通道要么从缓冲区中拉取数据，要么向缓冲区中放入数据。

本章将较为深入地探讨 nio 包。I/O 流这样的高级库使用了 nio，但是大部分时间你并不需要用到这一层的 I/O。如果使用的是 Java 7 和 8，那么除非特殊情况，否则你（在理想情况下）甚至不需要接触到 I/O 流。在理想情况下，你日常会用到的一切都已涵盖在基础卷第 17 章中了。只有在遇到性能问题（比如说在要用到内存映射文件的时候）或要实现自定义的 I/O 库时，才有必要理解 nio。

7.3.1　字节缓冲区 ByteBuffer

`ByteBuffer`（即保存原生字节的缓冲区）是唯一直接和通道通信的类型。如果你查阅 JDK 文档中关于 `java.nio.ByteBuffer` 部分的内容，便会发现它相当简单：只需要告诉它分配多少内存，就可以创建出一块字节缓冲区，并带有一些用于放入和取出数据（以原生字节的形式，或作为基本数据类型）的方法。但是并不能放入或取出对象，即使对象是字符串。正因为它可以在大多数操作系统中使（内存）映射更加高效，所以这是一种相当底层的处理方式。

"旧" I/O 库中有三个方法经过了修改，以用于生成 `FileChannel`：用于读的 `FileInputStream`、用于写的 `FileOutputStream`，以及既要读又要写的 `RandomAccessFile`。注意，这些都是操纵字节的流，和 nio 的底层性质一致。字符模式的 `Reader` 和 `Writer` 类不会生成通道，但是 `java.nio.channels.Channels` 类提供了工具方法来从通道生成 `Reader` 和 `Writer`。

下面我们练习一下这三个类型的流，生成可写、可读写，以及可读的通道：

```java
// newio/GetChannel.java
// 从流中得到通道
import java.nio.*;
import java.nio.channels.*;
import java.io.*;

public class GetChannel {
  private static String name = "data.txt";
  private static final int BSIZE = 1024;
  public static void main(String[] args) {
    // 写文件:
    try(
      FileChannel fc = new FileOutputStream(name)
        .getChannel()
    ) {
      fc.write(ByteBuffer
        .wrap("Some text ".getBytes()));
    } catch(IOException e) {
      throw new RuntimeException(e);
    }
    // 添加到文件末尾:
    try(
      FileChannel fc = new RandomAccessFile(
        name, "rw").getChannel()
    ) {
      fc.position(fc.size()); // 移动到尾部
      fc.write(ByteBuffer
        .wrap("Some more".getBytes()));
    } catch(IOException e) {
      throw new RuntimeException(e);
    }
    // 读文件:
    try(
      FileChannel fc = new FileInputStream(name)
        .getChannel()
    ) {
      ByteBuffer buff = ByteBuffer.allocate(BSIZE);
      fc.read(buff);
      buff.flip();
      while(buff.hasRemaining())
        System.out.write(buff.get());
    } catch(IOException e) {
      throw new RuntimeException(e);
    }
    System.out.flush();
  }
}
```

/* 输出：
略
*/

对于这里出现的任何流类，getChannel() 都会生成一个 FileChannel 通道。通道其实很简单：向其传入一个用于读写的 ByteBuffer，然后锁住文件区域以保证独占式访问（这部分稍后会讲述）。

将字节放入 ByteBuffer 的一种方法是，在一系列 put() 方法中选用一个，再用它将一个（或多个）字节或基本类型的值直接填充进去。不过如你所见，也可以用 wrap() 方法将已有的 byte 数组"包装"到 ByteBuffer 中。如果这么做，底层的数组就不再会被复制，而是充当所生成的 ByteBuffer 的存储。我们可以这么说：ByteBuffer 是由数组"支持"的。

RandomAccessFile 再次打开了 data.txt 文件。注意，可以在文件中来回移动 FileChannel，而在这里它被移动到了尾部，这样便可以附上（向末尾）追加的写操作了。

如要进行只读的访问，就需要用 static allocate() 方法显式地分配一个 ByteBuffer。nio 的目标就是为了快速地移动大量数据，因此需要特别重视 ByteBuffer 的大小——实际上，这里所用的 1KB 对于你日常的真实场景来说应该小了一点（你必须在实际运行的程序中验证，以找到最佳大小）。

甚至还可能达到更快的速度，方法是用 allocateDirect() 来取代 allocate()，以生成和操作系统结合度更高的"直接"缓冲区。不过这种分配方式的开销更大，实际的实现也因操作系统而不同，因此再次强调，你必须在实际运行的程序中验证，以确定直接缓冲区确实可以为你带来速度上的优势。

一旦调用了 read() 来让 FileChannel 将字节保存到 ByteBuffer，就必须在缓冲区上调用 flip()，使缓冲区做好被提取字节的准备（是的，这样操作看起来有点笨拙，但要记住这是非常底层的操作，目的是获得最快的速度）。并且如果要用缓冲区来做进一步的 read() 操作，就同样需要调用 clear() 来让缓冲区为后续的每次 read() 做好准备。下面这个简单的文件复制程序演示了上述操作：

```java
// newio/ChannelCopy.java
// 通过通道和缓冲来复制文件
// {java ChannelCopy ChannelCopy.java test.txt}
import java.nio.*;
import java.nio.channels.*;
import java.io.*;

public class ChannelCopy {
  private static final int BSIZE = 1024;
  public static void main(String[] args) {
    if(args.length != 2) {
      System.out.println(
        "arguments: sourcefile destfile");
      System.exit(1);
    }
```

```
    try(
      FileChannel in = new FileInputStream(
        args[0]).getChannel();
      FileChannel out = new FileOutputStream(
        args[1]).getChannel()
    ) {
      ByteBuffer buffer = ByteBuffer.allocate(BSIZE);
      while(in.read(buffer) != -1) {
        buffer.flip(); // 准备写
        out.write(buffer);
        buffer.clear();  // 准备读
      }
    } catch(IOException e) {
      throw new RuntimeException(e);
    }
  }
}
```

这里打开了两个 FileChannel，一个用于读，另一个用于写。然后分配了一个 ByteBuffer，如果 FileChannel.read() 返回了 -1（毫无疑问，这是继承了 UNIX 和 C 的惯例），就意味着到达了输入的尽头。在每次用 read() 将数据放入缓冲后，flip() 便会准备好缓冲区，这样缓冲区中的信息就可以通过 write() 提取出来了。在 write() 运行完后，信息仍然在缓冲区中，而 clear() 会重置所有的内部指针，这样通道就可以准备好在下一次 read() 运行中接收数据了。

不过上面这个程序并不是处理这种操作的理想方式。专用的 transferTo() 和 transferFrom() 方法可以将一个通道直接连接到另一个通道：

```
// newio/TransferTo.java
// 在通道之间使用 transferTo()
// {java TransferTo TransferTo.java TransferTo.txt}
import java.nio.channels.*;
import java.io.*;

public class TransferTo {
  public static void main(String[] args) {
    if(args.length != 2) {
      System.out.println(
        "arguments: sourcefile destfile");
      System.exit(1);
    }
    try(
      FileChannel in = new FileInputStream(
        args[0]).getChannel();
      FileChannel out = new FileOutputStream(
        args[1]).getChannel()
    ) {
      in.transferTo(0, in.size(), out);
```

```
      // 或者使用 out.transferFrom(in, 0, in.size());
    } catch(IOException e) {
      throw new RuntimeException(e);
    }
  }
}
```

这个方法你不会经常用到，但了解一下还是有好处的。

7.3.2 转换数据

在 GetChannel.java 中，为了将文件中的信息打印出来，我们将数据以一次一个 byte 的方式拉取出来，并将每个 byte 转型为 char。这种方式看起来很原始——如果仔细看看 java.nio.CharBuffer 类的内部，便会看到其中有个 toString() 方法，声明了"返回一个包含该缓冲区中字符的 String"。ByteBuffer 可以被看作一个带有 asCharBuffer() 方法的 CharBuffer，那么为什么不使用它呢？如下方输出语句中的第一行所示，这样是行不通的：

```
// newio/BufferToText.java
// 在 ByteBuffers 和文本间双向转换
import java.nio.*;
import java.nio.channels.*;
import java.nio.charset.*;
import java.io.*;

public class BufferToText {
  private static final int BSIZE = 1024;
  public static void main(String[] args) {
    try(
      FileChannel fc = new FileOutputStream(
        "data2.txt").getChannel()
    ) {
      fc.write(ByteBuffer.wrap("Some text".getBytes()));
    } catch(IOException e) {
      throw new RuntimeException(e);
    }
    ByteBuffer buff = ByteBuffer.allocate(BSIZE);
    try(
      FileChannel fc = new FileInputStream(
        "data2.txt").getChannel()
    ) {
      fc.read(buff);
    } catch(IOException e) {
      throw new RuntimeException(e);
    }
    buff.flip();
    // 这样不行:
    System.out.println(buff.asCharBuffer());
    // 用当前系统的默认字符集解码:
    buff.rewind();
```

```
/* 输出:
????
Decoded using windows-1252: Some text
Some text
Some textNULNULNUL
*/
```

```java
      String encoding =
        System.getProperty("file.encoding");
      System.out.println("Decoded using " +
        encoding + ": "
        + Charset.forName(encoding).decode(buff));
      // 对打印的内容进行编码:
      try(
        FileChannel fc = new FileOutputStream(
          "data2.txt").getChannel()
      ) {
        fc.write(ByteBuffer.wrap(
          "Some text".getBytes("UTF-16BE")));
      } catch(IOException e) {
        throw new RuntimeException(e);
      }
      // 现在再次试着读取:
      buff.clear();
      try(
        FileChannel fc = new FileInputStream(
          "data2.txt").getChannel()
      ) {
        fc.read(buff);
      } catch(IOException e) {
        throw new RuntimeException(e);
      }
      buff.flip();
      System.out.println(buff.asCharBuffer());
      // 用 CharBuffer 进行写操作:
      buff = ByteBuffer.allocate(24);
      buff.asCharBuffer().put("Some text");
      try(
        FileChannel fc = new FileOutputStream(
          "data2.txt").getChannel()
      ) {
        fc.write(buff);
      } catch(IOException e) {
        throw new RuntimeException(e);
      }
      // 读取并显示出来:
      buff.clear();
      try(
        FileChannel fc = new FileInputStream(
          "data2.txt").getChannel()
      ) {
        fc.read(buff);
      } catch(IOException e) {
        throw new RuntimeException(e);
      }
      buff.flip();
      System.out.println(buff.asCharBuffer());
  }
}
```

缓冲区中保存着简单的字节，为了将这些字节转换为字符，要么在将字节放入的时候进行**编码**（这样在读取出来的时候才会有意义），要么在将它们从缓冲区中读取出来的时候进行**解码**。可以通过 java.nio.charset.Charset 类来完成这些操作，这些类提供了将数据编码为多种不同字符集类型的工具：

```
// newio/AvailableCharSets.java
// 显示字符集和别名
import java.nio.charset.*;
import java.util.*;

public class AvailableCharSets {
  public static void main(String[] args) {
    SortedMap<String,Charset> charSets =
      Charset.availableCharsets();
    for(String csName : charSets.keySet()) {
      System.out.print(csName);
      Iterator aliases = charSets.get(csName)
        .aliases().iterator();
      if(aliases.hasNext())
        System.out.print(": ");
      while(aliases.hasNext()) {
        System.out.print(aliases.next());
        if(aliases.hasNext())
          System.out.print(", ");
      }
      System.out.println();
    }
  }
}
/* 输出（前7行）：
Big5: csBig5
Big5-HKSCS: big5-hkscs, big5hk, Big5_HKSCS, big5hkscs
CESU-8: CESU8, csCESU-8
EUC-JP: csEUCPkdFmtjapanese, x-euc-jp, eucjis,
Extended_UNIX_Code_Packed_Format_for_Japanese, euc_jp,
eucjp, x-eucjp
EUC-KR: ksc5601-1987, csEUCKR, ksc5601_1987, ksc5601,
5601, euc_kr, ksc_5601, ks_c_5601-1987, euckr
GB18030: gb18030-2000
GB2312: gb2312, euc-cn, x-EUC-CN, euccn, EUC_CN,
gb2312-80, gb2312-1980
        ...
*/
```

因此，让我们回到 BufferToText.java，如果对缓冲区执行 rewind()（回到数据的起始位置），然后使用该平台的默认字符集来 decode()（解码）数据，那么得到的 CharBuffer 便会正确地显示在控制台上。如果要检测出默认是哪种字符集，就可以使用 System.getProperty("file.encoding") 得到字符串类型的字符集名称。将该名称传入 Charset.forName()，就能生成可对字符串进行解码的 Charset 对象了。

另一种可选方案是用某个可在读取文件时生成可打印内容的字符集来 encode()（编码），如你在 BufferToText.java 中的第三部分所见。该处使用了 UTF-16BE 来将文本写入到文件，而在读取时，你需要做的就是将它转换为 CharBuffer，然后就能生成符合预期的文本内容了。

最后你可以看到，如果通过 CharBuffer 来对 ByteBuffer 进行**写入**（你会在后面学到更多相关内容），会发生什么情况。注意，这里为 ByteBuffer 分配了 24 字节，而每个 char 需要 2 字节，因此这足够容纳 12 个 char 了，但是"Some text"只有 9 个 char。剩余的空内容字节仍然会出现在由 CharBuffer 的 toString() 所展示的内容中，如你在输出中所见。

7.3.3　获取基本类型

虽然 ByteBuffer 只能保存字节类型，但其中包含的方法可以从所保存的字节生成各种不同类型的基本类型值。下面这个示例演示了如何通过这些方法来插入和提取各种类型的值：

```java
// newio/GetData.java
// 生成 ByteBuffer 的不同表示
import java.nio.*;

public class GetData {
  private static final int BSIZE = 1024;
  public static void main(String[] args) {
    ByteBuffer bb = ByteBuffer.allocate(BSIZE);
    // 该赋值会自动将 ByteBuffer 清零:
    int i = 0;
    while(i++ < bb.limit())
      if(bb.get() != 0)
        System.out.println("nonzero");
    System.out.println("i = " + i);
    bb.rewind();
    // 保存并读取 char 数组:
    bb.asCharBuffer().put("Howdy!");
    char c;
    while((c = bb.getChar()) != 0)
      System.out.print(c + " ");
    System.out.println();
    bb.rewind();
    // 保存并读取 short:
    bb.asShortBuffer().put((short)471142);
    System.out.println(bb.getShort());
    bb.rewind();
    // 保存并读取 int:
    bb.asIntBuffer().put(99471142);
    System.out.println(bb.getInt());
    bb.rewind();
```

```
/* 输出:
i = 1025
H o w d y !
12390
99471142
99471142
9.9471144E7
9.9471142E7
*/
```

```
    // 保存并读取 long：
    bb.asLongBuffer().put(99471142);
    System.out.println(bb.getLong());
    bb.rewind();
    // 保存并读取 float：
    bb.asFloatBuffer().put(99471142);
    System.out.println(bb.getFloat());
    bb.rewind();
    // 保存并读取 double：
    bb.asDoubleBuffer().put(99471142);
    System.out.println(bb.getDouble());
    bb.rewind();
  }
}
```

在为 ByteBuffer 分配内存后，会对它的值进行检查，以确认该分配是否自动清零了内容——可以看到，确实清零了。全部 1024 个值都经过了检查（由缓冲区的上限决定，该上限可通过 limit() 得到），并且全都为零。

将基本类型值插入 ByteBuffer 的最简单的方法是通过 asCharBuffer()、asShortBuffer() 等方法在该缓冲区上得到合适的"视图"，然后使用该视图的 put() 方法。该方法针对每种基本数据类型都有相应的实现。唯一看起来比较奇怪的是针对 ShortBuffer 的 put() 方法，它需要进行类型转换（该转型会截断并改变生成的值）。其他所有视图缓冲区的 put() 方法都不需要进行转型。

7.3.4 视图缓冲区

"视图缓冲区"（view buffer）相当于透过某个特定基本类型的视角来看底层的 ByteBuffer（因此称为"视图"）。实际存储数据的仍然是 ByteBuffer，它"支撑"着该视图，因此对视图所做的任何修改都会映射为 ByteBuffer 中数据的修改。如你在前面的示例中所见，这种方式非常便于向 ByteBuffer 中插入基本类型。视图还可以用一次一个（ByteBuffer 支持的方式）或批量（以数组形式）的方式从 ByteBuffer 中读取基本类型的值。下面这个示例便是通过 IntBuffer 来操纵 ByteBuffer 中的 int：

```
// newio/IntBufferDemo.java
// 通过 IntBuffer 来操纵 ByteBuffer 中的 int
import java.nio.*;

public class IntBufferDemo {
  private static final int BSIZE = 1024;
  public static void main(String[] args) {
    ByteBuffer bb = ByteBuffer.allocate(BSIZE);
    IntBuffer ib = bb.asIntBuffer();
    // 保存 int 数组：
```

```
    ib.put(new int[]{ 11, 42, 47, 99, 143, 811, 1016 });
    // 通过绝对地址读写：
    System.out.println(ib.get(3));
    ib.put(3, 1811);
    // 在回退缓冲区之前设置新的限制
    ib.flip();
    while(ib.hasRemaining()) {
      int i = ib.get();
      System.out.println(i);
    }
  }
}
/* 输出：
99
11
42
47
1811
143
811
1016
*/
```

首先用重载后的 put() 方法保存 int 数组。随后调用 get() 和 put() 方法直接访问了底层 ByteBuffer 中某个 int 的位置。注意此处是通过绝对地址访问的，也可以直接通过 ByteBuffer 以相同方式访问基本类型。

一旦通过视图缓冲区向底层的 ByteBuffer 填入了 int 或其他基本类型后，该 ByteBuffer 就可以直接写入通道中了。你可以同样轻松地从通道中读取数据，并使用视图缓冲区将所有内容都转换为指定的基本类型。下面这个示例通过在同一个 ByteBuffer 上建立不同的视图缓冲区，将同一个字节序列分别解析为 short、int、float、long 以及 double：

```java
// newio/ViewBuffers.java
import java.nio.*;

public class ViewBuffers {
  public static void main(String[] args) {
    ByteBuffer bb = ByteBuffer.wrap(
      new byte[]{ 0, 0, 0, 0, 0, 0, 0, 'a' });
    bb.rewind();
    System.out.print("Byte Buffer ");
    while(bb.hasRemaining())
      System.out.print(
        bb.position()+ " -> " + bb.get() + ", ");
    System.out.println();
    CharBuffer cb =
      ((ByteBuffer)bb.rewind()).asCharBuffer();
    System.out.print("Char Buffer ");
    while(cb.hasRemaining())
      System.out.print(
        cb.position() + " -> " + cb.get() + ", ");
    System.out.println();
    FloatBuffer fb =
      ((ByteBuffer)bb.rewind()).asFloatBuffer();
    System.out.print("Float Buffer ");
    while(fb.hasRemaining())
      System.out.print(
        fb.position()+ " -> " + fb.get() + ", ");
```

```
      System.out.println();
      IntBuffer ib =
        ((ByteBuffer)bb.rewind()).asIntBuffer();
      System.out.print("Int Buffer ");
      while(ib.hasRemaining())
        System.out.print(
          ib.position()+ " -> " + ib.get() + ", ");
      System.out.println();
      LongBuffer lb =
        ((ByteBuffer)bb.rewind()).asLongBuffer();
      System.out.print("Long Buffer ");
      while(lb.hasRemaining())
        System.out.print(
          lb.position()+ " -> " + lb.get() + ", ");
      System.out.println();
      ShortBuffer sb =
        ((ByteBuffer)bb.rewind()).asShortBuffer();
      System.out.print("Short Buffer ");
      while(sb.hasRemaining())
        System.out.print(
          sb.position()+ " -> " + sb.get() + ", ");
      System.out.println();
      DoubleBuffer db =
        ((ByteBuffer)bb.rewind()).asDoubleBuffer();
      System.out.print("Double Buffer ");
      while(db.hasRemaining())
        System.out.print(
          db.position()+ " -> " + db.get() + ", ");
  }
}
/* 输出:
Byte Buffer 0 -> 0, 1 -> 0, 2 -> 0, 3 -> 0, 4 -> 0, 5
-> 0, 6 -> 0, 7 -> 97,
Char Buffer 0 -> NUL, 1 -> NUL, 2 -> NUL, 3 -> a,
Float Buffer 0 -> 0.0, 1 -> 1.36E-43,
Int Buffer 0 -> 0, 1 -> 97,
Long Buffer 0 -> 97,
Short Buffer 0 -> 0, 1 -> 0, 2 -> 0, 3 -> 97,
Double Buffer 0 -> 4.8E-322,
*/
```

ByteBuffer 是由一个 8 字节的数组"包装"而成的，随后通过各种不同基本类型的视图缓冲区将该数组显示出来。从图 7-1 可以看出，当从不同类型的缓冲区中读取数据时，这些数据是如何以不同方式呈现的。

0	0	0	0	0	0	0	97	bytes
							a	chars
0		0		0		97		shorts
0				97				ints
0.0				1.36E-43				floats
97								longs
4.8E-322								doubles

图 7-1

图 7-1 与上面的程序输出是相对应的。

字节序

不同的机器可以使用不同的字节排序方式保存数据。"大端"（big endian，即高位优先）方式将最高位的字节放在最低的内存地址（即内存起始地址），而"小端"（little endian，即低位优先）方式将最高位的字节放在最高的内存地址（即内存末尾地址）。在存储大于一个字节的值（如 int、float 等）时，可能就需要考虑字节序问题了。ByteBuffer 以大端方式存储数据，数据在网络中传输时用的也都是大端序。可以通过向 order() 方法传入参数 ByteOrder.BIG_ENDIAN 或 ByteOrder.LITTLE_ENDIAN 来改变 ByteBuffer 中的字节序。

下面来看图 7-2 中包含两个字节的 ByteBuffer：

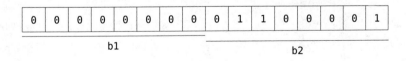

图 7-2

如果以 short 形式（ByteBuffer.asShortBuffer()）读取数据，会得到数字 97（二进制为 00000000 01100001）。改为小端方式后读取，还是这两个字节，却会得到数字 24832（二进制为 01100001 00000000）。

从下面这个示例可以看出，字节的顺序是由字节序的设置决定的：

```java
// newio/Endians.java
// 字节序的区别和数据存储
import java.nio.*;
import java.util.*;

public class Endians {
  public static void main(String[] args) {
    ByteBuffer bb = ByteBuffer.wrap(new byte[12]);
    bb.asCharBuffer().put("abcdef");
    System.out.println(Arrays.toString(bb.array()));
    bb.rewind();
    bb.order(ByteOrder.BIG_ENDIAN);
    bb.asCharBuffer().put("abcdef");
    System.out.println(Arrays.toString(bb.array()));
    bb.rewind();
    bb.order(ByteOrder.LITTLE_ENDIAN);
    bb.asCharBuffer().put("abcdef");
    System.out.println(Arrays.toString(bb.array()));
  }
}
/* 输出:
[0, 97, 0, 98, 0, 99, 0, 100, 0, 101, 0, 102]
[0, 97, 0, 98, 0, 99, 0, 100, 0, 101, 0, 102]
[97, 0, 98, 0, 99, 0, 100, 0, 101, 0, 102, 0]
*/
```

ByteBuffer 分配了一定内存空间,将 charArray 中的所有字节保存为一块扩展缓冲区,这样就可以调用 array() 方法来显示底层的字节了。array() 方法是"可选"(非必需)的,只能在基于数组的缓冲区上调用,否则便会抛出 UnsupportedOperationException 异常。

charArray 通过 CharBuffer 视图被插入 ByteBuffer 中。在底层字节显示出来后,可以看到默认的顺序和随后的大端序相同,而小端序则交换了字节的顺序。

7.3.5 用缓冲区操纵数据

图 7-3 表明了各种 nio 类之间的关系,并演示了如何移动和转换数据。例如,如果要将 byte 数组写入文件,就需要用 ByteBuffer.wrap() 方法将 byte 包装起来,并用 getChannel() 方法在 FileOutputStream 上打开一个通道,然后将 ByteBuffer 中的数据写入 FileChannel 中。

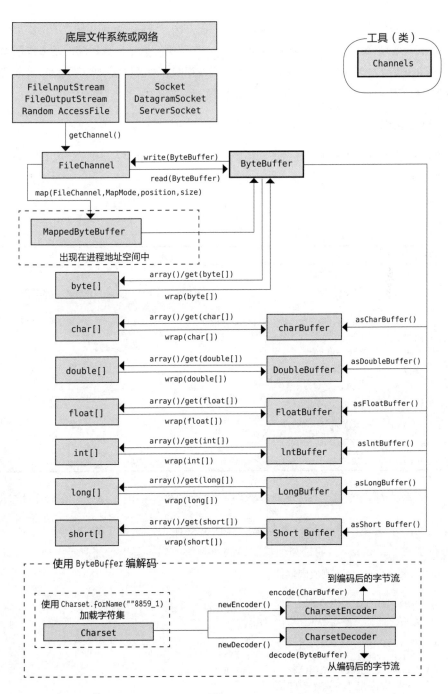

图 7-3

ByteBuffer 是唯一可将数据移入或移出通道的方法，我们只能创建独立的基本类型缓冲区，或者通过 "as" 方法从 ByteBuffer 生成一个该缓冲区。也就是说，无法将基本类型的缓冲区转换为 ByteBuffer。不过由于可以通过视图缓冲区将基本类型数据移入或移出 ByteBuffer，因此这实际上也就不构成什么限制了。

缓冲区的细节

Buffer 由数据和 4 个用于高效访问和操纵该数据的索引构成：mark（标记）、position（位置）、limit（界限）以及 capacity（容量）。表 7-8 列出了用于设置、重置及查询以上索引的方法：

表 7-8

方法	作用
capacity()	返回缓冲区的 capacity
clear()	清空缓冲区，将 position 设置为 0，将 limit 设置为 capacity。可以调用该方法来覆盖已有的缓冲区
flip()	将 limit 设置为 position，position 设置为 0。该方法用于让缓冲区在数据写入完成后，进入准备读取的就绪状态
limit()	返回 limit 的值
limit(int lim)	设置 limit 的值
mark()	将 mark 设置为 position
position()	返回 position 的值
position(int pos)	设置 position 的值
remaining()	返回 limit - position 后的值（即"剩余多少元素"）
hasRemaining()	如果在 position 和 limit 之间存在任何元素（即"还有剩余元素"），则返回 true

用于插入或提取数据的方法会同步更新表 7-8 中的这些索引，以反映出所发生的变化。

下面的示例使用了一种非常简单的算法（交换相邻字符）对 CharBuffer 中的字符进行加密（scramble，通过某种特殊算法对字符编码，可用于加密等用途）和解密（unscramble，scramble 的反向操作）：

```
// newio/UsingBuffers.java
import java.nio.*;

public class UsingBuffers {
  private static
  void symmetricScramble(CharBuffer buffer) {
    while(buffer.hasRemaining()) {
      buffer.mark();
      char c1 = buffer.get();
```

```
      char c2 = buffer.get();
      buffer.reset();
      buffer.put(c2).put(c1);
    }
  }
  public static void main(String[] args) {
    char[] data = "UsingBuffers".toCharArray();
    ByteBuffer bb =
      ByteBuffer.allocate(data.length * 2);
    CharBuffer cb = bb.asCharBuffer();
    cb.put(data);
    System.out.println(cb.rewind());
    symmetricScramble(cb);
    System.out.println(cb.rewind());
    symmetricScramble(cb);
    System.out.println(cb.rewind());
  }
}
/* 输出:
UsingBuffers
sUniBgfuefsr
UsingBuffers
*/
```

虽然可以通过调用 wrap() 方法并带上作为参数的 char 数组,由此直接生成 CharBuffer,但本例中并未这么做,而是分配了一个底层的 ByteBuffer,CharBuffer 只是作为视图在其上生成。由此凸显出我们的目标永远是操纵 ByteBuffer,因为它才是和通道交互的对象。

在进入 symmetricScramble() 方法时,缓冲区看起来如图 7-4 所示:

图 7-4

position(pos)指向缓冲区中的第一个元素,capacity(cap)和 limit(lim)则指向紧跟在最后一个元素之后的位置。

在 symmetricScramble() 中,while 循环会一直迭代执行,直到 position 等于 limit。缓冲区的 position 会随着相关的 get() 或 put() 方法的调用而指向不同的位置。也可以调用绝对(位置)的 get() 和 put() 方法,并传入索引参数,即 get() 或 put() 要作用的位置索引。不过这些方法并不会修改缓冲区中的 position 值。

在流程进入 while 循环后,便开始通过调用 mark() 设置 mark 的值。此时缓冲区的状态如图 7-5 所示:

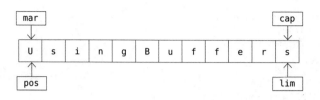

图 7-5

连续的两次 get() 调用将前两个字符的值分别保存到变量 c1 和 c2 中。在这之后,缓冲区看起来如图 7-6 所示:

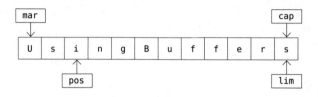

图 7-6

为了实现字符交换,我们在 position 为 0 处写入 c2,在 position 为 1 处写入 c1。可以用绝对 put() 方法实现这个目的,也可以通过 reset() 将 position 的值设置为 mark,如图 7-7 所示:

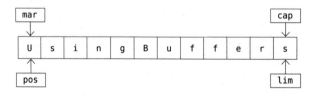

图 7-7

两次 put() 调用先后写入 c2 和 c1,如图 7-8 所示:

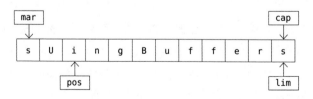

图 7-8

在循环的下一轮迭代中,将 mark 设置为 position 的当前值,如图 7-9 所示:

图 7-9

该过程一直持续到整个缓冲区遍历完成。在 while() 循环的最后，position 指向了缓冲区的末尾。如果将该缓冲区显示出来，则只会显示出位于 position 和 limit 之间的字符。因此，如果要显示出缓冲区的完整内容，就必须用 rewind() 方法将 position 设置为缓冲区的起始位置。

再次调用 symmetricScramble() 功能时，会对 CharBuffer 进行同样的处理，并将其恢复到初始状态。

7.3.6　内存映射文件

内存映射文件（Memory-mapped file）让你可以创建和修改那些因为太大而无法加载到内存中的文件。有了内存映射文件，便可以假装整个文件都已加载在内存中了，你可以将它当作一个非常大的数组来访问。这种方法极大地简化了实现文件修改所需的代码：

```java
// newio/LargeMappedFiles.java
// 使用（内存）映射创建超大文件
import java.nio.*;
import java.nio.channels.*;
import java.io.*;

public class LargeMappedFiles {
  static int length = 0x8000000; // 128MB
  public static void
  main(String[] args) throws Exception {
    try(
      RandomAccessFile tdat =
        new RandomAccessFile("test.dat", "rw")
    ) {
      MappedByteBuffer out = tdat.getChannel().map(
        FileChannel.MapMode.READ_WRITE, 0, length);
      for(int i = 0; i < length; i++)
        out.put((byte)'x');
      System.out.println("Finished writing");
      for(int i = length/2; i < length/2 + 6; i++)
        System.out.print((char)out.get(i));
    }
  }
}
/* 输出：
Finished writing
xxxxxx
*/
```

为了能同时读写，首先要用到 RandomAccessFile，先为该文件生成管道，然后调用 map() 生成 MappedByteBuffer，这是一种特殊类型的直接缓冲区。你需要指定文件的起始位置和映射区域的长度——这意味着你可以选择只映射大文件中的一块较小区域。

MappedByteBuffer 继承自 ByteBuffer，因此它拥有 ByteBuffer 的全部方法。此处只演示了最简单的 put() 和 get() 的用法，但也可以使用诸如 asCharBuffer() 等方法。

本示例创建的文件大小是 128MB，可能比你的操作系统所允许的单次可分配最大内存还要大。但该文件似乎可以一次性全部访问到，这是因为只有其中一部分被加载到了内存中，而其他部分则被交换了出去。通过这种方式，即使是非常巨大的文件（最大可到 2G），也可以轻松地进行修改。注意，底层操作系统的文件映射能力的用途就是将性能提升最大化。

性能

虽然"旧"流式 I/O 的性能在用 nio 重新实现后得到了改进，但是映射文件访问往往可以更大程度地提升速度。下面这段程序对性能做了简单的对比：

```
// newio/MappedIO.java
// {ExcludeFromGradle} Runs too long under WSL2
import java.util.*;
import java.nio.*;
import java.nio.channels.*;
import java.io.*;

public class MappedIO {
  private static int numOfInts = 4_000_000;
  private static int numOfUbuffInts = 100_000;
  private abstract static class Tester {
    private String name;
    Tester(String name) {
      this.name = name;
    }
    public void runTest() {
      System.out.print(name + ": ");
      long start = System.nanoTime();
      test();
      double duration = System.nanoTime() - start;
      System.out.format("%.3f%n", duration/1.0e9);
    }
    public abstract void test();
  }
  private static Tester[] tests = {
    new Tester("Stream Write") {
      @Override public void test() {
        try(
```

```
/* 输出：
Stream Write: 0.615
Mapped Write: 0.050
Stream Read: 0.577
Mapped Read: 0.015
Stream Read/Write: 4.069
Mapped Read/Write: 0.013
*/
```

```java
        DataOutputStream dos =
          new DataOutputStream(
            new BufferedOutputStream(
              new FileOutputStream(
                new File("temp.tmp"))))
    ) {
        for(int i = 0; i < numOfInts; i++)
          dos.writeInt(i);
      } catch(IOException e) {
        throw new RuntimeException(e);
      }
    }
  }
},
new Tester("Mapped Write") {
  @Override public void test() {
    try(
      FileChannel fc =
        new RandomAccessFile("temp.tmp", "rw")
          .getChannel()
    ) {
        IntBuffer ib =
          fc.map(FileChannel.MapMode.READ_WRITE,
            0, fc.size()).asIntBuffer();
        for(int i = 0; i < numOfInts; i++)
          ib.put(i);
      } catch(IOException e) {
        throw new RuntimeException(e);
      }
    }
  }
},
new Tester("Stream Read") {
  @Override public void test() {
    try(
      DataInputStream dis =
        new DataInputStream(
          new BufferedInputStream(
            new FileInputStream("temp.tmp")))
    ) {
        for(int i = 0; i < numOfInts; i++)
          dis.readInt();
      } catch(IOException e) {
        throw new RuntimeException(e);
      }
    }
  }
},
new Tester("Mapped Read") {
  @Override public void test() {
    try(
      FileChannel fc = new FileInputStream(
        new File("temp.tmp")).getChannel()
    ) {
        IntBuffer ib =
          fc.map(FileChannel.MapMode.READ_ONLY,
```

```
          0, fc.size()).asIntBuffer();
        while(ib.hasRemaining())
          ib.get();
      } catch(IOException e) {
        throw new RuntimeException(e);
      }
    }
  },
  new Tester("Stream Read/Write") {
    @Override public void test() {
      try(
        RandomAccessFile raf =
          new RandomAccessFile(
            new File("temp.tmp"), "rw")
      ) {
        raf.writeInt(1);
        for(int i = 0; i < numOfUbuffInts; i++) {
          raf.seek(raf.length() - 4);
          raf.writeInt(raf.readInt());
        }
      } catch(IOException e) {
        throw new RuntimeException(e);
      }
    }
  },
  new Tester("Mapped Read/Write") {
    @Override public void test() {
      try(
        FileChannel fc = new RandomAccessFile(
          new File("temp.tmp"), "rw").getChannel()
      ) {
        IntBuffer ib =
          fc.map(FileChannel.MapMode.READ_WRITE,
            0, fc.size()).asIntBuffer();
        ib.put(0);
        for(int i = 1; i < numOfUbuffInts; i++)
          ib.put(ib.get(i - 1));
      } catch(IOException e) {
        throw new RuntimeException(e);
      }
    }
  }
};
public static void main(String[] args) {
  Arrays.stream(tests).forEach(Tester::runTest);
}
}
```

Tester 使用了**模板方法**（Template Method）设计模式，为匿名内部子类中定义的 test() 的各种实现创建了一套测试框架。这些子类中的每一个都实现了一种测试，因此这些 test() 方法也算提供了一套实现各种 I/O 操作的原型参考。

虽然映射写操作看起来似乎应该使用 FileOutputStream，但是文件映射中的所有输出都必须使用 RandomAccessFile，正如前面的程序中的读写操作所做的一样。

注意，test() 方法的计时包括了各种 I/O 对象的初始化耗时，因此尽管映射文件的建立过程开销很大，但总的收益相较于流式 I/O 来说还是明显要大很多的。

7.3.7 文件加锁

对文件加锁会对文件的访问操作加上同步处理，这样文件才可以作为共享资源（而不会出现并发问题）。不过争用同一个文件的两个线程可能分别处在不同的 JVM 中，或者可能一个是 Java 线程，而另一个是操作系统中的某个本地线程。由于 Java 的文件加锁直接映射到了本地操作系统的加锁能力，因此文件锁对其他的操作系统进程也是可见的。

下面这段代码演示了基本的文件加锁：

```java
// newio/FileLocking.java
import java.nio.channels.*;
import java.util.concurrent.*;
import java.io.*;

public class FileLocking {
  public static void main(String[] args) {
    try(
      FileOutputStream fos =
        new FileOutputStream("file.txt");
      FileLock fl = fos.getChannel().tryLock()
    ) {
      if(fl != null) {
        System.out.println("Locked File");
        TimeUnit.MILLISECONDS.sleep(100);
        fl.release();
        System.out.println("Released Lock");
      }
    } catch(IOException | InterruptedException e) {
      throw new RuntimeException(e);
    }
  }
}
```

```
/* 输出:
Locked File
Released Lock
*/
```

通过在 FileChannel 上调用 tryLock() 或 lock()，便可以获得整个文件上的 FileLock（文件锁）。（SocketChannel、DatagramChannel 以及 ServerSocketChannel 并不需要上锁，因为它们天然就是单进程实体。你一般不会在两个进程间共享一个网络套接字）。

- tryLock() 是非阻塞操作。它会试图获取锁，但如果不成功（如果某个其他进程已经持有了同一个锁，并且该锁不可共享），便会直接从方法调用中返回。

- `lock()` 则会一直阻塞，直到成功获得锁，或者调用 `lock()` 的线程被中断，抑或是调用 `lock()` 的通道被关闭。可以通过 `FileLock.release()` 释放锁。

也可以用以下方法对文件的一部分上锁：

```
tryLock(long position, long size, boolean shared)
```

或者

```
lock(long position, long size, boolean shared)
```

这样会对 (size-position) 的区域上锁。第三个参数指定了锁是否可以共享。

虽然无参数的加锁方法会根据文件大小的变化自动调整，但是固定大小的锁不会随着文件大小的变化而变化。如果获得的是从 `position` 到 `position + size` 范围的锁，而文件增大到超出了 `position + size` 的范围，那么超出 `position + size` 的范围便不会被锁住。而无参数的加锁方法会锁住整个文件，即使文件后来增大了。

排他锁或共享锁必须由底层操作系统提供支持。如果操作系统不支持共享锁，却请求共享锁，那么就会使用排他锁来代替。可以通过 `FileLock.isShared()` 查询锁的类型（是共享锁还是排他锁）。

对部分映射文件加锁

文件映射一般用于非常大的文件。你可能会只想对这类文件中的某些部分加锁，这样其他的进程就可以修改未加锁的部分。比如，数据库就必须允许很多用户同时访问。

下面的示例中可以看到两个线程，每一个都对文件的不同部分加上了锁：

```java
// newio/LockingMappedFiles.java
// 对映射文件中的部分加锁
import java.nio.*;
import java.nio.channels.*;
import java.io.*;

public class LockingMappedFiles {
  static final int LENGTH = 0x8FFFFFF; // 128MB
  static FileChannel fc;
  public static void
  main(String[] args) throws Exception {
    fc = new RandomAccessFile("test.dat", "rw")
      .getChannel();
    MappedByteBuffer out = fc.map(
      FileChannel.MapMode.READ_WRITE, 0, LENGTH);
```

```
      for(int i = 0; i < LENGTH; i++)
        out.put((byte)'x');
      new LockAndModify(out, 0, 0 + LENGTH/3);
      new LockAndModify(
        out, LENGTH/2, LENGTH/2 + LENGTH/4);
    }
    private static class LockAndModify extends Thread {
      private ByteBuffer buff;
      private int start, end;
      LockAndModify(ByteBuffer mbb, int start, int end) {
        this.start = start;
        this.end = end;
        mbb.limit(end);
        mbb.position(start);
        buff = mbb.slice();
        start();
      }
      @Override public void run() {
        try {
          // 没有重叠部分的排他锁:
          FileLock fl = fc.lock(start, end, false);
          System.out.println(
            "Locked: "+ start +" to "+ end);
          // 执行修改:
          while(buff.position() < buff.limit() - 1)
            buff.put((byte)(buff.get() + 1));
          fl.release();
          System.out.println(
            "Released: " + start + " to " + end);
        } catch(IOException e) {
          throw new RuntimeException(e);
        }
      }
    }
  }
}
/* 输出：
Locked: 75497471 to 113246206
Locked: 0 to 50331647
Released: 75497471 to 113246206
Released: 0 to 50331647
*/
```

LockAndModify 线程类设置了缓冲区域，并创建了要修改的 slice()（片段），而 run() 中则获取到了文件通道上的锁（无法获取缓冲区上的锁，只能获得通道上的锁）。对 lock() 的调用和获取对象上的线程锁非常相似——这样便在文件上得到了一个具备排他访问权限的"临界区"。[①]

如果 JVM 退出，或获得锁的通道关闭，锁也会随即被自动释放，但也可以显式地在 FileLock 对象上调用 release()（释放锁），如以上示例所示。

---

① 本书第 6 章描述了关于多线程的更多细节。

08

设计模式

> 面向对象设计模式的演变历史参见《设计模式：可复用面向对象软件的基础》（以下简称《设计模式》）中的相关记载。[1]

《设计模式》这本书中演示了 23 种针对特定问题类型的不同解决方案。本章将通过各种示例来讲解设计模式的基本概念。这会激发你阅读《设计模式》的兴趣，这本书已经成为面向对象程序员的宝典之一。

本章最后引入了一个关于设计演进过程的例子，模拟了垃圾分类场景，从最开始的设计开始，逐步改进逻辑和流程，最终演进到更好的设计。可以将该演进过程看作某种演进原型——将满足某个特定问题的方案发展为可满足一类问题的灵活方案。

8.1 设计模式的概念

一开始，你可以将模式视为一种解决特定类型问题的兼

---

[1] 请注意，这本书中的示例是用 C++ 实现的。

具巧妙和洞察力的方法。模式看起来就像很多人已经解决了问题的所有细节部分，然后想出了最通用和灵活的方案。你可能以前遇到过这个问题，并且解决了，但你的方案很可能并不如模式中的那么完善。

它们虽然被称为"设计模式"，但实际上与设计领域并无关联。模式似乎和传统的分析、设计和实现的思维方式不同。模式在程序中体现了完整的思想，因此它有时会出现在分析或高层设计的阶段。由于模式在代码中有着直接的实现，你可能并不希望它出现在低层设计或实现阶段，甚至是维护阶段之前。通常，在进入这些阶段之前，你不会意识到你需要一个特定的模式。

模式的基本概念也可以看作程序设计的基本概念：增加一层抽象。无论何时，如果你要抽象某个事物，实际上就是在隔离某个特定的细节，而最有说服力的背后动机之一就是：
将会变化的事物和不会变化的事物分开。

还有一种说法，一旦你发现程序的某个部分可能会由于某种原因而发生变化，抽象就可以防止这些变化引发整个代码中的其他变化。

通常，要实现一个优雅且易维护的设计，其中最困难的部分是如何发现我所说的**变化的向量**（the vector of change，"向量"指的是最大的变化率，而不是某个集合类[①]）。这意味着要找到系统中最重要的会变化的事物。换言之，是要找到最大的（改动）成本。一旦找到了变化的向量，你就掌握了构建设计的关键点。

因此设计模式的目标就是隔离代码中的变化。如果能从这个角度看待设计模式，那就应该能意识到本书中已经出现过很多设计模式了。举例来说，继承可以看成一种设计模式（尽管是由编译器实现的）。它使你可以在具有相同接口的对象（保持不变的事物）中表现出行为的差异（变化的事物）。组合也可以被认为是一种模式，因为它使你可以动态或静态地改变实现类的对象，并由此改变对象的行为方式。

你也已经看到了另一种出现在《设计模式》中的模式：**迭代器**（Java1.0 和 1.1 任性地称其为"枚举"，Java 2 集合则称其为"迭代器"）。这会在遍历并逐个选择元素时隐藏集合的特定实现。迭代器使你可以在无须关心某个序列的构造方式的情况下，实现对该序列所有元素的某种操作。由此你的代码可以用于任何实现了迭代器的集合。

虽然设计模式很有用，但有些人断言：
设计模式代表着语言的失败之处。

---

① vector 既有"向量"的意思，也有"载体"的意思，此处解释了这个词并非指的后者。——译者注

这个看法很重要。比如说，某模式只是在 C++ 中很合理，但在 Java 或其他语言中却可能并无必要存在。出于这个原因，不能只是因为一个模式出现在了《设计模式》中，就认为它在你的语言中很有用。

我发现"语言的失败之处"这个看法很有用，但我也认为这过于简单化了。如果你在试图解决某个特定问题，而所用的语言又并未直接支持你所用的技术，那么你可以争辩说这是语言的一个失败之处。但是你的这种特定技术实际真的会经常用到吗？也许这种平衡才正是正确的：如果使用这种技术会使你耗费更多精力，那么你对该技术的需求程度可能并不足以证明提供语言级支持是合理的。另一方面，如果没有语言级支持，日常使用这种技术可能会太麻烦，但如果有了语言级支持，你可能会改变你的编程方式（例如 Java 8 的流就达到了这种效果）。

8.2 单例模式

最简单的设计模式可能就是**单例**（singleton）了，这是一种仅会提供唯一一个对象实例的方法。下面考虑一个参数化的 Resource（资源）：

```java
// patterns/Resource.java

public interface Resource<T> {
  T get();
  void set(T x);
}
```

假设我们想让每个不同的 Resource<T> 都仅能有一个实例。一种方法是为每个 T 都创建一个自定义的单例 Resource 类：

```java
// patterns/SingletonPattern.java

final class IntegerSingleton
  implements Resource<Integer> {
  private static IntegerSingleton value =
    new IntegerSingleton();
  private Integer i = Integer.valueOf(0);
  private IntegerSingleton() {
    System.out.println("IntegerSingleton()");
  }
  public static IntegerSingleton instance() {
    return value;
  }
  @Override public synchronized
  Integer get() { return i; }
  @Override public synchronized
```

```
/* 输出：
Inside main()
IntegerSingleton()
0
9
*/
```

```
    void set(Integer x) { i = x; }
  }
  public class SingletonPattern {
    public static <T> void show(Resource<T> r) {
      T val = r.get();
      System.out.println(val);
    }
    public static <T> void put(Resource<T> r, T val) {
      r.set(val);
    }
    public static void main(String[] args) {
      System.out.println("Inside main()");
      Resource<Integer> ir =
        IntegerSingleton.instance();
      Resource<Integer> ir2 =
        IntegerSingleton.instance();
      show(ir);
      put(ir2, Integer.valueOf(9));
      show(ir);
    }
  }
```

为了保证所控制的类型仅有一个对象被创建，我们将 IntegerSingleton 的构造器设为私有的，因此它只能在类中可用。

因为 value 对象是静态的，所以它在调用方程序员首次调用静态方法 instance() 时被创建，此时类被加载，并执行 value 的静态初始化。main() 中 ir2 的创建不会再次引发构造器调用。

JVM 的工作方式使得静态初始化是线程安全的。为了达到完全的线程安全，IntegerSingleton 的 getter 和 setter 都是 synchronized（同步）的。这很重要，因为多个线程可以分别持有指向同一个共享 IntegerSingleton 对象的引用。正如本书第 5 章中所描述的，即使我们自己并没有在并发程序中使用该类，也必须考虑并发问题。

Java 还允许通过克隆来创建对象（参见本书第 2 章）。本例将类修饰为 final 的，以防止克隆。 IntegerSingleton 并没有显式的基类，因此它直接继承自 Object。clone() 方法仍旧是 protected 的，因此无法调用（如果调用，则会引发编译器错误）。不过，如果是从某个以 public 权限重写了 clone() 方法，并实现了 Cloneable 的类层次结构中继承，那么阻止克隆就要重写 clone() 方法，并抛出本书第 2 章中描述过的 CloneNotSupportedException 异常（也可以重写 clone() 方法，并简单地返回 this，但这是自欺欺人，因为虽然调用方程序员认为他们是在克隆对象，但实际上处理的仍然是原本的对象）。

从 show() 和 put() 函数可以看出，我们可以向上转型为 Resource<T>，并使用多态。

也可以用这种方法来创建一个有限的对象池，尽管这样就必须管理池中的对象。如果这是个问题，那么我们的解决方案可以检查进出的共享对象。

继承如 Resource<T> 这样的基类可能看起来是不必要的额外工作。我们难道不应该仅仅创建一个实现单例概念的纯泛型类吗？下面是实现办法：

```java
// patterns/Single.java

@SuppressWarnings("unchecked")
public final class Single<T> {
  private static Object single;        // [1]
  public Single(T val) {
    if(single != null)
      throw new RuntimeException(
        "Attempt to reassign Single<" +
        val.getClass().getSimpleName() + ">"
      );
    single = val;
  }
  public T get() { return (T)single; }
}
```

在 [1] 处我们实际想声明的是：

```java
private static T single;
```

这样会导致编译器错误消息："非静态类型变量 T 不能从静态上下文中引用（non-static type variable T cannot be referenced from a static context）。"静态和泛型在 Java 中基本上是无法和谐共存的，因此需要将 Single 定义为 Object，然后在调用 get() 时对它进行转型。同样，不能将 get() 定义为静态方法（如果尝试这么做的话，则会得到很多关于静态泛型方法的教训）。

首次创建 Single 对象的时候，构造器将 static single 的值作为构造器参数。同一个 T 的第二次构造器调用会导致编译器错误消息。下面是个基本的测试：

```java
// patterns/TestSingle.java

public class TestSingle {
  public static void main(String[] args) {
    Single<String> ss = new Single<>("hello");
    System.out.println(ss.get());
    try {
      Single<String> ss2 = new Single<>("world");
    } catch(Exception e) {
```

```
      System.out.println(e.getMessage());
    }
  }
}
```

```
/* 输出：
hello
Attempt to reassign Single<String>
*/
```

用 Single 来创建一个 Double 单例对象，执行正常：

```
// patterns/SingletonPattern2.java

public class SingletonPattern2 {
  public static void main(String[] args) {
    Single<Double> pi =
      new Single<>(Double.valueOf(3.14159));
    Double x = pi.get();
    System.out.println(x);
  }
}
```

```
/* 输出：
3.14159
*/
```

我们无法修改 Double 的内容。但是如果所讨论的对象是可修改的，又会怎样呢？

```
// patterns/SingletonPattern3.java

class MyString {
  private String s;
  public MyString(String s) {
    this.s = s;
  }
  public synchronized
  void change(String s) {
    this.s = s;
  }
  @Override public synchronized
  String toString() {
    return s;
  }
}

public class SingletonPattern3 {
  public static void main(String[] args) {
    Single<MyString> x =
      new Single<>(new MyString("Hello"));
    System.out.println(x.get());
    x.get().change("World!");
    System.out.println(x.get());
  }
}
```

```
/* 输出：
Hello
World!
*/
```

如果 Single 管理着一个可修改的对象，在并发场景下便会导致竞态条件。在 MyString 中，change() 和 toString() 都是 synchronized 的，以防止发生竞态条件。

8.3 设计模式的分类

《设计模式》一书中讨论了 23 种不同的设计模式,并根据不同的目标将它们分为以下 3 类,每一类都围绕着某个可能发生变化的方面展开。

1. **创建类**:即创建对象的方式。这通常涉及隔离对象创建的细节,以使代码不依赖于对象的类型,这样在增加新对象类型时就不必做任何修改。单例模式可归为创建类模式,本章稍后你还会看到**工厂模式**的例子。

2. **结构类**:即如何设计满足特定项目约束的对象。这类设计主要围绕着这些对象和其他对象间的关联方式,以保证系统的变化不会导致这些关联方式的变化。

3. **行为类**:指处理程序中特定类型操作的对象。这些对象封装了要执行的流程,例如解释某种语言,满足某个请求,在序列中移动(比如通过迭代器),或者实现某种算法。本章列举了**观察者模式**(Observer)和**访问者模式**(Visitor)的相关示例。

《设计模式》中的 23 个模式各自都包含一节带有示例的内容,一般由 C++ 实现,但有时用的是 SmallTalk。本章不会把《设计模式》中的所有模式都复述一遍,因为该书自成一体,应该单独去学习(线上也有很多不错的资源,很多都是用 Java 讲解的)。相反,你会看到一些例子,它们会让你很好地领悟到模式究竟是什么,以及它们为什么很重要。

和设计模式打了多年交道之后,我开始觉得,模式本身使用的是基本的组织原则,而不是《设计模式》中描述的那些原则(而且比这些原则更基础)。这些原则基于实现的结构,从中我看到了模式间显著的相似性(比《设计模式》中展现出来的更多)。虽然我们通常会避免使用接口实现,但我发现根据这些结构原则来考虑模式,尤其是学习模式,往往会更容易。本章会试图基于模式自身的结构——而不是通过《设计模式》中的分类——来呈现模式。

8.4 模板方法

应用程序框架可以帮助我们,通过从提供的框架类中复用大部分代码,来创建新的应用程序,以及重写各种方法,以根据我们的需求自定义应用程序。

应用程序框架的重要概念之一是**模板方法模式**(Template Method),它通常是隐藏的,并通过调用各种基类方法来驱动该应用程序,我们可以重写这些基类方法来创建应用程序。

模板方法被定义在基类中,并且无法被改变。它有时是私有方法,但实际上通常是

final 的。它会调用其他的基类方法（我们会重写这些基类方法）来执行任务，但是通常只会作为初始化过程中的某个步骤被调用，调用方程序员不一定能够直接调用它。

```java
// patterns/TemplateMethod.java
// 基本的模板方法模式
import java.util.stream.*;

abstract class ApplicationFramework {
  ApplicationFramework() {
    templateMethod();
  }
  abstract void customize1(int n);
  abstract void customize2(int n);
  // private 意味着自动为 final 的:
  private void templateMethod() {
    IntStream.range(0, 5).forEach(
      n -> { customize1(n); customize2(n); });
  }
}

// 创建一个新应用程序:
class MyApp extends ApplicationFramework {
  @Override void customize1(int n) {
    System.out.print("customize1 " + n);
  }
  @Override void customize2(int n) {
    System.out.println(" customize2 " + n);
  }
}

public class TemplateMethod {
  public static void main(String[] args) {
    new MyApp();
  }
}
/* 输出:
customize1 0 customize2 0
customize1 1 customize2 1
customize1 2 customize2 2
customize1 3 customize2 3
customize1 4 customize2 4
*/
```

基类的构造器负责执行必要的初始化，然后启动"引擎"（即模板方法），运行应用程序（在 GUI 程序中，"引擎"指主要事件的循环）。调用方程序员只需简单地提供 customize1() 和 customize2() 的定义，"应用程序"就做好了运行的准备。

8.5　封装实现

设计模式中有种常见的结构，即引入一个代理[1]类，该代理类"封装"了实际执行工作的实现类。当在代理类中调用一个方法时，它会转而调用实现类中的某个方法。接下来

---

[1] surrogate，替代、代理的意思，和本章其他地方提到的 proxy（代理模式）不是同一个词，但意思相近。——译者注

会具体探讨 3 种使用了这种结构的模式：**代理模式**（Proxy）、**状态模式**（State）以及**状态机**（State Machine）。

代理类伴随着提供具体实现的一个或多个类，从基类中派生而来（见图 8-1）：

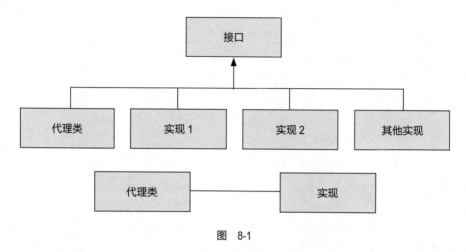

图 8-1

代理对象会连接到具体实现，它会在连接处转发所有的方法调用。

在结构上，**代理模式**只有一个实现，而**状态模式**有多个实现。（在《设计模式》中）一般认为这两个模式的应用应该是各不相同的：代理模式用来控制对自身实现的访问，而状态模式则可以让我们动态地改变实现。不过，如果将"对实现的访问控制"的概念延伸一下，那么这两者其实可以很好地结合在一起。

8.5.1 代理模式

如果根据图 8-1 来实现代理模式，看起来应该是这样：

```java
// patterns/ProxyDemo.java
// 代理模式的基本示例

interface ProxyBase {
  void f();
  void g();
  void h();
}

class Proxy implements ProxyBase {
  private ProxyBase implementation =
    new Implementation();
  // 将方法调用传递给实现：
  @Override
```

```
/* 输出：
Implementation.f()
Implementation.g()
Implementation.h()
*/
```

```
  public void f() { implementation.f(); }
  @Override
  public void g() { implementation.g(); }
  @Override
  public void h() { implementation.h(); }
}

class Implementation implements ProxyBase {
  @Override public void f() {
    System.out.println("Implementation.f()");
  }
  @Override public void g() {
    System.out.println("Implementation.g()");
  }
  @Override public void h() {
    System.out.println("Implementation.h()");
  }
}

public class ProxyDemo {
  public static void main(String[] args) {
    Proxy p = new Proxy();
    p.f();
    p.g();
    p.h();
  }
}
```

Implementation 并不需要和 Proxy 拥有相同的接口。只要 Proxy 能以某种方式"代言"实现类，就能符合基本的思路。不过，如果能有一个公共的接口，以强制 Implementation 必须实现 Proxy 需要调用的所有方法，会更方便。

8.5.2 状态模式

状态模式在代理模式的基础上增加了更多的实现，以及在代理类的生命周期内切换实现的方法。下面是一种状态模式的实现：

```
// patterns/StateDemo.java
// 基本的状态模式示例

class State {
  private State implementation;
  protected State() {}
  public State(State imp) {
    implementation = imp;
  }
  public void change(State newImp) {
    implementation = newImp;
  }
  // 将方法调用转发到实现类：
```

```
/* 输出：
Implementation1.f()
Implementation1.g()
Implementation1.h()
Changing implementation
Implementation2.f()
Implementation2.g()
Implementation2.h()
*/
```

```java
  public void f() { implementation.f(); }
  public void g() { implementation.g(); }
  public void h() { implementation.h(); }
}

class Implementation1 extends State {
  @Override public void f() {
    System.out.println("Implementation1.f()");
  }
  @Override public void g() {
    System.out.println("Implementation1.g()");
  }
  @Override public void h() {
    System.out.println("Implementation1.h()");
  }
}

class Implementation2 extends State {
  @Override public void f() {
    System.out.println("Implementation2.f()");
  }
  @Override public void g() {
    System.out.println("Implementation2.g()");
  }
  @Override public void h() {
    System.out.println("Implementation2.h()");
  }
}

public class StateDemo {
  static void test(State s) {
    s.f();
    s.g();
    s.h();
  }
  public static void main(String[] args) {
    State s = new State(new Implementation1());
    test(s);
    System.out.println("Changing implementation");
    s.change(new Implementation2());
    test(s);
  }
}
```

main() 方法一开始用的是第一个实现（类），然后切换到了第二种。Implementation1 和 Implementation2 都继承自 State，但这并不是必需的。只要你实现了一个对象——它可以被转发调用给成员对象，并且该成员对象可以被动态地替换——就可以认为你在使用状态模式。下面是第二个例子，其中 State 控制器对象和调用转发的目标对象并不是相同的类型：

```java
// patterns/CleanTheFloor.java
// State with different compositional interface.
//     由不同类型的接口组成的状态（模式）
class FloorCleaner {
  private Attachment attachment = new Vacuum();
  public void change(Attachment newAttachment) {
    attachment = newAttachment;
  }
  public void clean() { attachment.action(); }
}

interface Attachment {
  void action();
}

class Vacuum implements Attachment {
  @Override public void action() {
    System.out.println("Vacuuming");
  }
}

class Mop implements Attachment {
  @Override public void action() {
    System.out.println("Mopping");
  }
}

public class CleanTheFloor {
  public static void main(String[] args) {
    FloorCleaner floorCleaner = new FloorCleaner();
    floorCleaner.clean();
    floorCleaner.change(new Mop());
    floorCleaner.clean();
  }
}
/* 输出：
Vacuuming
Mopping
*/
```

代理模式和状态模式的不同之处在于要解决何种问题。《设计模式》中，代理模式常见的应用场景描述如下。

1. **远程代理**（remote proxy）。它用来代理处于不同地址空间的对象。例如，Java 远程方法调用（RMI）编译器 rmic 可以自动创建一个远程代理。

2. **虚拟代理**（virtual proxy）。提供了"延迟加载"的能力，可以按需创建开销较大的对象。如果永远不需要创建该对象，那么就无须花费相应的开销。可以等到需要的时候再初始化该对象，以减少启动耗时。

3. **保护代理**（protection proxy）。在你并不希望赋予调用方程序员访问被代理对象的完整权限时使用。

4. **智能引用**（smart reference）。在访问被代理的对象时执行额外的操作，例如：
   - 持续跟踪某个特定对象持有的引用数；
   - 实现**写时复制**（copy-on-write），以防止对象引用别名（减少系统开销和空间占用）；
   - 对指定方法的调用进行计数。

可以将 Java 的引用看作某种保护性代理，因为其控制了在堆上对实际对象的访问（并且保证了——比如说——你不会在 null 引用上调用方法）。

在《设计模式》中，代理模式和状态模式看起来彼此并无关联。正如我们所看到的，状态模式可以使用独立的实现层。类似地，代理模式不需要为了自身的实现而使用相同的基类，只需要代理（proxy）对象是其他对象的代理（surrogate）。抛开细节不论，在代理模式和状态模式中，都存在一个代理（surrogate）来将方法调用传递给某个实现对象。

8.5.3 状态机模式

这种模式在本书第 1 章中也出现过。

状态模式使得调用方程序员可以改变具体的实现，而状态机模式则通过一种强加的结构来自动改变具体实现。具体的实现表示系统的当前状态。在不同的状态下，系统有着不同的行为，这是因为它的内部包含状态模式。

将系统从一种状态改变为另一种状态的代码通常使用了模板方法，如下例所示：

```java
// patterns/state/StateMachineDemo.java
// 状态机模式
// {java patterns.state.StateMachineDemo}
package patterns.state;
import java.util.*;
import onjava.Nap;

interface State {
  void run();
}

abstract class StateMachine {
  protected State currentState;
  protected abstract boolean changeState();
  // 模板方法:
  protected final void runAll() {
    while(changeState())
      currentState.run();
  }
}

// 为每种状态实现一个不同的子类:
class Wash implements State {
```

```
/* 输出:
Washing
Spinning
Rinsing
Spinning
*/
```

```java
  @Override public void run() {
    System.out.println("Washing");
    new Nap(0.5);
  }
}

class Spin implements State {
  @Override public void run() {
    System.out.println("Spinning");
    new Nap(0.5);
  }
}

class Rinse implements State {
  @Override public void run() {
    System.out.println("Rinsing");
    new Nap(0.5);
  }
}

class Washer extends StateMachine {
  private int i = 0;
  private Iterator<State> states =
    Arrays.asList(
      new Wash(), new Spin(),
      new Rinse(), new Spin()
    ).iterator();
  Washer() { runAll(); }
  @Override public boolean changeState() {
    if(!states.hasNext())
      return false;
    // 将代理引用指向一个新的 State 对象：
    currentState = states.next();
    return true;
  }
}

public class StateMachineDemo {
  public static void main(String[] args) {
    new Washer();
  }
}
```

此处 StateMachine 控制着当前状态，并决定下一个状态是什么。不过状态对象自身可能也会决定下一个状态，一般是基于某种对系统的输入。这通常是一种更灵活的办法。

8.6 工厂模式：封装对象的创建

如果一个已有系统必须接受新的类型，那么明智的第一步便是使用多态性为这些新类型创建一个公共接口。这样可以在系统中将代码和所增加的具体类型隔离开（不必知道具体是什么类型）。添加新的类型可以不影响已有代码——至少看起来是这样。

它会让你感觉，代码中唯一要做的改动就是在继承出新类型的地方进行调整，但事实并非如此。你仍然要创建新类型的对象，并且在创建的时候必须指定具体要使用的构造器。如果这样的类型创建操作遍布于程序中，增加新类型便意味着需要找出所有要注意类型的地方。

工厂会强制通过某个公共节点来创建对象，以防止用于创建的代码在系统中四处扩散。系统中的所有代码都必须通过该工厂创建某个对象。这样在向系统增加新类的时候，只需要修改工厂即可。

每个面向对象的程序都会创建对象。由于有时你编写的程序会通过增加新类型的方式进行扩展，因此你可能会发现工厂很有用。

接下来会通过一系列 Shape（形状）层次结构来演示工厂模式。如果某个 Shape 无法成功创建，则需要使用合适的异常：

```java
// patterns/shapes/BadShapeCreation.java
package patterns.shapes;

public class BadShapeCreation
extends RuntimeException {
  public BadShapeCreation(String msg) {
    super(msg);
  }
}
```

然后是 Shape 的基类：

```java
// patterns/shapes/Shape.java
package patterns.shapes;

public class Shape {
  private static int counter = 0;
  private int id = counter++;
  @Override public String toString() {
    return
      getClass().getSimpleName() + "[" + id + "]";
  }
  public void draw() {
    System.out.println(this + " draw");
  }
  public void erase() {
    System.out.println(this + " erase");
  }
}
```

Shape 会自动为每个对象创建一个唯一的 id。toString() 方法则通过反射检测出具体的 Shape 子类名称。

现在可以创建一些 Shape 类了：

```java
// patterns/shapes/Circle.java
package patterns.shapes;

public class Circle extends Shape {}
```

```java
// patterns/shapes/Square.java
package patterns.shapes;

public class Square extends Shape {}
```

```java
// patterns/shapes/Triangle.java
package patterns.shapes;

public class Triangle extends Shape {}
```

工厂包含一个用于创建对象的方法：

```java
// patterns/shapes/FactoryMethod.java
package patterns.shapes;

public interface FactoryMethod {
  Shape create(String type);
}
```

create() 的参数用于指定要创建的 Shape 的具体类型。本例中的参数正好是 String，但也可以是任何一组数据。这个初始化数据可能来自系统外部。

下面是一个可复用的，对 FactoryMethod 实现进行的测试：

```java
// patterns/shapes/FactoryTest.java
package patterns.shapes;
import java.util.stream.*;

public class FactoryTest {
  public static void test(FactoryMethod factory) {
    Stream.of("Circle", "Square", "Triangle",
      "Square", "Circle", "Circle", "Triangle")
      .map(factory::create)
      .peek(Shape::draw)
      .peek(Shape::erase)
      .count(); // 终结操作
  }
}
```

记住，只有在最后加上终结操作，Stream 才会真的执行。此处丢弃了 count() 的值。

还有一种实现工厂的方法，即显式地创建每种类型：

```java
// patterns/ShapeFactory1.java
// 一个基本的静态工厂方法
import java.util.*;
import java.util.stream.*;
import patterns.shapes.*;

public class ShapeFactory1 implements FactoryMethod {
  @Override public Shape create(String type) {
    switch(type) {
      case "Circle": return new Circle();
      case "Square": return new Square();
      case "Triangle": return new Triangle();
      default:
        throw new BadShapeCreation(type);
    }
  }
  public static void main(String[] args) {
    FactoryTest.test(new ShapeFactory1());
  }
}
/* 输出:
Circle[0] draw
Circle[0] erase
Square[1] draw
Square[1] erase
Triangle[2] draw
Triangle[2] erase
Square[3] draw
Square[3] erase
Circle[4] draw
Circle[4] erase
Circle[5] draw
Circle[5] erase
Triangle[6] draw
Triangle[6] erase
*/
```

这样 create() 便成了在添加一种新的 Shape 类型时，系统中唯一需要修改的另一处代码了。

8.6.1 动态工厂模式

由于 ShapeFactory1.java 封装了对象的创建过程，因此它是一种合理的解决方案。不过，如果可以在增加新类时完全不需要修改任何代码，甚至是工厂本身，则会更好。ShapeFactory2.java 使用了反射（在基础卷第 19 章中介绍过），在首次需要时，将某种具体 Shape 的 Constructor 动态加载到 factories map 中：

```java
// patterns/ShapeFactory2.java
import java.util.*;
import java.lang.reflect.*;
import java.util.stream.*;
import patterns.shapes.*;

public class ShapeFactory2 implements FactoryMethod {
  private Map<String, Constructor> factories =
    new HashMap<>();
  private static Constructor load(String id) {
    System.out.println("loading " + id);
    try {
      return Class.forName("patterns.shapes." + id)
        .getConstructor();
    } catch(ClassNotFoundException |
            NoSuchMethodException e) {
      throw new BadShapeCreation(id);
    }
  }
  @Override public Shape create(String id) {
/* 输出:
loading Circle
Circle[0] draw
Circle[0] erase
loading Square
Square[1] draw
Square[1] erase
loading Triangle
Triangle[2] draw
Triangle[2] erase
Square[3] draw
Square[3] erase
Circle[4] draw
Circle[4] erase
Circle[5] draw
Circle[5] erase
Triangle[6] draw
Triangle[6] erase
*/
```

```
      try {
        return (Shape)factories
          .computeIfAbsent(id, ShapeFactory2::load)
          .newInstance();
      } catch(Exception e) {
        throw new BadShapeCreation(id);
      }
    }
    public static void main(String[] args) {
      FactoryTest.test(new ShapeFactory2());
    }
}
```

create() 方法基于传入的 String 参数生成各种新的 Shape，它会将该参数作为键，在 factories HashMap 中进行查询，返回值为 Constructor，然后通过调用其上的 newInstance() 创建新的 Shape 对象。

不过在刚开始运行该程序的时候，factories map 是空的。如果 Constructor 已存在于 Map 中，则 create() 会通过 Map 的 computeIfAbsent() 方法找到 Constructor。如果不在其中，则 computeIfAbsent() 会通过 load() 计算出 Constructor，并将其插入 Map。从输出可以看出，每个具体的 Shape 类型都只会在初次请求时进行加载，之后便可以直接从 Map 中获取了。

8.6.2 多态工厂模式

工厂方法模式可以从基类工厂中子类化出不同类型的工厂：

```
// patterns/ShapeFactory3.java
// 多态工厂方法
import java.util.*;
import java.util.function.*;
import java.util.stream.*;
import patterns.shapes.*;

interface PolymorphicFactory {
  Shape create();
}

class RandomShapes implements Supplier<Shape> {
  private final PolymorphicFactory[] factories;
  private Random rand = new Random(42);
  RandomShapes(PolymorphicFactory... factories) {
    this.factories = factories;
  }
  @Override public Shape get() {
    return
      factories[rand.nextInt(factories.length)]
```

```
/* 输出:
Triangle[0] draw
Triangle[0] erase
Circle[1] draw
Circle[1] erase
Circle[2] draw
Circle[2] erase
Triangle[3] draw
Triangle[3] erase
Circle[4] draw
Circle[4] erase
Square[5] draw
Square[5] erase
*/
```

```
      .create();
  }
}
public class ShapeFactory3 {
  public static void main(String[] args) {
    RandomShapes rs = new RandomShapes(      // [1]
      Circle::new, Square::new, Triangle::new
    );
    Stream.generate(rs)
      .limit(6)
      .peek(Shape::draw)
      .peek(Shape::erase)
      .count();
  }
}
```

由于 RandomShapes 是一个 Supplier<Shape>，因此它可以通过 Stream.generate() 生成 Stream。RandomShapes 的构造器接收 PolymorphicFactory 对象组成的可变参数列表。可变参数列表是以数组的形式传入的，因此这也是该列表在内部的存储形式。get() 方法随机索引到该数组中，并对结果调用 create()，以生成新的 Shape。

RandomShapes 构造器调用（[1]）是唯一需要在增加新 Shape 类型时修改的地方。

虽然 ShapeFactory2.java 可能引发异常，但是在这个方法中没有引发——这在编译期是确定的。

8.6.3　抽象工厂模式

抽象工厂模式看起来很像我们之前见过的工厂对象，只不过它带有多个工厂方法，每个工厂方法都会生成一种不同的对象。工厂类型也有多种。选定了一种工厂对象，即确定了该工厂所创建出的所有对象的具体类型。《设计模式》中给出的示例在多种图形用户界面（GUI）间实现了可移植性。只需创建适用于所用 GUI 的工厂对象即可。如果向该工厂对象请求菜单、按钮、滑块等组件，它便会生成适合该 GUI 版本的组件。这样便隔离了在不同 GUI 间切换的影响。

假设你要创建一套通用的游戏环境，用来支持不同类型的游戏。用抽象工厂来实现，便可能是下面这样：

```
// patterns/abstractfactory/GameEnvironment.java
// 抽象工厂模式的示例
// {java patterns.abstractfactory.GameEnvironment}
package patterns.abstractfactory;
import java.util.function.*;
```

```
interface Obstacle {
  void action();
}

interface Player {
  void interactWith(Obstacle o);
}

class Kitty implements Player {
  @Override
  public void interactWith(Obstacle ob) {
    System.out.print("Kitty has encountered a ");
    ob.action();
  }
}

class Fighter implements Player {
  @Override
  public void interactWith(Obstacle ob) {
    System.out.print("Fighter now battles a ");
    ob.action();
  }
}

class Puzzle implements Obstacle {
  @Override public void action() {
    System.out.println("Puzzle");
  }
}

class Weapon implements Obstacle {
  @Override public void action() {
    System.out.println("Weapon");
  }
}

// 抽象工厂:
class GameElementFactory {
  Supplier<Player> player;
  Supplier<Obstacle> obstacle;
}

// 具体工厂:
class KittiesAndPuzzles
extends GameElementFactory {
  KittiesAndPuzzles() {
    player = Kitty::new;
    obstacle = Puzzle::new;
  }
}

class Melee
```

/* 输出:
Kitty has encountered a Puzzle
Fighter now battles a Weapon
*/

```
  extends GameElementFactory {
    Melee() {
      player = Fighter::new;
      obstacle = Weapon::new;
    }
  }

public class GameEnvironment {
  private Player p;
  private Obstacle ob;
  public
  GameEnvironment(GameElementFactory factory) {
    p = factory.player.get();
    ob = factory.obstacle.get();
  }
  public void play() {
    p.interactWith(ob);
  }
  public static void main(String[] args) {
    GameElementFactory
      kp = new KittiesAndPuzzles(),
      ml = new Melee();
    GameEnvironment
      g1 = new GameEnvironment(kp),
      g2 = new GameEnvironment(ml);
    g1.play();
    g2.play();
  }
}
```

Player 和 Obstacle 相互作用。每个游戏都有着不同类型的 Player 和 Obstacle。通过选择具体的 GameElementFactory 来确定具体的游戏。GameEnvironment 则会设置并运行游戏。在本例中，游戏的设置和运行都非常简单，但这些行为（**初始条件**和**状态变化**）可以很大程度上决定游戏的结果。这里的 GameEnvironment 并不是为继承而设计的，虽然继承也是一个可选项。

本例中也用到了**双路分发**（Double Dispatching）模式，稍后会讲解。

8.7 函数对象模式

一个**函数对象**（Function Object）封装一个方法，其目的是将函数的选择和调用函数的位置解耦。

《设计模式》中对 "函数对象" 这个术语有所提及，但并未使用。不过函数对象的核心思想在该书的若干模式中反复出现过。

8.7.1 命令模式

这种模式是函数对象最纯粹的形式：一个身为对象的方法。将一个函数对象作为参数传递，它便会生成不同的行为。

命令（Command）模式在本书前面的若干地方描述过，主要包括以下几章：

- 基础卷第 11 章
- 基础卷第 19 章
- 进阶卷第 1 章

在下面这个示例中，show() 的代码保持不变，改变的是传给 show() 的 Comand 对象：

```java
// patterns/CommandPattern.java
import java.util.*;

class Command {
  public final String msg;
  public Command(String msg) {
    this.msg = msg;
  }
}

public class CommandPattern {
  public static void show(Command cmd) {
    System.out.println(cmd.msg);
  }
  public static void main(String[] args) {
    show(new Command("First Command"));
    show(new Command("Second Command"));
  }
}
/* 输出：
First Command
Second Command
*/
```

下面是另一个示例，创建了一个命令对象的列表，形成了一个"宏指令"。在 Java 8 之前的版本中，如果要实现独立函数的效果，就需要额外的繁文缛节，显式地将方法包装到一个对象中。而 Java 8 的 lambda 表达式可以创建函数对象，因此命令模式就变得相当简单了：

```java
// patterns/Macro.java
import java.util.*;

public class Macro {
  public static void main(String[] args) {
    List<Runnable> macro = new ArrayList<>(
      Arrays.asList(
        () -> System.out.print("Hello "),
        () -> System.out.println("World! ")
```

/* 输出：
Hello World!
Hello World!
I'm the command pattern!
*/

```
    ));
    macro.forEach(Runnable::run);
    macro.add(
      () -> System.out.print("I'm the command pattern!")
    );
    macro.forEach(Runnable::run);
  }
}
```

命令模式最主要的一点在于，可以让我们将想要的行为传给一个方法或对象。macro 会将一组要集体执行的行为排成一队，这样就可以动态地创建行为了，而这本来一般都需要编写新的代码来实现。

Macro.java 经过修改后，还可以用来解释脚本，参见**解释器**（Interpreter）模式。

《设计模式》中指出："命令模式是回调的面向对象形式的替代品。"然而我认为"回"字是回调概念的核心本质。也就是说，回调实际上会指回到回调的创建者，命令模式则一般只将命令对象创建出来，然后传给某个方法或对象，而不会随着时间的推移（往回）连接到命令对象上。不过这只是我的个人看法。稍后我会将一组设计模式结合起来，放到本章 8.10 节中。

8.7.2 策略模式

策略模式（Strategy）看起来就像命令模式的同胞兄弟，都是从同一个基类继承出来的一组函数对象。区别在于这些对象的使用方式。命令模式用于解决一类特定的问题——例如，通过传入一个描述所需文件的命令对象，从文件列表中选择一个文件。所调用的方法体即"保持不变的事物"，而变化的部分则被隔离在命令对象中。

策略模式包含一段可作为代理类的"上下文"，该代理类用于控制文件的选择以及对特定策略对象的使用——和状态模式很像。它看起来就像下面这样：

```
// patterns/strategy/StrategyPattern.java
// {java patterns.strategy.StrategyPattern}
package patterns.strategy;
import java.util.function.*;
import java.util.*;

// 公共策略基类:
class FindMinima {
  protected
  Function<List<Double>, List<Double>> algorithm;
}

// 各种策略，每一个都会生成无意义的数据:
class LeastSquares extends FindMinima {
  LeastSquares() {
```

```
    // Line is a sequence of points:
    // 直线是点的序列:
    algorithm = (line) -> Arrays.asList(1.1, 2.2);
  }
}

class Perturbation extends FindMinima {
  Perturbation() {
    algorithm = (line) -> Arrays.asList(3.3, 4.4);
  }
}

class Bisection extends FindMinima {
  Bisection() {
    algorithm = (line) -> Arrays.asList(5.5, 6.6);
  }
}

// "上下文"控制着策略:
class MinimaSolver {
  private FindMinima strategy;
  MinimaSolver(FindMinima strategy) {
    this.strategy = strategy;
  }
  List<Double> minima(List<Double> line) {
    return strategy.algorithm.apply(line);
  }
  void changeAlgorithm(FindMinima newAlgorithm) {
    strategy = newAlgorithm;
  }
}

public class StrategyPattern {
  public static void main(String[] args) {
    MinimaSolver solver =
      new MinimaSolver(new LeastSquares());
    List<Double> line = Arrays.asList(
      1.0, 2.0, 1.0, 2.0, -1.0,
      3.0, 4.0, 5.0, 4.0 );
    System.out.println(solver.minima(line));
    solver.changeAlgorithm(new Perturbation());
    System.out.println(solver.minima(line));
    solver.changeAlgorithm(new Bisection());
    System.out.println(solver.minima(line));
  }
}
```

```
/* 输出:
[1.1, 2.2]
[3.3, 4.4]
[5.5, 6.6]
*/
```

MinimaSolver 中的 changeAlgorithm() 将一个不同的策略插入 strategy 字段, 然后 minima() 就会使用不同的算法了。

可以通过将上下文合并到 FindMinima 中来简化该方案:

```java
// patterns/strategy/StrategyPattern2.java
// {java patterns.strategy.StrategyPattern2}
package patterns.strategy;
import java.util.function.*;
import java.util.*;

// "上下文"现在合并了:
class FindMinima2 {
  private
  Function<List<Double>, List<Double>> algorithm;
  FindMinima2() { leastSquares(); } // 默认
  // 各种策略:
  void leastSquares() {
    algorithm = (line) -> Arrays.asList(1.1, 2.2);
  }
  void perturbation() {
    algorithm = (line) -> Arrays.asList(3.3, 4.4);
  }
  void bisection() {
    algorithm = (line) -> Arrays.asList(5.5, 6.6);
  }
  List<Double> minima(List<Double> line) {
    return algorithm.apply(line);
  }
}

public class StrategyPattern2 {
  public static void main(String[] args) {
    FindMinima2 solver = new FindMinima2();
    List<Double> line = Arrays.asList(
      1.0, 2.0, 1.0, 2.0, -1.0,
      3.0, 4.0, 5.0, 4.0 );
    System.out.println(solver.minima(line));
    solver.perturbation();
    System.out.println(solver.minima(line));
    solver.bisection();
    System.out.println(solver.minima(line));
  }
}
/* 输出:
[1.1, 2.2]
[3.3, 4.4]
[5.5, 6.6]
*/
```

FindMinima2 封装了各种不同的算法,现在还同时包含了上下文,因此可以在单个类中控制算法的选择了。

8.7.3 职责链模式

这种模式在本书第 1 章中介绍过。

职责链模式(Chain of Responsibility)大概可以被看作策略对象实现的动态泛化版本的递归。先产生一个调用,然后一系列策略逐个尝试处理该调用。当某个策略处理成功,或者试过所有策略后(都不成功),整个过程结束。在递归过程中,一个方法不断重复调用自身，直到满足某个终结条件。而在职责链模式中，一个方法会调用相同基类的方法的不同实现，后者又会调用该基类方法的另一个实现，以此类推，直到满足终结条件。

不同于通过调用单个方法来处理请求，职责链中的多个方法都有机会处理该请求，因此职责链模式很适用于实现专家系统。职责链是高效的链表结构，因此可以动态地进行创建或修改。也可以将它看作一种更通用的、动态构建的 switch 语句。

你可能想让 StrategyPattern.java 自动尝试不同的算法，直到命中合适的那个。职责链模式提供了实现该功能的途径。

Result 是包含一个 success 标签的信使，这样收件人就可以知道该算法是否成功，line 则用来承载实际的数据。如果某个算法失败，则返回 Result.fail：

```java
// patterns/chain/Result.java
// 承载结果或表示失败
package patterns.chain;
import java.util.*;

public class Result {
  public final boolean success;
  public final List<Double> line;
  public Result(List<Double> data) {
    success = true;
    line = data;
  }
  private Result() {
    success = false;
    line = Collections.<Double>emptyList();
  }
  public static final Result fail = new Result();
}
```

由于失败是预期结果，因此在某个策略失败时，返回 Result.fail 比抛出异常更加合适。

```java
// patterns/chain/ChainOfResponsibility.java
// {java patterns.chain.ChainOfResponsibility}
package patterns.chain;
import java.util.*;
import java.util.function.*;

interface Algorithm {
  Result algorithm(List<Double> line);
}

class FindMinima {
  public static Result test(
    boolean success, String id, double d1, double d2) {
```

```
/* 输出：
LeastSquares
Perturbation
Bisection
[5.5, 6.6]
*/
```

```java
      System.out.println(id);
      if(success) // 实际的测试/计算:
        return new Result(Arrays.asList(d1, d2));
      else // 尝试职责链中的下一个:
        return Result.fail;
  }
  public static Result leastSquares(List<Double> line) {
    return test(false, "LeastSquares", 1.1, 2.2);
  }
  public static Result perturbation(List<Double> line) {
    return test(false, "Perturbation", 3.3, 4.4);
  }
  public static Result bisection(List<Double> line) {
    return test(true, "Bisection", 5.5, 6.6);
  }
  static List<Function<List<Double>, Result>>
    algorithms = Arrays.asList(
      FindMinima::leastSquares,
      FindMinima::perturbation,
      FindMinima::bisection
    );
  public static Result minima(List<Double> line) {
    for(Function<List<Double>, Result> alg :
        algorithms) {
      Result result = alg.apply(line);
      if(result.success)
        return result;
    }
    return Result.fail;
  }
}

public class ChainOfResponsibility {
  public static void main(String[] args) {
    FindMinima solver = new FindMinima();
    List<Double> line = Arrays.asList(
      1.0, 2.0, 1.0, 2.0, -1.0,
      3.0, 4.0, 5.0, 4.0);
    Result result = solver.minima(line);
    if(result.success)
      System.out.println(result.line);
    else
      System.out.println("No algorithm found");
  }
}
```

每个 Algorithm 的实现针对 algorithm() 方法都有不同的方法。在 FindMinima 中，创建了一组 algorithm 的 List（这便是职责链的那条"链"），而 minima() 方法会遍历该 List，并找出执行成功的算法。

8.8 改变接口

有时你要解决的问题可能就像"我没有想要的接口"这么简单。有两种设计模式可以解决这个问题：**适配器模式**（Adapter）接收一种类型为参数，并生成面向另一种类型的接口；**外观模式**（Facade）可以创建一个面向一组类的接口，这只是提供了一种更为简便的方式，来处理一个库或一批资源。

8.8.1 适配器模式

当你有了 A，而且想要 B 的时候，适配器模式便可以解决你的问题。这里唯一的需求就是生成 B，有多种方法可以完成该适配工作：

```java
// patterns/adapt/Adapter.java
// 多种适配器模式
// {java patterns.adapt.Adapter}
package patterns.adapt;

class WhatIHave {
  public void g() {}
  public void h() {}
}

interface WhatIWant {
  void f();
}

class ProxyAdapter implements WhatIWant {
  WhatIHave whatIHave;
  ProxyAdapter(WhatIHave wih) {
    whatIHave = wih;
  }
  @Override public void f() {
    // 用 WhatIHave 中的方法实现行为:
    whatIHave.g();
    whatIHave.h();
  }
}

class WhatIUse {
  public void op(WhatIWant wiw) {
    wiw.f();
  }
}

// 方法 2: 构造传入 op() 中使用的适配器
class WhatIUse2 extends WhatIUse {
  public void op(WhatIHave wih) {
    new ProxyAdapter(wih).f();
  }
```

```
  }

  // 方法 3：在 WhatIHave 内部构造适配器
  class WhatIHave2 extends WhatIHave
  implements WhatIWant {
    @Override public void f() {
      g();
      h();
    }
  }

  // 方法 4：使用内部类
  class WhatIHave3 extends WhatIHave {
    private class InnerAdapter implements WhatIWant{
      @Override public void f() {
        g();
        h();
      }
    }
    public WhatIWant whatIWant() {
      return new InnerAdapter();
    }
  }

  public class Adapter {
    public static void main(String[] args) {
      WhatIUse whatIUse = new WhatIUse();
      WhatIHave whatIHave = new WhatIHave();
      WhatIWant adapt= new ProxyAdapter(whatIHave);
      whatIUse.op(adapt);
      // 方法 2：
      WhatIUse2 whatIUse2 = new WhatIUse2();
      whatIUse2.op(whatIHave);
      // 方法 3：
      WhatIHave2 whatIHave2 = new WhatIHave2();
      whatIUse.op(whatIHave2);
      // 方法 4：
      WhatIHave3 whatIHave3 = new WhatIHave3();
      whatIUse.op(whatIHave3.whatIWant());
    }
  }
```

在 ProxyAdapter 的应用上，我对术语"代理"的使用没有那么严格。《设计模式》一书的作者主张代理必须具有和其代理的对象完全相同的接口。将代理和适配器这两个词合并为代理适配器（ProxyAdapter）来使用也许更合理。

8.8.2 外观模式

这是我在将需求转化为对象设计的开始阶段所遵循的原则：

如果有一段逻辑很丑陋，那么就把它藏到对象里。

这基本上就是外观模式所完成的任务。如果你有一堆相当杂乱的类和交互逻辑，并且不需要让调用方程序员看到，就可以将这些藏到一个接口后面，以一种更易懂的形式（只）暴露出外部需要知道的信息。

外观模式通常被实现为单例抽象工厂。可以使用包含静态工厂方法的类来实现这个效果：

```java
// patterns/Facade.java

class A { A(int x) {} }
class B { B(long x) {} }
class C { C(double x) {} }

// 其他不需要由外观暴露出去的类都放到这里

public class Facade {
  static A makeA(int x) { return new A(x); }
  static B makeB(long x) { return new B(x); }
  static C makeC(double x) { return new C(x); }
  public static void main(String[] args) {
    // 调用方程序员通过调用静态方法得到对象：
    A a = Facade.makeA(1);
    B b = Facade.makeB(1);
    C c = Facade.makeC(1.0);
  }
}
```

《设计模式》中给出的示例并不是真正的外观模式，只是一个用到了其他类的类而已。

可作为外观模式变体的 Java 包

对我来说，外观模式给人一种相当"面向过程"（而非面向对象）的感觉：我们只是在调用一些生成对象的函数而已。而这实际上和抽象工厂又有多大区别呢？外观模式的重点在于将一组类（以及它们之间的交互）形成的库部分隐藏起来，不让调用方程序员看到，使得这一组类对外暴露出的接口更好用、更易懂。

而这正是 Java 的包特性所实现的功能。在库以外只能创建和使用 public 的类，而所有非 public 的类都只能在包范围内访问。这就好像外观模式是 Java 的内建特性。

公平地讲，《设计模式》面向的主要是使用 C++ 的群体。虽然 C++ 有命名空间来防止全局冲突和类名冲突，但这并不能提供由 Java 中的非 public 类带来的类隐藏机制。大部分时候，Java 的包似乎就可以解决外观模式所要解决的问题了。

8.9 解释器模式：运行时的灵活性

在一些应用程序中，用户需要在运行时拥有更高的灵活性，例如创建用来描述系统所需行为的脚本。**解释器**设计模式可以用来创建语言解释器，并将其嵌入这样的程序中。

在构建应用程序的过程中，开发自定义语言以及为其构建解释器是一件费时费力的事。最好的方案是复用代码：嵌入一个已构建好并且调试过的解释器。Python 语言可以免费地嵌入到营利性质的应用程序中，而无须任何许可协议、版税或其他产权负担。另外，Python 还有一个称为 Jython 的版本（已经过时了，但仍然能用），它是彻底的 Java 字节码，因此将它加入 Java 程序是相对比较简单的。Python 是一种脚本语言，易于学习，编写和阅读的逻辑性很强，支持函数和对象，拥有大量可用的库，并且几乎可以在所有的平台上运行。你可以通过 Python 网站进一步学习。

8.10 回调

回调可以将代码和行为解耦。这其中包含**观察者模式**（Observer），以及一类称为**多路分发**（Multiple Dispatching）的回调，其中包含《设计模式》中的**访问者模式**（Visitor）。

8.10.1 观察者模式

观察者模式用于解决一个相当常见的问题：如果一组对象必须在某些其他对象改变状态时更新自身，该怎么做？这个场景可以在 SmallTalk 的"模型 – 视图 – 控制器"（model-view-controller, MVC）[1] 设计中的"模型 – 视图"部分，或者在几乎完全相同的"文档 - 视图架构"（Document-View Architecture，可视为 MVC 的一种变体）中看到。假设你有一些数据（即"文档"）以及多个视图，比如一个绘图视图和一个文本视图。修改数据的时候，这两个视图都必须更新自身，而这时观察者模式就能派上用场。这是个比较常见的问题，因此相应的解决方案早就是标准 `java.util` 库的一部分了。

观察者模式使我们可以将调用的来源和被调用的代码以完全动态的方式解耦。正如同其他的回调形式，观察者模式包含一个钩子（hook）点，我们可以在此处改变代码。不同之处在于观察者模式那完全动态的本质。它常被用于处理"由其他对象的状态改变所触发的变更"的特殊场景，但它也是事件管理的基础。

观察者模式中实际上有两种"会改变的事物"：观察对象的数量和更新发生的方式。

---

[1] 一种经典的三层架构设计，常见于 Web 服务中。——译者注

观察者模式使我们可以在不影响周围代码的情况下修改这两者。

Observer 是一个只有一个方法（update(Observable o, Object arg)）的接口。被观察的对象（即 Observable）认为是时候更新所有的 Observer 时，便会调用该方法。第一个参数是导致更新发生的 Observable 对象，第二个参数提供了 Observer 更新自身时可能需要的任何额外信息。

一个 Observer 可以注册多个 Observable。

Observable 是决定何时及如何进行更新的"被观察的对象"。Observable 类会跟踪每个希望在变更发生时得到通知的 Observer。当发生任何应该让每个 Observer 更新自身的变更时，Observable 会调用自身的 notifyObservers() 方法，该方法会调用每个 Observer 上的 update() 方法。

Observable 中有个标志，用于表示它是否被改变。该标志允许我们一直等待，直到我们认为时机成熟，才会通知 Observer。标志状态的控制是 protected 的，因此只有继承者才能决定怎样才算是变更，而不是由此得出的 Observer 子类的终端用户。

大部分的工作是在 notifyObservers() 中完成的。如果 changed 标志没有设置，则调用 notifyObservers() 不会引发任何动作。否则该方法会首先清空 changed 标志，这样对 notifyObservers() 的反复调用就不会浪费时间。该操作会在通知观察者之前完成，以防对 update() 的调用引发对这个 Observable 对象的变更。然后它会调用每个 Observer 上的 update() 方法。

一开始，似乎可以用一个普通的 Observable 对象来管理更新。但是这样做行不通，要达到效果，就必须继承 Observable，并在子类代码中的某处调用 setChanged()。该方法用于设置"changed"标志，意味着在调用 notifyObservers() 的时候，所有的观察着都会被通知到。应该在何处调用 setChanged() 取决于我们的程序逻辑。

注意，Observer 和 Observable（Java 1.0 后已成为语言中的一部分）已在 Java 9 中被弃用，取而代之的是 Java 9 所引入的 Flow 类。

8.10.2 示例：观察花朵

这是一个观察者模式的示例：

```
// patterns/observer/ObservedFlower.java
// 观察者模式的示例
// {java patterns.observer.ObservedFlower}
```

```
package patterns.observer;
import java.util.*;
import java.util.function.*;

@SuppressWarnings("deprecation")
class Flower {
  private boolean isOpen = false;
  private boolean alreadyOpen = false;
  private boolean alreadyClosed = false;
  Observable opening = new Observable() {
    @Override public void notifyObservers() {
      if(isOpen && !alreadyOpen) {
        setChanged();
        super.notifyObservers();
        alreadyOpen = true;
      }
    }
  };
  Observable closing = new Observable() {
    @Override public void notifyObservers() {
      if(!isOpen && !alreadyClosed) {
        setChanged();
        super.notifyObservers();
        alreadyClosed = true;
      }
    }
  };
  public void open() { // 花瓣打开（开花）
    isOpen = true;
    opening.notifyObservers();
    alreadyClosed = false;
  }
  public void close() { // 花瓣闭合（合拢）
    isOpen = false;
    closing.notifyObservers();
    alreadyOpen = false;
  }
}

@SuppressWarnings("deprecation")
class Bee {
  private String id;
  Bee(String name)  { id = name; }
  // 观察开花：
  public final Observer whenOpened = (ob, a) ->
    System.out.println(
      "Bee " + id + "'s breakfast time!");
  // 观察合拢：
  public final Observer whenClosed = (ob, a) ->
    System.out.println(
      "Bee " + id + "'s bed time!");
}
```

```
/* 输出：
Bee B's breakfast time!
Bee A's breakfast time!
Hummingbird A's breakfast time!
---------------
Bee B's bed time!
Hummingbird B's bed time!
Hummingbird A's bed time!
+++++++++++++++
===============
###############
Bee B's bed time!
Hummingbird B's bed time!
Hummingbird A's bed time!
*/
```

```java
@SuppressWarnings("deprecation")
class Hummingbird {
  private String id;
  Hummingbird(String name) { id = name; }
  public final Observer whenOpened = (ob, a) ->
    System.out.println("Hummingbird " +
      id + "'s breakfast time!");
  public final Observer whenClosed = (ob, a) ->
    System.out.println("Hummingbird " +
      id + "'s bed time!");
}

public class ObservedFlower {
  public static void main(String[] args) {
    Flower f = new Flower();
    Bee
      ba = new Bee("A"),
      bb = new Bee("B");
    Hummingbird
      ha = new Hummingbird("A"),
      hb = new Hummingbird("B");
    f.opening.addObserver(ha.whenOpened);
    f.opening.addObserver(hb.whenOpened);
    f.opening.addObserver(ba.whenOpened);
    f.opening.addObserver(bb.whenOpened);
    f.closing.addObserver(ha.whenClosed);
    f.closing.addObserver(hb.whenClosed);
    f.closing.addObserver(ba.whenClosed);
    f.closing.addObserver(bb.whenClosed);
    // 蜂鸟B决定睡觉
    // 移除whenOpened会停止开花的更新操作
    f.opening.deleteObserver(hb.whenOpened);
    // 一个会引起观察者注意的变更：
    f.open();
    f.open(); // 没有效果：花瓣已经打开
    System.out.println("---------------");
    // Bee A doesn't want to go to bed.
    // Removing whenClosed stops close updates.
    // 蜜蜂A不想睡觉
    // 移除whenClosed会停止合拢的更新操作
    f.closing.deleteObserver(ba.whenClosed);
    f.close();
    System.out.println("+++++++++++++++");
    f.close(); // 没有效果：花瓣已经闭合
    System.out.println("===============");
    f.opening.deleteObservers();
    f.open(); // 没有观察者会更新
    System.out.println("###############");
    f.close(); // 合拢的观察者仍然在这里
  }
}
```

Flower（花）可以 open()（开花）或 close()（闭拢）。通过匿名内部类，这两个事件成为可被分别独立观察的现象。opening 和 closing 都继承自 Observable，因此它们可以访问 setChanged()。

Observer 是基础卷第 13 章中介绍过的函数式接口（也称为 SAM 类型），因此 Bee 和 Hummingbird 中的 whenOpened 和 whenClosed 是使用 lambda 表达式定义的。whenOpened 和 whenClosed 各自独立观察着 Flower 的开花和闭拢。

在 main() 中可以看到观察者模式一个主要的好处：通过 Observable 动态地注册/取消注册 Observer 来获得在运行时改变行为的能力。

注意，我们可以创建完全不同的其他观察对象，Observer 唯一和 Flower 有关联的地方就是 Observer 接口。

8.10.3 一个可视化的观察者示例

下面这个示例通过使用 Swing 库创建各种图形来达到效果。有一些方框被放置在屏幕上的网格中，每个方框都被初始化为随机颜色。另外，每个方框都 implements（实现）了 Observer 接口，并注册了一个 Observable 对象。当你单击一个方框时，所有其他的方框都会得到通知，这是因为 Observable 对象会自动调用每个 Observer 对象的 update() 方法。在该方法内部，方框会检查其是否和被单击的方框相邻，如果相邻，则会改变自身的颜色，以匹配被单击的方框。

java.awt.event 库中有一个带有多个方法的 MouseListener 类，但是我们只对 mouseClicked() 方法感兴趣。如果只是想实现 mouseClicked()，则无法通过编写 lambda 表达式实现，这是因为 MouseListener 有多个方法，所以它并不是函数式接口。Java 9 允许我们通过 default 关键字简化代码，以创建一个辅助接口，从而解决这个问题：

```
// onjava/MouseClick.java
// Helper interface to allow lambda expressions.
// 用于支持 lambda 表达式的辅助接口
package onjava;
import java.awt.event.*;

// Default everything except mouseClicked():
// 将一切都设为 default，除了 mouseClicked() :
public interface MouseClick extends MouseListener {
    @Override
    default void mouseEntered(MouseEvent e) {}
    @Override
    default void mouseExited(MouseEvent e) {}
```

```
    @Override
    default void mousePressed(MouseEvent e) {}
    @Override
    default void mouseReleased(MouseEvent e) {}
}
```

现在可以成功地将 lambda 表达式转型为 MouseClick,并将其传给 addMouseListener()。

```
// patterns/BoxObserver.java
// 用 Java 内建的 Observer 类演示观察者模式
// {ExcludeFromGradle} // 不适用于WSL2
import javax.swing.*;
import java.awt.*;
import java.awt.event.*;
import java.util.*;
import onjava.*;
import onjava.MouseClick;

@SuppressWarnings("deprecation")
class Boxes extends JFrame {
  Observable notifier = new Observable() {
    @Override
    public void notifyObservers(Object b) {
      setChanged();
      super.notifyObservers(b);
    }
  };
  public Boxes(int grid) {
    setTitle("Demonstrates Observer pattern");
    Container cp = getContentPane();
    cp.setLayout(new GridLayout(grid, grid));
    for(int x = 0; x < grid; x++)
      for(int y = 0; y < grid; y++)
        cp.add(new Box(x, y, notifier));
    setSize(500, 400);
    setVisible(true);
    setDefaultCloseOperation(DISPOSE_ON_CLOSE);
  }
}

@SuppressWarnings("deprecation")
class Box extends JPanel implements Observer {
  int x, y; // 网格中的位置
  Color color = COLORS[
    (int)(Math.random() * COLORS.length)
  ];
  static final Color[] COLORS = {
    Color.black, Color.blue, Color.cyan,
    Color.darkGray, Color.gray, Color.green,
    Color.lightGray, Color.magenta,
    Color.orange, Color.pink, Color.red,
    Color.white, Color.yellow
```

```java
    };
    Box(int x, int y, Observable notifier) {
      this.x = x;
      this.y = y;
      notifier.addObserver(this);
      addMouseListener((MouseClick)
        e -> notifier.notifyObservers(Box.this));
    }
    @Override public void paintComponent(Graphics g) {
      super.paintComponent(g);
      g.setColor(color);
      Dimension s = getSize();
      g.fillRect(0, 0, s.width, s.height);
    }
    @Override
    public void update(Observable o, Object arg) {
      Box clicked = (Box)arg;
      if(nextTo(clicked)) {
        color = clicked.color;
        repaint();
      }
    }
    private boolean nextTo(Box b) {
      return Math.abs(x - b.x) <= 1 &&
             Math.abs(y - b.y) <= 1;
    }
  }

public class BoxObserver {
  public static void main(String[] args) {
    int grid = 8;
    if(args.length > 0)
      grid = Integer.parseInt(args[0]);
    new Boxes(grid);
  }
}
```

当你第一眼看到 Observable 的在线文档时，会觉得有点困惑，因为似乎可以用一个普通的 Observable 对象来管理更新操作。但这样做是行不通的，如果想实现这种效果，就必须继承自 Observable，然后在子类代码中的某个地方调用 setChanged()，如你在 Boxes.notifier 中所见。setChanged() 设置了"changed"标志，这意味着在调用 notifyObservers() 的时候，所有的观察者都会得到通知。本例中的 setChanged() 是在 notifyObservers() 中调用的，但你可以使用任何标准来决定何时调用 setChanged()。

Boxes 包含一个名为 notifier 的 Observable 对象，并且每创建出一个 Box，都会将该 Box 绑定到 notifier 上。在 Box 中，不论你何时单击鼠标，notifyObservers() 都会被调用，并将被单击的对象作为参数传递，因此所有（在自身的 update() 方法中）收到消息的方框都能知道谁被单击了，并能决定是否要改变自身的颜色。通过将 notifyObservers() 和 update() 中的代码组合使用，就能实现某些相当复杂的方案。

8.11 多路分发

在处理多个相互作用的类型时，程序很容易变得格外混乱。举个例子，考虑一个解析并执行数学表达式的系统。我们想声明 Number + Number、Number * Number 等运算，其中 Number 是一系列数值对象的基类。但如果我们想声明 a + b，又不知道 a 和 b 的具体类型，那怎样才能让它们正确地交互呢？

答案来自你很可能没有想到的地方：Java 只能实现单路分发。也就是说，在操作多个类型未知的对象时，Java 只能在其中一个类型上应用动态绑定机制。这并不能解决使用多个向上转型对象的问题，因此我们最终需要手动检测某些类型，并有效地生成自定义的动态绑定行为。

这个问题的解决方案称为**多路分发**（通常只会有两个分发，这称为**双路分发**）。因为多态性只能通过方法调用来生效，所以如果想实现双路分发，就必须有两次方法调用：第一次用于确定第一个未知类型，第二次用于确定第二个未知类型。

通常来说，需要创建一个生成多次动态方法调用的方法，从而确定流程中的多个类型。

本书第 1 章中的 RoShamBo1.java 示例介绍了"猜拳"游戏。我们会用该框架实现一个该游戏的变体，但用的是独立的 ItemFactory 和 Compete.match() 函数。compete() 和 eval() 方法都是同一个 Item 类型的成员（如果有两个相互作用的不同层次结构，则每个层次结构上都需要一次多态方法调用）。

```
// patterns/PaperScissorsRock.java
// 多路分发的示例
import java.util.*;
import java.util.function.*;
import java.util.stream.*;
import enums.Outcome;
import static enums.Outcome.*;
import enums.Item;
import enums.Paper;
import enums.Scissors;
import enums.Rock;
import onjava.*;
import static onjava.Tuple.*;
```

```java
class ItemFactory {
  static List<Supplier<Item>> items =
    Arrays.asList(
      Scissors::new, Paper::new, Rock::new);
  static final int SZ = items.size();
  private static SplittableRandom rand =
    new SplittableRandom(47);
  public static Item newItem() {
    return items.get(rand.nextInt(SZ)).get();
  }
  public static Tuple2<Item,Item> newPair() {
    return tuple(newItem(), newItem());
  }
}

class Compete {
  public static Outcome match(Tuple2<Item,Item> p) {
    System.out.print(p.a1 + " -> " + p.a2 + " : ");
    return p.a1.compete(p.a2);
  }
}

public class PaperScissorsRock {
  public static void main(String[] args) {
    Stream.generate(ItemFactory::newPair)
      .limit(20)
      .map(Compete::match)
      .forEach(System.out::println);
  }
}
/* 输出:
Scissors -> Rock : LOSE
Scissors -> Paper : WIN
Rock -> Paper : LOSE
Rock -> Rock : DRAW
Rock -> Paper : LOSE
Paper -> Scissors : LOSE
Rock -> Paper : LOSE
Scissors -> Scissors : DRAW
Scissors -> Rock : LOSE
Scissors -> Paper : WIN
Scissors -> Rock : LOSE
Paper -> Scissors : LOSE
Rock -> Rock : DRAW
Scissors -> Scissors : DRAW
Paper -> Paper : DRAW
Scissors -> Paper : WIN
Scissors -> Rock : LOSE
Scissors -> Paper : WIN
Rock -> Paper : LOSE
Rock -> Scissors : WIN
*/
```

Item 接口中包含双路分发的结构：compete() 实现了第一路分发，第二路分发则发生在 eval() 的调用过程中。

假设有两个 Item——a 和 b，并且我们不知道这两者的类型，那么在调用 a.compete(b) 的时候过程如下。compete() 方法是自动分发的，因此会在 a 类型的 compete() 方法体内被唤醒。如果 a 是 Paper 类型，那就是在 Paper 的 compete() 方法体中被唤醒，这样就通过第一路分发确定了第一个未知对象的类型。但现在 compete() 转而调用了第二个未知类型 b 的 eval() 方法，同时将 a 作为参数传入，因此会调用类型 b 上的重载后的 eval() 方法——这便是第二路分发。此时程序正运行至 eval() 方法中，而该方法已经知道了两个对象的类型。

访问者模式：多路分发中的一种类型

访问者模式的使用场景是这样的：存在一个不可改变的主类层次结构——该层次结构也许来自某个软件供应商，而你无法对其做任何修改。不过你希望为该层次结构加上某种多态操作，这通常意味着必须在基类接口中增加一个方法。其中的两难之处在于，必须为基类增加方法，但又无法直接修改基类。那么该如何解决这个问题呢？

访问者模式（《设计模式》中的最后一个设计模式）可以基于之前介绍的双路分发机制解决这个问题。

访问者模式使你可以通过创建一套独立的 Visitor 类型的类层次结构来扩展主类型的接口，由此模拟在主类型上执行的操作。主类型的对象只需"接受"访问者，然后调用访问者的动态绑定方法。

```java
// patterns/visitor/BeeAndFlowers.java
// 访问者模式的示例
// {java patterns.visitor.BeeAndFlowers}
package patterns.visitor;
import java.util.*;
import java.util.function.*;
import java.util.stream.*;

interface Visitor {
  void visit(Gladiolus g);
  void visit(Ranunculus r);
  void visit(Chrysanthemum c);
}

// Flower 的层次结构无法被更改:
interface Flower {
  void accept(Visitor v);
}

class Gladiolus implements Flower {
  @Override
  public void accept(Visitor v) { v.visit(this);}
}

class Ranunculus implements Flower {
  @Override
  public void accept(Visitor v) { v.visit(this);}
}

class Chrysanthemum implements Flower {
  @Override
  public void accept(Visitor v) { v.visit(this);}
}

// 增加生成字符串的能力:
class StringVal implements Visitor {
  private String s;
  @Override public String toString() { return s; }
  @Override public void visit(Gladiolus g) {
    s = "Gladiolus";
  }
  @Override public void visit(Ranunculus r) {
```

```
    s = "Ranunculus";
  }
  @Override public void visit(Chrysanthemum c) {
    s = "Chrysanthemum";
  }
}

// 增加执行"Bee"（蜜蜂）的各种行为的能力：
class Bee implements Visitor {
  @Override public void visit(Gladiolus g) {
    System.out.println("Bee and Gladiolus");
  }
  @Override public void visit(Ranunculus r) {
    System.out.println("Bee and Ranunculus");
  }
  @Override public void visit(Chrysanthemum c) {
    System.out.println("Bee and Chrysanthemum");
  }
}

class FlowerFactory {
  static List<Supplier<Flower>> flowers =
    Arrays.asList(Gladiolus::new,
      Ranunculus::new, Chrysanthemum::new);
  static final int SZ = flowers.size();
  private static SplittableRandom rand =
    new SplittableRandom(47);
  public static Flower newFlower() {
    return flowers.get(rand.nextInt(SZ)).get();
  }
}

public class BeeAndFlowers {
  public static void main(String[] args) {
    List<Flower> flowers =
      Stream.generate(FlowerFactory::newFlower)
        .limit(10)
        .collect(Collectors.toList());
    StringVal sval = new StringVal();
    flowers.forEach(f -> {
      f.accept(sval);
      System.out.println(sval);
    });
    // 在所有的 Flower 上执行"Bee"操作：
    Bee bee = new Bee();
    flowers.forEach(f -> f.accept(bee));
  }
}
/* 输出：
Gladiolus
Chrysanthemum
Gladiolus
Ranunculus
Chrysanthemum
Ranunculus
Chrysanthemum
Chrysanthemum
Chrysanthemum
Ranunculus
Bee and Gladiolus
Bee and Chrysanthemum
Bee and Gladiolus
Bee and Ranunculus
Bee and Chrysanthemum
Bee and Ranunculus
Bee and Chrysanthemum
Bee and Chrysanthemum
Bee and Chrysanthemum
Bee and Ranunculus
*/
```

注意本示例和之前示例的相似性：Flower 接收 Visitor 参数，作为第一路分发，然后转而调用 visit()（并将自身作为参数，以最终进入与 Flower 的类型对应的重载方法），作为第二路分发。

main()中基本上相当于为Flower添加了一系列方法，生成了String类型的描述，以及执行了Bee的各种行为。这便是Visitor的意义所在：为不可改变的层次结构增加方法。

8.12 模式重构

本章的剩余部分将探讨如何以一种逐渐演进的方式来应用设计模式，从而解决问题。首先会选择一种设计用于实现最初的方案，然后对该方案进行验证，随后会尝试更多的设计模式来解决问题——有些有效，有些行不通。寻找解决方案的过程中，最关键的问题永远是"哪些部分是会改变的"。

这个过程和Martin Fowler在其著作《重构：改善既有代码的设计》中的论述非常相似，虽然他更倾向于讨论代码级的优化（相较于模式级别的设计）。一开始我们会选择一种解决方案，然后渐渐发现该方案无法满足需求的持续变化，便要做出相应的改进。这是一种自然的趋势，但是在计算机编程中，很难用过程式的程序来完成。接受"我们可以重构代码和设计"这种理念，是迈向系统改进的第一步。

我们用来重构的示例是一个垃圾收集系统。垃圾以未分类的状态到达垃圾收集厂，随后我们建立了垃圾分类和评估的模型。在最开始的方案中，反射（参见基础卷第19章）会接收匿名的垃圾分块，并检测出它们的类型以进行分类。

8.12.1 Trash和它的子类

Trash（垃圾）的基类含有weight和price()等信息，并带有一个通过反射来生成Trash子类精确名称的toString()方法。其中还包含了accept()方法，稍后会用它实现访问者模式，不过在此之前，你可以先暂时忽略它。

```java
// patterns/trash/Trash.java
// 垃圾收集示例的基类
package patterns.trash;

public abstract class Trash {
  public final double weight;
  public Trash(double weight) {
    this.weight = weight;
  }
  public abstract double price();
  @Override public String toString() {
    return String.format(
      "%s weight: %.2f * price: %.2f = %.2f",
      getClass().getSimpleName(),
      weight, price(), weight * price());
```

```
    }
    // 暂时可以忽略，稍后会用到它：
    public abstract void accept(Visitor v);
}
```

price() 是一个会返回材料当前价格的 abstract 方法。价格并不会随着每块不同的 Trash 而变化——比如 Aluminum（铝）的价格就是固定的（由材质决定）。将所有的价格都放在一个地方非常方便，这样改变价格就很容易了。通过 interface，每个字段都被自动分配了 public、static 和 final 权限。

```
// patterns/trash/Price.java
package patterns.trash;

public interface Price {
  double
    ALUMINUM = 1.67,
    PAPER = 0.10,
    GLASS = 0.23,
    CARDBOARD = 0.11;
}
```

每种不同类型的 Trash 的 price() 方法都会返回 Price 中适当的字段：

```
// patterns/trash/Aluminum.java
package patterns.trash;

public class Aluminum extends Trash {
  public Aluminum(double wt) { super(wt); }
  @Override public double price() {
    return Price.ALUMINUM;
  }
  // 暂时可以忽略，稍后会用到它：
  @Override public void accept(Visitor v) {
    v.visit(this);
  }
}
```

```
// patterns/trash/Paper.java
package patterns.trash;

public class Paper extends Trash {
  public Paper(double wt) { super(wt); }
  @Override public double price() {
    return Price.PAPER;
  }
  // 暂时可以忽略，稍后会用到它：
  @Override public void accept(Visitor v) {
    v.visit(this);
  }
}
```

```
// patterns/trash/Glass.java
package patterns.trash;

public class Glass extends Trash {
  public Glass(double wt) { super(wt); }
  @Override public double price() {
    return Price.GLASS;
  }
  // 暂时可以忽略，稍后会用到它：
  @Override public void accept(Visitor v) {
    v.visit(this);
  }
}
```

```
// patterns/trash/Cardboard.java
package patterns.trash;

public class Cardboard extends Trash {
  public Cardboard(double wt) { super(wt); }
  @Override public double price() {
    return Price.CARDBOARD;
  }
  // 暂时可以忽略，稍后会用到它：
  @Override public void accept(Visitor v) {
    v.visit(this);
  }
}
```

TrashValue 是一个带有 static 函数 sum() 的工具类。这个类接收一个由 Trash 组成的 List，并显示其中的每块 Trash，最后显示出该 List 中所有 Trash 的总价值。

```
// patterns/trash/TrashValue.java
// 累加一个垃圾箱中所有垃圾的价值
package patterns.trash;
import java.util.*;

public class TrashValue {
  private static double total;
  public static void
  sum(List<? extends Trash> bin, String type) {
    total = 0.0;
    bin.forEach( t -> {
      System.out.println(t);
      total += t.weight * t.price();
    });
    System.out.printf(
      "Total %s value = %.2f%n", type, total);
  }
}
```

看起来并无必要将 total 定义在 sum() 外部，因为永远不会在 sum() 之外的地方用到它。不过如果试图将其定义为 sum() 内部的本地变量，便会报出错误消息："lambda 表达式引用的本地变量必须定义为 final，或具有 final 的效果（local variables referenced from a lambda expression must be final or effectively final）。"

一旦 List<Trash> 装满了 Trash 对象，Bins（垃圾箱）构造器便会通过 instanceof 将该 List 分类到其类型化的垃圾箱中：

```java
// patterns/trash/Bins.java
package patterns.trash;
import java.util.*;

public class Bins {
  final List<Trash> bin;
  final List<Aluminum> aluminum = new ArrayList<>();
  final List<Paper> paper = new ArrayList<>();
  final List<Glass> glass = new ArrayList<>();
  final List<Cardboard> cardboard = new ArrayList<>();
  public Bins(List<Trash> source) {
    bin = new ArrayList<>(source); // 复制
    bin.forEach( t -> {
      // 通过反射发现 Trash 的类型：
      if(t instanceof Aluminum)
        aluminum.add((Aluminum)t);
      if(t instanceof Paper)
        paper.add((Paper)t);
      if(t instanceof Glass)
        glass.add((Glass)t);
      if(t instanceof Cardboard)
        cardboard.add((Cardboard)t);
    });
  }
  public void show() {
    TrashValue.sum(aluminum, "Aluminum");
    TrashValue.sum(paper, "Paper");
    TrashValue.sum(glass, "Glass");
    TrashValue.sum(cardboard, "Cardboard");
    TrashValue.sum(bin, "Trash");
  }
}
```

增加一种新类型的垃圾意味着在 Bins 中增加一个新的 List 和 instanceof，以及 show() 中的另一行。

现在我们已经准备好创建一个简单工厂了，该工厂持有一个名为 constructors 的 List，其中包含了用于创建新 Trash 对象的构造器。

```java
// patterns/recyclea/RecycleA.java
// 用反射实现的垃圾收集
// {java patterns.recyclea.RecycleA}
package patterns.recyclea;
```

```java
import java.util.*;
import java.util.function.*;
import java.util.stream.*;
import patterns.trash.*;

class SimpleFactory {
  static final
  List<Function<Double, Trash>> constructors =
    Arrays.asList(
      Aluminum::new, Paper::new, Glass::new);
  static final int SIZE = constructors.size();
  private static SplittableRandom rand =
    new SplittableRandom(42);
  public static Trash random() {
    return constructors
      .get(rand.nextInt(SIZE))
      .apply(rand.nextDouble());
  }
}

public class RecycleA {
  public static void main(String[] args) {
    List<Trash> bin =
      Stream.generate(SimpleFactory::random)
        .limit(10)
        .collect(Collectors.toList());
    Bins bins = new Bins(bin);
    bins.show();
  }
}
/* 输出：
Aluminum weight: 0.34 * price: 1.67 = 0.57
Aluminum weight: 0.62 * price: 1.67 = 1.03
Aluminum weight: 0.49 * price: 1.67 = 0.82
Aluminum weight: 0.50 * price: 1.67 = 0.83
Total Aluminum value = 3.26
Paper weight: 0.69 * price: 0.10 = 0.07
Total Paper value = 0.07
Glass weight: 0.16 * price: 0.23 = 0.04
Glass weight: 0.87 * price: 0.23 = 0.20
Glass weight: 0.80 * price: 0.23 = 0.18
Glass weight: 0.52 * price: 0.23 = 0.12
Glass weight: 0.20 * price: 0.23 = 0.05
Total Glass value = 0.59
Total Cardboard value = 0.00
Glass weight: 0.16 * price: 0.23 = 0.04
Aluminum weight: 0.34 * price: 1.67 = 0.57
Glass weight: 0.87 * price: 0.23 = 0.20
Glass weight: 0.80 * price: 0.23 = 0.18
Aluminum weight: 0.62 * price: 1.67 = 1.03
Aluminum weight: 0.49 * price: 1.67 = 0.82
Glass weight: 0.52 * price: 0.23 = 0.12
Glass weight: 0.20 * price: 0.23 = 0.05
Aluminum weight: 0.50 * price: 1.67 = 0.83
Paper weight: 0.69 * price: 0.10 = 0.07
Total Trash value = 3.91
*/
```

注意 SimpleFactory 不会生成 Cardboard，因为 constructors 中并没有 Cardboard::new。

constructors 类型并不是一个 List<Constructor> 类型。相反，该 List 持有的是 Function<Double, Trash>。这说明 Function 对象接收 Double 类型的参数并返回 Trash 对象——这正是 Trash 的单参数构造器的行为。然而这些单参数构造器接收的是 double，而不是包装类 Double，因此 Java 会进行自动装箱。

random() 为随机的 Trash 类型取出 Function，通过 apply() 方法调用 Function，并传入随机生成的 double 参数。

在 main() 中，Stream.generate() 的参数并不是 Supplier<T>，后者是 generate() 的具体参数类型。即使生成 Trash 的函数名是 random() 而不是 get()，也是可以运行的。

这个程序满足了设计需求：能够运行。如果这是一次性解决问题的长久方案，便不会有问题。但是，一个有用的程序往往会随着时间的推移而发展，因此你需要问自己："万一情况有变化呢？"比如，塑料是一种有价值的可回收商品，那么应该怎样将其集成到系统中呢（特别是在程序代码很庞杂的情况下）？虽然 SimpleFactory 确实封装了创建过程，但程序的其余部分散落着若干类型检查的代码，因此每次增加新类型时，都必须找到这些代码。如果遗漏了某一处，则编译器并不会产生任何有帮助的错误消息。

如果**每种类型都经过了验证**，你就会知道自己是在误用反射。如果只是因为需要特殊对待类型的某个子集而要找出该子集，那么这样做大概没什么问题。但是如果要找出 switch 语句中的所有类型，那么可能会有办法来改进设计，使其具有更好的可维护性。本章剩余部分会通过多个阶段逐步演化该程序，使其变得更加灵活。这种方式应该可以为如何设计程序提供有一个有价值的案例。

8.12.2　信使对象

面向对象的设计有一个原则（我最初是从 Grady Booch 那里听到的）："如果觉得设计太复杂，那就生成更多对象。"这种说法既违反直觉，又简单得可笑，但我发现它很有用（"生成更多对象"通常等同于"再增加一层抽象"）。总的来说，如果发现有些地方代码很乱，就要考虑用哪种类可以清理代码。通常清理代码带来的副作用是使系统更灵活并且结构更好。

SimpleFactory 是个合理的首选方案，但是如果派生的 Trash 构造器需要不同的或更多的参数呢？"生成更多对象"可以解决该问题。为了隐藏用于创建的数据，TrashInfo 包含了工厂为创建合适的 Trash 对象的所有必要信息：

```java
// patterns/trash/TrashInfo.java
// 携带 Trash 创建时的数据的信使类
package patterns.trash;

public class TrashInfo {
  public final String type;
  public final double data;
  public TrashInfo(String type, double data) {
    this.type = type;
    this.data = data;
  }
  @Override public String toString() {
    return "TrashInfo(" + type + ", " + data + ")";
  }
}
```

TrashInfo 对象的唯一职责就是持有并传递信息，因此它被称为**信使对象**（Messenger Object）或者**数据传输对象**（Data Transfer Object, DTO）。信使一般是不可变的，因此这两个字段都是 final 并且 public 的。

如果某些事物发生了变化，导致工厂需要不同的或更多的信息来创建新的 Trash 对象，则此时无须改变工厂的参数。可以通过增加新的数据和构造器，或者通过子类化来直接修改 TrashInfo。

8.12.3　使工厂通用化

如果有类型被加入 RecycleA.java 中，就必须为每个新类型修改 SimpleFactory。在 DynaFactory 中，我们会通过反射（如 ShapeFactory2.java 所示）来为该类型的 Trash 生成一个 Constructor。现在再向系统中增加新类型时，工厂就无须变动了：

```java
// patterns/trash/DynaFactory.java
// Trash 类型的动态发现
package patterns.trash;
import java.util.*;
import java.util.function.*;
import java.lang.reflect.*;

public class DynaFactory {
  private Map<String, Constructor> constructors =
    new HashMap<>();
  private String packageName;
  public DynaFactory(String packageName) {
    this.packageName = packageName;
  }
  @SuppressWarnings("unchecked")
  public
  <T extends Trash> T create(TrashInfo info) {
```

```
      try {
        String typename =
          "patterns." + packageName + "." + info.type;
        return (T)constructors.computeIfAbsent(
          typename, this::findConstructor
        ).newInstance(info.data);
      } catch(Exception e) {
        throw new RuntimeException(
          "Cannot create() Trash: " + info, e);
      }
    }
    private
    Constructor findConstructor(String typename) {
      try {
        System.out.println("Loading " + typename);
        return Class.forName(typename)
          .getConstructor(double.class);
      } catch(Exception e) {
        throw new RuntimeException(
          "Trash(double) Constructor Not Found: " +
          typename, e);
      }
    }
  }
```

constructors 会将一个 String（Trash 子类的名称）映射到它对应的 Constructor。具体的 Trash 子类名称来自 TrashInfo type 属性，并用来构造 typename。typename 用于在 constructors 中查找 Constructor。如果不存在 typename 的实体，computeIfAbsent() 就会调用 findConstructor() 来生成与 typename 对应的 Constructor，后者会被添加到 constructors 中。最后，newInstance(info.data) 会调用 Constructor，向其传入重量（weight）。

可以通过很多种方法来决定要使用哪个对象——这里使用了 info.type 的方法，因此可以将从某个文件中读取的文本信息转化为对象。

DynaFactory 会假定所有的 Trash 类型都有一个以 double（注意 double.class 和 Double.class 是不同的）为参数的构造器。还可能有另一种更灵活的方案，即调用 getConstructors()，该方法会返回包含所有可能的构造器的数组。

你会看到，这种设计的美妙之处在于代码是无须变动的，不论该方案是用于哪种场景的。

8.12.4 从文件解析 Trash

接下来不再随机生成 Trash 对象，而是通过从文本文件中读取信息的方式来创建它们。该文件包含了每块垃圾需要的所有信息，以 Trash:weight 的格式保存：

```
// patterns/trash/Trash.dat
Cardboard:4.4
Paper:8.0
Aluminum:1.8
Glass:5.4
Aluminum:3.4
Cardboard:2.2
Glass:4.3
Cardboard:1.2
Paper:6.6
Aluminum:2.7
Paper:9.1
Glass:3.6
```

ParseTrash.fillBin() 会打开 Trash.dat，并针对每一行创建一块 Trash，并添加到一个 Fillable 中：

```
// patterns/trash/ParseTrash.java
// 打开一个文件，将其中的内容解析为 Trash 对象，并将这些对象放入一个 Fillable 中
// {java patterns.trash.ParseTrash}
package patterns.trash;
import java.util.*;
import java.util.stream.*;
import java.io.*;
import java.nio.file.*;
import java.nio.file.Files;

public class ParseTrash {
  public static String source = "Trash.dat";
  public static <T extends Trash> void
  fillBin(String packageName, Fillable<T> bin) {
    DynaFactory factory =
      new DynaFactory(packageName);
    try {
      Files.lines(Paths.get("trash", source))
        // 移除注释和空行:
        .filter(line -> line.trim().length() != 0)
        .filter(line -> !line.startsWith("//"))
        .forEach(line -> {
          String type = line.substring(
            0, line.indexOf(':')).trim();
          double weight = Double.valueOf(
            line.substring(line.indexOf(':') + 1)
              .trim());
          bin.addTrash(factory.create(
            new TrashInfo(type, weight)));
        });
    } catch(IOException | NumberFormatException e) {
      throw new RuntimeException(e);
    }
  }
}
```

```java
// 处理 List 的专用操作：
public static <T extends Trash> void
fillBin(String packageName, List<T> bin) {
  fillBin(packageName, new FillableList<>(bin));
}
// 基本的测试：
public static void main(String[] args) {
  List<Trash> bin = new ArrayList<>();
  fillBin("trash", bin);
  bin.forEach(System.out::println);
}
/* 输出：
Loading patterns.trash.Cardboard
Loading patterns.trash.Paper
Loading patterns.trash.Aluminum
Loading patterns.trash.Glass
Cardboard weight: 4.40 * price: 0.11 = 0.48
Paper weight: 8.00 * price: 0.10 = 0.80
Aluminum weight: 1.80 * price: 1.67 = 3.01
Glass weight: 5.40 * price: 0.23 = 1.24
Aluminum weight: 3.40 * price: 1.67 = 5.68
Cardboard weight: 2.20 * price: 0.11 = 0.24
Glass weight: 4.30 * price: 0.23 = 0.99
Cardboard weight: 1.20 * price: 0.11 = 0.13
Paper weight: 6.60 * price: 0.10 = 0.66
Aluminum weight: 2.70 * price: 1.67 = 4.51
Paper weight: 9.10 * price: 0.10 = 0.91
Glass weight: 3.60 * price: 0.23 = 0.83
*/
```

文件路径是相对于父目录（patterns）而言的，因为相关的示例也会从该目录打包。Trash.dat 被转化为一个多行的流，并会被过滤，移除掉注释和空行。然后 String 的 indexOf() 方法得到了 : 的索引。String 的 substring() 方法提取出了垃圾类型的名称。接下来使用 substring() 找到了 weight，后者通过 Double.valueOf() 转化为 double。trim() 方法移除了 String 两端的空白。

RecycleA.java 中用一个 ArrayList 持有 Trash 对象。为了可以使用其他类型的集合，fillBin() 将 Fillable 的引用作为参数，Fillable 只是一个支持调用 addTrash() 方法的接口：

```java
// patterns/trash/Fillable.java
// 一个可以用 Trash 填充的对象
package patterns.trash;

public interface Fillable<T extends Trash> {
  void addTrash(T t);
}
```

任何支持该接口的类都可以用于 fillBin()。不过 List 并未实现 Fillable，因此需要一些帮助才能使用。大多数示例中使用了 List，因此有理由增加另一个以 List<Trash> 为参数的重载 fillBin() 方法。通过适配器类，可以将 List<Trash> 当作 Fillable 使用。

```
// patterns/trash/FillableList.java
// 使 List 具备 Fillable 能力的适配器
package patterns.trash;
import java.util.*;

public class FillableList<T extends Trash>
implements Fillable<T> {
  private List<T> list;
  public FillableList(List<T> list) {
    this.list = list;
  }
  @Override public void addTrash(T t) {
    list.add(t);
  }
}
```

该类的唯一职责是将 Fillable 的 addTrash() 方法与 List 的 add() 方法联系起来。有了这个类，重载的 fillBin() 方法就可以和 List 在 ParseTrash.java 中结合使用了。

这种方法对于任何常用的集合类都有效。或者，集合类可以提供自己的实现了 Fillable 接口的适配器（稍后你会在 TypeMapTrash.java 中看到）。

8.12.5　用 DynaFactory 实现回收

下面是 RecycleA.java 的另一个版本，使用了 ParseTrash.fillBin()，继而用 DynaFactory 执行了 Trash 的创建：

```
// patterns/recycleb/RecycleB.java
// {java patterns.recycleb.RecycleB}
package patterns.recycleb;
import patterns.trash.*;
import java.util.*;

public class RecycleB {
  public static void main(String[] args) {
    List<Trash> bin = new ArrayList<>();
    ParseTrash.fillBin("trash", bin);
    Bins bins = new Bins(bin);
    bins.show();
  }
}
```

```
/* 输出:
Loading patterns.trash.Cardboard
Loading patterns.trash.Paper
Loading patterns.trash.Aluminum
Loading patterns.trash.Glass
Aluminum weight: 1.80 * price: 1.67 = 3.01
Aluminum weight: 3.40 * price: 1.67 = 5.68
Aluminum weight: 2.70 * price: 1.67 = 4.51
Total Aluminum value = 13.19
Paper weight: 8.00 * price: 0.10 = 0.80
Paper weight: 6.60 * price: 0.10 = 0.66
Paper weight: 9.10 * price: 0.10 = 0.91
Total Paper value = 2.37
Glass weight: 5.40 * price: 0.23 = 1.24
Glass weight: 4.30 * price: 0.23 = 0.99
Glass weight: 3.60 * price: 0.23 = 0.83
Total Glass value = 3.06
Cardboard weight: 4.40 * price: 0.11 = 0.48
Cardboard weight: 2.20 * price: 0.11 = 0.24
Cardboard weight: 1.20 * price: 0.11 = 0.13
Total Cardboard value = 0.86
Cardboard weight: 4.40 * price: 0.11 = 0.48
Paper weight: 8.00 * price: 0.10 = 0.80
Aluminum weight: 1.80 * price: 1.67 = 3.01
Glass weight: 5.40 * price: 0.23 = 1.24
Aluminum weight: 3.40 * price: 1.67 = 5.68
Cardboard weight: 2.20 * price: 0.11 = 0.24
Glass weight: 4.30 * price: 0.23 = 0.99
Cardboard weight: 1.20 * price: 0.11 = 0.13
Paper weight: 6.60 * price: 0.10 = 0.66
Aluminum weight: 2.70 * price: 1.67 = 4.51
Paper weight: 9.10 * price: 0.10 = 0.91
Glass weight: 3.60 * price: 0.23 = 0.83
Total Trash value = 19.48
*/
```

解析 Trash.dat 的过程包装到了 ParseTrash.fillBin() 中，因此不再是我们设计重点的一部分。在本章的剩余部分，不论添加了什么新类，ParseTrash.fillBin() 都无须做任何改变。

8.12.6 将用法抽象化

解决了创建问题后，我们就可以解决剩余部分的设计了：在什么位置使用这些类。将垃圾分类到垃圾箱的操作尤为丑陋，并且暴露在外，因此我们应用**将丑陋的隐藏到方法或者类中**这个原则进行优化，如图 8-2 所示：

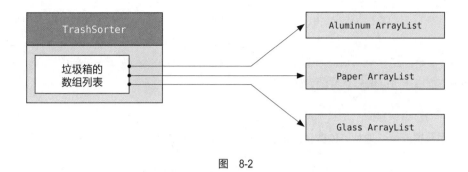

图 8-2

TrashSorter 类大概可以这样实现：

```
class TrashSorter extends ArrayList<ArrayList<Trash>> {
  void sort(Trash t) { /* ... */ }
}
```

也就是说，TrashSorter 是个由 Trash 组成的 List 组成的 List，而通过 List 的 add() 方法，可以向其中添加其他的 List<Trash> 对象。因此，必须改变 TrashSorter 的初始化方式，以应对向模型中添加新类型 Trash 的情况。

然而现在 sort() 成了一个问题。静态代码的方法如何能够处理添加新类型的情况呢？要解决这个问题，就必须从 sort() 中去除类型信息。新版本的 sort() 方法调用了一个处理类型细节的通用方法。这是另一种实现动态绑定方法效果的方式。因此 sort() 只需在序列中移动，并为每个 List<Trash> 调用一个动态绑定的方法。该方法的任务是抓取它感兴趣的垃圾块，因此方法名是 grab(Trash)。最后的结构看起来如图 8-3 所示：

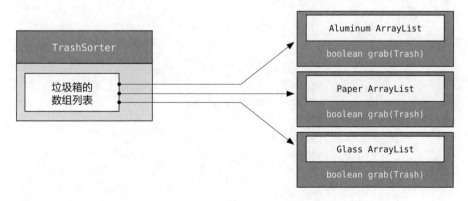

图 8-3

TrashSorter 会调用每个 grab() 方法,并根据当前 List<Trash> 持有的 Trash 类型得到不同的结果。也就是说,每个 List 都必须知道它所持有的是什么类型。解决这个问题的经典方法是创建一个 TrashBin 基类,并为每个要持有的不同类型继承一个新的子类。通过进一步观察还可以得到更好的方法。

面向对象编程的一个基本设计原则是:

用数据成员处理状态的变化,用多态处理行为的变化。

你最初可能会认为 grab() 方法对待 List<Paper> 和 List<Glass> 的行为肯定不同。但是该方法的行为完全取决于类型,和其他因素无关。这可以解释为一种不同的状态。在 Java 中,Class 是类型在运行时的表现形式,而这可以用来确定某个特定 TrashBin 会持有的 Trash 类型。

可以为 TrashBin 的构造器提供你选择的 Class,这会告诉 List 它应该持有什么类型。然后 grab() 会通过 binType 和反射确认传递给它的 Trash 对象是否和它应该捕获的类型相匹配:

```java
// patterns/recyclec/RecycleC.java
// {java patterns.recyclec.RecycleC}
package patterns.recyclec;
import patterns.trash.*;
import java.util.*;

// 只允许正确类型的 List :
class
TrashBin<T extends Trash> extends ArrayList<T> {
  final Class<T> binType;
  TrashBin(Class<T> binType) {
    this.binType = binType;
  }
  @SuppressWarnings("unchecked")
  boolean grab(Trash t) {
    // 比较类的类型:
    if(t.getClass().equals(binType)) {
      add((T)t); // 向下转型为这个 TrashBin 类型
      return true; // 捕获到了 Trash
    }
    return false; // 未捕获到 Trash
  }
}

class TrashBinList<T extends Trash>
extends ArrayList<TrashBin<? extends T>> {    // [1]
  @SuppressWarnings("unchecked")
  public TrashBinList(Class<? extends T>... types) {
    for(Class<? extends T> type : types)
```

```
      add(new TrashBin<>(type));
  }
  public boolean sort(T t) {
    for(TrashBin<? extends T> ts : this)
      if(ts.grab(t))
        return true;
    return false; // 未找到 t 的 TrashBin
  }
  public void sortBin(TrashBin<T> bin) {     // [2]
    for(T trash : bin)
      if(!sort(trash))
        throw new RuntimeException(
          "Bin not found for " + trash);
  }
  public void show() {
    for(TrashBin<? extends T> tbin : this) {
      String typeName = tbin.binType.getSimpleName();
      TrashValue.sum(tbin, typeName);
    }
  }
}

public class RecycleC {
  public static void main(String[] args) {
    TrashBin<Trash> bin =
      new TrashBin<>(Trash.class);
    ParseTrash.fillBin("trash", bin);
    @SuppressWarnings("unchecked")
    TrashBinList<Trash> trashBins =
      new TrashBinList<>(
        Aluminum.class, Paper.class, Glass.class,
        // 增加一项：
        Cardboard.class                   // [3]
      );
    trashBins.sortBin(bin);               // [4]
    trashBins.show();
    TrashValue.sum(bin, "Trash");
  }
}
/* 输出：
Loading patterns.trash.Cardboard
Loading patterns.trash.Paper
Loading patterns.trash.Aluminum
Loading patterns.trash.Glass
Aluminum weight: 1.80 * price: 1.67 = 3.01
Aluminum weight: 3.40 * price: 1.67 = 5.68
Aluminum weight: 2.70 * price: 1.67 = 4.51
Total Aluminum value = 13.19
Paper weight: 8.00 * price: 0.10 = 0.80
Paper weight: 6.60 * price: 0.10 = 0.66
Paper weight: 9.10 * price: 0.10 = 0.91
```

```
Total Paper value = 2.37
Glass weight: 5.40 * price: 0.23 = 1.24
Glass weight: 4.30 * price: 0.23 = 0.99
Glass weight: 3.60 * price: 0.23 = 0.83
Total Glass value = 3.06
Cardboard weight: 4.40 * price: 0.11 = 0.48
Cardboard weight: 2.20 * price: 0.11 = 0.24
Cardboard weight: 1.20 * price: 0.11 = 0.13
Total Cardboard value = 0.86
Cardboard weight: 4.40 * price: 0.11 = 0.48
Paper weight: 8.00 * price: 0.10 = 0.80
Aluminum weight: 1.80 * price: 1.67 = 3.01
Glass weight: 5.40 * price: 0.23 = 1.24
Aluminum weight: 3.40 * price: 1.67 = 5.68
Cardboard weight: 2.20 * price: 0.11 = 0.24
Glass weight: 4.30 * price: 0.23 = 0.99
Cardboard weight: 1.20 * price: 0.11 = 0.13
Paper weight: 6.60 * price: 0.10 = 0.66
Aluminum weight: 2.70 * price: 1.67 = 4.51
Paper weight: 9.10 * price: 0.10 = 0.91
Glass weight: 3.60 * price: 0.23 = 0.83
Total Trash value = 19.48
*/
```

[1] TrashBinList 包含一个由 TrashBin 引用组成的 List，因此 sort() 在为每个 Trash 对象查找匹配项时，可以遍历 TrashBin。

[2] sortBin() 从它的 TrashBin 参数中选出每一块 Trash，并将其分类到相应的具体 TrashBin 中。注意，在添加新类型的时候，这段代码完全无须修改。如果在添加新类型（或发生其他变更）时大部分代码不需要修改，那么我们就拥有了一个易于扩展的系统。

[3] 加上这行代码，就会增加 Cardboard 类型。

[4] 一次方法调用会触发操作，将 bin 的内容分别放到各自特定类型的垃圾箱（bin）中。

8.12.7 用多路分发重新设计

上述设计肯定满足了需求。通过增加或修改不同的类即可增加新的类型，而不会在系统中产生额外的代码变更。反射没有像在 RecycleA.java 中那样被"误用"。然而，还有可能更进一步——可以用一种更纯粹的观点来看待反射，也就是说，应该将反射从垃圾分类的操作中完全去掉。

为了达到这个目的，应该首先采用这样的观点：所有依赖类型的活动（例如检测一块垃圾的类型，并将其放入合适的垃圾箱）都应该通过多态来控制。

前面的示例首先根据类型分类，然后在由某个特定类型的元素组成的序列上进行操作。如果你发现自己在挑选特定类型，就要停下来想想。多态（动态绑定方法调用）的整个思想就是为你处理特定类型的行为。所以为什么还要寻找类型呢？

你会寻找类型，是因为 Java 只会实现单路分发。也就是说，如果一个操作要作用于一个以上的未知类型的对象，Java 只会在其中一个类型上应用动态绑定机制。这样并不能解决问题，因此你最终需要手动检测某些类型，并实现自己的动态绑定行为。

解决的方案是使用**多路分发**。我们将代码配置为让单个方法生成多个动态方法调用，从而确定多个类型。要实现这个效果，一般需要用到多种层次结构：每路分发使用一个层次结构。下面这个示例使用了两个层次结构：已有的 Trash 家族，以及要放入垃圾的垃圾箱的层次结构。如果只有两路分发，就称为双路分发。

为了实现双路分发，我们在 Trash 的层次结构中增加了一个名为 addToBin() 的新方法，该方法以 List<TypedBin> 为参数。一块垃圾 Trash 会调用 addToBin() 来遍历 List，并使用双路分发，将自身加入到正确的 TypedBin 中（见图 8-4）。

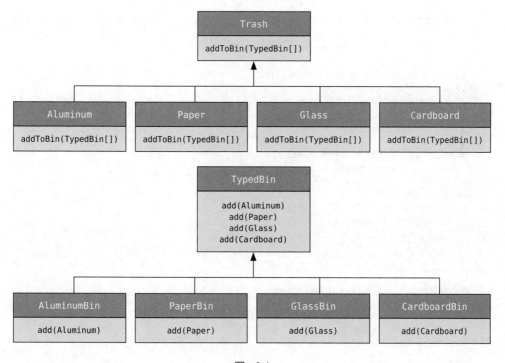

图 8-4

新的层次结构是 TypedBin，它包含一个同样以多态方式使用的 add() 方法。但是这里有个别扭的地方：为了能够接收不同类型的 Trash，add() 被**重载**了。因此双路分发方案中很重要的部分也会涉及重载。

重新设计该程序会产生一个难题：Trash 类现在必须包含一个 addToBin() 方法。一种方法是从已有的 Trash 层次结构中复制代码，并引入 addToBin() 方法，由此创建一个新的层次结构。另一种方法则可以在你没有源代码的控制权，或者不希望向已有代码中引入新 bug 风险的时候采用。这种方法是将 addToBin() 方法放入一个接口，留下 Trash，并继承出新的 Aluminum、Paper、Glass 和 Cardboard 的具体类型。下面便是这个接口：

```java
// patterns/doubledispatch/TypedBinMember.java
// 在不修改原有层次结构的情况下
// 将双路分发方法适配到 trash 的层次结构中
package patterns.doubledispatch;
import java.util.*;

public interface TypedBinMember {
  boolean addToBin(List<TypedBin> bins);
}
```

每一个适配后的 Aluminum、Paper、Glass 和 Cardboard 的子类中都实现了 addToBin() 方法，但是每种情况下的 addToBin() 代码看起来都完全相同：

```java
// patterns/doubledispatch/Aluminum.java
// 使用了双路分发的 Aluminum
package patterns.doubledispatch;
import patterns.trash.*;
import java.util.*;

public class Aluminum extends patterns.trash.Aluminum
implements TypedBinMember {
  public Aluminum(double wt) { super(wt); }
  @Override
  public boolean addToBin(List<TypedBin> tbins) {
    return tbins.stream()
      .anyMatch(tb -> tb.add(this));
  }
}
```

```java
// patterns/doubledispatch/Paper.java
// 使用了双路分发的 Paper
package patterns.doubledispatch;
import patterns.trash.*;
import java.util.*;

public class Paper extends patterns.trash.Paper
implements TypedBinMember {
  public Paper(double wt) { super(wt); }
  @Override
```

```java
    public boolean addToBin(List<TypedBin> tbins) {
      return tbins.stream()
        .anyMatch(tb -> tb.add(this));
    }
}

// patterns/doubledispatch/Glass.java
// 使用了双路分发的 Glass
package patterns.doubledispatch;
import patterns.trash.*;
import java.util.*;

public class Glass extends patterns.trash.Glass
implements TypedBinMember {
  public Glass(double wt) { super(wt); }
  @Override
  public boolean addToBin(List<TypedBin> tbins) {
    return tbins.stream()
      .anyMatch(tb -> tb.add(this));
  }
}
// patterns/doubledispatch/Cardboard.java
// 使用了双路分发的 Cardboard
package patterns.doubledispatch;
import patterns.trash.*;
import java.util.*;

public class Cardboard extends patterns.trash.Cardboard
implements TypedBinMember {
  public Cardboard(double wt) { super(wt); }
  @Override
  public boolean addToBin(List<TypedBin> tbins) {
    return tbins.stream()
      .anyMatch(tb -> tb.add(this));
  }
}
```

anyMatch(tb -> tb.add(this)) 调用了数组中每个 TypedBin 上的 add() 方法。add() 方法返回一个 boolean，表示该 Trash 是否成功加入到了特定的 TypedBin 中，而 anyMatch() 则会找出这些 add() 调用中是否有执行成功的。如果有，那么这块 Trash 就被分到了一个垃圾箱中。

每个 addToBin() 方法的代码似乎都是一样的。但是要注意 tb.add(this) 中的 this 参数。每个 Trash 子类的 this 的类型都是不同的，因此代码实际上也是不同的。这边是双路分发的第一个部分，在 addToBin() 中，你知道你持有的 Trash 是明确的 Aluminum 或 Paper 等类型。在调用 add() 的过程中，该信息是通过 this 类型传递的。编译器会将调用解析到恰当的 add() 重载版本。但是 tb 生成的是指向基类 TypedBin 的引用，因此该动态调用最终会根据当前所选择的 TypedBin 类型，调用到不同的方法上。这便是第二路分发。

下面便是 TypedBin 的基类。注意 private 的字段 typedBin 永远都不会直接对外暴露——在调用 bin() 时，会复制一份 ArrayList 的副本。由此，typedBin Arraylist 就不会被外部的代码所修改：

```java
// patterns/doubledispatch/TypedBin.java
// 可以捕获到正确类型的 List
package patterns.doubledispatch;
import patterns.trash.*;
import java.util.*;

public class TypedBin {
  private List<Trash> typedBin =
    new ArrayList<>();
  public final String type;
  public TypedBin(String type) {
    this.type = type;
  }
  public List<Trash> bin() {
    // 返回 typedBin 的副本：
    return new ArrayList<Trash>(typedBin);
  }
  protected boolean addIt(Trash t) {
    typedBin.add(t);
    return true;
  }
  public boolean add(Aluminum a) {
    return false;
  }
  public boolean add(Paper a) {
    return false;
  }
  public boolean add(Glass a) {
    return false;
  }
  public boolean add(Cardboard a) {
    return false;
  }
}
```

这些重载后的 add() 方法全都会返回 false。add() 方法如果没有在子类中被重写，就会一直返回 false，而调用者（addToBin()）会假设当前的 Trash 对象还没有被加入到集合中。

每个 TypedBin 的子类只会重写一个 add() 方法，即和要定义的 TypedBin 类型相对应。例如，CardboardBin 重写了 add(Cardboard)。重写后的方法将 Trash 对象加入自身的集合并返回 true，而 CardboardBin 中其余的 add() 方法都会继续返回 false，因为它们并没有被重写。

下面是这个程序剩余的部分：

```java
// patterns/doubledispatch/DoubleDispatch.java
// 在一次方法调用中，用多路分发处理多个未知类型
// {java patterns.doubledispatch.DoubleDispatch}
package patterns.doubledispatch;
import patterns.trash.*;
import java.util.*;

class AluminumBin extends TypedBin {
  public AluminumBin() { super("Aluminum"); }
  @Override public boolean add(Aluminum a) {
    return addIt(a);
  }
}

class PaperBin extends TypedBin {
  public PaperBin() { super("Paper"); }
  @Override public boolean add(Paper a) {
    return addIt(a);
  }
}

class GlassBin extends TypedBin {
  public GlassBin() { super("Glass"); }
  @Override public boolean add(Glass a) {
    return addIt(a);
  }
}

class CardboardBin extends TypedBin {
  public CardboardBin() { super("Cardboard"); }
  @Override public boolean add(Cardboard a) {
    return addIt(a);
  }
}

class TrashBinSet {
  public final List<TypedBin> binSet =
    Arrays.asList(
      new AluminumBin(), new PaperBin(),
      new GlassBin(), new CardboardBin()
    );
  @SuppressWarnings("unchecked")
  public void sortIntoBins(List bin) {
    bin.forEach( aBin -> {
      TypedBinMember t = (TypedBinMember)aBin;
      if(!t.addToBin(binSet))
        throw new RuntimeException(
          "sortIntoBins() couldn't add " + t);
    });
  }
}
```

```java
}

public class DoubleDispatch {
  public static void main(String[] args) {
    List<Trash> bin = new ArrayList<>();
    ParseTrash.fillBin("doubledispatch", bin);
    TrashBinSet bins = new TrashBinSet();
    // 将主 bin 中分类到单个类型的 bin 中:
    bins.sortIntoBins(bin);
    // 累加每个 bin 的值……
    bins.binSet.forEach(tb ->
      TrashValue.sum(tb.bin(), tb.type));
    // ……累加主 bin 的值
    TrashValue.sum(bin, "Trash");
  }
}
/* 输出:
Loading patterns.doubledispatch.Cardboard
Loading patterns.doubledispatch.Paper
Loading patterns.doubledispatch.Aluminum
Loading patterns.doubledispatch.Glass
Aluminum weight: 1.80 * price: 1.67 = 3.01
Aluminum weight: 3.40 * price: 1.67 = 5.68
Aluminum weight: 2.70 * price: 1.67 = 4.51
Total Aluminum value = 13.19
Paper weight: 8.00 * price: 0.10 = 0.80
Paper weight: 6.60 * price: 0.10 = 0.66
Paper weight: 9.10 * price: 0.10 = 0.91
Total Paper value = 2.37
Glass weight: 5.40 * price: 0.23 = 1.24
Glass weight: 4.30 * price: 0.23 = 0.99
Glass weight: 3.60 * price: 0.23 = 0.83
Total Glass value = 3.06
Cardboard weight: 4.40 * price: 0.11 = 0.48
Cardboard weight: 2.20 * price: 0.11 = 0.24
Cardboard weight: 1.20 * price: 0.11 = 0.13
Total Cardboard value = 0.86
Cardboard weight: 4.40 * price: 0.11 = 0.48
Paper weight: 8.00 * price: 0.10 = 0.80
Aluminum weight: 1.80 * price: 1.67 = 3.01
Glass weight: 5.40 * price: 0.23 = 1.24
Aluminum weight: 3.40 * price: 1.67 = 5.68
Cardboard weight: 2.20 * price: 0.11 = 0.24
Glass weight: 4.30 * price: 0.23 = 0.99
Cardboard weight: 1.20 * price: 0.11 = 0.13
Paper weight: 6.60 * price: 0.10 = 0.66
Aluminum weight: 2.70 * price: 1.67 = 4.51
Paper weight: 9.10 * price: 0.10 = 0.91
Glass weight: 3.60 * price: 0.23 = 0.83
Total Trash value = 19.48
*/
```

TrashBinSet 封装了所有不同类型的 TypedBin，以及用于启动双路分发的 sortIntoBins()

方法。一旦构建好了结构，分类到不同的 TypedBin 中就是很简单的事了。

添加新类型所需的修改相对独立：继承新的 Trash 类型和其中的 addToBin() 方法，然后继承一个新的 TypedBin，最后在 TrashBinSet.binSet 的初始化过程中添加新类型。

8.12.8 访问者模式

接下来考虑应用一种设计目标与之前完全不同的设计模式。

对于这种设计模式，我们不再关心如何优化"向系统中添加新的 Trash 类型"这一动作。在某些方面，访问者模式反而会使新 Trash 类型的添加变得更复杂。假设现在有一个不可改动的主类层次结构——也许来自供应商提供的系统。由于它不可改动，因此无法对其做任何修改。然而我们希望向该层次结构中添加新的多态方法，一般情况下，我们会在基类中增加方法，但是现在整个层次结构你都动不了。怎样才能解决这个问题呢？

访问者模式是建立在双路分发的基础上的。访问者使我们可以通过创建独立的 Visitor 类层次结构，来扩展主类型的接口，以模拟对主类型执行的操作。主类的对象只需简单地"接受"访问者，然后调用访问者的动态绑定方法，如图 8-5 所示。

假设 v 是一个指向某个 Aluminum 对象的引用，则以下代码：

```
PriceVisitor pv = new PriceVisitor();
v.accept(pv);
```

会产生两次多态方法调用：第一次用来选择 Aluminum 的 accept() 的版本，第二次发生在 accept() 中，即通过 Visitor 的引用 pv 动态调用具体版本的 visit() 的时候。

这种设置意味着新功能可以用新的 Visitor 子类的形式添加到系统中，而 Trash 的层次结构则不需要修改。这便是访问者模式最大的好处：可以在不修改原有类层次结构的情况下，向其中增加新的多态功能。

要注意一点：这个不可改动的层次结构的提供者需要能预见到访问者模式的使用并做好支持，可以通过显式地引入 accept() 方法（就像在 Trash 层次结构中所做的那样），也可以通过使层次结构中的类具有足够的可访问性来覆盖包含 accept() 的接口，类似于前面双路分发的示例。

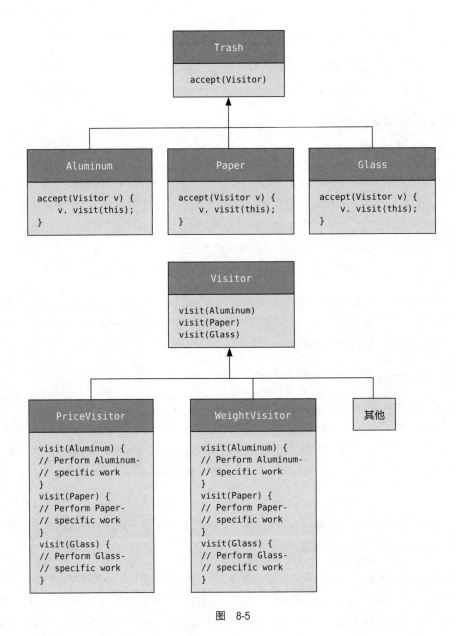

图 8-5

访问者是有帮助的,但它并不是本来想实现的,因此你可能会认为这并不是想要的方案。但是看看它做到的一件事:访问者的方案避免了从主 Trash 序列分类到单个类型的容器中。因此,可以将所有内容都保留在单个主序列中,只需使用合适的访问者对它进行遍历,即可实现目标。

访问者模式中的双路分发在没有使用反射的情况下，确定了 Trash 和 Visitor 这两者的类型。在随后的示例中，实现了两个 Visitor：用来显示和累加价格的 PriceVisitor，以及用来显示和累加重量的 WeightVisitor。

现在你应该理解为什么要在 Trash 的层次结构中加上 accept( ) 方法（即之前让你暂时忽略的方法）了。现在回过头来看一下用来表示各种类型的 Trash 的类：每个类中的 accept( ) 都只是简单的 v.visit(this);。正如在双路分发的示例中所看到的，所有 accept( ) 方法看起来都一样，但其实每个都不同，因为 this 的类型是不同的。

Visitor 是个抽象类而不是接口，因为它使用了 double 来持有不同类型垃圾的总数，以及用 show() 和 total() 相关的公共代码来显示结果。descriptor 由各个子类设置为"price"（价格）和"重量"（weight），因此 show() 可以生成有意义的输出：

```java
// patterns/trash/Visitor.java
// 访问者 Vistor 的基类
package patterns.trash;

public abstract class Visitor {
  protected double alTotal; // Aluminum
  protected double pTotal;  // Paper
  protected double gTotal;  // Glass
  protected double cTotal;  // Cardboard
  protected String descriptor;
  protected Visitor(String descriptor) {
    this.descriptor = descriptor;
  }
  protected void show(String type, double value) {
    System.out.printf(
      "%s %s: %.2f%n", type, descriptor, value);
  }
  public void total() {
    show("Total Aluminum", alTotal);
    show("Total Paper", pTotal);
    show("Total Glass", gTotal);
    show("Total Cardboard", cTotal);
  }
  abstract void visit(Aluminum a);
  abstract void visit(Paper p);
  abstract void visit(Glass g);
  abstract void visit(Cardboard c);
}
```

注意，在添加新的 Trash 类型时，必须修改 Visitor 和它的子类。这实际上并不太糟糕，因为修改是隔离于 Visitor 层次结构的。比如，如果你向 Visitor 添加的是类似 visit(Plastic p) 这样的内容，则编译器会确保所有的 Visitor 子类都会实现新的重载版本的 visit()。

该程序的剩余部分创建了具体的 Visitor 类，并通过一个 Trash 对象的 list 来传递：

```java
// patterns/trash/TrashVisitor.java
// 包装在每个 Visitor 实现中的相关算法
// {java patterns.trash.TrashVisitor}
package patterns.trash;
import java.util.*;

class PriceVisitor extends Visitor {
  public PriceVisitor() { super("price"); }
  @Override public void visit(Aluminum al) {
    double price = al.weight * al.price();
    show("Aluminum", price);
    alTotal += price;
  }
  @Override public void visit(Paper p) {
    double price = p.weight * p.price();
    show("Paper", price);
    pTotal += price;
  }
  @Override public void visit(Glass g) {
    double price = g.weight * g.price();
    show("Glass", price);
    gTotal += price;
  }
  @Override public void visit(Cardboard c) {
    double price = c.weight * c.price();
    show("Cardboard", price);
    cTotal += price;
  }
}

class WeightVisitor extends Visitor {
  public WeightVisitor() { super("weight"); }
  @Override public void visit(Aluminum al) {
    show("Aluminum", al.weight);
    alTotal += al.weight;
  }
  @Override public void visit(Paper p) {
    show("Paper", p.weight);
    pTotal += p.weight;
  }
  @Override public void visit(Glass g) {
    show("Glass", g.weight);
    gTotal += g.weight;
  }
  @Override public void visit(Cardboard c) {
    show("Cardboard", c.weight);
    cTotal += c.weight;
  }
```

```
/* 输出：
Loading patterns.trash.Cardboard
Loading patterns.trash.Paper
Loading patterns.trash.Aluminum
Loading patterns.trash.Glass
Cardboard price: 0.48
Cardboard weight: 4.40
Paper price: 0.80
Paper weight: 8.00
Aluminum price: 3.01
Aluminum weight: 1.80
Glass price: 1.24
Glass weight: 5.40
Aluminum price: 5.68
Aluminum weight: 3.40
Cardboard price: 0.24
Cardboard weight: 2.20
Glass price: 0.99
Glass weight: 4.30
Cardboard price: 0.13
Cardboard weight: 1.20
Paper price: 0.66
Paper weight: 6.60
Aluminum price: 4.51
Aluminum weight: 2.70
Paper price: 0.91
Paper weight: 9.10
Glass price: 0.83
Glass weight: 3.60
Total Aluminum price: 13.19
Total Paper price: 2.37
Total Glass price: 3.06
Total Cardboard price: 0.86
Total Aluminum weight: 7.90
Total Paper weight: 23.70
Total Glass weight: 13.30
Total Cardboard weight: 7.80
*/
```

```
}
public class TrashVisitor {
  public static void main(String[] args) {
    List<Trash> bin = new ArrayList<>();
    ParseTrash.fillBin("trash", bin);
    List<Visitor> visitors = Arrays.asList(
      new PriceVisitor(), new WeightVisitor());
    bin.forEach(
      trash -> visitors.forEach(trash::accept)
    );
    visitors.forEach(Visitor::total);
  }
}
```

在 `main()` 中只有一个 Trash bin（垃圾箱）。两个 Visitor 对象被该序列中的所有元素所接受。访问者保留了自己的内部数据，以计算总的重量和价格。

在双路分发的方案中，在每个子类的创建过程中，只有一个重载的 `add()` 方法在被重写了。而此处，每个重载的 `visit()` 方法都在 Visitor 的所有子类中被重写了。

8.12.9 反射是有害的？

本章中的各种设计都在试图去除反射，这可能会给你"反射是有害的"的印象。实际上并不是这样的，问题在于对反射的**滥用**。我们的设计要去除反射，因为对反射的误用会阻碍程序的可扩展性——我们的预期目标是在对周围代码的影响较小的情况下向系统中添加新类型。反射常常被拿来在系统中挨个查找每个类型，这是一种滥用，会产生无扩展性的代码，因为在添加新类型的时候，必须找到所有使用了反射的代码，如果有遗漏的话，编译器是帮不上忙的。

不过，反射并不会自动创建无扩展性的代码。我们来回顾一下垃圾收集器，并引入一个新工具，我称之为 TypeMap，它包含一个 Map<Class, List<T>>。接口很简单：可以 `add()`（添加）一个新对象，`values()` 会生成一个 List 的 Stream，每个 List 都包含了某个具体类型的全部对象。只要向系统中添加新类型，TypeMap 都会动态地进行适配。

```
// patterns/TypeMap.java
// 泛型的 TypeMap 可用于任何类型
package patterns;
import java.util.*;
import java.util.stream.*;

public class TypeMap<T> {
  public final Map<Class, List<T>> map =
    new HashMap<>();
```

```java
  public void add(T o) {
    Class type = o.getClass();
    map.computeIfAbsent(type,
      k -> new ArrayList<T>()).add(o);
  }
  public Stream<List<T>> values() {
    return map.values().stream();
  }
}
```

add() 方法完成了所有的任务。在用 add() 添加一个新对象的时候，会提取出该对象的 Class，该 Class 会作为键，用来确认（查询）持有该类型对象的 List 在 Map 中是否已经存在。如果已经存在，则会返回该 List。如果不存在，computeIfAbsent() 会添加一个由该 Class 对象和一个新 ArrayList 组成的键值对。不论哪种情况，对象都会被添加到 List 中。

接下来的两个示例会通过 show() 方法展示出 Map<Class, List<Trash>> 所包含的内容，因此会将此实现为一个工具函数：

```java
// patterns/trash/ClassToListOfTrashMap.java
// 显示 Map<Class, List<Trash>> 的内容
package patterns.trash;
import java.util.*;

public class ClassToListOfTrashMap {
  public static void
  show(Map<Class, List<Trash>> map) {
    map.values().forEach( bin -> {
      String typeName = "Trash";
      if(!bin.isEmpty())
        typeName =
          bin.get(0).getClass().getSimpleName();
      TrashValue.sum(bin, typeName);
    });
  }
}
```

每次向 bin 中插入一个 Trash 对象时，ParseTrash.fillBin() 都会进行分类：

```java
// patterns/trash/TypeMapTrash.java
// 使用一个由 List 组成的 Map 和反射来将 Tash 分类到不同的 List 中
// 尽管使用了反射，该功能仍然具有可扩展性
// {java patterns.trash.TypeMapTrash}
package patterns.trash;
import patterns.TypeMap;

// ParseTrash.fillBin() 的适配类:
class TypeMapAdapter implements Fillable {
```

```
  private TypeMap<Trash> map;
  TypeMapAdapter(TypeMap<Trash> map) {
    this.map = map;
  }
  @Override
  public void addTrash(Trash t) { map.add(t); }
}

public class TypeMapTrash {
  @SuppressWarnings("unchecked")
  public static void main(String[] args) {
    TypeMap<Trash> bin = new TypeMap<>();
    ParseTrash.fillBin(
      "trash", new TypeMapAdapter(bin));
    ClassToListOfTrashMap.show(bin.map);
  }
}
/* 输出:
Loading patterns.trash.Cardboard
Loading patterns.trash.Paper
Loading patterns.trash.Aluminum
Loading patterns.trash.Glass
Paper weight: 8.00 * price: 0.10 = 0.80
Paper weight: 6.60 * price: 0.10 = 0.66
Paper weight: 9.10 * price: 0.10 = 0.91
Total Paper value = 2.37
Glass weight: 5.40 * price: 0.23 = 1.24
Glass weight: 4.30 * price: 0.23 = 0.99
Glass weight: 3.60 * price: 0.23 = 0.83
Total Glass value = 3.06
Aluminum weight: 1.80 * price: 1.67 = 3.01
Aluminum weight: 3.40 * price: 1.67 = 5.68
Aluminum weight: 2.70 * price: 1.67 = 4.51
Total Aluminum value = 13.19
Cardboard weight: 4.40 * price: 0.11 = 0.48
Cardboard weight: 2.20 * price: 0.11 = 0.24
Cardboard weight: 1.20 * price: 0.11 = 0.13
Total Cardboard value = 0.86
*/
```

有了前面的示例，你对 class TypeMapTrash 的大部分内容应该很熟悉。这一次没有将新 Trash 对象放入 List 类型的 bin，而是使用了 TypeMap 类型的 bin。在将垃圾放入 bin 中的时候，TypeMap 的内部排序机制会立刻对该垃圾进行分类。之后再对每个独立的 List 进行操作就是很简单的事情了。

向系统中添加新类型完全不会影响到这部分代码，也不会影响到 Trash 中的代码。这个方案确实很依赖反射，但要注意到，Map 中的每个键值对都只会查找一种类型。此外，在增加新类型的时候，你无须担心忘记添加或修改代码，因为根本就无须任何改动。

Java 8 提供了 Stream 的函数 groupingBy()，这个函数同样也可以得到 Map<Class, List<Trash>> ：

```java
// patterns/trash/GroupingBy.java
// {java patterns.trash.GroupingBy}
package patterns.trash;
import java.util.*;
import java.util.stream.*;

public class GroupingBy {
  public static void main(String[] args) {
    List<Trash> bin = new ArrayList<>();
    ParseTrash.fillBin("trash", bin);
    Map<Class, List<Trash>> m =
      bin.stream().collect(
        Collectors.groupingBy(Object::getClass));
    ClassToListOfTrashMap.show(m);
  }
}
/* 输出：
Loading patterns.trash.Cardboard
Loading patterns.trash.Paper
Loading patterns.trash.Aluminum
Loading patterns.trash.Glass
Glass weight: 5.40 * price: 0.23 = 1.24
Glass weight: 4.30 * price: 0.23 = 0.99
Glass weight: 3.60 * price: 0.23 = 0.83
Total Glass value = 3.06
Paper weight: 8.00 * price: 0.10 = 0.80
Paper weight: 6.60 * price: 0.10 = 0.66
Paper weight: 9.10 * price: 0.10 = 0.91
Total Paper value = 2.37
Cardboard weight: 4.40 * price: 0.11 = 0.48
Cardboard weight: 2.20 * price: 0.11 = 0.24
Cardboard weight: 1.20 * price: 0.11 = 0.13
Total Cardboard value = 0.86
Aluminum weight: 1.80 * price: 1.67 = 3.01
Aluminum weight: 3.40 * price: 1.67 = 5.68
Aluminum weight: 2.70 * price: 1.67 = 4.51
Total Aluminum value = 13.19
*/
```

这种方式产生的效果和 TypeMapTrash 相同，但是不再需要 TypeMap 或 TypeMapAdapter 类了。此外也不再需要屏蔽 "unchecked" 警告了。

为了帮助你理解本章中 8.10 节之后的内容，可以试着增加一个叫 Plastic 的 Trash 子类，看看它是怎样融入所有这些示例中的。

8.13　总结

　　找出变化的向量并不是件简单的事情。通常在程序的初版设计完成前，仅靠分析是很难找出来的。可能要到工程的最后阶段，才会积累出所需的洞察力。有时只有在实现阶段，才能发现某些更深层次或更隐晦的系统需求。在添加新类型这个场景中（如本章最后所探讨的），可能只有到了维护阶段并开始扩展系统的时候，才会意识到我们还需要什么（例如某种特定的继承层次结构）。

　　设计模式告诉我们，面向对象并不仅仅是关于多态，而是"将会变化的事物和不会变化的事物分开"。多态是实现该目标的一种重要方法，如果编程语言直接支持多态，它就会很有帮助（这样就不必亲自实现多态了，否则往往会使实现成本过高）。但设计模式通常可以告诉我们能够实现基本目标的其他方法，一旦建立了这种意识，我们就会开始寻找更具创造性的设计。

A 编程指南

> 本附录包含的建议有助于为你的底层程序设计和代码编写提供指引。

当然,这些只是指南,并不是规范。我的本意是让它们为你提供启发。要记住,这些建议在某些场景下可能并不完美,甚至并不适用。

A.1 设计建议

1. **优雅总会带来回报**。从短期来看,找到一个问题真正优雅的解决方案似乎需要很长的时间,但是一旦它开始发挥作用,系统就可以很容易地适应新的情况,而不再需要耗费数小时、数天,甚至数月的磨合时间。这时你便可以看到回报(即使没人能够量化衡量)。它让你的程序不仅更易于构建和调试,而且更易于理解和维护,这也正是它在运营成本上的价值所在。这一点需要有些经验后才能理解,因为将一段代码变得更优雅需要一段过程,其间可能会显得效率不高。要克制赶工的冲动,欲速则不达。

2. **先运行起来,再优化性能**。即使你确信某段代码真的非常重要,是系统的主要瓶颈,也应该坚持这个理念。不要着急优化。让系统以尽可能简单的设计先运行起来。然后,如果速度不够快,再对其进行分析。你几乎总是会发现"你以为的"瓶颈并不是问题所在。要把时间留给真正重要的事情。

3. **要牢记"分而治之"的原则**。如果你要解决的问题太过复杂,那么可以想象一下程序的基本操作,设想一个能处理这些复杂部分的神奇"模块"。该"模块"就是一个对象——先编写使用该对象的代码,然后回头深入这个对象,并将其中复杂的部分再封装到其他对象中，以此类推。

4. **将类的创建者和使用者（调用方程序员）分开**。类的使用者就是"客户"，他们既不需要也不想了解类的内部细节逻辑。类的创建者则必须精通类的设计，编写出来的类既能供最缺乏经验的程序员使用，又仍然可以稳健地运行在程序中。可以将一个类看作其他类的**服务提供者**。库只有在简单透明的情况下才易于使用。

5. **在创建类的时候，尽量让类名清晰易懂，甚至不用添加注释**。你的目标应该是使调用方程序员的接口在概念上尽量简单。为此，需要在适当的时候使用方法重载来创建直观、易用的接口。

6. **分析和设计的产出应该至少包含系统中的类、它们的公共接口，以及它们和其他类（特别是基类）之间的关系**。如果你的设计方法所产出的不止这些，就要问问自己，产出的所有内容在程序的生命周期内是否都有价值。如果没有，维护它们将会耗费额外的精力。开发团队的成员们倾向于不去维护任何对生产力没有贡献的东西，很多设计方法都没有考虑到这个现实问题。

7. **使一切自动化**。先编写测试代码（如果可以，在编写类本身之前就做），并保持它和类的同步，再通过构建工具（可以使用 Gradle，它是事实上的标准 Java 构建工具）实现测试的自动化运行。通过这种方式，任何变更都可以通过运行测试代码自动验证，从而让你可以在第一时间发现错误。由于你知道测试框架提供了安全保障，因此在发现新需求时，就可以更大胆地进行大范围的修改。要记住，尽管由类型检查和异常处理等能力所带来的内建测试，为语言带来了巨大的提升，但这些能力仅限于此。你必须依靠自己继续前行，通过填充用于验证类或程序相关功能的测试，创建出稳健的系统。

8. **先编写测试代码（在编写类本身之前）来验证类的设计是否完善**。如果不写出测试代码，你就不知道类实际是什么样子的。此外，编写测试代码的行为常常可以暴露出类中额外所需的功能或者约束——这些功能或约束有时不会在设计和开发中显露出来。同时，测试还可作为示例代码，演示类该如何使用。

9. **所有的软件设计问题都可以通过额外引入一层概念上的中间层来简化**。这一软件工程上的基本规则[①]是抽象的基础，而抽象则是面向对象编程的主要特性。在面向对象编程中，也可以这么说："如果代码太过复杂，就引入更多的对象。"

10. **中间层应该是有意义的**（与建议 (9) 一致）。这条指南的含义可以简单到如同"将

[①] Andrew Koenig 为我解释了这一点。

常用的代码放进一个方法中"一样。如果你增加了一层没有任何意义的中间层（抽象、封装等），那就几乎和没有中间层一样糟糕。

11. 使类尽量原子化。要为每个类赋予单一、清晰的目的，即为其他类提供内聚性服务。如果类或系统的设计变得过于复杂，就将复杂的类拆分为更简单的类。最明显的效果就是代码量的减少。如果某个类很大，那么它做的事很可能太多了，应该进行拆分。一个类如果出现了下列迹象，就应该考虑重新设计了。

- 复杂的 switch 语句：可以考虑使用多态。
- 大量方法，广泛覆盖了许多不同类型的操作：可以考虑使用（拆分为）多个类。
- 大量成员变量，广泛涉及不同的特征：可以考虑使用（拆分为）多个类。
- 其他建议可以在 Martin Fowler 所著的《重构：改善既有代码的设计》一书中找到。

12. 小心冗长的参数列表。这会导致方法调用变得难以编写、阅读和维护。可以试着将该方法移动到（更加）合适的类中，并/或传入对象作为参数。

13. 不要自我重复。如果一段代码出现在了子类的多个方法中，就要将这段代码放入基类的一个方法中，并从子类方法中调用该方法。这样不仅节省了代码空间，还能更简单地进行变更传播。有些时候，找出这些通用的代码，可以为接口带来有价值的功能。本建议还有个更简单、不需要使用继承的版本：如果类中的多个方法含有重复的代码，就要将这些代码拆分到一个通用方法中，然后在其他方法中调用该方法。

14. 小心 switch 语句或者链式的 if-else 子句。这类语句可以作为**类型检查式编程**的信号，意味着你是在根据某种类型信息选择要执行哪些代码（精确的类型可能在一开始并不明显）。通常，你可以将这类代码替换为继承和多态。多态方法调用会为你执行类型检查，并带来更可靠、更简单的可扩展性。

15. 从设计的角度出发，寻找会变化的事物，并将其和不会变化的事物分离。也就是说，寻找系统中你希望不用修改设计就能改变的那些元素，然后将它们封装到类中。

16. 不要通过子类化来扩展基础功能。如果一个接口的元素对于类来说非常重要，那就应该把它放在基类中，而不应该在派生的过程中再加入。如果你正在继承的过程中增加方法，就要考虑是否要重新设计了。

17. 少即是多。先从一个最小的类接口开始，使其尽可能小而简单，能解决手头的问题就行，但不要试图预测这个类所有**可能**的使用方式。随着对类的使用，你会逐渐清楚需要怎样扩展这个接口。不过，一旦开始使用一个类，你就无法在不破坏调用方代码的情况下收缩接口了。如果需要增加更多方法，那么没有问题，这不会破坏代码。但是即使用新方法取代了旧方法的功能，也要保留已有接口不变（如果需要，可以在内部实现中组合该功能）。如果必须通过增加更多的参数来扩展已有方法的接口，就要创建应用新参数的重载方法，这样就不会影响任何对已有方法的调用了。

18. **确保类之间关系的合理性**。基类和子类之间是"is-a"的关系，和成员对象之间是"has-a"的关系。

19. **在选择是继承还是组合的时候，问问自己是否必须向上转型为基类型**。如果不需要，优先考虑组合（成员对象），而不是继承。这样可以消除对多个基类型的需求。如果使用继承，使用者会认为他们应该向上转型。

20. **小心重载**。方法不应该根据参数的值（的不同）有条件地执行代码。此时，应该创建两个或更多的重载方法。

21. **使用异常体系**——最好从 Java 标准异常体系中选取合适的类，并由其派生出来。捕获了异常的人可以为具体的异常类型编写处理器，然后为基类型编写处理程序。如果要增加新的派生异常，已有的调用方代码仍然可以通过基类型捕获异常。如果使用异常来报告程序错误，错误报告会更易于理解。

22. **简单的聚合有时就能达成目的**。例如，飞机上的"乘客舒适系统"由以下互不相连的元素构成：座椅、空调、视频系统，等等。如果必须在飞机上创建很多这样的元素，你会生成私有成员并构建一个新接口吗？不会。在本场景中，这些组件也是公共接口的一部分，所以要创建公共成员对象。这些对象都有自己的私有实现，仍然是安全的。要小心，简单聚合并不是常用的解决方案，但有时确实会用到。

23. **从客户程序员和代码维护人员的视角考虑问题**。要将类设计得尽可能浅显易用，还要预见到可能发生的变化，并将类设计得易于适应这些变化。

24. **小心"巨型对象综合征"**。这常常让那些刚从面向过程编程转向面向对象编程的程序员痛苦不堪，他们最终写出的仍然是面向过程的程序，然后将其塞进一两个巨大的对象中。除了应用程序框架之外，对象代表的是应用程序中的概念，而不是应用程序本身。

25. **如果必须使用一些丑陋的设计，那么至少把它们限制在一个类中**。

26. **如果必须实现一套不具备移植性的服务，就对该服务进行抽象，并限制在一个类中**。这个额外的中间层可以防止不可移植性在程序中四处散播。（这种用法和其他一些用法都体现在**桥接模式**中。）

27. **对象不应只是简单地持有数据**。它们也应该具有定义明确的行为。不过，在通用的集合不适用的时候，"数据传输对象"（又称"信使"）适合明确地用来打包并传输一组条目数据。

28. **在基于已有的类创建新类时，组合优于继承**。只有在设计需要时，才使用继承。如果你在用组合就能解决问题的地方使用继承，就会使设计变得不必要的复杂。

29. **用继承和方法重写来表达行为的区别，用字段来表达状态的不同**。如果一个方法是基于类中的字段来改变自身的行为，就要考虑重新设计，通过在子类中实现不同的重写方法来表现不同的行为。

30. **小心差异**。两个语义不同的对象可能具有相同的动作或职责。出于对继承的迷信，人们会不自觉地试图使其中一个类成为另一个类的子类，认为这样会有好处。这称为**差异**（variance），但是如果实际上超类/子类的关系并不存在，并没有理由强行引入。更好的办法是创建一个通用的基类，该基类为两个类都生成一个接口，作为子类。这样仍然能够得到继承的好处，且可能会在设计上催生重要的发现。

31. **小心继承过程的限制**。最清晰的设计会为被继承的类添加新的功能。可疑的设计则会在继承过程中移除旧的功能，而不是增加新的功能。但是规则是用来打破的，如果你在使用旧的库，那么将现有的类限制在它的子类中可能会比重新构建继承层次结构更有效。这样新类就可以在旧类之上，适应它应该在的位置。

32. **使用设计模式消除"裸功能"**。也就是说，如果只需要创建类的一个对象，就不要匆忙地去程序中编写注释"只创建一个"，而是要将其包装到一个单例中。如果主程序中有很多混乱的代码会创建对象，就要寻找一种创建模式（如工厂方法），让你可以将创建过程封装在其中。消除"裸功能"不仅能使你的代码更易于理解和维护，也会使代码更安全，不容易被善意的维护人员无意破坏。

33. **检查工具类**。工具类（utility class）大体上是一种用于创建独立函数的方法，其中的方法都是静态的。有时静态字段会悄悄进入工具类中。这意味着该类也会开始隐藏自身状态，可能最好作为普通类来使用。可以试着移除所有的静态字段，看看是否效果更好。

34. **小心"分析瘫痪"**。记住，你通常会在掌握足够信息之前进入工程开发。要了解那些未知的因素，最好、最快的方法通常就是进入下一阶段，而不是试图在脑中凭空构思方案。你只能一步步摸索出解决方案，而难以提前知道方案是什么样的。Java 有内建的防火墙，要好好利用。你在一个类或一组类中出的错不会毁掉整个系统的完整性。

35. **如果你认为自己得出了满意的分析成果、设计或实现，请实际演练一遍**。引入团队外的人——不一定是专业顾问，公司其他团队中的人也可以。用一双崭新的眼睛来重新审视你的成果，可以在一个更容易解决问题的阶段发现问题，意义远大于为演练过程花费的时间和金钱开销。

A.2 实现建议

1. **遵循编程惯例**。业界有着大量不同的编程惯例，例如 Google Java 规范（本书中的代码就尽量遵循了该规范）。如果你顽固地坚持自己习惯的其他语言的代码风格，会给阅读代码的人带来困难。不论你决定使用哪种编程惯例，都要确保在整个项目中保持一致。集成开发环境通常会有内建的格式化工具和检查工具。

2. **不论使用哪种代码风格，如果你的团队（更好的是，整个公司）以其为标准，真的会带来好处**。这意味着每个人都认为，如果某人的编程风格不符合标准，那么修改它就是合理的。标准化的价值在于，它可以减少解析代码所耗费的精力，使你可以聚焦于代码的含义。

3. **遵循标准大小写规范**。类名的首字母要大写，字段、方法和对象（引用）的首字母应小写。所有的标识符都应该将其单词放在一起，并将第一个单词之后所有单词的首字母大写。例如：

    - ThisIsAClassName

    - thisIsAMethodOrFieldName
      对于在定义中具有常量初始值设定项的 static final 的基本类型标识符，**所有字母都要大写**（并使用下划线分隔符）。这表明它们是编译时常量。

    - **包名比较特殊**：字母应该全部小写。域名后缀（com、org、net、edu 等）也应该小写。（这是 Java 1.1 到 Java 2 之间的变化）。

4. **不要创建自定义的私有"装饰"字段名**。这通常以下划线和字符前缀的形式出现。匈牙利命名法便是最糟糕的例子，它（在各种命名前）附加了表示数据类型、用途、位置等信息的额外字符，就好像在编写汇编语言，并且编译器完全没有提供额外帮助一样。这些符号令人困惑、难以阅读，对执行和维护十分不友好。让类和包为你管理命名的范围界定。如果你觉得必须通过修饰命名来防止混淆，那么你的代码很可能太不清晰，应该简化。

5. 在创建通用的类时，**遵循"经典形式"**。要带上 equals()、hashCode()、toString()、clone()（实现 Cloneable，或选用其他的对象复制方式，如序列化）的定义，并实现 Comparable 和 Serializable。对于**信使**类，Java 17 中的 record 特性会为你完成以上大部分工作。

6. 对于读取或修改 private 字段的方法，**使用 get、set 和 is 命名惯例**。这样不仅使类更易于使用，而且是此类方法的标准命名方式，使阅读代码的人更容易理解。

7. **对你创建的每个类，都引入 JUnit 测试**。（参见 JUnit 网站，以及基础卷第 16 章中的示例。）在工程中使用类时，无须移除测试代码，而且如果你修改了代码，还可以很轻松地重新运行测试。测试代码也可以成为代码使用方法的示例。

8. **有时，你需要使用继承来访问基类中 protected 的成员**。这可能会引发感知多种基类型的需求。如果不需要向上转型，就要首先派生出一个新类来实现对 protected 权限的访问。然后将该新类作为要使用它的任意类的成员对象，而不使用继承。

9. **避免为了提升性能而使用 final 方法**。只有在通过分析发现方法调用成为性能瓶颈时，才应该使用 final。

10. **如果两个类以某种函数式的方式相互关联（例如集合和迭代器），那么试着将其中一个作为另一个的内部类**。这不仅强调了类之间的关联，而且可以通过将类名内嵌在另一个类中，实现对类名的复用。Java 集合库通过在每个集合类中定义一个内部 Iterator 类实现了这一点，从而为集合提供了一个公共接口。使用内部类的另一个原因是，这是作为 private 实现的一部分。此处，内部类的作用是为了隐藏实现，而不是上面提到的类关联和防止命名空间污染。

11. **一旦你注意到类之间有耦合过重的迹象，就要考虑用内部类来获得编程和维护上的改进**。使用内部类不会使类解耦，而是可以使耦合更显式、更方便。

12. **不要成为过早优化的牺牲品**。这条道路充满了疯狂。特别是，在你初次构建系统的时候，不要担心编写（或避免）本地方法，不要将某些方法设为 final 的，也不要为了提升效率而调整代码。你的主要目标应该是验证设计。即使你的设计需要一定的性能，也**要先运行起来，再来优化**。

13. **将作用域保持得尽量小**，这样对象的可视性和生命周期就会尽量小、尽量短。这样可以减少在错误的上下文中使用对象，以及产生难以发现的隐藏 bug 的概率。举例来说，假设有一个集合和一段迭代遍历该集合的代码。如果你将这段代码复制了一份，并用在了一个新的集合上，最终可能会意外地将原有集合的大小作为新集合的上限。但是，如果原有集合在作用域之外，就可以在编译期捕获错误。

14. **使用 Java 标准库中的集合**。熟练掌握这些集合的使用，将会极大提升你的生产效率。对于序列，首选 `ArrayList`；对于集（set），首选 `HashSet`；对于关联数组，首选 `HashMap`；对于栈和队列，优先使用 `LinkedList`（栈的首选并不是 `Stack`，尽管你可以创建一个适配器来提供栈接口；队列也可能需要使用适配器，如本书中所示）。在使用前三者的时候，要分别向上转型为 `List`、`Set` 和 `Map`，这样就可以在需要的时候改用其他的具体实现了。

15. **程序的整体稳健性需要依赖每个组件的稳健性**。在创建的每个类中充分使用 Java 提供的所有工具（访问控制、异常、类型检查、同步，等等），这样你就可以在构建系统时安全地进入下一层的抽象工作了。

16. **相较于运行时错误，优先利用编译时错误**。要在尽量靠近错误发生的地方处理错误。要在最临近、拥有足够信息来处理错误的异常处理器中捕获任何异常。要在当前层尽可能地处理异常，如果这样解决不了问题，就将异常抛出。

17. **小心冗长的方法定义**。方法应该是用来描述和实现类接口中某个离散部分的简短、函数式单元。冗长且复杂的方法会给维护带来困难和额外开销，而且很可能独自做了太多的事情。如果看见了这样的方法，就代表至少应该将它拆分为多个方法，也可能代表着需要创建一个新类。短小的方法也会促进该方法在类中的复用性。（有时方法必须很长，但仍然应该只做一件事。）

18. **让一切尽可能"私密"**。一旦开放（publicize）库的某些方面（某个方法、类或字段），就永远无法将它们移除。如果移除，就会破坏其他人的已有代码，迫使他们重新设计和编写。如果只开放必要的部分，就可以安全地改变其他任何部分了。由于设计会逐渐演进，这也带来了很大的自由度。通过这种方式，实现的变更对子类的影响会减到最小。私密性在处理多线程时格外重要——只有 `private` 的字段可以受到保护，免于非 `synchronized` 的使用。具有包访问权限的类仍然应该包含 `private` 的字段，但是对于其中的方法，合理的做法通常是赋予其包访问权限，而不是设为 `public` 权限。

19. **充分使用注释，并且使用 Javadoc 的注释文档语法来生成程序文档**。不过注释应该为代码增加真正的含义，只将代码已经清晰表达出的意图复述一遍的注释会令人厌烦。注意，Java 类典型的冗长细节信息和方法命名减少了某些注释的必要性。

20. **避免使用"魔法数字"**。"魔法数字"指的是硬编码到代码中的数字。如果你需要修改这些数字，它们就会如噩梦般可怕，因为你永远不知道"100"指的是"数组大小"还是"完全另一码事"。相反，应该创建一个名称具有明确含义的常量，并在程序中全部使用该常量的标识符。这样可以使程序更易理解、更易维护。

21. **在创建构造器时考虑异常**。在最好的情况下，构造器不会做任何会抛出异常的事。在次优的情况下，类只会由稳健的类组合和继承而来，所以在抛出异常时不需要进行清理。除此以外，都必须在 `finally` 子句中对组合而成的类进行清理。如果构造器必须失败，妥当的举动是抛出异常，以免调用者什么都不知道，以为对象已经成功创建了。

22. **在构造器内，只做将对象设置到合适状态所必需的事**。主动避免调用其他方法（除了 `final` 的方法），因为这些方法可能会被其他人重写，在构造过程中产生预期外的结果。（更多细节参见基础卷第 6 章。）越简短的构造器越不容易抛出异常或导致问题。

23. **在客户程序员完成对象的使用后，如果你的类需要任何清理，就要将用于清理的代码放在一个定义明确的类中**。这个类需要使用诸如 `dispose()` 这样能够清晰表达意图的名称。此外，可以在类中放置一个 `boolean` 标志，来表示 `dispose()` 方法是否已被调用，这样 `finalize()` 方法就可以检查"终结条件"（参见基础卷第 6 章）了。

24. `finalize()` 的职责仅限于验证调试对象的"终结条件"。（参见基础卷第 6 章。）在特殊情况下，可能需要该方法来释放垃圾收集器无法释放的内存。JVM 有可能不会为某个对象调用垃圾收集器，因此无法使用 `finalize()` 执行必要的清理。为此，必须创建自己的 `dispose()` 方法。在类的 `finalize()` 方法中，需要进行检查，以确保对象被清理了。如果没有清理，则抛出 `RuntimeException` 的子类，以指出编程错误。在依赖此方案之前，要确保 `finalize()` 在系统上可用（可能需要调用 `System.gc()` 来确保该行为的执行）。

25. 如果必须在特定作用域内手动清理一个对象（而不是用垃圾收集器清理），需要使用如下方法：初始化该对象；如果成功，就立刻进入 `try` 语块，并在 `finally` 子句中执行清理。

26. **在继承的过程中重写 `finalize()` 时，要调用 `super.finalize()`**。（如果当前的父类就是 `Object`，则不需要这么做。）要将 `super.finalize()` 放在重写的 `finalize()` 的**最末尾**处调用，而不是最前面，以确保基类的组件在需要时仍然有效。

27. **在创建固定大小的对象集合时，将它们转换为数组**，特别是在从方法中返回该集合时。这样就可以得到数组的运行时类型检查所带来的好处，而且数组的接收方在使用数组中的对象时可能不需要进行类型转换。注意，集合库的基类 java.util.Collection 中有两个 toArray() 方法可以达到这个目的。

28. **选择接口而不是抽象类**。如果你知道这是一个基类，第一选择应该是将其实现为接口，并且只在需要方法定义或成员变量时，才将其改为抽象类。接口关注的是调用者的需求，而类则倾向于关注（或者达成）实现细节。

29. **为了避免糟糕的体验，请确保不要在类路径中的任何位置上出现同名且均未放在包中的类（一个名称只能有一个不在包中的类）**。否则编译器会先找到同名的其他类，并报告毫无意义的错误消息。如果你怀疑类路径有问题，可以试着在类路径的每一个起始节点位置用相同的名称查找 .class 文件。在理想情况下，所有的类都应该放在包中。

30. **一定要使用 @Override**。如果你想要重写某个基类方法，却写错了方法名或使用了错误的参数，那么你实际上是增加了一个新的方法，而不是重写了某个已有方法。这不会引发编译器或运行时系统的错误信息——你的代码只是不会按你预期那样的运行而已。要防止这种情况，就一定要使用 @Override 注解，它会在你重写出错的时候提示你。

31. **小心过早优化**。先运行起来，再优化性能——但是只在必要且已证明代码中某部分出现了性能瓶颈时再优化。除非确实使用分析工具发现了瓶颈，否则就是在浪费时间。调整性能带来的额外潜在成本在于，会使代码变得不易于理解和维护。

32. **要牢记，代码被阅读的时间要远远长于编写所花的时间**。简洁的设计可以创造出易于理解的程序，但是注释、详细的解释、测试和示例同样是非常宝贵的，它们会帮助你和之后的所有人。即使没有其他问题，在 JDK 文档中苦寻信息的那些挫折经历也会驱使你这么做。

B

Javadoc

为代码编写文档，最大的问题可能就是对文档的维护了。如果文档和代码是分离的，那么每次更新代码都必须更新文档，会非常麻烦。解决的办法似乎很简单：将代码和文档关联起来。最简单的办法就是将所有东西都放在同一个文件里。不过，如果要完成这个构想，就需要特殊的注释语法来为文档打上标签，以及一个专门的工具，来将这些注释提取出来并转换为某种有用的格式。

这个提取注释的工具称为 Javadoc，可通过安装 JDK 获得。它使用了 Java 编译器的某些技术来寻找特殊的注释标签。除了提取出用这些标签标记的信息外，它还会拉取紧挨注释的类名或方法名。通过这种方式，你可以用最少的工作量生成优雅的程序文档。

Javadoc 的输出是 HTML 文件，可以用 Web 浏览器查看。有了 Javadoc，你就有了一套创建文档的简单标准，因此就具备了为所有 Java 库生成文档的能力。

此外，还可以编写称为 doclet 的自定义 Javadoc 处理程序，用来对 Javadoc 所处理的信息执行特殊操作（例如生成不同格式的输出）。

本附录随后的内容只是对 Javadoc 的基本介绍和概述。你可以在 JDK 文档中找到更详细的描述。

B.1 语法

所有的 Javadoc 指令都出现在以 /** 开头的注释内（但仍然以 */ 结束）。Javadoc 有两种主要的使用方式：一是嵌入式 HTML，二是使用"doc 标签"。**独立 doc 标签**是以 @ 开头的指令，放在一行注释的开头。（不过开头的 * 会被忽略。）**行内 doc 标签**可以出现在

Javadoc 注释内的任何地方，同样以 @ 开头，但必须用花括号包围起来。

一共有三种类型的注释文档，分别对应于注释前面的元素：类、字段或方法。也就是说，类的注释出现在类定义的正前方，字段的注释出现在字段定义的正前方，而方法的注释则出现在方法定义的正前方。下面是一个简单的例子：

```
// javadoc/Documentation1.java
/** 类的注释 */
public class Documentation1 {
  /** 字段的注释 */
  public int i;
  /** 方法的注释 */
  public void f() {}
}
```

Javadoc 只会为有 public 和 protected 权限的成员处理注释文档。private 和包权限成员的注释（见基础卷第 7 章）默认会被忽略，你不会看到相关输出。这很合理，因为以调用方程序员的视角来看，只有 public 和 protected 权限的成员在文件外是可用的。可以使用 -private 标志将 private 成员也纳入处理范围。

如果通过 Javadoc 处理前面的代码，则代码如下：

```
javadoc Documentation1.java
```

这会生成一组 HTML 文件。如果在浏览器中打开 index.html，就会看到结果和所有其他 Java 文档具有相同的标准格式，因此使用者可以很轻松地浏览各个类。

B.2 嵌入式 HTML

Javadoc 会将 HTML 代码原样传给生成的 HTML 文档。这使你可以充分地使用 HTML，不过主要目的是为了格式化代码，例如：

```
// javadoc/Documentation2.java
/** <pre>
 * System.out.println(new Date());
 * </pre>
 */
public class Documentation2 {}
```

你也可以像在其他任何 Web 文档中那样，使用 HTML 来格式化描述中的常规文本：

```
// javadoc/Documentation3.java
/** 你 <em>甚至 </em>还能插入列表：
```

```
 * <ol>
 * <li> 元素 1
 * <li> 元素 2
 * <li> 元素 3
 * </ol>
 */

public class Documentation3 {}
```

注意，在文档注释中，行首的星号和空格会被 Javadoc 丢弃。Javadoc 会重新格式化所有内容，使其符合标准的文档外观。不要使用诸如 <h1> 或 <hr> 之类的标题来作为嵌入式 HTML，因为 Javadoc 会插入自己的标题，这样两者就冲突了。

所有类型的注释文档（类、字段和方法）都可以支持嵌入式 HTML。

B.3 部分示例标签

下面是一些可用于代码文档的 Java 标签。在正式尝试使用 Javadoc 之前，请认真查阅 JDK 文档中对 Javadoc 的说明，了解 Javadoc 的所有使用方式。

B.3.1 @see

此标签用于引用其他类中的文档。Javadoc 会生成通过 @see 标签超链接到其他文档的 HTML。形式如下：

```
@see classname
@see fully-qualified-classname
@see fully-qualified-classname#method-name
```

每个 @see 都会向生成的文档中增加一个"See Also"（另请参阅……）超链接入口。Javadoc 不会验证该超链接是否可用。

B.3.2 {@link package.class\#member label}

它和 @see 非常相似，只不过还可以用在行内（@see 必须用在行首，{@link} 则无此限制），并且使用 label 作为超链接文本，而不是"See Also"。

B.3.3 {@docRoot}

生成相对文档根目录的相对路径。在显式超链接到文档树中的页面时很有用。

B.3.4 {@inheritDoc}

将文档从当前类最近的基类继承到当前的文档注释中。

B.3.5 @version

此标签形式如下：

```
@version version-information
```

其中 version-information（版本信息）是你认为适合引入的任意重要信息。如果在 Javadoc 命令行中使用 -version 标志，会在生成的 HTML 文档中专门调出版本信息。

B.3.6 @author

此标签形式如下：

```
@author author-information
```

其中 author-information（作者信息）通常是你的名字，但也可以包括你的电子邮箱地址或其他合适的任意信息。如果在 Javadoc 命令行中使用 -author 标签，会在生成的 HTML 文档中专门调出作者信息。

你可以使用多个作者标签来表示一组作者，但是必须将其连续排列。在生成的 HTML 中，所有的作者信息都会被集中放在一个段落里。

B.3.7 @since

此标签表示本代码是从哪个版本开始使用某个特定功能的。例如，它会出现在 HTML 的 Java 文档中，表示某功能首次出现的 JDK 版本。

B.3.8 @param

此标签会为方法参数生成文档：

```
@param parameter-name description
```

其中 parameter-name（参数名）是方法参数列表中的标识符，description（描述）则是可以继续在后面的行中显示的文本。此描述文本在遇到一个新的文档标签时即被认为结束。该标签的使用数量没有限制，通常每个参数用一个。

B.3.9 @return

此标签会记录返回值：

```
@return description
```

其中 description 用于表示返回值的意义。它支持换行。

B.3.10 @throws

方法可以产生任意数量的不同类型的异常，这些异常都需要描述。异常标签的形式如下：

```
@throws fully-qualified-class-name description
```

其中 fully-qualified-class-name（完整的合格类名）用于为异常类提供一个明确的名字，description（它支持换行）则会告诉你为什么该类型的异常会在方法调用中出现。

B.3.11 @deprecated

此标签表示该功能已被改进后的功能取代。deprecated（弃用）标签建议你不要继续使用该功能，因为它可能会在未来某个时候被移除。如果调用了被标记为 @deprecated 的方法，会导致编译器产生警告。在 Java 5 中，Javadoc 标签 @deprecated 被 @Deprecated 注解取代了（见本书第 4 章）。

B.4 文档示例

下面是带有文档注释的 objects/HelloDate.java：

```java
// javadoc/HelloDateDoc.java
import java.util.*;

/** 本书的第一个示例程序。
 * 显示一个字符串和今天的日期
 * @author Bruce Eckel
 * @author www.MindviewInc.com
 * @version 5.0
 */
public class HelloDateDoc {
  /** 类或应用的进入点
   * @param args array of String arguments
   * @throws exceptions No exceptions thrown
   */
  public static void main(String[] args) {
    System.out.println("Hello, it's: ");
    System.out.println(new Date());
  }
}
/* Output:
Hello, it's:
Sun Jan 24 08:49:10 MST 2021
*/
```

你可以在 Java 标准库的源代码中找到 Javadoc 注释文档的很多例子。

理解 equals() 和 hashCode()

在基于哈希实现的容器里，如果想使用某个类，不仅需要定义它的 hashCode() 方法（我们将在本附录稍后详细探讨），还需要定义 equals() 方法。这两个方法一起使用，才能实现对哈希容器的正确查找。

C.1 经典的 equals()

当创建一个新类时，它会自动继承 Object 类。如果该类没有重写 equals() 方法，它会使用 Object 类中的 equals() 方法。这个方法默认比较内存地址，所以只有比较的是同一个对象时，它才会返回 true。这个默认实现是"最具辨别力"的。

```java
// equalshashcode/DefaultComparison.java

class DefaultComparison {
  private int i, j, k;
  DefaultComparison(int i, int j, int k) {
    this.i = i;
    this.j = j;
    this.k = k;
  }

  public static void main(String[] args) {
    DefaultComparison
      a = new DefaultComparison(1, 2, 3),
      b = new DefaultComparison(1, 2, 3);
    System.out.println(a == a);
    System.out.println(a == b);
  }
}
```

```
/* 输出:
true
false
*/
```

通常情况下，我们希望能放宽这个限制。如果两个对象的类型相同并且字段值也相等，一般就可以认为这两个对象相等了。但也可能存在我们不想包含在 equals() 中进行比较的字段，这属于类设计过程的一部分。

一个适当的 equals() 方法必须满足以下五个条件。

1. 自反性：对于任意的 x，调用 x.equals(x) 时应该返回 true。
2. 对称性：对于任意的 x 和 y，当且仅当 y.equals(x) 返回 true 时，x.equals(y) 返回 true。
3. 传递性：对于任意的 x、y 和 z，如果 x.equals(y) 和 y.equals(z) 都返回 true，那么 x.equals(z) 也应该返回 true。
4. 一致性：对于任意的 x 和 y，如果对象中用于相等性比较的信息没有被修改过，那么多次调用 x.equals(y) 应该始终返回 true 或 false。
5. 对于任意非空的 x，调用 x.equals(null) 应该返回 false。

下面是满足这些条件的一系列测试，它可以确定自身与所比较的对象（在这里称为 rval）是否相等。

1. 如果 rval 为 null，则两个对象不相等。
2. 如果 rval 为 this 对象（即自己和自己比较），则两个对象相等。
3. 如果 rval 不是 this 对象所属的类或子类，则两个对象不相等。
4. 如果以上测试都通过，接下来就需要确定 rval 中哪些字段是重要的（并且是一致的），然后对它们进行比较。

Java 7 中引入了 Objects 类来帮助完成上述流程，它可以帮助我们编写出更好的 equals() 方法。

下面的示例对不同版本的 Equality 类进行了比较。为了避免代码冗余，我们将使用**工厂方法**设计模式（参见本书第 8 章 8.6 节）来构建示例。EqualityFactory 接口提供了 make() 方法来创建 Equality 对象，这样不同的 EqualityFactory 就可以创建不同的 Equality 子类型了：

```
// equalshashcode/EqualityFactory.java
import java.util.*;

interface EqualityFactory {
  Equality make(int i, String s, double d);
}
```

理解 equals() 和 hashCode()

现在我们将定义一个 Equality 类，它包含了三个字段（在比较时我们认为所有这些字段都很重要），还有一个满足了上述四项检查的 equals() 方法。构造器会打印类型的名称，以确保程序执行的是我们想要的测试：

```java
// equalshashcode/Equality.java
import java.util.*;

public class Equality {
  protected int i;
  protected String s;
  protected double d;
  public Equality(int i, String s, double d) {
    this.i = i;
    this.s = s;
    this.d = d;
    System.out.println("made 'Equality'");
  }
  @Override public boolean equals(Object rval) {
    if(rval == null)
      return false;
    if(rval == this)
      return true;
    if(!(rval instanceof Equality))
      return false;
    Equality other = (Equality)rval;
    if(!Objects.equals(i, other.i))
      return false;
    if(!Objects.equals(s, other.s))
      return false;
    if(!Objects.equals(d, other.d))
      return false;
    return true;
  }
  public void
  test(String descr, String expected, Object rval) {
    System.out.format("-- Testing %s --%n" +
      "%s instanceof Equality: %s%n" +
      "Expected %s, got %s%n",
      descr, descr, rval instanceof Equality,
      expected, equals(rval));
  }
  public static void testAll(EqualityFactory eqf) {
    Equality
      e = eqf.make(1, "Monty", 3.14),
      eq = eqf.make(1, "Monty", 3.14),
      neq = eqf.make(99, "Bob", 1.618);
    e.test("null", "false", null);
    e.test("same object", "true", e);
    e.test("different type",
           "false", Integer.valueOf(99));
    e.test("same values", "true", eq);
```

```
    e.test("different values", "false", neq);
  }
  public static void main(String[] args) {
    testAll( (i, s, d) -> new Equality(i, s, d));
  }
}
/* 输出:
made 'Equality'
made 'Equality'
made 'Equality'
-- Testing null -
null instanceof Equality: false
Expected false, got false
-- Testing same object --
same object instanceof Equality: true
Expected true, got true
-- Testing different type --
different type instanceof Equality: false
Expected false, got false
-- Testing same values --
same values instanceof Equality: true
Expected true, got true
-- Testing different values --
different values instanceof Equality: true
Expected false, got false
*/
```

testAll() 方法比较了我们想要比较的所有不同类型的对象。它使用工厂方法来创建 Equality 对象。

在 main() 方法中，可以看到调用 testAll() 方法非常简单。EqualityFactory 接口中只有一个方法，所以可以使用 lambda 表达式来实现 make() 方法。

不过上面的 equals() 方法冗长得令人生厌，可以简化成经典形式。我们可以看到：

1. 使用了 instanceof 进行检查之后，就不必再检查是否为 null 了；
2. 与 this 进行比较是多余的。只要能正确编写 equals() 方法，自我比较肯定不会有问题。

&& 操作符可以进行短路比较，它在第一次遇到失败时就会退出并返回 false。因此，用 && 操作符将这些检查连接起来，这样实现的 equals() 方法会更简洁：

```java
// equalshashcode/SuccinctEquality.java
import java.util.*;

public class SuccinctEquality extends Equality {
  public SuccinctEquality(int i, String s, double d) {
    super(i, s, d);
    System.out.println("made 'SuccinctEquality'");
  }
```

```
  @Override public boolean equals(Object rval) {
    return rval instanceof SuccinctEquality &&
      Objects.equals(i, ((SuccinctEquality)rval).i) &&
      Objects.equals(s, ((SuccinctEquality)rval).s) &&
      Objects.equals(d, ((SuccinctEquality)rval).d);
  }
  public static void main(String[] args) {
    Equality.testAll( (i, s, d) ->
      new SuccinctEquality(i, s, d));
  }
}
/* 输出：
made 'Equality'
made 'SuccinctEquality'
made 'Equality'
made 'SuccinctEquality'
made 'Equality'
made 'SuccinctEquality'
-- Testing null --
null instanceof Equality: false
Expected false, got false
-- Testing same object --
same object instanceof Equality: true
Expected true, got true
-- Testing different type --
different type instanceof Equality: false
Expected false, got false
-- Testing same values --
same values instanceof Equality: true
Expected true, got true
-- Testing different values --
different values instanceof Equality: true
Expected false, got false
*/
```

对于 SuccinctEquality 类而言，基类构造器会在子类构造器之前调用。输出显示我们仍然得到了正确的结果。我们可以判断短路的确发生了，否则在 equals() 的比较列表更下方部分进行强制类型转换时，null 测试和 "不同类型" 测试都会抛出异常。

当使用另一个类来组合新类时，Objects.equals() 更是能大显身手：

```
// equalshashcode/ComposedEquality.java
import java.util.*;

class Part {
  String ss;
  double dd;
  Part(String ss, double dd) {
    this.ss = ss;
    this.dd = dd;
  }
```

```java
  @Override public boolean equals(Object rval) {
    return rval instanceof Part &&
      Objects.equals(ss, ((Part)rval).ss) &&
      Objects.equals(dd, ((Part)rval).dd);
  }
}

public class ComposedEquality extends SuccinctEquality {
  Part part;
  public ComposedEquality(int i, String s, double d) {
    super(i, s, d);
    part = new Part(s, d);
    System.out.println("made 'ComposedEquality'");
  }
  @Override public boolean equals(Object rval) {
    return rval instanceof ComposedEquality &&
      super.equals(rval) &&
      Objects.equals(part,
        ((ComposedEquality)rval).part);
  }
  public static void main(String[] args) {
    Equality.testAll( (i, s, d) ->
      new ComposedEquality(i, s, d));
  }
}
/* 输出：
made 'Equality'
made 'SuccinctEquality'
made 'ComposedEquality'
made 'Equality'
made 'SuccinctEquality'
made 'ComposedEquality'
made 'Equality'
made 'SuccinctEquality'
made 'ComposedEquality'
-- Testing null --
null instanceof Equality: false
Expected false, got false
-- Testing same object --
same object instanceof Equality: true
Expected true, got true
-- Testing different type --
different type instanceof Equality: false
Expected false, got false
-- Testing same values --
same values instanceof Equality: true
Expected true, got true
-- Testing different values --
different values instanceof Equality: true
Expected false, got false
*/
```

注意这里对 super.equals() 的调用——不需要重复造轮子（而且你也并不总是能够访

问基类的必要部分）。

子类型之间的相等性

继承的特性告诉我们，两个不同子类型的对象在向上强制转型后可以是"相同的"。假设现在有一个 Animal 对象的集合，这个集合只能接受 Animal 的子类型——本例中为 Dog 和 Pig。每个 Animal 都有一个 name 和一个 size，以及一个唯一的内部 id。

Objects 类提供了 equals() 和 hashCode() 的经典形式定义来供我们使用。

我们可以通过 Objects 类来定义经典形式的 equals() 和 hashCode() 方法，但本例中我们只在基类 Animal 中定义它们，并且这两个方法都不包含唯一的 id。从 equals() 的角度来看，这意味着我们只关心某物是否是 Animal，而不在乎它是否是特定类型的 Animal：

```java
// equalshashcode/SubtypeEquality.java
import java.util.*;

enum Size { SMALL, MEDIUM, LARGE }

class Animal {
  private static int counter = 0;
  private final int id = counter++;
  private final String name;
  private final Size size;
  Animal(String name, Size size) {
    this.name = name;
    this.size = size;
  }
  @Override public boolean equals(Object rval) {
    return rval instanceof Animal &&
      // Objects.equals(id, ((Animal)rval).id) && // [1]
      Objects.equals(name, ((Animal)rval).name) &&
      Objects.equals(size, ((Animal)rval).size);
  }
  @Override public int hashCode() {
    return Objects.hash(name, size);
    // return Objects.hash(name, size, id);       // [2]
  }
  @Override public String toString() {
    return String.format("%s[%d]: %s %s %x",
      getClass().getSimpleName(), id,
      name, size, hashCode());
  }
}

class Dog extends Animal {
  Dog(String name, Size size) {
    super(name, size);
  }
}
```

```
/* 输出：
Dog[0]: Ralph MEDIUM 931523a9
*/
```

```
class Pig extends Animal {
  Pig(String name, Size size) {
    super(name, size);
  }
}

public class SubtypeEquality {
  public static void main(String[] args) {
    Set<Animal> pets = new HashSet<>();
    pets.add(new Dog("Ralph", Size.MEDIUM));
    pets.add(new Pig("Ralph", Size.MEDIUM));
    pets.forEach(System.out::println);
  }
}
```

如果我们只考虑类型的话，有时只从基类型的角度出发来理解类是有道理的，这也是**里氏替换原则**（Liskov Substitution Principle）的基础。这段代码就非常符合该原则，因为和基类相比，子类型并没有添加任何额外的功能（方法）。子类型仅在行为上不同，但在接口上并没有什么不同（这当然不是一般情况）。

但是当我们提供具有相同数据的两种不同类型的对象时，如果将它们都放进 HashSet<Animal> 中，结果只有一个对象添加成功。这里强调了 equals() 不是一个完美的数学概念，而是（至少部分是）一个机制上的概念。 如果想让类型在哈希数据结构中正常工作，那么 hashCode() 和 equals() 方法必须同时定义。

在本例中，Dog 和 Pig 都哈希到了 HashSet 中的同一个桶里。此时，HashSet 会回退到使用 equals() 方法来区分对象，但是本例的 equals() 方法也认为两者相等。HashSet 不会添加 Pig 对象，因为它已经有了一个相同的对象 。

即使对象的其他方面都相同，我们还是可以通过给它们强制添加唯一性来使示例正常工作。本例中，每个 Animal 已经有了一个唯一的 id，因此你可以取消 equals() 中第 [1] 行的注释，或者在 hashCode() 中将代码切换到第 [2] 行。在经典形式中一般会两者都执行，这样就可以在这两个操作中包含所有"不变"的字段（"不变"意味着 equals() 和 hashCode() 在哈希数据结构里存储和检索时不会产生不同的值。我之所以将"不变"放在引号中，是因为你必须自行评估某个字段是否可能会被修改）。

旁注：在自己实现的 hashCode() 方法中，如果只涉及单个字段，请使用 Objects.hashCode() 方法。如果有多个字段，请使用 Objects.hash() 方法。

我们也可以通过遵循标准形式来解决这个问题：在子类中定义 equals() 方法（但仍然不包括这个唯一的 id）：

```java
// equalshashcode/SubtypeEquality2.java
import java.util.*;

class Dog2 extends Animal {
  Dog2(String name, Size size) {
    super(name, size);
  }
  @Override public boolean equals(Object rval) {
    return rval instanceof Dog2 &&
      super.equals(rval);
  }
}

class Pig2 extends Animal {
  Pig2(String name, Size size) {
    super(name, size);
  }
  @Override public boolean equals(Object rval) {
    return rval instanceof Pig2 &&
      super.equals(rval);
  }
}

public class SubtypeEquality2 {
  public static void main(String[] args) {
    Set<Animal> pets = new HashSet<>();
    pets.add(new Dog2("Ralph", Size.MEDIUM));
    pets.add(new Pig2("Ralph", Size.MEDIUM));
    pets.forEach(System.out::println);
  }
}
/* 输出：
Dog2[0]: Ralph MEDIUM 931523a9
Pig2[1]: Ralph MEDIUM 931523a9
*/
```

请注意这里的 hashCode() 是相同的，但是因为对象不再是 equals() 的，所以现在两者都出现在了 HashSet 中。此外，super.equals() 意味着我们不需要访问基类中的私有字段。

这个问题的一种解读方式是，Java 分离了可替代性与 equals() 和 hashCode() 的定义。我们仍然可以将 Dog 和 Pig 放入 Set<Animal> 中，而不去理会 equals() 和 hashCode() 是如何定义的。但是在定义这些方法时如果不想着哈希结构，这些对象在哈希数据结构中就不会表现正常。不幸的是，equals() 并不仅仅与 hashCode() 结合使用。当你试图避免为特定类定义它时，会使事情变得复杂，这也是应该遵循经典形式的原因。而使情况更加复杂的是，有些时候你只需要实现这两个方法中的一个。

C.2 哈希和哈希码

在基础卷第 12 章的示例中，我们使用了预定义的类来作为 HashMap 的键。这些示例之所以有效，是因为预定义的类实现了所有必要的功能，这使得它可以作为键来正常工作。

如果想用自定义的类作为 HashMap 的键，一个经常会犯的错误是，忘记实现必要的功能。例如，考虑一个天气预测系统，它用 Groundhog 对象去关联匹配 Prediction 对象。这看起来很简单——使用 Groundhog 作为键，Prediction 作为值：

```java
// equalshashcode/Groundhog.java
// 看起来可行，但是作为 HashMap 的键时不能正常工作

public class Groundhog {
  protected int number;
  public Groundhog(int n) { number = n; }
  @Override public String toString() {
    return "Groundhog #" + number;
  }
}
```

```java
// equalshashcode/Prediction.java
// 预测天气
import java.util.*;

public class Prediction {
  private static Random rand = new Random(47);
  @Override public String toString() {
    return rand.nextBoolean() ?
      "Six more weeks of Winter!" : "Early Spring!";
  }
}
```

```java
// equalshashcode/SpringDetector.java
// 未来的天气怎么样?
import java.util.*;
import java.util.stream.*;
import java.util.function.*;
import java.lang.reflect.*;

public class SpringDetector {
  public static <T extends Groundhog>
  void detectSpring(Class<T> type) {
    try {
      Constructor<T> ghog =
        type.getConstructor(int.class);
      Map<Groundhog, Prediction> map =
        IntStream.range(0, 10)
          .mapToObj(i -> {
            try {
              return ghog.newInstance(i);
            } catch(Exception e) {
              throw new RuntimeException(e);
            }
          })
          .collect(Collectors.toMap(
            Function.identity(),
```

```
        gh -> new Prediction()));
    map.forEach((k, v) ->
      System.out.println(k + ": " + v));
    Groundhog gh = ghog.newInstance(3);
    System.out.println(
      "Looking up prediction for " + gh);
    if(map.containsKey(gh))
      System.out.println(map.get(gh));
    else
      System.out.println("Key not found: " + gh);
  } catch(NoSuchMethodException |
          IllegalAccessException |
          InvocationTargetException |
          InstantiationException e) {
    throw new RuntimeException(e);
  }
}
public static void main(String[] args) {
  detectSpring(Groundhog.class);
}
}
/* 输出:
Groundhog #5: Six more weeks of Winter!
Groundhog #2: Early Spring!
Groundhog #1: Six more weeks of Winter!
Groundhog #0: Early Spring!
Groundhog #8: Early Spring!
Groundhog #6: Six more weeks of Winter!
Groundhog #4: Early Spring!
Groundhog #3: Early Spring!
Groundhog #7: Six more weeks of Winter!
Groundhog #9: Six more weeks of Winter!
Looking up prediction for Groundhog #3
Key not found: Groundhog #3
*/
```

每个 Groundhog 都有一个标识号，因此如果想要在 HashMap 中查找某个 Prediction 的时候，可以这样表示："请给我与 Groundhog #3 关联的 Prediction。"Prediction 通过随机生成一个布尔值来选择天气情况。detectSpring() 方法使用反射来实例化，并且接收 Groundhog 类或其子类作为参数。这个特性稍后会派上用场，我们会继承一种新类型的 Groundhog 来解决这里演示的问题。

HashMap 中存储了 Groundhog 及其对应的 Prediction。遍历 HashMap 后的输出显示了它已经被填充。然后用编号 Groundhog #3 作为键去查询相应的预测结果（这个结果肯定是在 Map 中的）。

这看起来很简单，但不起作用——我们并没有找到值为 #3 的键。问题在于 Groundhog 自动继承了 Object 类，并且调用了 Object 的 hashCode() 方法为每个对象生成哈希码。默认情况下，该方法使用对象的内存地址。因此，Groundhog(3) 的第一个实例的哈希码，与用于查询的第二个实例的哈希码并不相同。

我们需要适当地重写 hashCode() 方法。但这仍然不会起作用，还需要完成一件事：重写 Object 的 equals() 方法，这是因为 HashMap 使用了 equals() 方法来判定你的键和列表中的其他键是否相等。

默认的 Object.equals() 方法比较对象的内存地址，而两个 Groundhog(3) 对象的内存地址并不相等。因此，想要让自定义的类作为 HashMap 的键，就必须重写 hashCode() 和 equals() 这两个方法，下面这个方案就解决了 Groundhog 的问题：

```java
// equalshashcode/Groundhog2.java
// 使用一个类作为 HashMap 的键时，必须重写它的 hashCode() 和 equals() 方法

import java.util.*;

public class Groundhog2 extends Groundhog {
  public Groundhog2(int n) { super(n); }
  @Override
  public int hashCode() { return number; }
  @Override public boolean equals(Object o) {
    return o instanceof Groundhog2 &&
      Objects.equals(
        number, ((Groundhog2)o).number);
  }
}
```

```java
// equalshashcode/SpringDetector2.java
// 能正常工作的键

public class SpringDetector2 {
  public static void main(String[] args) {
    SpringDetector.detectSpring(Groundhog2.class);
  }
}
/* 输出：
Groundhog #0: Six more weeks of Winter!
Groundhog #1: Early Spring!
Groundhog #2: Six more weeks of Winter!
Groundhog #3: Early Spring!
Groundhog #4: Early Spring!
Groundhog #5: Six more weeks of Winter!
Groundhog #6: Early Spring!
Groundhog #7: Early Spring!
Groundhog #8: Six more weeks of Winter!
Groundhog #9: Six more weeks of Winter!
Looking up prediction for Groundhog #3
Early Spring!
*/
```

Groundhog2.hashCode() 方法返回了 Groundhog 的编号作为哈希值。在本例中，程序员负责确保 Groundhog 编号的唯一性。hashCode() 方法并没有被要求一定要返回唯一的标识（本附录后面会讲解），但是 equals() 方法必须严格判断两个对象是否相等。这里的 equals() 方法基于 Groundhog 的编号，所以如果两个 Groundhog 对象的编号相同，它们就不能同时作为键保存在 HashMap 中。

关于如何定义 equals() 方法，C.1 节已经介绍过。上面示例的输出显示，现在功能已经正常。

C.2.1 理解 hashCode()

前面的示例只是正确解决问题的开始。它表明，如果作为键的对象没有重写 hashCode() 和 equals() 方法，那么哈希数据结构（比如 HashSet、HashMap、LinkedHashSet 和 LinkedHashMap）就可能无法正确地处理键。而如果想要很好地解决这个问题，就必须了解哈希数据结构的内部原理。

首先，考虑一下使用哈希背后的动机：用一个对象去查找另一个对象。但你也可以使用 TreeMap 来完成此操作，甚至可以实现一个自己的 Map。与哈希实现相反，下面的例子使用了一对 ArrayList 来实现一个 Map。与 AssociativeArray.java 不同的是，它完整地实现了 Map 接口，因此也包含了 entrySet() 方法：

```java
// equalshashcode/SlowMap.java
// 用 ArrayLists 实现的 Map
import java.util.*;
import onjava.*;

public class SlowMap<K, V> extends AbstractMap<K, V> {
  private List<K> keys = new ArrayList<>();
  private List<V> values = new ArrayList<>();

  @Override public V put(K key, V value) {
    V oldValue = get(key); // 旧的值或 null
    if (!keys.contains(key)) {
      keys.add(key);
      values.add(value);
    } else
      values.set(keys.indexOf(key), value);
    return oldValue;
  }

  @Override
  public V get(Object key) { // key 是 Object 类型的，而非参数化类型 K
    if (!keys.contains(key))
      return null;
```

```java
      return values.get(keys.indexOf(key));
    }
    @Override public Set<Map.Entry<K, V>> entrySet() {
      Set<Map.Entry<K, V>> set = new HashSet<>();
      Iterator<K> ki = keys.iterator();
      Iterator<V> vi = values.iterator();
      while (ki.hasNext())
        set.add(new MapEntry<>(ki.next(), vi.next()));
      return set;
    }
    public static void main(String[] args) {
      SlowMap<String, String> m = new SlowMap<>();
      m.putAll(Countries.capitals(8));
      m.forEach((k, v) -> System.out.println(k + "=" + v));
      System.out.println(m.get("BENIN"));
      m.entrySet().forEach(System.out::println);
    }
}
/* 输出:
CAMEROON=Yaounde
ANGOLA=Luanda
BURKINA FASO=Ouagadougou
BURUNDI=Bujumbura
ALGERIA=Algiers
BENIN=Porto-Novo CAPE
VERDE=Praia
BOTSWANA=Gaberone
Porto-Novo
```
（转右栏）

```
CAMEROON=Yaounde
ANGOLA=Luanda
BURKINA FASO=Ouagadougou
BURUNDI=Bujumbura
ALGERIA=Algiers
BENIN=Porto-Novo
CAPE VERDE=Praia
BOTSWANA=Gaberone
*/
```

put() 方法将键和值放在相应的 ArrayList 中。与 Map 接口一样，它返回了旧键，如果找不到相应的键，则返回 null。

根据 Map 的规范，如果要查找的键不在 SlowMap 中，get() 方法就会返回 null。如果键存在，它会被用来查找对应的数字索引值（这个值表示它在列表 keys 中的位置），然后再以这个数字值为索引，在列表 values 中查找对应的值。请注意，get() 方法的参数 key 是 Object 类型，而不是 K 类型（而在 AssociativeArray.java 中就是这样用的）。这是因为 Java 语言在发展了很长时间后才引入泛型——如果泛型是该语言的原始特性，那么 get() 就可以指定其参数的类型了 [1]。

SlowMap 的字符串表示形式是由 AbstractMap 中定义的 toString() 方法自动生成的。

---

[1] 具体原因可参考 Gilad Bracha 关于泛型的文章 "Converting Legacy Code to Use Generics"，以及 HashMap 的作者 Josh Bloch 相关的演讲 "Advanced Topics in Programming Languages: Java Puzzlers"。——译者注

在 SlowMap.main() 中，我们先加载了一个 SlowMap，然后打印了它的内容。我们还调用了 get() 方法，验证了它的确是有效的。

Map.entrySet() 会生成一组 Map.Entry 对象。但 Map.Entry 只是一个接口，它所描述结构依赖于具体实现，因此要实现自己的 Map 类型，还必须实现 Map.Entry：

```java
// equalshashcode/MapEntry.java
// Map 实现里的一个简单 Map.Entry
import java.util.*;

public class MapEntry<K, V> implements Map.Entry<K, V> {
  private K key;
  private V value;
  public MapEntry(K key, V value) {
    this.key = key;
    this.value = value;
  }
  @Override public K getKey() { return key; }
  @Override public V getValue() { return value; }
  @Override public V setValue(V v) {
    V result = value;
    value = v;
    return result;
  }
  @Override public int hashCode() {
    return Objects.hash(key, value);
  }
  @SuppressWarnings("unchecked")
  @Override public boolean equals(Object rval) {
    return rval instanceof MapEntry &&
      Objects.equals(key,
        ((MapEntry<K, V>)rval).getKey()) &&
      Objects.equals(value,
        ((MapEntry<K, V>)rval).getValue());
  }
  @Override public String toString() {
    return key + "=" + value;
  }
}
```

equals() 方法遵循了 C.1 节里的要求。Objects 类提供了一个类似的方法来帮助创建 hashCode() 方法：Objects.hash()。在定义 hashCode() 时，如果涉及多个字段就使用该方法。如果只涉及单个字段，请改用 Objects.hashCode()。

上面这个简单的解决方案可以通过 SlowMap.main() 中的小测试，但它不是一个正确的实现，因为它复制了键和值。entrySet() 的正确实现需要提供一个 Map 的视图，而不是一个副本，并且这个视图允许我们修改原始的 Map（副本则不行）。

C.2.2 用哈希来提高速度

SlowMap.java 表明创建一种新类型的 Map 并不难。但顾名思义，SlowMap 并不快，所以如果有其他选择，我们就不会使用它。它的问题出现在查找过程中：键没有按任何特定的顺序来保存，所以搜索的话只能使用简单的线性搜索。而线性搜索是最慢的查找方式。

我们之所以用哈希就是为了提高速度，因为通过哈希来查找值的速度非常快。由于瓶颈在于键的查找速度，因此一种解决方案是保证键是有序的，然后使用 Collections.binarySearch() 来执行查找。

哈希则更进一步，它通过一种可以快速定位的方式来存储键。存储一组元素的最快结构是数组，因此数组用来表示键的信息（注意我说的是"键的信息"，而不是键本身）。但因为数组不能调整大小，所以就遇到了一个问题：我们想在 Map 中存储不确定数量的值，而键的数量被数组大小所固定，那该怎么办呢？

这里的数组并不保存键。我们会根据键对象来生成一个数，然后通过这个数来索引数组。这个数就是**哈希码**（hash code），它可以通过 Object 中定义的 hashCode() 方法（使用一个哈希函数）生成，也可以通过自己类里重写的 hashCode() 方法生成。

为了解决数组固定大小的问题，多个键可以生成相同的索引。也就是说，可能存在**碰撞**（collision）。因此，数组有多大并不重要，任何键对象的哈希码都能在数组中找到自己对应的位置。

因此，为了查找值，我们首先计算哈希码并使用它来索引数组。如果能保证没有冲突（值的数量固定的话是有可能的），你就会有一个**完美的哈希函数**，但这是特殊情况[①]。在一般情况下，冲突由外部的链表来处理。数组并不直接指向一个值，而是指向一个值的链表。我们通过 equals() 方法以线性方式这些值中搜索（因此，equals() 对于哈希也是必不可少的）。这个搜索要慢很多，但是如果哈希函数足够好，每个槽中只会有几个值。这样就无须搜索整个链表，而是快速跳到一个槽里，然后在该槽中只需比较几个条目即可找到值。这就是 HashMap 能够进行快速查找的原因。

了解了哈希的基础知识，你就可以用哈希实现一个简单的 Map 了：

```
// equalshashcode/SimpleHashMap.java
// 一个哈希实现的演示版本 Map
import java.util.*;
import onjava.*;
```

[①] EnumMap 和 EnumSet 实现了完美的哈希函数，这是因为 enum 定义了固定数量的实例，请参阅本书第 1 章。

理解 equals() 和 hashCode()

```java
public
class SimpleHashMap<K, V> extends AbstractMap<K, V> {
  // 为哈希表大小选择一个质数,以实现均匀分布:
  static final int SIZE = 997;
  // 你不能定义一个实际的泛型数组(如K[]),但可以通过向上转型来获得
  @SuppressWarnings("unchecked")
  LinkedList<MapEntry<K, V>>[] buckets =
    new LinkedList[SIZE];
  @Override public V put(K key, V value) {
    V oldValue = null;
    int index = Math.abs(key.hashCode()) % SIZE;
    if(buckets[index] == null)
      buckets[index] = new LinkedList<>();
    LinkedList<MapEntry<K, V>> bucket = buckets[index];
    MapEntry<K, V> pair = new MapEntry<>(key, value);
    boolean found = false;
    ListIterator<MapEntry<K, V>> it =
      bucket.listIterator();
    while(it.hasNext()) {
      MapEntry<K, V> iPair = it.next();
      if(iPair.getKey().equals(key)) {
        oldValue = iPair.getValue();
        it.set(pair); // 用新值代替旧值
        found = true;
        break;
      }
    }
    if(!found)
      buckets[index].add(pair);
    return oldValue;
  }
  @Override public V get(Object key) {
    int index = Math.abs(key.hashCode()) % SIZE;
    if(buckets[index] == null) return null;
    for(MapEntry<K, V> iPair : buckets[index])
      if(iPair.getKey().equals(key))
        return iPair.getValue();
    return null;
  }
  @Override public Set<Map.Entry<K, V>> entrySet() {
    Set<Map.Entry<K, V>> set= new HashSet<>();
    for(LinkedList<MapEntry<K, V>> bucket : buckets) {
      if(bucket == null) continue;
      for(MapEntry<K, V> mpair : bucket)
        set.add(mpair);
    }
    return set;
  }
  public static void main(String[] args) {
    SimpleHashMap<String,String> m =
      new SimpleHashMap<>();
    m.putAll(Countries.capitals(8));
```

```
/* 输出:
CAMEROON=Yaounde
ANGOLA=Luanda
BURKINA FASO=Ouagadougou
BURUNDI=Bujumbura
ALGERIA=Algiers
BENIN=Porto-Novo
CAPE VERDE=Praia
BOTSWANA=Gaberone
Porto-Novo
CAMEROON=Yaounde
ANGOLA=Luanda
BURKINA FASO=Ouagadougou
BURUNDI=Bujumbura
ALGERIA=Algiers
BENIN=Porto-Novo
CAPE VERDE=Praia
BOTSWANA=Gaberone
*/
```

```
    m.forEach((k, v) ->
      System.out.println(k + "=" + v));
    System.out.println(m.get("BENIN"));
    m.entrySet().forEach(System.out::println);
  }
}
```

因为哈希表中的"槽"通常被称为**桶**（bucket），所以示例中代表实际的表的数组被命名为 buckets。为了得到更均匀的分布，桶的数量通常是质数[①]。注意它是一个 LinkedList 的数组，能自动为碰撞提供解决方案——每个新项都会被添加到特定桶里的 LinkedList 末尾。尽管 Java 不允许直接创建泛型数组，但可以对此类数组进行引用。示例中我们向上转型到一个这样的数组，这用起来会很方便，不需要在后面的代码里进行额外的转换。

在 put() 方法中，我们会在键上调用 hashCode() 方法，返回结果只能为正数。为了使结果数值在 buckets 数组中有效，我们会将它与该数组的大小进行模运算。如果该位置为 null，则表示没有哈希到该位置的元素，因此需要先将刚刚哈希到该位置的对象保存到一个新的 LinkedList 里，然后将这个 LinkedList 保存到该位置。但正常情况下一般会查看列表是否有重复项，如果有，则将旧值放入 oldValue 中，然后用新值替换旧值。found 标志会跟踪是否找到了旧的键值对，如果没有，则将新的键值对附加到列表的末尾。

get() 方法计算 buckets 数组索引的方式与 put() 方法相同（这对于能否保证它们定位到同一位置很重要）。如果已经存在一个 LinkedList 对象，则会对它进行搜索来寻找匹配项。

这个特定的实现并没有进行性能调优，它仅用于显示哈希映射时所执行的操作。如果你查看 java.util.HashMap 的源代码，会看到一个经过调优的实现。此外，为了简单起见，SimpleHashMap 使用了与 SlowMap 相同的 entrySet() 方法，它过于简单并且不适用于通用的 Map。

C.2.3 重写 hashCode()

了解了哈希的工作原理后，编写一个合适的 hashCode() 方法就变得更合情理了。

首先，我们无法控制用于索引存储桶数组的这个实际值的生成。它取决于这个特定 HashMap 的容量，而容量的变化取决于这个集合有多满，以及**负载因子**（load factor，该术语稍后描述）的情况。因此，hashCode() 产生的值会被进一步处理以创建桶索引（在 SimpleHashMap 中，只是以桶数组大小为模数进行了计算）。

[①] 质数实际上并不是哈希桶的理想大小，Java 中的哈希实现使用的大小是 2 的幂（经过了大量测试）。除法或余数是现代处理器上最慢的运算。如果哈希表的长度为 2 的幂，就可以使用掩码代替除法。get() 是迄今为止最常见的操作，因此计算成本的很大一部分是 % 操作，而使用 2 的幂消除了这一点（但也可能影响一些 hashCode() 方法）。

创建 hashCode() 方法时，要考虑的最重要因素是，无论何时调用 hashCode()，它每次都会为特定对象生成相同的值。如果你有一个对象，当它被 put() 方法放入 HashMap 时产生一个 hashCode() 值，而在 get() 方法中产生另一个值，你将无法重新得到这个对象。因此，如果你的 hashCode() 依赖于对象中的可变数据，就必须让用户意识到更改数据会生成不同的键，因为它会生成不同的 hashCode。

另外，你可能不会根据唯一的对象信息来生成一个 hashCode()——特别值得一提的是，在 hashCode() 里使用 this 的值就不好，因为那样就不能在生成一个新键时，保证它与之前 put() 时用到的原始键值对的键相同。这也是在 SpringDetector.java 中出现的问题，因为 hashCode() 的默认实现就使用了对象的地址。因此，请使用对象中有意义的信息来标识对象。

在 String 类中可以看到这样的一个示例。字符串有一个特殊的特性——如果程序里有多个包含相同字符序列的字符串对象，那么这些字符串对象就都会映射到同一个内存地址上。因此，字符串 "hello" 的两个单独实例生成的 hashCode() 应该是相同的，如下所示：

```
// equalshashcode/StringHashCode.java

public class StringHashCode {
  public static void main(String[] args) {
    String[] hellos = "Hello Hello".split(" ");
    System.out.println(hellos[0].hashCode());
    System.out.println(hellos[1].hashCode());
  }
}
/* 输出：
69609650
69609650
*/
```

String 类的 hashCode() 方法显然是基于字符串内容的。

因此，要使 hashCode() 有效，它必须既快速又有意义。也就是说，它必须根据对象的内容来生成值。记住这个值不需要是唯一的——我们应该倾向于速度而不是唯一性——但是对于 hashCode() 和 equals()，必须能确保它们定位到相同的对象。

hashCode() 会在桶索引生成之前进一步处理，因此它的取值范围并不重要，只需要保证它是 int 就可以了。

还有另一个因素：一个好的 hashCode() 应该保证值的均匀分布。如果值趋于聚集，那么 HashMap 或 HashSet 在某些区域的负载就会更重，并且速度也没有均匀分布的哈希函数那么快。

在 *Effective Java Programming Language Guide* 一书中，Joshua Bloch 给出了一个

生成比较好的 hashCode() 的基本准则。

1. 把一个非零常数，比如 17，保存在一个叫 result 的变量中。
2. 对于对象中的每个重要字段 f（即 equals() 方法会考虑的每个字段），计算出该字段的 int 类型的哈希码 c，具体如表 C-1 所示。
3. 组合上面计算出来的哈希码：result = 37 * result + c;。
4. 返回 result 值。
5. 观察得到的 hashCode()，确保相等的实例具有相同的哈希码。

表 C-1

字段类型	计算方法
boolean	c = (f ? 0 : 1)
byte、char、short 或 int	c = (int)f
long	c = (int)(f ^ (f>>>32))
float	c = Float.floatToIntBits(f);
double	long l = Double.doubleToLongBits(f); c = (int)(l ^ (l >>> 32))
Object，其中比较时调用的是这个字段自身的 equals()	c = f.hashCode()
Array	对每个元素应用上述规则

下面是一个遵循了这些准则的示例。请注意，实际编码时并不需要这样——相反，使用 Objects.hash() 对多个字段进行哈希（如本例中），或使用 Objects.hashCode() 对单个字段进行哈希即可。

```
// equalshashcode/CountedString.java
// 实现一个好的 hashCode()
import java.util.*;

public class CountedString {
  private static List<String> created =
    new ArrayList<>();
  private String s;
  private int id = 0;
  public CountedString(String str) {
    s = str;
    created.add(s);
    // id 表示 CountedString 里使用的这个字符串的实例总数:
    for(String s2 : created)
      if(s2.equals(s))
        id++;
  }
  @Override public String toString() {
    return "String: " + s + " id: " + id +
```

```
      " hashCode(): " + hashCode();
  }
  @Override public int hashCode() {
    // 最简单的实现:
    // return s.hashCode() * id;
    // 使用 Joshua Bloch 提供的方法:
    int result = 17;
    result = 37 * result + s.hashCode();
    result = 37 * result + id;
    return result;
  }
  @Override public boolean equals(Object o) {
    return o instanceof CountedString &&
      Objects.equals(s, ((CountedString)o).s) &&
      Objects.equals(id, ((CountedString)o).id);
  }
  public static void main(String[] args) {
    Map<CountedString,Integer> map = new HashMap<>();
    CountedString[] cs = new CountedString[5];
    for(int i = 0; i < cs.length; i++) {
      cs[i] = new CountedString("hi");
      map.put(cs[i], i); // int 自动装箱为 Integer
    }
    System.out.println(map);
    for(CountedString cstring : cs) {
      System.out.println("Looking up " + cstring);
      System.out.println(map.get(cstring));
    }
  }
}
/* 输出:
{String: hi id: 4 hashCode(): 146450=3, String: hi id:
5 hashCode(): 146451=4, String: hi id: 2 hashCode():
146448=1, String: hi id: 3 hashCode(): 146449=2,
String: hi id: 1 hashCode(): 146447=0}
Looking up String: hi id: 1 hashCode(): 146447
0
Looking up String: hi id: 2 hashCode(): 146448
1
Looking up String: hi id: 3 hashCode(): 146449
2
Looking up String: hi id: 4 hashCode(): 146450
3
Looking up String: hi id: 5 hashCode(): 146451
4
*/
```

CountedString 包含了一个字符串和一个 id，id 表示包含相同字符串的 CountedString 对象的数量。我们通过在构造器中遍历存储了所有字符串的静态 ArrayList 来完成计数。

hashCode() 和 equals() 都基于两个字段来生成结果。如果它们仅基于字符串或单独的 id，就会重复匹配不同的值。

在 main() 中，我们使用相同的字符串来创建多个 CountedString 对象，这样就可以看到，由于计数 id 的存在，重复项也能生成唯一值。我们打印了 HashMap，这样就可以看到它在内部是如何存储的（没有可辨别的顺序）。我们还单独查找了每个键，来证实查找机制是能正常工作的。

第二个示例里，我们使用了基础卷第 19 章在 reflection.pet 库里定义的基类 Individual 类。在那一章里我们使用了 Individual 类，但定义被推迟到本附录中介绍，以便你能正确理解它的实现。

这里我们将合理地使用 Objects.hash()，而不是手动创建 hashCode() 方法：

```java
// reflection/pets/Individual.java
package reflection.pets;
import java.util.*;

public class
Individual implements Comparable<Individual> {
  private static long counter = 0;
  private final long id = counter++;
  private String name;
  public Individual(String name) { this.name = name; }
  // name 是可选的:
  public Individual() {}
  @Override public String toString() {
    return getClass().getSimpleName() +
      (name == null ? "" : " " + name);
  }
  public long id() { return id; }
  @Override public boolean equals(Object o) {
    return o instanceof Individual &&
      Objects.equals(id, ((Individual)o).id);
  }
  @Override public int hashCode() {
    return Objects.hash(name, id);
  }
  @Override public int compareTo(Individual arg) {
    // 先比较类名:
    String first = getClass().getSimpleName();
    String argFirst = arg.getClass().getSimpleName();
    int firstCompare = first.compareTo(argFirst);
    if(firstCompare != 0)
      return firstCompare;
    if(name != null && arg.name != null) {
      int secondCompare = name.compareTo(arg.name);
      if(secondCompare != 0)
        return secondCompare;
    }
    return (arg.id < id ? -1 : (arg.id == id ? 0 : 1));
```

```
    }
}
```

compareTo() 方法中的比较逻辑是有层次结构的：首先按实际类型来排序，然后按 name（如果有的话）排序，最后如果还没有结果的话就按创建顺序来排序。下面的示例展示了其工作机制：

```
// equalshashcode/IndividualTest.java
import collections.MapOfList;
import reflection.pets.*;
import java.util.*;

public class IndividualTest {
  public static void main(String[] args) {
    Set<Individual> pets = new TreeSet<>();
    for(List<? extends Pet> lp :
        MapOfList.petPeople.values())
      for(Pet p : lp)
        pets.add(p);
    pets.forEach(System.out::println);
  }
}
/* 输出：
Cat Elsie May
Cat Pinkola
Cat Shackleton
Cat Stanford
Cymric Molly
Dog Margrett
Mutt Spot
Pug Louie aka Louis Snorkelstein Dupree
Rat Fizzy
Rat Freckly
Rat Fuzzy
*/
```

所有的宠物都有名字，因此它们首先按类型排序，然后在同一类型中按名称排序。

C.3 HashMap 调优

我们可以对 HashMap 进行手动调优，来提高它在特定应用程序里的性能。在 HashMap 调优时，想要理解性能问题，需要先知道一些术语。

- **容量**：哈希表中桶的数量。
- **初始容量**：创建哈希表时桶的数量。HashMap 和 HashSet 都提供了允许指定初始容量的构造器。
- **大小**：哈希表中当前的条目数。
- **负载因子**：大小/容量。负载因子为 0 表示哈希表为空，0.5 表示哈希表半满，以此类推。轻负载的哈希表很少发生冲突，因此最适合插入和查找（但会减慢迭代器的遍历过程）。HashMap 和 HashSet 都提供了允许指定负载因子的构造器。当达到这个负载因子时，集合会自动增加容量（桶的数量），方法是将其直接加倍，并将现有对象重新分配到新的桶里 [这称为**再哈希**（rehashing）]。

HashMap 默认的负载因子是 0.75（即它直到哈希表的四分之三已满才会再哈希）。这似乎是一个在时间和空间上比较好的权衡。更高的负载因子会减少哈希表所需的空间，但会增加查找成本，而这很重要，因为我们大部分时间所做的工作就是查找（包括 get() 和 put()）。

如果你知道自己将在一个 HashMap 中存储大量数据，可以适当加大初始容量来创建它，这样就可以避免自动再哈希的开销[1]。

---

[1] 在一封私信中，Joshua Bloch 写道："……我认为允许将实现细节（例如哈希表大小和负载因子）添加到我们的 API 中是一个错误。客户或许应该只提供预期的最大容量，然后我们接手后续的所有处理。客户为这些参数提供选项的话，很容易就会变得弊大于利。举一个极端的例子，考虑一下 Vector 的 capacityIncrement。所有人都不应该设置这个参数，我们也不应该提供它。如果将其设置为任何非零值，那么序列添加操作的渐近成本将从线性变为平方。换句话说，它破坏了性能。随着时间的推移，我们在这类事情上的决策变得更为明智。如果你查看 IdentityHashMap，就会发现它没有任何底层的调优参数。"

D

数据压缩

Java I/O 库包含了以压缩格式读取和写入流的类。你可以将它们包装在其他 I/O 类中，从而提供压缩功能。

这些类并没有继承 Reader 和 Writer 类，而是 InputStream 和 OutputStream 继承层次结构的一部分。这是因为压缩库使用的是字节，而不是字符。但是，有时可能不得不混合使用这两种类型的流。（请记住，你可以使用 InputStreamReader 和 OutputStreamWriter 在两种类型之间轻松转换。）表 D-1 中介绍了压缩类及其功能。

表 D-1

压缩类	功能
CheckedInputStream	getCheckSum() 为任意 InputStream（不仅仅包括解压缩的流）生成校验和
CheckedOutputStream	getCheckSum() 为任意 OutputStream（不仅仅包括压缩过的流）生成校验和
DeflaterOutputStream	压缩类的基类
ZipOutputStream	将数据压缩为 Zip 文件格式的 DeflaterOutputStream
GZIPOutputStream	将数据压缩为 GZIP 文件格式的 DeflaterOutputStream
InflaterInputStream	解压缩类的基类
ZipInputStream	属于 InflaterInputStream，用于解压缩以 Zip 文件格式存储的数据
GZIPInputStream	属于 InflaterInputStream，用于解压缩以 GZIP 文件格式存储的数据

尽管有很多压缩算法，但 Zip 和 GZIP 可能是最常见的。Java 里有许多可用于读取和写入这些格式的工具，使用它们就能很轻松地处理压缩数据。

D.1 使用 GZIP 进行简单的压缩

GZIP 接口很简单，因此要压缩单个数据流（而不是不同数据块的容器）时，使用它可能更合适。下面的示例会压缩单个文件：

```java
// compression/GZIPcompress.java
// {java GZIPcompress GZIPcompress.java}
// {VisuallyInspectOutput}
import java.util.zip.*;
import java.io.*;

public class GZIPcompress {
  public static void main(String[] args) {
    if(args.length == 0) {
      System.out.println(
        "Usage: \nGZIPcompress file\n" +
        "\tUses GZIP compression to compress " +
        "the file to test.gz");
      System.exit(1);
    }
    try(
      InputStream in = new BufferedInputStream(
        new FileInputStream(args[0]));
      BufferedOutputStream out =
        new BufferedOutputStream(
          new GZIPOutputStream(
            new FileOutputStream("test.gz")))
    ) {
      System.out.println("Writing file");
      int c;
      while((c = in.read()) != -1)
        out.write(c);
    } catch(IOException e) {
      throw new RuntimeException(e);
    }
    System.out.println("Reading file");
    try(
      BufferedReader in2 = new BufferedReader(
        new InputStreamReader(new GZIPInputStream(
          new FileInputStream("test.gz"))))
    ) {
      in2.lines().forEach(System.out::println);
    } catch(IOException e) {
      throw new RuntimeException(e);
    }
  }
}
```

使用压缩类很简单：将输出流包装在 `GZIPOutputStream` 或 `ZipOutputStream` 中，并将输入流包装在 `GZIPInputStream` 或 `ZipInputStream` 中。其他都是普通的 I/O 读写。这是一个混合使用字符流和字节流的例子。`in` 使用了 `Reader` 类，`GZIPOutputStream` 的构造器只能接受一个 `OutputStream` 对象，而不是一个 `Writer` 对象。当文件被打开时，`GZIPInputStream` 被转换为一个 `Reader`。

D.2 使用 Zip 进行多文件存储

Java 库对 Zip 格式的支持力度比 GZIP 更大。你可以通过 Java 库轻松地存储多个文件，它甚至还提供了一个单独的类，来帮助我们更容易地读取 Zip 文件。该库使用标准的 Zip 格式，因此可以与当前因特网上下载的所有 Zip 工具无缝协作。下面的示例与前面的示例形式相同，但它可以根据需要处理任意数量的命令行参数。此外，它还使用了 Checksum 类来计算和验证文件的校验和。有两种 Checksum 类型：Adler32（更快）和 CRC32（更慢但更准确）。

```java
// compression/ZipCompress.java
// 使用 Zip 压缩功能来压缩命令行上给出的任意数量的文件:
// {java ZipCompress ZipCompress.java}
// {VisuallyInspectOutput}
import java.util.zip.*;
import java.io.*;
import java.util.*;

public class ZipCompress {
  public static void main(String[] args) {
    try(
      FileOutputStream f =
        new FileOutputStream("test.zip");
      CheckedOutputStream csum =
        new CheckedOutputStream(f, new Adler32());
      ZipOutputStream zos = new ZipOutputStream(csum);
      BufferedOutputStream out =
        new BufferedOutputStream(zos)
    ) {
      zos.setComment("A test of Java Zipping");
      // 不过没有对应的 getComment() 方法
      for(String arg : args) {
        System.out.println("Writing file " + arg);
        try(
          InputStream in = new BufferedInputStream(
            new FileInputStream(arg))
        ) {
          zos.putNextEntry(new ZipEntry(arg));
          int c;
          while((c = in.read()) != -1)
            out.write(c);
        }
        out.flush();
      }
      // 校验和只有在文件关闭后才有效:
      System.out.println(
        "Checksum: " + csum.getChecksum().getValue());
    } catch(IOException e) {
      throw new RuntimeException(e);
    }
    // 现在提取文件:
    System.out.println("Reading file");
    try(
```

```
        FileInputStream fi =
          new FileInputStream("test.zip");
        CheckedInputStream csumi =
          new CheckedInputStream(fi, new Adler32());
        ZipInputStream in2 = new ZipInputStream(csumi);
        BufferedInputStream bis =
          new BufferedInputStream(in2)
    ) {
      ZipEntry ze;
      while((ze = in2.getNextEntry()) != null) {
        System.out.println("Reading file " + ze);
        int x;
        while((x = bis.read()) != -1)
          System.out.write(x);
      }
      if(args.length == 1)
        System.out.println(
          "Checksum: "+csumi.getChecksum().getValue());
    } catch(IOException e) {
      throw new RuntimeException(e);
    }
    // 另一种打开并读取 Zip 文件的方法:
    try(
      ZipFile zf = new ZipFile("test.zip")
    ) {
      Enumeration e = zf.entries();
      while(e.hasMoreElements()) {
        ZipEntry ze2 = (ZipEntry)e.nextElement();
        System.out.println("File: " + ze2);
        // ……像以前一样提取数据
      }
    } catch(IOException e) {
      throw new RuntimeException(e);
    }
  }
}
```

对于要添加到存档中的每个文件，都必须调用 putNextEntry() 并传递给它一个 ZipEntry 对象。ZipEntry 对象提供了功能丰富的接口，用于获取和设置 Zip 文件中该特定条目的所有可用数据：名称、压缩和未压缩大小、日期、CRC 校验和、额外字段数据、注释、压缩方法，以及它是否是一个目录条目。不过，虽然 Zip 格式允许设置密码，但 Java 的 Zip 库并不支持这一功能。尽管 CheckedInputStream 和 CheckedOutputStream 支持 Adler32 和 CRC32 校验和，但 ZipEntry 类只支持 CRC 的接口。这是底层 Zip 格式的一个限制，这可能会限制我们使用更快的 Adler32。

为了提取文件，ZipInputStream 有一个 getNextEntry() 方法，如果有数据则返回下一个 ZipEntry。作为更简洁的替代方法，你可以使用 ZipFile 对象来读取文件，该对象提供了 entries() 方法，可以返回一个 Enumeration 给 ZipEntries。

如果想要读取校验和，必须以某种方式访问关联的 `Checksum` 对象。本示例中我们保留了 `CheckedOutputStream` 和 `CheckedInputStream` 对象的引用，但你也可以只保留对 `Checksum` 对象的引用。

Zip 流中有一个令人困惑的方法 `setComment()`。如 ZipCompress.java 所示，你可以在写文件的时候设置注释，但在 `ZipInputStream` 中没有办法恢复这个注释。似乎只能通过 `ZipEntry` 在每一项上设置注释。

`GZIP` 或 `Zip` 库的使用不仅限于文件——你可以压缩任何内容，包括通过网络连接发送的数据。

D.3 Java 档案（Jars）

JAR（Java ARchive）文件格式里也使用了 Zip 格式，它是一种将一组文件收集到单个压缩文件的方法，就像 Zip 一样。但是，与 Java 中的其他内容一样，JAR 文件是跨平台的，因此我们不需要担心平台问题。JAR 文件还可以包含音频和图像文件，以及类文件。

JAR 文件由单个文件组成，该文件包含了一组压缩文件以及描述它们的"清单文件（manifest）"。（你可以创建自己的清单文件，否则 jar 程序会自动为你创建。）你可以在 JDK 文档中找到有关 JAR 清单的更多信息。

JDK 附带的 jar 工具会自动压缩所选择的文件。可以在命令行上这样调用它：

jar [选项] 目的地 [清单] 输入文件

选项是一组字母（不需要连字符或任何其他指示符）。UNIX/Linux 用户可能会注意到它与 tar 选项的相似之处。这些选项如表 D-2 所示。

表 D-2

字母	效果
c	创建一个新的或空的存档
t	列出目录
x	提取所有文件
x file	提取指定文件
f	它表示"我将给你指定文件名"。如果不使用这个选项，jar 会假定输入来自标准输入；或者，如果正在创建一个文件，它的输出将转到标准输出
m	表示第一个参数是用户创建的清单文件的名称
v	生成详细的输出来描述 jar 正在做什么

（续）

字母	效果
0	只存储文件；不压缩文件（用于创建放在类路径中的 JAR 文件）
M	禁止自动创建清单文件

如果放入 JAR 文件的内容包含子目录，那么它会自动添加该子目录，包括该子目录的子目录，以此类推。同时路径信息也会被保留下来。

下面是调用 jar 的一些典型方式。下面这个命令创建了一个名为 myJarFile.jar 的 JAR 文件，其中包含当前目录下的所有类文件，以及一个自动生成的清单文件：

```
jar cf myJarFile.jar *.class
```

下一个命令与前面的类似，但它添加了一个名为 myManifestFile.mf 的用户创建的清单文件：

```
jar cmf myManifestFile.mf myJarFile.jar *.class
```

下面的命令会打印 myJarFile.jar 中文件的目录列表：

```
jar tf myJarFile.jar
```

添加一个"详细模式"的标志，以提供 myJarFile.jar 中文件的更多信息：

```
jar tvf myJarFile.jar
```

假设 audio、classes 和 image 是子目录，下面的命令将所有子目录合并到文件 myApp.jar 中。它还提供了一个"详细模式"的标志，用来在 jar 程序运行时提供额外的反馈：

```
jar cvf myApp.jar audio classes image
```

如果使用了 0（零）选项来创建 JAR 文件，则生成的文件可以放在你的 CLASSPATH 下：

```
CLASSPATH="lib1.jar;lib2.jar;"
```

然后 Java 可以搜索 lib1.jar 和 lib2.jar 来查找类文件。

jar 工具不像 Zip 工具那样通用。例如，不能向现有的 JAR 文件添加或更新文件，只能从头开始创建 JAR 文件。此外，无法将文件移动到 JAR 文件中，也不能在移出时将其删除。但是，在一个平台上创建的 JAR 文件可以在任何其他平台上用 jar 工具毫无阻碍地读取（Zip 工具有时会遇到麻烦）。

E

对象序列化

重要提示：几乎从诞生之日起，Java 就提供了对象序列化功能，但直到最近 Java 团队才承认，Java 序列化在设计上存在危险的缺陷，并且一直是这门语言里严重安全漏洞的来源。这个附录之所以被包含在本书中，主要是为了让你理解可能会遇到的序列化／反序列化旧代码。

JDK17 实现的 Java 特性增强提案（Java Enhancement Proposal, JEP）415 里描述了这个问题：

"对不受信任的数据进行反序列化操作，这根本就是一个危险的行为，因为传入数据流的内容决定了创建的对象、字段的值以及它们之间的引用。在许多典型的场景里，流的字节数据来源于未知的、不受信任的或未经验证的客户端。通过仔细地构建流数据，攻击者可以不怀好意地执行任意类中的代码。如果对象构建时会改变状态或调用其他操作，则这些操作可能会危及数据的完整性，这些数据包括应用程序对象、库对象甚至 Java 运行时。防止反序列化攻击的关键在于，禁止任意类的实例被反序列化，这样就能防止在反序列化过程中直接或间接地执行类的方法。"

JDK 9（JEP 290）添加了反序列化过滤器，这样应用程序和库代码就可以在反序列化之前验证传入的数据流。但这种方法无法扩展，并且过滤器也很难更新。此外，它不能对第三方库进行过滤。为了解决这个问题，JEP 290 提供了在 JVM 范围内起作用的反序列化过滤器。但这个方案还是有局限性，尤其是对包含多层库代码和多个执行上下文的复杂应用程序来说。JEP 415 提供了过滤器工厂来尝试完全解决这个问题，它可配置并在 JVM 范围内起作用。

本附录的其余部分描述了经典的序列化／反序列化，来帮助你理解旧代码。但出于安全目的，不应该再基于此来实现新系统，而是应该学习 JEP 415 的内容。

E.1 概述

当创建了一个对象后，只要你需要它，它就可以一直存在；但是如果程序退出，这个对象就不存在了。虽然这乍看起来说得通，但在某些情况下，如果程序不再运行，而对象仍然能够存在并且保留相关信息，会对我们非常有用。也就是说，在下次启动程序时，包含上次程序运行时信息的对象还会在那里。我们可以通过将信息写入文件或数据库来实现此效果，但本着一切皆对象的宗旨，如果能将对象声明为持久性的，然后让编程语言自动为你处理所有的细节，会方便很多。

Java 的**对象序列化**（object serialization）机制会接受实现了 Serializable 接口的任意对象，并将其转换成一个字节序列，便于以后重新生成原始对象。它甚至可以通过网络工作，这意味着序列化机制会自动消除操作系统之间的差异。也就是说，你可以在 Windows 机器上创建一个对象、将其序列化，然后通过网络发送到 UNIX 机器上，它会在那里被正确重建。你不需要担心不同机器上的数据表示、字节顺序或任何其他细节。

对象序列化可以实现**轻量级持久化**（lightweight persistence）。持久化意味着对象存活于程序调用**之间**，其生命周期不是由程序是否在执行决定的。通过获取一个可序列化的对象并将其写入磁盘，然后在重新调用程序时恢复该对象，这样就产生了持久化的效果。之所以称其为"轻量级"，是因为你不能使用某个"持久化"关键字定义一个对象，并让编程语言替你处理一切细节。相反，你必须在程序中显式地序列化和反序列化对象。如果需要更严格的持久化机制，请考虑使用像 Hibernate 这样的工具。

在语言中添加对象序列化，主要是为了支持两个功能。首先，Java 的**远程方法调用**（remote method invocation, RMI）可以让存在于远程机器上的对象表现得像存在本地机器上一样。当把消息发送给远程对象时，需要对象序列化来传输参数和返回值。

其次，JavaBeans 也需要对象序列化（在撰写本书时，JavaBeans 被认为是一项失败的技术）。当使用 Bean 时，一般会在设计时配置其状态信息。这个状态信息必须存储起来，然后在程序启动时恢复，对象序列化就被用来执行此任务。

序列化一个对象很简单，只要对象实现了 Serializable 接口就可以。Serializable 是一个标签接口，没有方法。当序列化被添加到语言中时，标准库里的许多类被更改以便可序列化，包括所有基本类型的包装类、所有容器类，等等。甚至 Class 对象也是可以序列化的。

要序列化对象，你需要创建某种 OutputStream 对象，并将其包装在 ObjectOutputStream 对象中。然后只需要调用 writeObject() 就可以了，这样你的对象就会被序列化并发送到 OutputStream（对象序列化是面向字节的，因此使用 InputStream 和 OutputStream 的层次结构）。为了逆转这个过程，你需要将一个 InputStream 包装在 ObjectInputStream 中并调用 readObject() 方法。像往常一样，这返回的是一个向上转型了的 Object 引用，因此必须向下转型才能得到正确结果。

对象序列化里一个特别出彩的地方是，它不仅保存了对象的镜像，而且会跟踪该对象包含的所有引用并保存这些引用对象的镜像，然后跟踪每个对象所包含的全部引用，以此类推。这有时称为单个对象可以连接到的"对象网络"，包括成员对象以及存储了对象引用的数组。如果你必须维护自己的对象序列化方案，那么遍历所有这些链接的维护代码可能会极为复杂。不过，Java 的对象序列化似乎完美地实现了这个目标。毫无疑问，它使用了一种遍历对象网络的优化算法。下面的示例创建了一个"蠕虫"（worm）来测试序列化机制。蠕虫里的每个关联对象都包含一个指向蠕虫下一段的链接，还包含一个存储了 Data 对象引用的数组：

```java
// serialization/Worm.java
// 演示对象序列化
import java.io.*;
import java.util.*;

class Data implements Serializable {
  private int n;
  Data(int n) { this.n = n; }
  @Override public String toString() {
    return Integer.toString(n);
  }
}

public class Worm implements Serializable {
  private static Random rand = new Random(47);
  private Data[] d = {
    new Data(rand.nextInt(10)),
    new Data(rand.nextInt(10)),
    new Data(rand.nextInt(10))
  };
  private Worm next;
  private char c;
  // Value of i == number of segments
  public Worm(int i, char x) {
    System.out.println("Worm constructor: " + i);
    c = x;
```

```java
    if(--i > 0)
      next = new Worm(i, (char)(x + 1));
  }
  public Worm() {
    System.out.println("Zero-argument constructor");
  }
  @Override public String toString() {
    StringBuilder result = new StringBuilder(":");
    result.append(c);
    result.append("(");
    for(Data dat : d)
      result.append(dat);
    result.append(")");
    if(next != null)
      result.append(next);
    return result.toString();
  }
  public static void
  main(String[] args) throws ClassNotFoundException,
    IOException {
    Worm w = new Worm(6, 'a');
    System.out.println("w = " + w);
    try(
      ObjectOutputStream out = new ObjectOutputStream(
        new FileOutputStream("worm.dat"))
    ) {
      out.writeObject("Worm storage\n");
      out.writeObject(w);
    }
    try(
      ObjectInputStream in = new ObjectInputStream(
        new FileInputStream("worm.dat"))
    ) {
      String s = (String)in.readObject();
      Worm w2 = (Worm)in.readObject();
      System.out.println(s + "w2 = " + w2);
    }
    try(
      ByteArrayOutputStream bout =
        new ByteArrayOutputStream();
      ObjectOutputStream out2 =
        new ObjectOutputStream(bout)
    ) {
      out2.writeObject("Worm storage\n");
      out2.writeObject(w);
      out2.flush();
      try(
        ObjectInputStream in2 = new ObjectInputStream(
          new ByteArrayInputStream(
            bout.toByteArray()))
      ) {
        String s = (String)in2.readObject();
        Worm w3 = (Worm)in2.readObject();
```

```
                System.out.println(s + "w3 = " + w3);
            }
        }
    }
}
```

```
/* 输出：
Worm constructor: 6
Worm constructor: 5
Worm constructor: 4
Worm constructor: 3
Worm constructor: 2
Worm constructor: 1
w = :a(853):b(119):c(802):d(788):e(199):f(881)
Worm storage
w2 = :a(853):b(119):c(802):d(788):e(199):f(881)
Worm storage
w3 = :a(853):b(119):c(802):d(788):e(199):f(881)
*/
```

Worm 里的数组所包含的每个 Data 对象都使用一个随机数来初始化。（这样，你就不用担心编译器会保留某种元信息了。）我们使用一个 char 字段来标记 Worm 的每一段，这个 char 是在递归生成 Worm 链表的过程中自动生成的。当创建一个 Worm 时，你在构造器里指定它的长度。为了生成 next 引用，它会再次调用自己的构造器，但长度会减一，以此类推。最后的 next 引用为 null，表示 Worm 的结尾。

这一切的目的就是让生成的结果复杂一些，从而不易序列化。不过，序列化的行为很简单。一旦从其他流中创建了 ObjectOutputStream，调用 writeObject() 方法就可以序列化该对象。注意我们是如何调用 writeObject() 来序列化字符串的。还可以使用 DataOutputStream（它们共享相同的接口）里相同的方法来写入所有的原始数据类型。

示例里的两块代码看起来很相似。第一块代码写入和读取一个文件，第二块代码写入和读取一个 ByteArray。你可以通过序列化在任何 DataInputStream 或 DataOutputStream 上读取和写入对象，甚至可以通过网络。

输出显示，反序列化后得到的对象确实包含了原始对象中的所有链接。

注意，在反序列化 Serializable 对象的过程中，我们没有调用任何构造器，连无参构造器也没有调用。整个对象的数据都是从 InputStream 里恢复的。

E.2 查找对应的类

你可能想知道，如果要从序列化状态恢复一个对象，到底需要些什么。例如，假设你序列化了一个对象并将其作为文件或通过网络发送到另一台机器，那么另一台机器上的程序能否仅根据文件的内容来重建对象呢？

回答这个问题的最好方法（像往常一样）是进行实验。以下文件位于本附录的子目录中：

```
// serialization/Alien.java
// 一个可序列化的类
import java.io.*;
public class Alien implements Serializable {}
```

创建和序列化 Alien 对象的文件位于同一个目录中：

```
// serialization/FreezeAlien.java
// 创建一个序列化输出文件
import java.io.*;

public class FreezeAlien {
  public static void
  main(String[] args) throws Exception {
    try(
      ObjectOutputStream out = new ObjectOutputStream(
        new FileOutputStream("X.file"));
    ) {
      Alien quellek = new Alien();
      out.writeObject(quellek);
    }
  }
}
```

一旦程序被编译并运行，它就会在 serialization 目录中生成一个名为 X.file 的文件。下面的代码位于 xfiles 子目录中：

```
// serialization/xfiles/ThawAlien.java
// 从序列化文件中恢复
// {java serialization.xfiles.ThawAlien}
// {RunFirst: FreezeAlien}
package serialization.xfiles;
import java.io.*;

public class ThawAlien {
  public static void
  main(String[] args) throws Exception {
    ObjectInputStream in = new ObjectInputStream(
      new FileInputStream(new File("X.file")));
    Object mystery = in.readObject();
    System.out.println(mystery.getClass());
  }
}
```

```
/* 输出
class Alien
*/
```

如果想让示例正常运行，JVM 就必须能够找到相应的 .class 文件。

E.3 控制序列化

默认的序列化机制使用起来很简单。但是如果有特殊需求呢？也许你有特殊的安全问题，不想序列化对象的某些部分；也许某个子对象在反序列化时需要重新创建，这样的话，序列化这个子对象就没有意义了。

你可以通过实现 Externalizable 接口而不是 Serializable 接口来控制序列化过程。Externalizable 接口扩展了 Serializable 接口并添加了两个方法：writeExternal() 和 readExternal()。它们在序列化和反序列化期间会自动调用，这样你就可以执行自己的特殊操作了。

下面的示例简单实现了 Externalizable 接口里的方法。注意，Blip1 和 Blip2 几乎相同，除了一个细微的差异（看看你能否通过阅读代码发现它）：

```java
// serialization/Blips.java
// Externalizable 的简单使用和一个陷阱
import java.io.*;

class Blip1 implements Externalizable {
  public Blip1() {
    System.out.println("Blip1 Constructor");
  }
  @Override
  public void writeExternal(ObjectOutput out)
      throws IOException {
    System.out.println("Blip1.writeExternal");
  }
  @Override
  public void readExternal(ObjectInput in)
      throws IOException, ClassNotFoundException {
    System.out.println("Blip1.readExternal");
  }
}

class Blip2 implements Externalizable {
  Blip2() {
    System.out.println("Blip2 Constructor");
  }
  @Override
  public void writeExternal(ObjectOutput out)
      throws IOException {
    System.out.println("Blip2.writeExternal");
  }
  @Override
  public void readExternal(ObjectInput in)
      throws IOException, ClassNotFoundException {
    System.out.println("Blip2.readExternal");
  }
}
```

```
/* 输出:
Constructing objects:
Blip1 Constructor
Blip2 Constructor
Saving objects:
Blip1.writeExternal
Blip2.writeExternal
Recovering b1:
Blip1 Constructor
Blip1.readExternal
*/
```

```java
}
public class Blips {
  public static void main(String[] args) {
    System.out.println("Constructing objects:");
    Blip1 b1 = new Blip1();
    Blip2 b2 = new Blip2();
    try(
      ObjectOutputStream o = new ObjectOutputStream(
        new FileOutputStream("Blips.serialized"))
    ) {
      System.out.println("Saving objects:");
      o.writeObject(b1);
      o.writeObject(b2);
    } catch(IOException e) {
      throw new RuntimeException(e);
    }
    // 现在恢复它们：
    System.out.println("Recovering b1:");
    try(
      ObjectInputStream in = new ObjectInputStream(
        new FileInputStream("Blips.serialized"))
    ) {
      b1 = (Blip1)in.readObject();
    } catch(IOException | ClassNotFoundException e) {
      throw new RuntimeException(e);
    }
    // 哎呀！抛出异常：
    //- System.out.println("Recovering b2:");
    //- b2 = (Blip2)in.readObject();
  }
}
```

Blip2 对象没有被恢复，这是因为在尝试这样做时抛出了异常。你能看出 Blip1 和 Blip2 之间的区别吗？ Blip1 的构造器是 public 的，而 Blip2 的构造器不是，这在恢复时导致了异常。尝试将 Blip2 的构造器设为 public 的，然后删除 //- 注释来查看正确结果。

当恢复 b1 时，Blip1 的无参构造器被调用。这与恢复 Serializable 对象不同，后者完全从其存储的字节位里构建对象，没有调用构造器。在使用 Externalizable 对象时，所有正常的默认构造行为都会发生（包括在字段定义处的初始化），之后 readExternal() 被调用。请注意这一点，特别是所有默认的构造总是会发生，这样才能在 Externalizable 对象中产生正确的行为。

下面的示例显示了完全存储和恢复 Externalizable 对象必须执行的操作：

```java
// serialization/Blip3.java
// 重新构建一个 Externalizable 对象
import java.io.*;
```

```java
public class Blip3 implements Externalizable {
  private int i;
  private String s; // 没有初始化
  public Blip3() {
    System.out.println("Blip3 Constructor");
    // s 和 i 没有被初始化
  }
  public Blip3(String x, int a) {
    System.out.println("Blip3(String x, int a)");
    s = x;
    i = a;
    // s 和 i 只在有参数的构造器里进行了初始化
  }
  @Override public String toString() { return s + i; }
  @Override public void writeExternal(ObjectOutput out)
  throws IOException {
    System.out.println("Blip3.writeExternal");
    // 必须这样做:
    out.writeObject(s);
    out.writeInt(i);
  }
  @Override public void readExternal(ObjectInput in)
  throws IOException, ClassNotFoundException {
    System.out.println("Blip3.readExternal");
    // 必须这样做:
    s = (String)in.readObject();
    i = in.readInt();
  }
  public static void main(String[] args) {
    System.out.println("Constructing objects:");
    Blip3 b3 = new Blip3("A String ", 47);
    System.out.println(b3);
    try(
      ObjectOutputStream o = new ObjectOutputStream(
        new FileOutputStream("Blip3.serialized"))
    ) {
      System.out.println("Saving object:");
      o.writeObject(b3);
    } catch(IOException e) {
      throw new RuntimeException(e);
    }
    // 现在进行反序列化:
    System.out.println("Recovering b3:");
    try(
      ObjectInputStream in = new ObjectInputStream(
        new FileInputStream("Blip3.serialized"))
    ) {
      b3 = (Blip3)in.readObject();
    } catch(IOException | ClassNotFoundException e) {
      throw new RuntimeException(e);
    }
    System.out.println(b3);
  }
}
/* 输出:
Constructing objects:
Blip3(String x, int a)
A String 47
Saving object:
Blip3.writeExternal
Recovering b3:
Blip3 Constructor
Blip3.readExternal
A String 47
*/
```

字段 s 和 i 仅在第二个构造器中进行了初始化，在无参数构造器中没有初始化。这意味着，如果你没有在 readExternal() 中初始化 s 和 i，那么 s 就是 null 且 i 是零（对象的存储在其创建的第一步被擦除为零）。如果注释掉"必须这样做："后面的两行代码并运行该程序，你将看到当对象恢复时，s 为 null、i 为零。

如果你继承了一个 Externalizable 对象，通常需要调用 writeExternal() 和 readExternal() 的基类版本，从而提供基类组件的正确存储和恢复。

因此，为了使序列化正常，不仅需要在 writeExternal() 方法里写入对象的重要数据（序列化机制不会默认为 Externalizable 对象写入任何成员对象），而且需要在 readExternal() 方法中恢复该数据。起初这可能有点令人困惑，因为 Externalizable 对象的默认构造行为有可能使它看起来像存在某种自动存储和恢复行为，而这实际上是没有的。

E.3.1　transient 关键字

当你控制序列化时，可能存在不希望 Java 序列化机制自动保存和恢复的特定子对象。如果该子对象表示你不想序列化的敏感信息（例如密码），就会出现这种情况。即使该信息在对象中是 private 的，一旦它被序列化，其他人就有可能通过读取文件或拦截网络传输来访问它。

防止对象的敏感部分被序列化的一种方法是将你的类实现为 Externalizable，如前所示。这样就不会自动序列化任何内容了，你可以仅显式序列化 writeExternal() 中的必要部分。

但是，如果你正在使用 Serializable 对象，则所有序列化都会自动发生。为了控制这一点，可以使用 transient 关键字逐个字段地关闭序列化，它表示"不要费心保存或恢复这个字段——我会处理它的"。

例如，考虑一个保存了有关特定登录会话信息的 Logon 对象。假设在验证完登录后，你想存储这个登录数据，但不包括密码。最简单的方法是实现 Serializable 并将 password 字段标记为 transient，如下所示：

```
// serialization/Logon.java
// 演示 transient 关键字
import java.util.concurrent.*;
import java.io.*;
import java.util.*;
import onjava.Nap;
```

```java
public class Logon implements Serializable {
  private Date date = new Date();
  private String username;
  private transient String password;
  public Logon(String name, String pwd) {
    username = name;
    password = pwd;
  }
  @Override public String toString() {
    return "logon info: \n   username: " +
      username + "\n   date: " + date +
      "\n   password: " + password;
  }
  public static void main(String[] args) {
    Logon a = new Logon("Hulk", "myLittlePony");
    System.out.println("logon a = " + a);
    try(
      ObjectOutputStream o =
        new ObjectOutputStream(
          new FileOutputStream("Logon.dat"))
    ) {
      o.writeObject(a);
    } catch(IOException e) {
      throw new RuntimeException(e);
    }
    new Nap(1);
    // 下面进行反序列化:
    try(
      ObjectInputStream in = new ObjectInputStream(
        new FileInputStream("Logon.dat"))
    ) {
      System.out.println(
        "Recovering object at " + new Date());
      a = (Logon)in.readObject();
    } catch(IOException | ClassNotFoundException e) {
      throw new RuntimeException(e);
    }
    System.out.println("logon a = " + a);
  }
}
/* 输出:
logon a = logon info:
   username: Hulk
   date: Sun Jan 24 08:49:30 MST 2021
   password: myLittlePony
Recovering object at Sun Jan 24 08:49:31 MST 2021
logon a = logon info:
   username: Hulk
   date: Sun Jan 24 08:49:30 MST 2021
   password: null
*/
```

date 和 username 是普通字段（不是 transient 的），因此会自动序列化。然而 password 是 transient 的，所以不会被存储到磁盘，序列化机制也不会尝试恢复它。当对象被恢复时，password 字段为 null。注意，toString() 使用重载的 + 运算符组装了一个字符串对象，而其中的 null 引用被自动转换成了字符串 "null"。

你还可以看到 date 字段被存储到磁盘并从磁盘恢复，而不是重新生成。

Externalizable 对象默认不存储自身的任何字段，因此 transient 关键字仅适用于 Serializable 对象。

E.3.2　Externalizable 的替代方案

如果不想实现 Externalizable 接口，还有另一种方案。你可以实现 Serializable 接口并添加（注意我说的是"添加"，而不是"重写"或"实现"）名为 writeObject() 和 readObject() 的方法，它们会在对象序列化和反序列化时分别被调用。也就是说，如果你提供了这两个方法，它们就会被调用，来代替默认的序列化。

这两个方法必须具有以下确切的签名：

```
private void writeObject(ObjectOutputStream stream)
throws IOException;

private void readObject(ObjectInputStream stream)
throws IOException, ClassNotFoundException
```

从设计的角度来看，这里的事情变得非常奇怪。首先，你可能认为，因为这些方法不是基类或 Serializable 接口的一部分，所以它们应该在自己的接口中定义。但是请注意，它们被定义为 private 的，这意味着它们只能被此类的其他成员调用。然而，这个类的其他成员实际上并没有调用它们，是 ObjectOutputStream 和 ObjectInputStream 对象的 writeObject() 和 readObject() 方法调用了你的对象的 writeObject() 和 readObject() 方法。（请注意，我在这里非常克制，没有对使用相同方法名称的问题长篇大论。一句话：这令人困惑。）你可能想知道 ObjectOutputStream 和 ObjectInputStream 对象如何访问你的类的 private 方法。我们只能假设这是序列化魔法的一部分[1]。

接口中定义的任何成员都自动是 public 的，所以如果 writeObject() 和 readObject() 必须是 private 的，那么它们就不能是接口的一部分。你必须完全遵循方法签名，才能有如同实现了接口一样的效果。

---

[1] 基础卷 19.9 节展示了如何从类的外部访问 private 方法。

当调用 ObjectOutputStream.writeObject() 时，传递给它的 Serializable 对象看起来好像会被检查（使用反射，毫无疑问），看它是否实现了自己的 writeObject()。如果是，则跳过正常的序列化过程并调用自定义的 writeObject()。readObject() 的情况也一样。

这里还有一个意外之处。在 writeObject() 中，你可以选择调用 defaultWriteObject() 来执行默认的 writeObject() 操作。同样，在 readObject() 中也可以调用 defaultReadObject()。下面是一个简单的示例，演示了如何控制 Serializable 对象的存储和恢复：

```java
// serialization/SerialCtl.java
// 通过添加自己的writeObject()和readObject()方法来控制序列化
import java.io.*;

public class SerialCtl implements Serializable {
  private String a;
  private transient String b;
  public SerialCtl(String aa, String bb) {
    a = "Not Transient: " + aa;
    b = "Transient: " + bb;
  }
  @Override public String toString() {
    return a + "\n" + b;
  }
  private void writeObject(ObjectOutputStream stream)
  throws IOException {
    stream.defaultWriteObject();
    stream.writeObject(b);
  }
  private void readObject(ObjectInputStream stream)
  throws IOException, ClassNotFoundException {
    stream.defaultReadObject();
    b = (String)stream.readObject();
  }
  public static void main(String[] args) {
    SerialCtl sc = new SerialCtl("Test1", "Test2");
    System.out.println("Before:\n" + sc);
    try (
      ByteArrayOutputStream buf =
        new ByteArrayOutputStream();
      ObjectOutputStream o =
        new ObjectOutputStream(buf);
    ) {
      o.writeObject(sc);
      // 下面进行反序列化:
      try (
        ObjectInputStream in =
          new ObjectInputStream(
            new ByteArrayInputStream(
              buf.toByteArray()));
```

```
/* 输出:
Before:
Not Transient: Test1
Transient: Test2
After:
Not Transient: Test1
Transient: Test2
*/
```

```
    ) {
      SerialCtl sc2 = (SerialCtl)in.readObject();
      System.out.println("After:\n" + sc2);
    }
  } catch(IOException | ClassNotFoundException e) {
    throw new RuntimeException(e);
  }
  }
}
```

在这个例子里，两个字符串字段中的一个是普通字段，另一个是 transient 的，用来证明非 transient 字段会被 defaultWriteObject() 方法保存，而 transient 字段则需要显式保存和恢复。这些字段在构造器内部初始化，而不是在定义处，以证明它们在反序列化过程中没有被某种自动机制初始化。

如果使用默认机制来序列化对象的非 transient 部分，那么在 writeObject() 中的第一个操作必须是调用 defaultWriteObject()，同样，readObject() 中的第一个操作必须是调用 defaultReadObject()。这些方法调用很奇怪。这看起来好像是你在调用 ObjectOutputStream 的 defaultWriteObject()，并且没有给它传递任何参数，但它不知怎么知道了你的对象引用，反而转过来序列化你的对象的所有非 transient 部分。这挺吓人的。

示例中 transient 对象的存储和检索使用了我们熟悉的代码。然而，想想这里发生了什么吧。在 main() 中，我们创建了一个 SerialCtl 对象并将其序列化为一个 ObjectOutputStream。（注意这里使用的是缓冲区而不是文件，不过对 ObjectOutputStream 来说它们没什么不同。）序列化发生在下面这一行代码：

```
o.writeObject(sc);
```

writeObject() 方法肯定检查了 sc，查看它是否有自己的 writeObject() 方法。[不是通过检查接口——这里没有接口——或类类型（class type），而是使用反射去查询方法。] 如果有，这个方法就会被调用。readObject() 也采用类似的方案。这也许是解决问题的唯一可行方法，但确实让人觉得很奇怪。

E.3.3 版本控制

你可能会更改可序列化类的版本（例如，老版本的类的对象可能存储在数据库中）。这是可行的，但应该只在特殊情况下执行此操作。此外，这需要对序列化有更深入的理解，这里不再介绍。JDK 文档对这个主题进行了透彻的讲解。

E.4 使用持久性

我们可以使用序列化技术来存储程序的一些状态，方便以后轻松地将程序恢复到当前状态，这很有吸引力。但在执行此操作之前，你必须回答一些问题。如果序列化的两个对象都具有对第三个对象的引用，会发生什么呢？当你从两个对象的序列化状态恢复它们时，第三个对象是否只生成一次？如果将两个对象序列化为单独的文件，并在代码的不同部分反序列化它们，又会怎样呢？

下面是一个演示了这些问题的示例：

```java
// serialization/MyWorld.java
import java.io.*;
import java.util.*;

class House implements Serializable {}

class Animal implements Serializable {
  private String name;
  private House preferredHouse;
  Animal(String nm, House h) {
    name = nm;
    preferredHouse = h;
  }
  @Override public String toString() {
    return name + "[" + super.toString() +
      "], " + preferredHouse + "\n";
  }
}

public class MyWorld {
  public static void main(String[] args) {
    House house = new House();
    List<Animal> animals = new ArrayList<>();
    animals.add(
      new Animal("Bosco the dog", house));
    animals.add(
      new Animal("Ralph the hamster", house));
    animals.add(
      new Animal("Molly the cat", house));
    System.out.println("animals: " + animals);
    try(
      ByteArrayOutputStream buf1 =
        new ByteArrayOutputStream();
      ObjectOutputStream o1 =
        new ObjectOutputStream(buf1)
    ) {
      o1.writeObject(animals);
      o1.writeObject(animals); // 再次序列化
      // 写到另一个流里:
      try(
```

```
      ByteArrayOutputStream buf2 =
        new ByteArrayOutputStream();
      ObjectOutputStream o2 =
        new ObjectOutputStream(buf2)
    ) {
      o2.writeObject(animals);
      // 下面进行反序列化：
      try(
        ObjectInputStream in1 =
          new ObjectInputStream(
            new ByteArrayInputStream(
              buf1.toByteArray()));
        ObjectInputStream in2 =
          new ObjectInputStream(
            new ByteArrayInputStream(
              buf2.toByteArray()))
      ) {
        List
          animals1 = (List)in1.readObject(),
          animals2 = (List)in1.readObject(),
          animals3 = (List)in2.readObject();
        System.out.println(
          "animals1: " + animals1);
        System.out.println(
          "animals2: " + animals2);
        System.out.println(
          "animals3: " + animals3);
      }
    }
  } catch(IOException | ClassNotFoundException e) {
    throw new RuntimeException(e);
  }
 }
}
/* 输出：
animals: [Bosco the dog[Animal@19e0bfd], House@139a55
, Ralph the hamster[Animal@1db9742], House@139a55
, Molly the cat[Animal@106d69c], House@139a55
]
animals1: [Bosco the dog[Animal@1ee12a7], House@10bedb4
, Ralph the hamster[Animal@103dbd3], House@10bedb4
, Molly the cat[Animal@167cf4d], House@10bedb4
]
animals2: [Bosco the dog[Animal@1ee12a7], House@10bedb4
, Ralph the hamster[Animal@103dbd3], House@10bedb4
, Molly the cat[Animal@167cf4d], House@10bedb4
]
animals3: [Bosco the dog[Animal@a987ac], House@a3a380
, Ralph the hamster[Animal@1453f44], House@a3a380
, Molly the cat[Animal@ad8086], House@a3a380
]
*/
```

我们能将对象序列化到 byte 数组或从 byte 数组中反序列化，这样就可以对任意 Serializable 对象进行**深拷贝**（deep copy）了。（深拷贝意味着要复制整个对象网络，而不仅仅是基本对象及其引用。）本书第 2 章对对象的复制进行了深入讲解。

Animal 对象包含类型为 House 的字段。在 main() 中，我们创建并序列化了含有这些 Animal 的 List，两次序列化到同一个流，然后序列化到另一个单独的流。当它们被反序列化并打印时，你可以看到某次运行时显示的输出（每次运行时对象位于不同的内存位置）。

你可能期望反序列化的对象与原始对象具有不同的地址。但是请注意，在 animals1 和 animals2 中出现了相同的地址，包括两者共享的 House 对象引用。另外，当 animals3 恢复时，系统无法知道这个流中的对象是第一个流中对象的别名，因此它构建了一个完全不同的对象网络。

只要将所有内容序列化到同一个流，你就可以恢复自己序列化时的对象网络，而不会出现意外的重复对象。可以在序列化第一个和最后一个对象之间更改对象的状态，但这就要你自己负责了，对象在序列化它们时是以其当前状态（以及它们与其他对象的任何连接）写入的。

保存系统状态的最安全方法是"原子"地序列化该状态。如果你序列化了一些东西，然后做一些其他的工作，再序列化一些，等等，那么就不能安全地存储系统。相反，要将构成系统状态的所有对象放在一个容器中，然后在一次操作中序列化该容器。这样，你就可以通过一个方法调用来恢复它了。

下面的示例演示了使用该方案的一个虚构的计算机辅助设计（computer-aided design, CAD）系统。此外，它还抛出了 static 字段的问题。如果查看 JDK 文档，你会看到 Class 也实现了 Serializable 接口，因此应该很容易通过序列化 Class 对象来存储 static 字段。无论如何，这似乎是一个合理的做法。

```
// serialization/AStoreCADState.java
// 保存一个虚构的 CAD 系统的状态
import java.io.*;
import java.util.*;
import java.util.stream.*;

enum Color { RED, BLUE, GREEN }

abstract class Shape implements Serializable {
  private int xPos, yPos, dimension;
  Shape(int xVal, int yVal, int dim) {
    xPos = xVal;
```

```java
      yPos = yVal;
      dimension = dim;
    }
    public abstract void setColor(Color newColor);
    public abstract Color getColor();
    @Override public String toString() {
      return "\n" + getClass() + " " + getColor() +
        " xPos[" + xPos + "] yPos[" + yPos +
        "] dim[" + dimension + "]";
    }
    private static Random rand = new Random(47);
    private static int counter = 0;
    public static Shape randomFactory() {
      int xVal = rand.nextInt(100);
      int yVal = rand.nextInt(100);
      int dim = rand.nextInt(100);
      switch(counter++ % 2) {
        default:
        case 0: return new Circle(xVal, yVal, dim);
        case 1: return new Line(xVal, yVal, dim);
      }
    }
  }

  class Circle extends Shape {
    Circle(int xVal, int yVal, int dim) {
      super(xVal, yVal, dim);
    }
    private static Color color = Color.RED;
    @Override public void setColor(Color newColor) {
      color = newColor;
    }
    @Override public Color getColor() { return color; }
  }

  class Line extends Shape {
    Line(int xVal, int yVal, int dim) {
      super(xVal, yVal, dim);
    }
    private static Color color = Color.RED;
    @Override public void setColor(Color newColor) {
      color = newColor;
    }
    @Override public Color getColor() { return color; }
    public static void
    serializeStaticState(ObjectOutputStream os)
    throws IOException {
      os.writeObject(color);
    }
    public static void
    deserializeStaticState(ObjectInputStream os)
    throws IOException, ClassNotFoundException {
      color = (Color)os.readObject();
```

```
    }
  }

public class AStoreCADState {
  public static void main(String[] args) {
    List<Class<? extends Shape>> shapeTypes =
      Arrays.asList(Circle.class, Line.class);
    List<Shape> shapes = IntStream.range(0, 5)
      .mapToObj(i -> Shape.randomFactory())
      .collect(Collectors.toList());
    // 将所有的静态字段 color 设置为 GREEN：
    shapes.forEach(s -> s.setColor(Color.GREEN));
    // 将所有信息序列化到 CADState.dat：
    try(
      ObjectOutputStream out =
        new ObjectOutputStream(
          new FileOutputStream("CADState.dat"))
    ) {
      out.writeObject(shapeTypes);
      Line.serializeStaticState(out);
      out.writeObject(shapes);
    } catch(IOException e) {
      throw new RuntimeException(e);
    }
    // 展示形状：
    System.out.println(shapes);
  }
}
/* 输出：
[
class Circle GREEN xPos[58] yPos[55] dim[93],
class Line GREEN xPos[61] yPos[61] dim[29],
class Circle GREEN xPos[68] yPos[0] dim[22],
class Line GREEN xPos[7] yPos[88] dim[28],
class Circle GREEN xPos[51] yPos[89] dim[9]]
*/
```

Shape 类实现了 Serializable 接口，因此它的子类自动是可序列化的。每个 Shape 都包含数据，Shape 的每个子类都包含一个 static 字段，该字段保存这种 Shape 类型所有对象的颜色。（将 static 字段放置在基类中的话，这个字段将只有一份，因为 static 字段是不会在子类里复制的。）可以覆盖基类中的方法来设置和获取不同子类的颜色。static 方法不能动态绑定，所以这些是普通方法。每次调用 randomFactory() 方法都会创建一个不同的 Shape，其包含的数据是随机的。（这个示例可以很方便地添加额外的 Shape 子类。）

在 main() 方法中，shapeTypes 列表用来保存 Class 对象，shapes 列表用来保存 Shape 对象。

恢复对象相当简单：

```java
// serialization/RecoverCADState.java
// 恢复这个虚构的 CAD 系统的状态
// {RunFirst: AStoreCADState}
import java.io.*;
import java.util.*;

public class RecoverCADState {
  @SuppressWarnings("unchecked")
  public static void main(String[] args) {
    try(
      ObjectInputStream in =
        new ObjectInputStream(
          new FileInputStream("CADState.dat"))
    ) {
      // 按照它们的写入顺序读取:
      List<Class<? extends Shape>> shapeTypes =
        (List<Class<? extends Shape>>)in.readObject();
      Line.deserializeStaticState(in);
      List<Shape> shapes =
        (List<Shape>)in.readObject();
      System.out.println(shapes);
    } catch(IOException | ClassNotFoundException e) {
      throw new RuntimeException(e);
    }
  }
}
/* 输出:
[
class Circle RED xPos[58] yPos[55] dim[93],
class Line GREEN xPos[61] yPos[61] dim[29],
class Circle RED xPos[68] yPos[0] dim[22],
class Line GREEN xPos[7] yPos[88] dim[28],
class Circle RED xPos[51] yPos[89] dim[9]]
*/
```

xPos、yPos 和 dim 的值都成功地存储并恢复了，但是 static 字段的恢复出了问题。我们写入的都是 GREEN，但读取出来并不是这样：Circle 的值为 RED。就好像 Circle 的静态数据根本没有被序列化一样！没错，即使类 Class 实现了 Serializable 接口，它也不会像你期望的那样进行序列化。因此，要序列化 static 字段，你必须自己做。

这就是 Line 的静态方法 serializeStaticState() 和 deserializeStaticState() 的作用。它们作为存储和恢复过程的一部分被显式调用。（注意，写入序列化文件和从中读取的顺序必须不变。）因此，为了使这些程序正常运行，你必须：

1. 向 Shape 里添加 serializeStaticState() 和 deserializeStaticState() 方法；
2. 移除 shapeTypes 列表和与之相关的所有代码；
3. 添加对 Shape 中新的序列化和反序列化静态方法的调用。

另一个你可能会考虑的问题是安全性，因为序列化还会保存 private 数据。为了防止它们序列化到 CADState.dat 中，应该将这些字段标记为 transient。你必须设计一种安全的方式来存储该信息，这样才能恢复这些 private 变量。

XML

对象序列化的一个重要限制是，它仅适用于 Java 平台：只有 Java 程序可以反序列化此类对象。一个更具互操作性的解决方案是将数据转换为 XML 格式，这使得它可以被各种平台和语言使用。

由于 XML 的流行，使用它编程有很多令人困惑的选项，包括与 JDK 一起分发的 javax.xml.* 库。我选择使用 Elliotte Rusty Harold 的开源 XOM 库（可从 XOM 网站下载并查看其文档），因为在使用 Java 生成并修改 XML 的各种方式中，它似乎是最简单、最直接的。此外，XOM 还强调了 XML 的正确性。

例如，假设有一个包含名字和姓氏的 APerson 对象，你希望将其序列化为 XML。下面的 APerson 类有一个 getXML() 方法，它使用 XOM 将 APerson 的数据抽取并转换为 XML 的 Element 对象，还有一个接受 Element 对象的构造器，可以提取相应的 APerson 数据（注意 XML 示例在它自己的子目录中）：

```java
// serialization/APerson.java
// 使用 XOM 库读写 XML
// nu.xom.Node 来自 XOM 网站
import nu.xom.*;
import java.io.*;
import java.util.*;

public class APerson {
  private String first, last;
  public APerson(String first, String last) {
    this.first = first;
    this.last = last;
  }
  // 从这个 APerson 对象生成一个 XML 元素:
  public Element getXML() {
    Element person = new Element("person");
    Element firstName = new Element("first");
    firstName.appendChild(first);
    Element lastName = new Element("last");
    lastName.appendChild(last);
    person.appendChild(firstName);
    person.appendChild(lastName);
    return person;
  }
```

```
// 构造器从 XML 中恢复 APerson：
public APerson(Element person) {
  first = person
    .getFirstChildElement("first").getValue();
  last = person
    .getFirstChildElement("last").getValue();
}
@Override public String toString() {
  return first + " " + last;
}
// 使其可读：
public static void
format(OutputStream os, Document doc)
throws Exception {
  Serializer serializer =
    new Serializer(os,"ISO-8859-1");
  serializer.setIndent(4);
  serializer.setMaxLength(60);
  serializer.write(doc);
  serializer.flush();
}
public static void
main(String[] args) throws Exception {
  List<APerson> people = Arrays.asList(
    new APerson("Dr. Bunsen", "Honeydew"),
    new APerson("Gonzo", "The Great"),
    new APerson("Phillip J.", "Fry"));
  System.out.println(people);
  Element root = new Element("people");
  for(APerson p : people)
    root.appendChild(p.getXML());
  Document doc = new Document(root);
  format(System.out, doc);
  format(new BufferedOutputStream(
    new FileOutputStream("People.xml")), doc);
  }
}
/* 输出：
[Dr. Bunsen Honeydew, Gonzo The Great, Phillip J. Fry]
<?xml version="1.0" encoding="ISO-8859-1"?>
<people>
    <person>
        <first>Dr. Bunsen</first>
        <last>Honeydew</last>
    </person>
    <person>
        <first>Gonzo</first>
        <last>The Great</last>
    </person>
    <person>
        <first>Phillip J.</first>
```

```
        <last>Fry</last>
    </person>
</people>
*/
```

XOM 的方法是不言自明的,你可以在 XOM 文档中找到它们。

XOM 还包含一个在 format() 方法中使用的 Serializer 类,它用于将 XML 转换为更具可读性的形式。如果调用 toXML() 方法 [①],你也可以把所有东西都运行起来,所以 Serializer 是一个给我们提供了方便的工具。

从 XML 文件反序列化为 APerson 对象也很简单:

```
// serialization/People.java
// nu.xom.Node 来自 XOM 网站
// {RunFirst: APerson}
import nu.xom.*;
import java.io.File;
import java.util.*;

/* 输出:
[Dr. Bunsen Honeydew, Gonzo The Great, Phillip J. Fry]
*/

public class People extends ArrayList<APerson> {
  public People(String fileName) throws Exception {
    Document doc =
      new Builder().build(new File(fileName));
    Elements elements =
      doc.getRootElement().getChildElements();
    for(int i = 0; i < elements.size(); i++)
      add(new APerson(elements.get(i)));
  }
  public static void
  main(String[] args) throws Exception {
    People p = new People("People.xml");
    System.out.println(p);
  }
}
```

People 的构造器使用 XOM 的 Builder.build() 方法打开并读取文件,getChildElements() 方法会生成一个 Elements 列表(它不是 Java 的标准 List,只有 size() 和 get() 方法——Harold 不想强迫人们使用特定版本的 Java,但仍然想要一个类型安全的容器)。这个列表中的每个 Element 代表着一个 APerson 对象,所以它会被传递给 APerson 的第二个构造器。注意,这要求你提前知道 XML 文件的确切结构,而对于此类问题,你通常是知道的。如果它的结构不符合预期,XOM 将抛出异常。如果传入的 XML 结构的具体信息较少,你也

[①] 该方法来自 Document 类。——译者注

可以编写更复杂的代码来探索 XML 文档，而不是对其进行假设。

要编译这些示例，可以将 XOM 发行版中的 JAR 文件放入你的类路径中。

本附录只简单介绍了如何使用 Java 和 XOM 库进行 XML 编程，更多信息请参阅 XOM 网站。

静态类型检查的利与弊

本附录中收集了我多年来撰写过的论文,并做了些编辑。这些论文试图从一种新视角来看待静态类型检查语言和动态类型检查语言之间的争论,而在本附录的两个前言中,我阐述了自己对该主题的最新见解。

F.1　2021 版前言

在本书首次出版后,我在静态类型检查领域体验到两个进展。

1. 在写 *Atomic Kotlin* 这本书的过程中,我意识到了声明风格的影响力。特别是在 C、C++、Java 以及类似语言中看到的"Algol 风格",它有别于我所说的"Haskell 风格"——后者出现在 Scala、Kotlin 和新版本 Python 等语言中。"Haskell 风格"(我认为)不仅更具可读性,而且还能提供更清晰的类型推断。

2. Python 3.5 版本增加了"类型提示"(type hint),这是一种可选的静态类型检查。我发现这种 Haskell 风格的类型声明语法几乎没有负面影响,而好处非常显著,比如开发环境可以据此提供更好的检查和建议。例如,当 PyCharm 看到一个类型声明时,它可以告诉我这个对象能做的所有事情,这节省了时间。它还可以在编辑时就提示我使用对象的方式不正确,这也节省了时间。由于语法很容易添加,它也改善了我作为 Python 程序员的生活。

我现在觉得自己关于静态类型检查的问题和抱怨是由于语法的影响——即使本书中演示了很多改进,Java 语法仍然包含很多我不想关心的"视觉噪声"。

阅读本附录时,请记住,这不是我现在的想法,而是我以前的想法。随着时间的推移,我的想法发生了变化。我目前的立场是,只要语法是低噪声的,而且过程不引人注目,我就认为静态类型检查利大于弊。

F.2 原版前言

我对软件开发的主要兴趣一直是程序员的生产力。程序员周期（Programmer cycle）昂贵,而 CPU 周期（CPU cycle）廉价,我认为不应该让前者为后者买单。

我自己使用静态类型检查的经验,首先来自汇编语言（根本没有任何类型检查）,其次是 ANSI C 之前的版本,它解决了很多问题。然后 C++ 出现了,检查了函数参数和返回值类型,发现了很多错误（并影响了 ANSI C,让它也采用了这种方法）。我成了这种错误发现机制的"粉丝"。时光飞逝,为了支持更多的类型检查,我们背负着越来越多的编码开销。与此同时,我开始尝试动态语言,结果发现了 Python 极其惊人的生产力。这让我开始怀疑,在追求完美静态检查的过程中,我们是否遗漏了什么?（在编写 *Atomic Scala* 时,我还尝试了静态类型检查的极端情况）。

我总是能听到诸如"更多的静态检查总是更好"和"动态语言只能用来解决简单的问题,它们不适用于现实世界的应用程序"之类的话。

我写这些文章的目的是解释动态语言的好处,并质疑那些认为静态类型检查是"免费"的、所有这些功能都没有相关成本的想法。有时这些操作带来的成本会阻碍我们的进步。

我原先以为人们之所以有这些想法只是因为不理解,如果我更努力地解释,他们最终会明白。

然后有趣的事情发生了。随着时间的流逝,人们使用 Python 和 Ruby 等语言创建了越来越多大型的、真实世界里的应用程序。JVM 上也出现了像 Groovy 和 Clojure 这样的动态语言,人们开始用这些语言编写令人赞叹的程序。

在看到这些语言可以做什么之后,如果再来辩解说某些事情只能用静态语言完成,就找不到合理论据了。这时,有些人又开始说他们只是"更喜欢静态类型检查"。

直到最近我才猛然醒悟:

这是一个文化问题,关乎信念。

这就是为什么我所有的解释都失败了。这群人已经习惯了这套信念,而我正在挑战这些信念,这让他们感到不舒服。所以他们反击了,并且更加努力地寻找论据。

编程语言创建了自己的文化社区。

事后看来这个新观点再平常不过，但接受了这个观点后，我再看相关的问题和人们的选择，感觉就大不相同了。有些问题可以通过理性和实验来争论，但另一些问题是社区核心信念体系的一部分，如果你加入这个社区，你就跟随了这个信念，无论它把你带到什么地方。

F.3　静态类型检查与测试

我们如何才能最大限度地掌握我们试图解决的问题？每当新工具（尤其是编程语言）出现时，它都会提供某种抽象，这个抽象可能会也可能不会对程序员隐藏不必要的细节。然而，我始终关注浮士德式的讨价还价，尤其是当它们试图要说服我，让我忽略为了实现这种抽象所需要绕过的各种圈套时。Perl 就是一个很好的例子——这个语言的即时性隐藏了编写程序时的无意义细节，但不可读的语法（这是因为要与 awk、sed 和 grep 等 UNIX 工具向后兼容导致的）是一个显著的代价。

多年来，我慢慢理解了更为传统的编程语言和它们对静态类型检查的态度。这开始于很久以前，那时我与 Perl 开始了两个月的恋情，它的快速改变能力提高了我的生产力（Perl 对引用和类的处理方式应受到谴责，这件事终止了我们的恋情，直到后来我才看到它语法的真正问题）。在 Perl 中看不到静态类型检查与动态类型检查之间的争论，因为你无法构建足够大的项目来发现这些问题，而它的语法掩盖了较小程序中的所有问题。

在使用 Python（一种可以构建大型复杂系统的语言）后，我注意到，尽管它显然没有在意类型检查，但 Python 程序似乎不需要太多努力就能运行得很好，并且没有那些我们都"知道"的编程问题。我们一直都认为没有静态类型检查的语言肯定会有这些问题，而静态类型检查是解决它们唯一的正确方法。

这对我来说成了一个谜题：如果静态类型检查如此重要，为什么人们能够在没有灾难的情况下构建大型、复杂的 Python 程序（而且比静态程序花费更短的时间和更少的努力）？而我本来确信这些灾难肯定会发生的。

这动摇了我对静态类型检查（从 C 迁移到 C++ 时获得的，因为当时的改进非常显著）无条件接受的态度，以至于后来遇到 Java 中关于检查型异常的问题时，我会问"为什么"。这引发了一场大讨论，有人告诉我，如果我继续提倡非检查型异常，就会有大麻烦、大灾难了。在 *Thinking in Java, 3rd Edition* 中，我提倡使用 RuntimeException 作为包装类来"关闭"检查型异常。每次我这样做的时候,感觉都很好(我注意到 Martin Fowler 大致在同一时间提出了相同的想法),但我仍然会偶尔收到电子邮件,来警告说我的言论荒唐透顶,会导致大灾难之类的。

但是,认为检查型异常带来的麻烦多于它们的价值[①],并没有回答这个问题:在传统观点都认为它应该产生大量失败的情况下,Python 为什么能工作得如此之好? Python 和类似的动态类型语言都会延迟类型检查。Ruby、SmallTalk 和 Python 等语言都不会尽早给对象类型施加尽可能强的约束(如 C++ 和 Java 所做的那样)。相反,它们会对类型施加尽可能宽松的约束,并且仅在必需时才检查类型。也就是说,你可以向任何对象发送任何消息,而语言只关心该对象是否可以接受该消息——它不要求该对象是特定类型,而 Java 和 C++ 则不是这样。例如,在 Java 中如果要实现会说话的宠物,代码如下所示:

```java
// staticchecking/petspeak/PetSpeak.java
// 用 Java 实现会说话的宠物
// {java staticchecking.petspeak.PetSpeak}
package staticchecking.petspeak;

interface Pet {
  void speak();
}

class Cat implements Pet {
  @Override public void speak() {
    System.out.println("meow!");
  }
}

class Dog implements Pet {
  @Override public void speak() {
    System.out.println("woof!");
  }
}

public class PetSpeak {
  static void command(Pet p) { p.speak(); }
  public static void main(String[] args) {
    Pet[] pets = { new Cat(), new Dog() };
    for(Pet pet : pets)
      command(pet);
  }
}
/* 输出:
meow!
woof!
*/
```

注意,command() 必须确切地知道它要接受的参数类型——Pet——并且它不会接受任

[①] 这里强调的是检查,而不是异常。我相信单一且一致的错误报告机制是必不可少的,尽管最终的错误报告可能并不与异常相关——就像 Go 语言那样。

何其他类型的参数。因此，我必须创建一个 Pet 层次结构，并使 Dog 和 Cat 来继承，这样才能让它们向上转型后传递给通用的 command() 方法。

在很长一段时间里，我认为向上转型是面向对象编程的固有部分，并且觉得用 SmallTalk 的人关于向上转型的问题很烦人。但是当我开始使用 Python 时，我发现了下面的这些可贵之处。上面的代码可以直接翻译成如下的 Python 代码：

```python
# staticchecking/PetSpeak.py
#- 用 Python 实现会说话的宠物
class Pet:
    def speak(self): pass

class Cat(Pet):
    def speak(self):
        print("meow!")

class Dog(Pet):
    def speak(self):
        print("woof!")

def command(pet):
    pet.speak()

pets = [ Cat(), Dog() ] # (1)
for pet in pets: # (2)
    command(pet)

output = """
meow!
woof!
"""
```

如果你以前从未用过 Python，你会注意到它以一种非常好的方式重新定义了语言的简洁性。你认为 C/C++ 简洁？扔掉那些花括号吧。缩进对人类的思维是有意义的，所以将用它来表示作用域。参数类型和返回类型？让语言自行解决吧。在创建类时，基类在括号中表示。def 表示正在创建一个函数或方法定义。另一方面，Python 在方法定义时如果用到 this 参数（按惯例称为 self），则需要显式使用。

注释以 # 开头，一直到行尾。

Pet 类的定义使用了 pass 关键字，类似于 Java 中的 abstract。它表示没有定义类的主体。

注意，command(pet) 只声明了它需要名为 pet 的对象，但没有要求该对象必须是什么类型的。这是因为它并不在乎这些，只要这个对象可以调用 speak() 方法就行，或者其他任何你自己的函数或方法想做的事情。稍后会更仔细地研究这方面内容。

另外，command(pet) 只是一个普通的函数，这在 Python 中是允许的。也就是说，Python 不坚持让所有东西都成为对象，因为有时你想要的就是函数。

(1) 在 Python 中，列表和映射 / 字典 / 关联数组都非常重要，它们被内置到语言的核心里，所以不需要导入任何特殊的库就能使用它们。上例中创建了一个列表，它包含了类型分别为 Cat 和 Dog 的两个新对象。构造器被调用了，但没有使用 new（现在你再回到 Java，就意识到它也不需要 new。这只是从 C++ 里继承来的多余关键字）。

(2) 序列遍历也十分重要，Python 对它提供了原生操作。for 会选择列表 pets 中的每一项，然后赋值给变量 pet。

上例的输出与 Java 版本相同，并展示在了 output 变量的定义里。三引号可以创建一个多行字符串。

Python 是一门优秀的伪代码语言，具有可以实际执行的奇妙属性。这意味着可以在 Python 中快速尝试想法，当找到可行的想法时，就可以用 Java、C++ 或自己选择的其他语言重写它。或者你可能意识到问题已经在 Python 中解决了，那为什么还要重写它呢（通常我就是这样做的）？在研讨会期间，我会用 Python 来指导练习，因为这样我就不会透露自己的整个想法，而人们又可以在解决方案中看到我想要使用的形式，这样他们就可以继续前进。我能够验证这种形式是否正确。这就是为什么 Python 通常被称为"可执行的伪代码"。

对于这个讨论，一个有趣的地方是，command(pet) 方法不在乎它得到的是什么类型，因此不必向上转型。这样我就可以在不使用基类的情况下重写 Python 程序：

```python
# staticchecking/NoBasePetSpeak.py
#- 不使用基类重写会说话的宠物程序

class Cat:
    def speak(self):
        print("meow!")

class Dog:
    def speak(self):
        print("woof!")

class Bob:
    def bow(self):
        print("thank you, thank you!")
```

```
        def speak(self):
            print("Welcome to the neighborhood!")
        def drive(self):
            print("beep, beep!")

    def command(pet):
        pet.speak()

    pets = [ Cat(), Dog(), Bob() ]

    for pet in pets:
        command(pet)

    output = """
    meow!
    woof!
    Welcome to the neighborhood!
    """
```

command(pet) 只在乎它是否可以将 speak() 消息发送给它的参数，所以我删除了基类 Pet，甚至添加了一个名为 Bob 的非宠物类，不过 Bob 有一个 speak() 方法，因此它也适用于 command(pet) 函数。

此时，静态类型语言怒不可遏，坚持认为这种草率的做法会造成混乱和灾难。显然，在某些时候，"错误"的类型会被传递给 command()，或溜进系统里。更简单、更清晰的概念表达所带来的好处并不值得我们冒这么大的险，即使这种好处是生产力比 Java 或 C++ 提高 5~10 倍。

那当 Python 程序出现这样的问题时会发生什么呢——一个对象不知何故出现在了它不应该出现的地方？Python 将所有的错误都报告为异常，所以你的确能发现问题，但它总是在运行时才报告。"啊哈！"你说，"这就是问题：你不能保证程序的正确性，因为你没有必需的编译时类型检查。"

这个程序甚至可以用 Go 语言重写，如下所示：

```
// staticchecking/petspeak.go
package main
import "fmt"

type Cat struct {}
func (this Cat) speak() { fmt.Printf("meow!\n")}

type Dog struct {}
func (this Dog) speak() { fmt.Printf("woof!\n")}

type Bob struct {}
```

```go
func (this Bob) bow() {
  fmt.Printf("thank you, thank you!\n")
}
func (this Bob) speak() {
  fmt.Printf("Welcome to the neighborhood!\n")
}
func (this Bob) drive() {
  fmt.Printf("beep, beep!\n")
}

type Speaker interface {
  speak()
}

func command(s Speaker) { s.speak() }

// 如果"Speaker"没有在其他地方使用过，可以匿名：
func command2(s interface { speak() }) { s.speak() }

func main() {
  command(Cat{})
  command(Dog{})
  command(Bob{})
  command2(Cat{})
  command2(Dog{})
  command2(Bob{})
}
/* 输出：
meow!
woof!
Welcome to the neighborhood!
meow!
woof!
Welcome to the neighborhood!
*/
```

Go 没有 class 关键字，但你可以使用上述形式来创建与基类等效的类：其他语言通常会定义类，而这里定义了一个 struct，它可以包含数据字段（本例中没有）。对于每个方法，从 func 关键字开始，然后——为了将方法附着到你的类上——跟着一对包含了对象引用的括号。对象引用可以用任何标识符，但我在这里使用 this 来提醒你，它就像 C++ 或 Java 中的 this 一样。然后定义函数的其余部分，这一部分和 Go 中其他任何函数的定义没有什么不同（请注意，Go 也没有继承，因此这种"面向对象"的形式相对原始，这可能是阻碍我在该语言上花更多时间的主要因素。不过，Go 里的组合倒是很直观）。

command() 和 command2() 函数都使用了结构化类型 / 鸭子类型：参数的确切类型并不重要，只要它包含一个 speak() 方法。我在这里展示了两种方法：command() 使用外部定义的 Speaker 接口，但如果该接口从未在其他任何地方使用过，你也可以匿名地、内联地定义它，如你在 command2() 中所见。

main() 表明 command() 和 command2() 确实不关心其参数的确切类型，只要有一个 speak() 方法就可以。但是，就像 C++ 模板函数一样，Go 会在编译时检查类型（语法 Cat{} 创建了一个匿名的 Cat 结构体）。

类型检查也只是一种测试

当编写《C++ 编程思想》第 1 版的时候，我采用了一种非常粗略的测试形式：我编写了一个程序，该程序会自动从书中提取所有代码，使用放置在代码中的注释标记来查找代码的开头和结尾。然后它会构建一个编译所有代码的 makefile。这样就能保证书中的所有代码都经过了编译，然后我就有理由确定地说"书中的代码都是正确的"。我忽略了"编译并不意味着正确执行"的唠叨声，因为这是迈向代码自动化验证的一大步（任何读过编程书的人都知道，许多作者现在仍然不会在验证代码正确性方面付出努力）。但这自然会导致一些示例运行不正确，当这些年来报告的出错示例足够多时，我意识到自己不能再忽视测试问题了。我开始相信：

如果没有经过测试，代码就不可能正常工作。

在静态类型语言中，如果一个程序编译通过，这只是意味着它通过了一些测试，由此保证语法是正确的（Python 在编译时也会检查一些语法，只是没有那么多的语法约束）。但不能仅仅因为编译器编译通过了你的代码，就认为这保证了代码的正确性。即使你的代码可以运行，也不能保证代码的正确性。

无论你用的语言是强类型还是"灵活"类型，正确性的唯一保证是，它是否通过了定义程序正确性的所有测试。你必须自己编写其中的一些测试。当然，这些都是单元测试。一旦成为"测试感染者①"（test infected），你就不会再走回头路了。

这非常像从旧的 C 迁移到 C++。忽然间，编译器会替你执行更多的测试，你可以更快地工作。但这些语法测试也就仅仅能做到这种程度。编译器无法知道你想要程序做什么，因此你必须通过添加单元测试来"扩展"编译器（无论使用的是哪种语言）。如果你这样做，就可以快速地进行彻底更改（重构代码或修改设计），因为你知道自己的测试套件会支持你，如果出现问题就会立即失败——就像在有语法问题的时候编译失败一样。

但是如果没有完整的单元测试（最低要求），就不能保证程序的正确性。声称 C++ 或 Java 中的静态类型检查约束能避免你写出糟糕的代码，这显然是一种错觉（从个人经验中就能知道这一点）。其实我们需要：

强测试，而不是强类型检查。

所以，我断言，这就是 Python 能够工作的一个很好的原因。C++ 测试发生在编译时（除了一些次要的特殊情况）。Java 测试有些发生在编译时（语法检查），有些发生在

---
① 即尝到了测试带来的好处。——译者注

运行时（例如数组边界检查）。大多数 Python 测试发生在运行时而不是编译时，但它们确实存在，这才是最重要的（而不是什么时候）。相较于编写等效的 C++ 或 Java 程序，我可以在更少的时间内启动和运行 Python 程序，所以可以更快地开始运行真正的测试：单元测试、假设性测试，以及替代方式的测试等。如果 Python 程序有足够的单元测试，它也可以像 C++ 或 Java 程序一样稳健，即使 C++ 或 Java 程序也有足够的单元测试（不过 Python 中的测试编写速度更快）。

Robert Martin 是 C++ 社区的长期参与者，他写过书和文章，做过咨询，也教过课。他是一个非常铁杆的、强静态类型检查的支持者——至少以前我是这么以为的，直到读了他的博文"动态语言会取代静态语言吗"（"Are Dynamic Languages Going to Replace Static Languages"）。Robert 或多或少地得出了与我相同的结论，但他先是成为"测试感染者"，然后意识到编译器只是一种（不完整的）测试形式，进而理解了动态类型语言可以通过提供足够的测试，来创建与静态类型语言一样稳健的程序。

当然，马丁也收到了司空见惯的"你怎么可以这样想"之类的评论。正是这个问题让我开始与静态/动态类型概念做斗争。当然，我们两个都是从倡导静态类型检查开始的。有时需要一次惊天动地的经历——比如被测试"感染"或学习一种不同的语言——才能重新评估自己所相信的东西。

F.4 如何讨论类型检查

许多人观察到，类型检查就像某种宗教话题一样，最好避免讨论。我大体上同意这种观点，鉴于我已经在这方面引发了太多的论战。然而，一些关于类型和类型检查的争论能够体现编程语言之间的基本区别。这种理解使我们值得冒这方面的险，所以在本节中——一篇我在 2004 年 Python 会议（PyCon）上的闭幕主题演讲——我将研究围绕类型概念的各种问题和论点，特别是"结构化类型"（structural typing），又名"潜在类型"[latent typing，不幸的是，有时也称为"弱类型检查"（weak typing）]的现象，为什么这个概念如此强大，以及它如何在不同的语言中表示。

背景 当我在南加州读初中时，老师们试图通过在图书馆里播放关于"E Man"（我记得好像是一只狗）和 albondigas（肉丸）的电影来教我们西班牙语，这就是我记得的全部。老师对我们讲西班牙语，然后期待我们也能用西班牙语回应，我对此感到头疼。我没有意识到学习它的动机，而且这种努力似乎很痛苦。弹吉他对我来说也是如此（它会让手指疼痛），任何未能让我掌握实际技能的学习过程都是如此。

所以我同情那些拒绝学习多种语言的人。通过学习外语，我可能只知道几个短语，而只有通过体验，我才能发现学习不止一种编程语言的价值难以置信。最大的收获是，我现在可以思考 Java 或 C++ 中的一些事情，而这些事情是我在学习 Python 之前无法想象的。从我读过的资料和代码来看，单一语言的用户很少能想到这些。你的语言的确会限制你的思想。

几年前我开始注意到，编程语言特性似乎以某些静态检查的哲学思想为中心。这些哲学思想的基础是这样一种想法：特定的做事方式总是会产生更好的结果。此外，人们在给出各种术语含义的时候，其上下文往往会受限于自己最了解和最喜欢的语言。

当我开始使用"弱类型检查"这个术语时 [我发现它与"弱类型"（weakly typed）完全不同]，我自己就深陷其中。当时，这个术语的出处似乎很权威，但现在我无法追溯它的起源。当我断言在 Java 中，检查型异常——即编译器强制要求编写代码来处理特定调用的异常，而不允许我们自行决定是否处理异常——是该语言的一个实验性功能，实际上弊大于利时（并不是说检查型异常没有用，但使用它会带来更多的麻烦），这引起了一片抗议之声。这些讨论似乎触动了人们在基本信仰体系中的深层神经。

我最初称为"弱类型检查"的类型机制，其实更应该称为"结构化类型"或"潜在类型"，无论如何称呼，它的价值都令人着迷。基本上它就是在说，"我不在乎这种类型是什么（在某些编程结构中），但我仍然关心该类型的行为是否正确……"。在 C++ 中，这个机制就是模板。而 Java 使用了术语"泛型"，这虽然意味着同样的事情，但实际上它最多能泛化到根类 Object。但是在 Python（还有 SmallTalk）中，这就是你定义任何函数的方式，就像这里对 speak() 的定义一样：

```python
# staticchecking/DogsAndRobots.py

def speak(anything):
    anything.talk()

class Dog:
    def talk(self):  print("Arf!")
    def reproduce(self): pass

class Robot:
    def talk(self): print("Click!")
    def oilChange(self): pass

a = Dog()
b = Robot()

speak(a)
```

```
  speak(b)

output = """
Arf!
Click!
"""
```

speak() 不关心它的参数类型。你可以传递任何支持 talk() 方法的对象。

这个例子很容易转换为 C++[1]：

```cpp
// staticchecking/DogsAndRobots.cpp
#include <iostream>
using namespace std;

class Dog {
public:
  void talk() { cout << "Arf!" << endl; }
  void reproduce() {}
};

class Robot {
public:
  void talk() { cout << "Click!" << endl; }
  void oilChange() {}
};

template<class T> void speak(T speaker) {
  speaker.talk();
}

int main() {
  Dog d;
  Robot r;
  speak(d);
  speak(r);
}
/* 输出:
Arf!
Click!
*/
```

下面是用 Go 编写的相同程序：

```go
// staticchecking/dogsandrobots.go
package main
import "fmt"

type Dog struct {}
func (this Dog) talk() { fmt.Printf("woof!\n")}
func (this Dog) reproduce() {}

type Robot struct {}
/* 输出:
woof!
Click!
*/
```

---

[1] 你可以在 C++ Shell 网站上体验一下 C++，不用费劲安装 C++ 编译器 。

```
func (this Robot) talk() { fmt.Printf("Click!\n") }
func (this Robot) oilChange() {}

func speak(speaker interface { talk() }) {
  speaker.talk();
}

func main() {
  speak(Dog{})
  speak(Robot{})
}
```

这几乎就像在发明一种新类型——"一种 speak() 可接受的类型",因为它肯定不是系统中存在的任何类型。你可能还会争辩说,speak() 定义了两种类型——speak() 可接受的类型和其他所有类型。这两个论点我都不认可。这里没有为 x 指定类型,或者可以说,对 x 的类型约束的削弱使其允许存在许多不同类型。

介绍完术语后,本节将介绍在 Python 和 C++ 中实现结构化类型/潜在类型的方式,以及 Java(不支持结构化类型)中的泛型是什么样的。

F.4.1 弱类型与弱类型检查

C++ 拥护者:"C++ 是强类型的,同时还具有静态类型检查。"

专家:"C/C++ 是弱类型的,因为它有 cast 和 union。"

C++ 拥护者:"这有点极端,不是吗?不如我们说 C++ 是强类型、静态类型的,但类型系统里有一个漏洞: cast(哪还有人使用 union 呢?)"

专家:"不。如果一种语言允许一个对象接受任何不正确的消息,它就是弱类型的。Java 也允许类型转换,但它会先进行检查,在编译时查看类型转换是否可行,并在运行时确保类型转换是正确的。所以 Java 是强类型的,而 C++ 是弱类型的。"

C++ 拥护者:"所以你是在说,即使一门语言在任何方面都是强类型的,但只要它的类型系统上有漏洞,这门语言就是弱类型的?"

专家:"绝对是弱类型的呀,老兄。"

显然,上面的问题在某些方面和定义以及对定义的遵守程度有关,所以首先必须定义"什么是类型"。这里有两种不同的方式来定义类型。一个对象可以符合如下两种定义之一。

1. 语法兼容的:对象提供所有预期的操作(类型名称、函数签名、接口)。
2. 语义兼容的:对象所有的操作都以预期的方式运行(状态语义、逻辑公理、证明)。

第一种通常被认为是**类型**（type，它提供了一个特定的接口），第二种通常被称为**类**（class，描述了实现约束）。我认为这与本节开头要处理的问题非常吻合：我们主要关注可以对对象执行哪些操作，唯一感兴趣的语义是这些操作是否安全，另外语言系统是否始终会报告不安全的操作（例如 C++ 允许旧式的未经检查的类型转换，如果你使用了它们，语言系统不会报告任何内容）。

- 语言将类型约束作用于符号，还是对象？
  我们可以把类型约束作用于符号，尤其是在运行时支持（runtime support）不足的情况下（C++）。也可以两者兼而有之，就像 Java 一样，它在编译时限制了所有符号的类型和用法，但也有足够的运行时支持来执行有限的动态检查。如果它不能静态检查某些东西（例如数组边界或空指针），则它会动态检查它们。

- 约束是在程序运行之前（静态类型检查）还是在程序运行时（动态类型检查）起作用？
  同样，Java 两者兼而有之。但是动态类型检查的语言似乎具有更大的灵活性，同时能更好地确保对象得到正确处理。它们可以对语法符号更随意一点，因为对象在运行时会保护自己免受不当使用，而不是依靠编译器来保护。

- 语言是根据用法"猜测"类型（类型推断，type inference），还是通过某处的声明（显式类型，manifest typing）获知类型的？
  人们有时会把类型推断与动态类型相混淆，因为语法符号可能看起来没有类型（或没有固定类型）。Scala[①] 是类型推断语言的一个很好的例子，在定义类型时使用类型推断要好得多。

- 如果一种语言是静态类型的，它是否会放宽类型约束以允许更灵活的编程？ [这有时称为**渐进类型**（gradual typing）。]
  这是我本章中最感兴趣的地方。注意，在上面的 dog-robot 示例中，许多人得出的结论是，因为 speak() 参数列表中的 anything 不受约束，因此不存在类型安全。但是，强类型安全是存在的——你不能向对象发送不正确的消息。

词典

- **静态类型**：在编译时检查类型。
- **动态类型**：在运行时检查类型。
- **强类型**：你不能在对象上执行不正确的操作（发送坏消息）。

---

① Scala 是静态类型语言，但类型推断很强大。——译者注

- **"弱类型"**：你可以在对象上执行不正确的操作。很多人可能会把弱类型和下面其他的术语混淆，但不管怎么样人们还是这样使用。
- **潜在类型 / 结构类型**：在少数特定情况下会放宽类型约束，使编程更加灵活和强大。
- **类型推断**：你不必指定类型，语言系统会根据它们的使用方式来自行推断类型（Scala 就是这样的）。
- **显式类型**：创建变量时必须指定类型。
- **鸭子类型**：这是 Ruby 采用的一个术语，用于描述结构化类型 / 潜在类型。"如果它走路像鸭子，说话像鸭子，那它就是鸭子。"鸭子类型的价值似乎在于能够直接创建适配器对象，而不需要创建新的适配器类（请参阅本书第 8 章 8.8.1 节）。

短语"隐式类型检查"似乎很好地描述了我在这里谈论的内容：当定义函数、模板或泛型时，会创建隐式类型。不幸的是，Fortran 已经采用了这个术语，表示"标识符的第一个字母意味着实数或整数"。我很想将它稍微更改为"隐式类型检查"（implied typing），因为它们表达的含义差不多，但我相信这样做的话最终也会导致与"弱类型与弱类型检查"（weakly vs. weak）相同的问题。[我还注意到人们也在使用"隐式类型"（implicitly typed）一词。]

latent（隐式）
1. 存在或潜在的，但不明显或活跃。存在但不显化。潜在人才。
2. 在心理学里，表示在无意识的头脑中存在且可见，但未有意识地表达出来。
3. 一种肉眼不明显但充分可见的指纹，但经过除尘或发烟（fuming）后，可以很清晰地识别出来。

上面的词义隐含了我们想要表达的意思，但并没有显式表达出来。毕竟这个词就是这样的，用在这里很恰当。或者：

"如果你只需要嘎嘎叫的东西，那么就不用关心它是否是鸭子。"

——Andrew Dalke

当然，对于潜在类型，我们甚至一开始就不想让它的命名里直接出现"鸭子"。我相信鸭子就在那里，即使只有很少的证据能证明它的存在。

F.4.2 模板、潜在类型和 Java 泛型

在将模板引入 C++ 之前，我们使用宏来创建通用容器。对于库设计者来说这很痛苦，对于调用者来说可能不那么难受。当 C++ 引入了模板时，我评论道：如果 C++ 有一个单根层次结构（所有东西最终都继承自 `Object`），那么这些就没有必要了，因为我们可以编写存储 `Object` 的容器，这样就可以自动存储所有的东西了。这也是 SmallTalk 的做法，而

且看起来挺成功的。

当然，SmallTalk 使用了潜在类型，因此不需要类型转换。当 Java 采用这个方案时，它实现了静态类型，因此我们会受益于单根层次结构创建的容器。但是对于 Java 来说，我们需要时时刻刻进行类型转换，包括向上转型和向下转型。

Java 5 引入的解决方案是所谓的**泛型**，泛型意味着"任何类型"，但实际上它是受到限制的。泛型的引入让集合框架更好用：

```java
// staticchecking/drc/DogAndRobotCollections.java
// {java staticchecking.drc.DogAndRobotCollections}
package staticchecking.drc;
import java.util.*;

class Dog {
  public void talk() {
    System.out.println("Woof!");
  }
  public void reproduce() { }
}

class Robot {
  public void talk() {
    System.out.println("Click!");
  }
  public void oilChange() { }
}

public class DogAndRobotCollections {
  public static void main(String[] args) {
    List<Dog> dogList = new ArrayList<>();
    List<Robot> robotList = new ArrayList<>();
    for(int i = 0; i < 10; i++)
      dogList.add(new Dog());
    //- dogList.add(new Robot()); // 编译时错误
    for(int i = 0; i < 10; i++)
      robotList.add(new Robot());
    //- robotList.add(new Dog()); // 编译时错误
    dogList.forEach(Dog::talk);
    robotList.forEach(Robot::talk);
  }
}
/* 输出:
Woof!
Woof!
Woof!
Woof!
Woof!
Woof!
Woof!
Woof!
Woof!
Woof!
Click!
Click!
Click!
Click!
Click!
Click!
Click!
Click!
Click!
Click!
*/
```

这绝对改进了集合框架以及其他方面。但是处理参数的时候，不能像在 C++ 中那样"不在乎是什么类型"。所以在 Java 中不能这样做：

```java
class Communicate {
  public static <T> void speak(T speaker) {
```

```
    speaker.talk();
  }
}
```

我们只能说："这个东西不能比 Object 更具体。"这样之所以能编译，是因为它调用了 Object 里的方法：

```
public class NothingButObject {
  public <T> String f(T anyObject) {
    return anyObject.toString();
  }
}
```

结果证明，当操作 Java 集合时它可以工作得很好，这些集合已经被限制为只包含 Object 的了（因此它们可以包含任何类型）。

但是要做比这更通用的事情，以及编写真正的"泛型代码"，还是不行的（不过 Java 8 提供了"辅助潜在类型"，在基础卷第 20 章的末尾进行了描述）。要将 communicate() 函数应用于 Dog 和 Robot，必须像下面这样做：

```
// staticchecking/dr/DogsAndRobots.java
// {java staticchecking.dr.DogsAndRobots}
package staticchecking.dr;

interface Speaks { void talk(); }
class Dog implements Speaks {
  @Override public void talk() {
    System.out.println("Woof!");
  }
  public void reproduce() { }
}

class Robot implements Speaks {
  @Override public void talk() {
    System.out.println("Click!");
  }
  public void oilChange() { }
}

class Communicate {
  public static <T extends Speaks>
  void speak(T speaker) {
    speaker.talk();
  }
}

public class DogsAndRobots {
  public static void main(String[] args) {
```

```
/* 输出:
Woof!
Click!
*/
```

```
    Dog d = new Dog();
    Robot r = new Robot();
    Communicate.speak(d);
    Communicate.speak(r);
  }
}
```

必须约束函数能接受的泛型类型，让这个泛型类型符合该函数内调用操作所属的类或接口。

这当然并不比直接使用接口作为参数更好，实际上这更令人困惑：

```
class Communicate {
  public static void speak(Speaks speaker) {
    speaker.talk();
  }
}
```

我希望 Java 泛型能实现真正的潜在类型：

```
<T> void f(T objectOfAnyType) {
  objectOfAnyType.anyOperation();
}
```

这里我得到的是精简版模板和其他一些功能，但没有得到真正的泛型（潜在）类型的力量（直到 Java 8，前面提到过）。反对 Java 实现潜在类型的主要论点是"它不是类型安全的"，我希望本附录后面的内容能证明这个论点是错误的。

C++ 模板始终允许实现真正的泛型代码。它的缺点是，早期编译器会为模板代码生成可怕的错误消息（后来有所改进），因此许多人得出结论：模板不好。

针对 C++ 模板的另一个反对意见是它们导致代码膨胀，因为每次将它实例化为新类型，都要复制模板代码。但是，这种代码膨胀通常是因为对功能的误解产生的，特别是将所有代码放在模板内，而不是继承或使用组合。基类或成员对象的代码不需要模板化（或复制）。

这句话非常有力量："我不关心对象是什么类型的，只要能够对其执行某些操作就可以了。"这也是 C++ 标准模板库（STL）算法的基础。C++ 容器实现了该算法，容器本身也被设计为尽可能少受限制，因此它们可以存储任何对象。这就好像 C++ 里有一个"隐藏的单根层次结构"（一个非常简单的根类）一样。

F.4.3 灵活性与（所认为的）安全性

Java 专家："静态类型检查对于安全性来说是必要的。"

我："但是 Java 在运行时也会做一些检查。这就是为什么会有 ClassCastException。"

Java 专家："好吧，那我这样说：为了安全性，尽可能多的静态类型检查总是可取的。"

我："如果有太多的类型检查要做，这难道不会阻碍我们对代码的理解？而对于'每个程序员每天只能写 20 行代码'的统计，我们迄今为止仍然没有什么改进。"

Java 专家："你只是在抱怨要多打几个字而已。像 IntelliJ 这样的工具可以帮我们打很多字①，所以我可以做到两全其美：少打字的同时还拥有强大的类型检查。"

我："是的，但是代码读得比写得多。让代码阅读者做更多的工作，会拖延开发进度。"

Java 专家："恰恰相反，这使代码更明确，因此更容易理解。"

我："我发现准确阅读 Python 代码时的速度要快得多，因为它更短、更直接，而我经常被 Java 代码里的冗余类型检查所打断。两者相比，Python 似乎更容易理解。"

Java 专家："我不懂 Python，所以我不相信你。Java 的静态类型检查保证了安全性，我很确定这正是我想要的。"

我："什么'保证'？软件开发项目的失败率至少有 50%，甚至可能高达 80%？"

Java 专家："这句话没法验证，并且 Java 项目的成功率比 C++ 要高。这肯定让我的生活更轻松了。"

我："是的，很少有人愿意承认他们的失败，所以我们无法得到一个确切的数字，只能说失败率非常高，而且开销也很大。当失败率如此之高时，你怎么能说这是可以保证的呢？与 C++ 相比，Java 看起来确实提高了项目的成功率，但是当大部分项目走向失败时，你怎么能说一切都正常呢？"

Java 专家："但是你想减少类型检查的数量。这会让情况变得更糟。"

我："不，类型检查的数量至少还是一样的，只是改变了它发生的阶段。"

Java 专家："如果你要在运行时而不是在编译时进行类型检查，错误可能就会从裂缝

---

① 指 IDE 可以帮我们生成很多模板代码。——译者注

里溜进来。"

我："来看一个在 Java 里进行运行时检查的示例吧。Java 5 之前的集合允许存储任何对象，然后在取出时必须将其转换回所需的类型，并且在运行时检查强制类型转换的正确性。"

Java 专家："是的，这种做法很糟糕。你可能会将错误的类型放入集合中。我甚至可以将'狗'放入'猫'的集合里。咦，这好像是你自己书中的一个示例呀。"

我："是的，这会是个问题，但是当它发生时，你会得到一个运行时异常，所以你会发现它。"

Java 专家："但是如果我们使用 Java 泛型那样的静态类型，那就根本不会发生。这解决了一个大麻烦。"

我："你经常遇到这种情况吗？"

Java 专家："呃……并不。实际上，我不记得上次发生这种情况是什么时候了。但我确信问题只是潜伏着，等待时机让一个重要的程序崩溃。"

我："只要它发生时你能知道，这就足够了。你也可以通过单元测试来做到这一点。"

Java 专家："静态检查比单元测试更好，因为我不必编写测试。"

我："你最终还是要编写一些测试。语言的语法和特定程序的语义之间总会有一条分界线，而我们不得不越过这条分界线来为特定程序的语义编写测试。如果没有自己的测试，就无法拥有可验证的程序。"

Java 专家："单元测试很有用，是的。"

我："所以我的建议是，承认无论如何都要编写自己的测试——从而对应用程序进行动态检查——这样就可以考虑移动一下这条分界线，使编程变得更加动态。我认为让代码更易于编写和阅读所带来的好处，将远远超过减少静态类型检查带来的损失。尽可能多地保留类型检查，但将它稍微移动一下，放到运行时阶段。"

Java 专家："静态类型检查太重要了。这也是 Java 的精华。更多的静态类型检查总是更好。"

F.4.4　是否需要潜在类型 / 结构化类型

最后，我们不得不考虑：潜在类型 / 结构化类型是不是一个必需的基本功能？它到底有什么作用？

它允许我们跨越类的层次结构，调用不属于公共接口的方法。

类只能共享公共基类中的方法。C++ 没有最基础的 Object 基类，因此对潜在类型的需求更加明显。Java 具有单根层次结构，并且在泛型代码里，T 的 className 能调用 Object 的方法。只要不超出泛化的范围（Object），它们就可以是"泛型"的。这适用于集合和任何其他只需要 Object 基本操作的代码。

但是，Java 中仍然可能存在这样一些层次结构，它们不相交，具有通用方法，但没有通用接口。不同类层次结构的原始作者没办法预测它们之间是否有交集，你也无法提前预测到可以编写一段适用于两者的代码。

这更有可能发生在现有的层次结构中，即层次结构是从别处引入的。要跨越多个不相交的层次结构执行常见的操作，就必须编写重复的代码。认为这没问题，实际上是在说"代码复制没问题"，尽管工作量看起来微不足道，而且也不是很频繁。

但是，如果你只是因为不知道有这个功能，所以才没有使用它呢？例如，过程化（procedural）程序员眼中的对象，或者可以大大简化一些架构的切面编程（aspect）。C++ 程序员不限于在容器类里使用模板，这让他们受益很多（不过模板元编程似乎还是超出了大多数人的理解）。

即使你的确认为所有的类都可以硬塞进一个层次结构，从而实现一个公共接口，但对于额外代码导致的维护开销也存在着争论。作为 Java 程序员，我们就像温水里的青蛙（一个迷思——显然当水太热时它们会跳出来），适应了用越来越多的代码做基本的事情。随着时间的推移，Java 增加了部分语法来减少一些代码，这是一个令人钦佩的进展。但它仍然需要更多的代码，这超出了它所需的。相反的观点是，有像 IntelliJ 这样的工具来帮我们打字，但是阅读和维护代码的开销——这是大部分金钱最终流向的地方——仍然在增加。

考虑一下这个相反的观点。在 2004 年的软件开发大会上，James Hobart 指出了"概率"和"可能性"之间的区别。与 Java 相比，可以认为 C++ 遇到不相交类层次结构的概率更大。Java 泛型似乎在集合里工作得很好。在 Java 里，你可能会遇到不相交的类层次结构，并希望用一段代码来跨越这些层次结构，但概率可能并不高，在实践中也不会经常遇到。这就是 Hobart 所说的"走在快乐通道中"。在 Java 中，我们别无选择，只能走这条路，但在其他语言中，我们可以体验一下潜在类型，看看自己到底错过了什么。

F.4.5 我们其实是在讨论测试

归根结底，这些都只是不同类型的测试。我们可以在同一个范围内考虑它们：

- 编译器测试（静态类型检查）；
- 运行时系统测试（动态类型检查及其他）；
- 类的单元测试；
- 程序的一致性测试。

编译器和运行时系统能做的就这么多。它们只能提供一组足够通用的测试，以此来检查每个程序。到了某个阶段，你就只能开始编写自己的测试，来验证特定程序的语义，以确保其正确性。

那么，我们谈论的不是测试是否重要（实际上它们很重要），而是这些测试在什么时候进行，以及如何进行。更重要的是，对于语言的表现力和清晰度来说，它们会受到测试发生在哪个阶段的影响，而我们需要对此进行权衡。我的观点是，程序就像散文，必须是可读的。如果静态检查不影响语言的清晰度、表达力和生产力，它就是好的。如果这些方面受到影响，则应该调整检查的阶段来减少这种影响。

这个观点基于我的看法和经验：不能依赖静态检查来验证程序的正确性。因此在某种更安全的错觉下，用过多类型检查语法来掩盖程序的含义是不合理的。为了做到万无一失，肯定需要运行时测试。鉴于此，对我来说，由于这种错觉而让代码被过多类型检查语法搞得混乱难懂，进而使工作变得更艰难，这是没有意义的。

F.4.6 决定论

阿尔伯特·爱因斯坦对海森堡不确定性原理（你不可能同时知道小粒子的位置和动量）感到非常不安。他相信"上帝不掷骰子"，并提出了一种"隐变量"理论，这种变量一旦被发现，我们就可以完全确定一切。

贝尔定理的结果并不支持隐变量理论。

斯蒂芬·霍金描述了对于事件而言，（在时空中）离得越远，不确定性就越强。不确定性原理的影响在时空中是累积的。

希腊人相信命运和自由意志是共存的，对此人们觉得很困惑，因为我们似乎只能选择其中之一。也许霍金的理论解释了这种二分法。

创建能被证明为正确的程序是可能的，但我猜测这些程序只占一般解决方案空间的一小部分。

静态类型检查只是一种测试,其效果非常有限——当你离开编译时空后,"软件不确定性原理"给结果增加了越来越多的噪声。只有通过全面的测试,才可能得到一个尽可能正确的程序。

F.5　生产力成本

我对潜在类型的探讨吸引了一些非常有趣的分析,尤其是一位名叫 Pixel 的匿名博主的分析。尽管 Pixel 在某些地方误解了我的意思(或者更有可能是我说得不够清楚),但总的来说,他 / 她对 "Java 案例" 的表述相当恰当。

这个分析包括了一些常规论点,比如"人们在 Java 中的生产力与在 Python 中是一样高的",这个论点几乎普遍来自想象,而不是来自使用这两种语言的直接经验。我经常使用这两种语言,对比之下的结果总是:若要完成很多工作,那就使用 Python。正如 Pixel 所说的,测试方法并不是什么特殊的人工测试,而是通过要操作的实际数据来实现的。用 Python 就是能更快地完成工作,并可以用真实数据更快地发现真正的问题。我一次又一次地体验着相同的结果。

我要解决的各种问题理论上只能使用 Java 来解决。这样做至少需要 5~10 倍的时间,并且还要假设你不会先放弃或迷失其中。例如,你可以从 Python 里移植一些内容,但是相比之下,在规划时就直接用 Python 做设计要容易得多,同时移植后生成的代码会更多、更混乱,并且(这是一个很大的争论点,因为人们经常声称 Java 代码越冗长就越容易维护,因为它更明确)更难维护。

Pixel 接着问道:"对于在编译时尽可能多检查,你认为有多大价值呢?"这就将问题过于简化了。如果类型检查是无须任何代价的,那么答案就像 Pixel 在这个问题中暗示的那样简单明了。我发现大多数认为必须不惜一切代价保留和增加静态类型检查的人,就从来没有体验过它的任何不利之处。如果没有看到任何不利的一面,那你就会认为反对它显然是疯狂的。

我知道这些是因为我体验过从 ANSI C 之前版本的 C 语言迁移到 C++ 后带来的好处,这让我从一开始就坚定地站在了静态类型检查的阵营。C++ 静态类型检查发现了很多(尤其是 ANSI C 之前版本)C 没有发现的错误,这是一个很明显的优势。因此,这种看法延伸到 Java 里也就顺理成章了。但那时我的体验是一维的,只有当我开始使用更动态、更简洁的语言（Python）时，我才体验到了新的维度。

大多数人的经历是这样的：他们对 Python（或任何你选择的动态语言）有足够的了解，并将其牢记在心。当出现一个可能需要临时解决方案的问题时，他们知道使用 Java 实现要花费多少时间，这种情况下正好可以尝试一下 Python，因为它可能会节省一些工作。这就成了他们解决问题体验的分水岭，Python 比 Java 花费的时间竟然少那么多。当下一次出现类似问题时，他们会更容易决定使用 Python，直到他们开始用 Python 作为首选工具，而 Java 则只有在没有其他选择的情况下才使用。

在这个过程中，使用 Python 能显著提高生产力的体验一遍又一遍地重复。这不是哲学论证（因为答案"显然"是静态类型语言更好），而是直接经验。最终生成的 Python 程序正确性和质量始终很高。

然后你就遇到了"知识与经验相矛盾"的危机：你"知道"Java"更好"，但你的经验是 Python"更好"。当你认为正确的所有论据都清楚地表明静态类型语言必然"更好"时，怎么会出现这种情况？这时，你就真正开始质疑自己的预设立场，并尝试解决这场危机。但是你听到的唯一论点就是，你自己（也包括我自己）以前"知道"的是正确的，而从经验得到的体会是不确定的。

根据我的经验，静态类型的价值及其对生产力的影响之间存在平衡。提出"静态类型显然更好"论点的那些人，通常没有使用替代语言获得更高生产力的经验。当你有这种经验时，会发现静态类型的开销并不总是有益的，因为它有时会减慢你的速度，最终对生产力产生重大影响。

我无法量化这一点。我无法从基本原则出发给出一个数学证明，这可能是因为它依赖人为因素，例如开发者需要记住如何打开文件并将 try 块放在正确的位置，如何读取行记录，以及读取文件后打算用它实现什么功能（尽管这种体验在 Java 8 中得到了极大的改善）。在 Python 中，我可以通过以下方式处理文件中的每一行：

```
for line in open("FileName.txt"):
    # 处理文件中的行
```

（Python 3 添加了一个上下文管理器来自动关闭文件，这里没有演示该功能。）我不需要去查什么文档，甚至不用去想应该怎么做，因为这太自然了。我总是不得不查找文档来看看如何在 Java 中打开文件和读取行记录。你可能会争辩说，Java 并不是为文本处理而设计的。我同意你的看法，但不幸的是，Java 似乎主要用于服务端，而其中一项非常常见的任务就是处理文本。

做事情时如果被打断，需要多长时间才能恢复？相关研究有很多，但我自己无法做出任何论断，只能说在我看来这肯定有很大的影响，无论是在编写代码时（是的，我知道其中很大一部分是用 IntelliJ 和类似编辑器自动生成的），还是在阅读代码时。

Java 该是什么就是什么。我在这里证明 Java 泛型不支持潜在类型，主要是想让一些人向我展示 Java 泛型是如何真正支持潜在类型的，以防我误解了什么。他们没能向我展示这些，我也没有误解什么，所以情况就是这样的。

最后，我意识到在任何地方都使用接口确实没什么大不了的。在 Python 这种简洁的语言里，这样做的确很扎眼，但在 Java 冗长的代码里却不是这样的，而且很多人似乎喜欢这种感觉，觉得它是一个好处。四处添加一些额外的接口，对生成的 Java 代码并没有什么太大的影响，而且也符合 Java 语言的哲学（如果 Java 真的直接支持了潜在类型，人们甚至可能会感到惊讶）。

所以我现在明白了，Java 泛型实际上只是为了更好地使用容器而提供的自动类型转换，当然它也很好、很有用，但泛型也就只能做到这一步了。

在 Java 中也可以实现潜在类型，但它需要更多的努力，所以你得真的很想这样做才行。让我们再重温一遍狗和机器人的程序，这次使用反射来生成一个实现了潜在类型的方法：

```java
// staticchecking/latent/Latent.java
// {java staticchecking.latent.Latent}
package staticchecking.latent;
import java.lang.reflect.*;

class Dog {
  public void talk() {
    System.out.println("Woof!");
  }
  public void reproduce() {}
}

class Robot {
  public void talk() {
    System.out.println("Click!");
  }
  public void oilChange() {}
}

class Mime {
  public void walkAgainstTheWind() {}
  @Override
  public String toString() { return "Mime"; }
}
```

```
/* 输出：
Woof!
Click!
Mime cannot talk
*/
```

```java
class Communicate {
  public static void speak(Object speaker) {
    try {
     Class<? extends Object> spkr =
       speaker.getClass();
     Method talk =
       spkr.getMethod("talk", (Class[])null);
     talk.invoke(speaker, new Object[]{});
    } catch(NoSuchMethodException e) {
     System.out.println(
       speaker + " cannot talk");
    } catch(IllegalAccessException e) {
     System.out.println(
       speaker + " IllegalAccessException");
    } catch(InvocationTargetException e) {
     System.out.println(
       speaker + " InvocationTargetException");
    }
  }
}

public class Latent {
  public static void main(String[] args) {
    Communicate.speak(new Dog());
    Communicate.speak(new Robot());
    Communicate.speak(new Mime());
  }
}
```

我可以给 speak() 方法传递任何东西，但它只会在实际具有 talk() 方法的对象上调用 talk()。

F.6 静态与动态

下面的内容节选自一封读者来信。我对它进行了一些编辑，并"更正"了部分术语，如信后所述。

我是三一大学（Trinity University）的一名教授，执教一门编程语言的课程。静态类型是我和学生经常讨论的话题，我的一个学生偶然发现了您的网站，它在我们的一节课里很有效地推动了课堂讨论。我个人倾向于支持静态类型，并且如果可以的话，我想向您请教两个问题。

第一个问题和我喜欢静态类型的一般原因有关。基本上，静态类型提供了证据，表明程序在某些方面是有效的。诚然，类型系统并不能证明一切正常，我们仍然需要测试，但它至少能证明某些方面是正确的。为什么这对我来说很重要呢？因为如果想要做完整的测试，测试的数量会随着代码量增加至少呈指数级增长。如果测试数量不够多，则会导致代码中的许多执行路径没有被覆盖。无论使用某种语言的工作效率如何，编写指数级的测试用例都不堪重负。鉴于此，让编译器证明至少某些方面是正确的，这种可靠性难道不值得吗？

我还有一个更有趣的问题：对于元语言及其衍生语言在强测试与静态类型参数方面的表现，您是怎么看的？这些语言具有动态类型语言的优点，因为它们很少需要用户指定类型，但是它们会推断所有表达式的类型，而且它们是静态类型的。在这方面，它们的类型系统比需要动态检查的语言（Java）或用户可以迫使类型系统出错的语言（C/C++）要安全得多。

当我阅读您关于强测试与强类型的文章时，我立刻想到了元语言与 Scheme 的比较，因为 Scheme 动态地进行所有类型检查，并在编写代码时为用户提供完全的灵活性。然而根据我的经验，用它编写大型程序非常困难，而 Scheme 里能接受大量代码也并不意味着这些代码都是正确的。元语言则相反，它对所有内容进行静态类型检查，即使用户没有指定类型。当我的元语言代码通过编译后，大部分时间它能工作得很好，残存的错误通过一两个简单的测试就能发现。

随着微软将 F# 语言纳入 .NET 框架，很可能会有更多的程序员接触到元语言风格的编码。我很想知道，您有没有觉得，在程序员没有指定类型的情况下就能拥有强大的静态类型，可以让我们拥有两全其美的结果？

首先，对于这些混淆的术语的传播，我至少担负部分责任，而且我没有时间也没有意愿再纠正以前的文章中的错误。在计算机世界中，这些术语似乎也没有明确定义。因此，为了帮助纠正这种错误，我在此特地做出如下说明。

- **弱类型检查**：允许向对象发送不正确的消息。C 和 C++ 允许用户转换为错误的类型，因此被视为弱类型（尽管 Stroustrup 曾经说过 "C++ 是强类型的，只是类型机制里里有几个漏洞"）。我过去曾认为 "弱类型"（我最初用来表示 "潜在类型" 的术语）与 "弱类型检查" 不同，但我认为这两个术语的差异太微妙了，不值得争论 ①。此外，潜在类型的概念也让这个问题令人迷惑。C++（是静态类型，还可以认为语言的某几个地方是弱类型的）、Python 和 Ruby（动态类型）都支持结构化类型 / 潜在类型（Ruby 称之为 "鸭子类型"）。

---

① 作者此处的观点与 F.4 节有些不一致，这是因为它们写于不同时期，具体可参考篇首说明。——译者注

- **动态类型检查**：类型检查仍然会发生，语言还可以是强类型，但类型检查发生在运行时而不是编译时。强类型动态语言仍然只允许向对象发送有效消息。

这里澄清一个重点：我不反对静态类型检查，只是觉得它存在如下一些问题。

1. 有一种错觉，认为静态类型检查可以解决所有的问题，随之而来的结论是，更多的静态类型检查总是更好。

2. 额外的各式各样的静态类型检查经常被添加到语言中，而不考虑实际成本。在极端情况下，你甚至会将所有时间都花在解决编译器问题上。

总体来说，我的态度是静态类型检查是可取的，只要它不会花费太多。正如这封信中所指出的那样，元语言和 Scala 中的类型推断提供了静态类型检查，不需要提供额外的信息，编译器就可以自行计算出结果。

正如这封信中所说的，我确实在寻求"世界上最好的"。问题是静态类型检查的某个特定实现带来的帮助是不是比阻碍更多。我经常被指责老是抱怨"多打了几个字"，但我观察到的不仅仅只是腕关节受苦（有人指出，其中大部分代码可以由 IntelliJ 等工具自动生成）。

这种现象背后真正的问题是人类思维管理复杂性时的局限。我们的极限是著名的"7±2"，即我们在同一时刻能记住的事物的最大数量。我认为，计算机编程的所有进步都来自对心智模型的改进，这使我们能够更轻松地操作基本概念。我觉得 Python 之所以如此迷人，是因为它似乎经常有新视角，能够抓住并结合"更简单"和"基本概念"这两点。我的大部分兴趣在于这是如何发生的，而我目前对它背后的秘密知之甚少或根本不了解。Python 语言的设计似乎融入了计算机编程的心理学。

尽管如此，我还是倾向于静态类型检查。正如这封信中所指出的那样，我们的目标是创建可靠的组件——问题是如何实现？在动态语言中，你可以灵活地进行高效的快速实验，但要确保代码无懈可击，就必须既专业又努力地进行单元测试。在倾向于静态类型检查的语言中，编译器会确保某些错误不会溜进来，这很有帮助，但这种语言通常会让你费更多力气才能获得所需的结果，并且代码的阅读者也要更努力才能理解你做了什么。我认为它的影响比我们所想象的要大得多。

此外，我认为静态类型语言会给人一种程序正确性的错觉。事实上，它们只能通过检查语法来确定程序的正确性。我认为这样的语言会鼓励人们认为一切都正常，而实际上它对开发人员进行测试的要求和动态语言一样高。我还怀疑运行编译器所需的额外工作消耗了开发人员进行测试所需的一些能量。这封信中提出了一个有趣的问题，即动态类型语言可能比静态类型语言需要更多的单元测试。我不相信这一点，我觉得数量可能大致相同。为了克服静态类型检查带来的阻碍，我们付出了额外的努力，如果我是正确的，这意味着这些努力可能没有想象的那么富有成效。

两全其美的是什么？根据我自己的经验，用动态语言创建模型非常有帮助，因为在实验过程中重新设计的障碍非常少。你可以快速尝试自己的想法，看看它们在处理实际数据时表现如何，获得一些有关模型准确性的真实反馈，并快速更改模型以适配你的新理解。

我认为这种方法比简单地用 UML 建模好处更大。用 UML 建模只会生成一个幻想中的模型，只有经过一些复杂的转换，你才能表达和测试你的想法。我对此类复杂转换的经验是，它们会给你带来压力，并阻碍你进行实验和更改。而另一方面，使用动态语言，模型就是代码，反之亦然，因此你可以不受惯性约束地进行实验。我认为这种轻便性非常重要，因为它更接近于我们大脑的工作方式。

一旦使用动态语言制定并测试了模型，是否值得将其转换为静态类型语言？我的 Python 使用经验并没有让我觉得需要这样做，原因有两个。

1. 一旦我用 Python 实现了某个东西，它似乎就能工作得挺好。此时将模型转换为静态类型语言的好处并没有多少。其他人对大型 Python 程序的体验似乎也支持这一点——大型的 Python 程序错误少到令人惊讶。

2. 我发现任何经常使用的程序都会被更改。在 Python 中更改程序比在静态类型语言中更改要容易得多，因此我有进一步的动力让它留在 Python 中。

我当然不是在暗示将调试过的 Python 模型转换为 Java 永远没有意义，只是我尚未遇到过这种情况。我可以很容易地想出许多符合此种场景的业务，如下所示。

1. 首先使用"UML lite"和 / 或 CRC 卡"勾勒出"一个初步模型。

2. 然后使用 Python 或 Ruby 实现此模型，开发人员习惯哪种语言就可以用哪种。此时，你就可以将纸质模型抛在脑后，代码成了你的模型。Python 经常被描述为"可执行的伪代码"，这在建模和实验过程中非常有用。

3. 尝试并改进这个"实时"模型，直到解决了各种难缠的问题，并且模型能够正常工作为止。

4. 最后将它移植到 Java、C++ 或项目规定的任何语言。

我的经验是，通过使用鼓励改变的语言开发模型，最终会得到一个更好的模型，当该模型被翻译成你的实现语言时,会带来明显的好处。

最后,我的主要兴趣在于生产力和可扩展性。Visual Basic(VB)对于小型项目来说是一种非常高效的语言(有许多这样的小项目,因此 VB 解决了很多问题),但它的可扩展性不好。Perl 也属于这一类,而明显继承了它的 Ruby,似乎相当好地实现了对象范式,并且可以扩展到更大的项目。Python 则已经被用于许多大型项目,尽管它缺乏静态类型检查,但结果似乎显示 bug 非常少。

最后这一点是困扰很多人的谜题——我们相信静态类型检查可以防止错误,然而一门动态类型语言竟然能产生这么好的结果。当我试图更深入地研究这个谜题时,我的许多先入为主之见——主要是静态类型检查是必不可少的——受到了挑战。对此的最初反应通常是简单地否认事实,但是一旦开始否认证据,你的理论很快就变成了不切实际的幻想。以我自己的经验,很难弄清楚 Python 为何如此高效。然而,在尝试这样做的过程中,我发现了很多新东西,并对其他语言有了更深入的了解。

我的猜测是 Python 能让我能够更清楚地思考试图要解决的问题。当我尝试生成问题空间的有效模型时,Python 不会那么分散注意力,因为它不会强迫我考虑语言强加的规则——这些规则基本上是任意的。通过这种让步,Python 和类似的动态语言让只能同时处理几件事的大脑更多地考虑问题本身,而更少地花在语言实现的细节上。我不止一次有过这种体验,例如从 C++ 迁移到 Java 时,我不再需要担心 operator= 和复制构造函数。让 Java 为我们处理更多事情似乎可以提高程序员的生产力,而我在 Python 里也有相同的体验。

最后,有的项目使用 Python 或 Ruby 对模型进行演进,并用 Java 重新实现它。我能看出采用这种方式的价值。这种方式最大的好处在于能适配现有的 Java 开发环境,不过似乎也可能会因为静态类型检查而发现一些错误。如果有人做过这个实验,听听他们的经验以及 bug 数量的变化,可能会很有趣。

这封信中还问到元语言的类型推断是否可以让我们获得两全其美的体验。我喜欢类型推断,因为它减少了程序员的开销,但它也是一种静态类型机制(除非有足够的上下文来推断静态类型,否则它不起作用)。此外,它仅仅给我们带来了动态语言提供的部分好处。

如果想要创建正确的程序,我们就需要帮助,因此很难找到理由来反对静态类型。如果说面向对象是鱼的话,类型系统就是水,允许类型系统存在漏洞和后门就像在说"对于类型来说,这些都很有道理,但是只有当人们知道自己在做什么且遵循规范时,类型系统才能被信赖"。这不是很让人放心。静态类型实际上通过确保类型的正确行为，来帮助我们思考我们的对象模型。

问题不在于类型检查是否是一件好事。我毫不含糊地说：这确实是一件好事。问题是，何时进行类型检查，以及静态检查与动态检查的成本是多少。在 C++ 中，实际上所有类型检查都是静态的。在 Java 中，类型检查既是静态又是动态的（我认为任何试图实现彻底类型检查的语言都需要某些动态类型检查），而在 Python 和其他动态语言中，类型检查主要是动态的。正如你所注意到的那样，在大多数动态语言中，并非所有的执行路径都经过编译器的测试，这可能是一个问题，但这个问题到底有多糟糕？我感觉如果对 Java 程序及其等效的 Python 版本执行相同数量的单元测试，我们就能测试所有的执行路径。我认为在这两种情况下，单元测试都处于足够高的水平，最终会得到相同的结果。换句话说，我不认为需要额外的单元测试来进行语法检查，因为单元测试在测试类和接口时，就自然而然地进行了语法检查。我没有直接的证据，不过我自己就不用做这些额外的工作。

Java 泛型提供的一个功能是集合类的静态类型检查。它可以防止在运行时出现 `ClassCastException`，这还是有点用处的，但如果这是引入 Java 泛型的唯一原因，那么引入它所带来的语法的复杂性，就让我们的付出不值得了（幸运的是，通过一些努力，我们可以创建泛型代码，正如之前所展示的那样）。由此产生的 `ClassCastException` 不会经常发生，并且在发生时也不难发现。在这种情况下，动态类型检查就足够了。

我还指出，检查型异常不能很好地扩展，即使对于小程序来说，这也很容易成为障碍。自 Java 以来新设计的任何语言都没有采用这个功能，我认为这是一个强有力的论据。不过，我发现 Java 1.4 `RuntimeException(Throwable cause)` 构造器消除了我对检查型异常的大部分抱怨——如果它妨碍到了我，我可以只用很少的编码就将它关闭。因此，我可以选择让它保持开启或关闭。

上面提到的两个例子都引入了过多的静态类型检查，无论初衷多么好，这都会妨碍代码的创建和阅读。尽管如此，我实际上还是喜欢静态类型检查的，前提是使用得当。Python 3 甚至添加了一个可选的类型注解机制，由外部静态检查工具强制执行。但我拒绝任何"所有静态检查都很好，所以越多越好"的暗示。如果你考虑一下这一点，很容易就能想象，这样会走向极端。而在许多情况下，Java 正是这样做的，而没有对结果进行成本效益分析。我反对的是认为静态类型检查没有成本的错觉。

·作者简介·

布鲁斯·埃克尔（Bruce Eckel），C++标准委员会的创始成员之一，知名技术顾问，专注于编程语言和软件系统设计方面的研究，常活跃于世界各大顶级技术研讨会。他自1986年以来，累计出版 *Thinking in C++*、*Thinking in Java*、*On Java 8* 等十余部经典计算机著作，曾多次荣获 Jolt 最佳图书奖（"被誉为软件业界的奥斯卡"），其代表作 *Thinking in Java* 被译为中、日、俄、意、波兰、韩等十几种语言，在世界范围内产生广泛影响。

·译者简介·

孙卓，现任职于百度健康研发中心，百度技术委员会成员。从业十余年，熟悉 Java、PHP 等语言体系，同时也是一名语言文字爱好者。

陈德伟，深耕软件研发十余年，目前专注于金融系统研发工作。

臧秀涛，InfoQ 前会议内容总编。现于涛思数据负责开源时序数据库 TDengine 的社区生态。代表译作有《Java 性能权威指南》《C++ API 设计》《Groovy 程序设计》等。

图灵教育

站在巨人的肩上
Standing on the Shoulders of Giants

图灵教育

站 在 巨 人 的 肩 上
Standing on the Shoulders of Giants